Lecture Notes in Management and Industrial Engineering

Series Editor

Adolfo López-Paredes, INSISOC, University of Valladolid, Valladolid, Spain

This book series provides a means for the dissemination of current theoretical and applied research in the areas of Industrial Engineering and Engineering Management. The latest methodological and computational advances that can be widely applied by both researchers and practitioners to solve new and classical problems in industries and organizations contribute to a growing source of publications written for and by our readership.

The aim of this book series is to facilitate the dissemination of current research in the following topics:

- Strategy and Entrepreneurship
- Operations Research, Modelling and Simulation
- Logistics, Production and Information Systems
- Quality Management
- Product Management
- Sustainability and Ecoefficiency
- Industrial Marketing and Consumer Behavior
- Knowledge and Project Management
- Risk Management
- Service Systems
- Healthcare Management
- Human Factors and Ergonomics
- Emergencies and Disaster Management
- Education

More information about this series at http://www.springer.com/series/11786

Fethi Calisir
Editor

Industrial Engineering in the Internet-of-Things World

Selected Papers from the Virtual Global Joint Conference on Industrial Engineering and Its Application Areas, GJCIE 2020, August 14–15, 2020

Springer

Editor
Fethi Calisir
Department of Industrial Engineering
Istanbul Technical University
Istanbul, Turkey

ISSN 2198-0772 ISSN 2198-0780 (electronic)
Lecture Notes in Management and Industrial Engineering
ISBN 978-3-030-76723-5 ISBN 978-3-030-76724-2 (eBook)
https://doi.org/10.1007/978-3-030-76724-2

This Springer imprint is published by the registered company Springer Nature Switzerland AG
The registered company address is: Gewerbestrasse 11, 6330 Cham, Switzerland

Preface

This book compiles extended versions of a selection of the best papers presented at the Global Joint Conference on Industrial Engineering and Its Application Areas (GJCIE) 2020 held online. They represent a good sample of the current state of the art in the field of industrial engineering and its application areas.

The papers presented in this book address methods, techniques, studies, and applications of industrial engineering with the theme of "Industrial Engineering in the World of IoT." The Internet of things (IoT) is a paradigm where objects can be equipped with identifying, sensing, networking, and processing capabilities that will let them connect, communicate, and exchange data with one another and with other devices and services over the Internet. IoT has been developed vastly due to the developments of a wide variety of technologies, such as wireless communications, sensors, and computing. There are a lot of applications of IoT, which range from simple home and personal appliances to large-scale systems. IoT is one of the critical technologies that will have an enormous impact on all organizations. For example, IoT may play an enabling role in the design of systems and products by making them more effective, better connected, and more intelligent and transparent. This challenges the way we educate industrial engineers and the way we manage organizations. But IoT research is still in its early stages, and related studies are not well integrated. This book will shed light on the role of industrial engineering in this endeavor. Contributions have been arranged in three parts:

- Industrial Engineering
- Engineering and Technology Management
- Healthcare Systems Engineering and Management

I would like to express our gratitude to all the contributors, reviewers, and international scientific committee members who have aided in the publication of this book. I would also like to express our gratitude to Springer for their full support during the publishing process. Last but not least, we gratefully acknowledge the sponsors (LC Waikiki, Elginkan Foundation, and Entertech) of GJCIE 2020.

Istanbul, Turkey
August 2020

Fethi Calisir
Conference Chair

Contents

Engineering and Technology Management

Healthcare Systems Engineering and Management

Industrial Engineering

Safety Risk Management in Complex Systems

Gulsum Kubra Kaya

Abstract Risk management has been applied in a wide range of industries to ensure safety by using risk management tools and techniques. Many of those techniques were developed long ago to analyze individual system components. In complex systems, however, accidents emerge from system interactions. Hence, traditional risk management tools and techniques have become inadequate to analyze risks in complex systems. The Functional Resonance Analysis Method (FRAM) was recently developed to address the limitations of traditional risk management methods. This study provides an example of the use of FRAM to demonstrate its use and to highlight its potential value in safety risk management.

Keywords Risk management · Safety management · FRAM · Complex systems

Introduction

Risk assessment plays an essential role in safety risk management; it aims to address three main questions: "*What could go wrong*?" "*What is the level of the risk*?" and "*Is there a need for actions*?" [28]. In other words, risk assessment involves the steps of risk identification, risk analysis, and risk evaluation [10].

In the risk identification step, risk sources are identified [7, 10]. Before that, it is essential to set a common understanding of the term 'risk' to prevent misuses and, in turn, inadequate risk identifications [43]. The British Standards Institution (BSI) identifies risk as an "*effect of uncertainty on objectives*" [10]. In the literature, the term 'risk' has been identified in varied ways. Aven [4] classified risk definitions into nine categories, which are "*expected value (loss),*" "*probability of an event,*" "*objective uncertainty,*" "*uncertainty,*" "*the possibility of a loss,*" "*probability and*

G. K. Kaya (✉)
Industrial Engineering Department, Faculty of Engineering and Natural Sciences, Istanbul Medeniyet University, Istanbul, Turkey
e-mail: kubra.kaya@medeniyet.edu.tr

F. Calisir (ed.), *Industrial Engineering in the Internet-of-Things World*,
Lecture Notes in Management and Industrial Engineering,
https://doi.org/10.1007/978-3-030-76724-2_1

consequences," "*consequence*," "*consequence and uncertainty*" and the BSI definition. While there is no single correct or wrong description of the risk, the risk level should be measured based on the definition followed.

In the risk analysis step, the level of risk is measured through the estimation of the likelihood of a risk and its potential consequences [10]. Likelihood can be measured as probability or frequency. The consequence can be measured by its impact on safety [3, 17, 29].

In the risk evaluation step, the level of risk is compared with risk criteria to prioritize risks and identify actions required for risk treatment [5, 11]. At this stage, the formulation of the risk criteria is key to making effective decisions. Many organizations adopt the As Low as Reasonably Practicable (ALARP) principle to prioritize risks. In addition, organizations conduct a cost-benefit analysis before deciding on risk treatment [5, 13, 19].

Numerous tools and techniques have been developed to support safety risk management. Most the well-known traditional techniques are Failure Mode and Effect Analysis (FMEA), Hazard and Operability (HAZOP) study, Fault Tree Analysis (FTA), Event Tree Analysis (ETA), Bow-tie Analysis (BA) and Risk Matrix (RM) [11, 15, 34]. Among these techniques, FMEA identifies each failure mode individually, and determines the likelihood, consequence and detectability of the failure mode [9]. HAZOP is similar to FMEA, but it provides guide words to identify the error modes within the system to be assessed [8]. FTA has a top-down approach. It starts with a top-level event, and, then, breaks this event down to identify causal factors [24]. ETA starts with an event and represents the sequence of events in different scenarios [11]. BA begins with a top-level event, and analyses the causes and consequences of this event by examining the existing control mechanisms [11, 41]. An RM is a tool to represent findings from risk analysis. RMs have two indexes: likelihood and consequence. Often, a score from 1 to 5 is used to assign the level of risk. RMs support the decisions made in the risk evaluation step [6, 17, 28, 33].

Such traditional tools and techniques, however, have been criticized in terms of their ability to assess complex systems [21, 22, 31]. These techniques follow a linear cause-effect way of thinking, and they solely focus on individual system components. As a result, they can be ineffective when considering the interactions of system components [20, 25, 32, 35].

In complex systems, accidents emerge from system interactions [21, 22, 31]. Therefore, safety risk management in complex systems (e.g., health-care, defense and aviation industries, and nuclear and chemical plants) requires the use of techniques that are built on systems theory. Researchers pioneered new techniques: the Functional Resonance Analysis Method (FRAM) [20] and the Systems Theoretic Process Analysis (STPA) [31] to address the limitations of traditional techniques. These two techniques have been used in various industries and have demonstrated their value in analyzing both system interactions and components [30, 37, 42].

Moreover, performance variability might be necessary to achieve success in complex systems. This might lead to deviation from the practice described, which is referred to as failure in traditional risk assessment techniques. While traditional techniques focus on the reliability of each system component, FRAM focuses on system resilience [20]. FRAM has been used in various industries to investigate accidents, to assess risks, to measure performance variability, to understand system complexity and, thus, to enhance system resilience [1, 27, 38].

This study applies FRAM to assess risks in a Turkish coffee-making process, and it discusses the applicability of FRAM in complex systems.

Methodology

The Principles of FRAM

FRAM is based on four principles. The first is "*the equivalence of successes and failures.*" Traditional risk assessment approaches follow the "*find and fix*" approach, which focuses on failures, whereas FRAM assumes that both success and failure are due to the performance adjustments of organizations, groups, and individuals [20, 23].

The second principle is "*approximate adjustments.*" Socio-technical systems describe activities in-detail to operate successfully at all times. However, adjustments of the activities might be required to respond to a change in working conditions (e.g., requirements, resources, and opportunities). As individuals might have a different degree of understanding of the event, their adjustments are approximate rather than precise [23].

The third principle is "*emergent outcomes.*" Understanding events tend to be based on causality credo, where things have a cause. Thus, adverse events are dealt with by finding problems and fixing them. FRAM recognizes that variability can be as a result of the combined variability of multiple functions [21, 23].

The fourth principle is "*functional resonance.*" The variability of a function might have an impact on the variability of other functions. Similarly, multiple functions might have variability simultaneously and, as a result, might affect others [23].

The FRAM Steps

FRAM starts with the identification of the purpose of the analysis. This study applies FRAM to assess risks in the Turkish coffee-making process. It must be noted that this application is only for illustrative purposes. It is better to use FRAM in a complex system rather than a simple process, as in this example.

FRAM involves four steps: identifying and describing functions, characterizing the variability of functions, the aggregation of performance variability, and responding to performance variability [20].

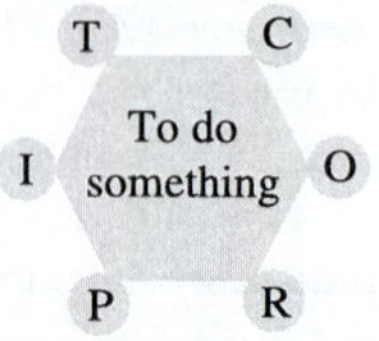

Fig. 1 A hexagon illustration of a function

A function represents an activity that is required to achieve a goal and describes everyday work [23]. In the first step, functions in a system or process are described in six aspects: input, output, precondition, resource, control, and time (Fig. 1).

Input can be seen as a form of instruction, information, or a change that starts the function. The output is the outcome of the function, and it may trigger other functions or may feed them. A precondition is the system state that should be in place before the function is carried out. Resources are needed to carry out the function, such as information, material, workforce, and software. Control represents the supervision and regulations to ensure desirable outcomes from the function. Time can be seen as a form of control when the function is completed before, after, or simultaneously with other functions. Similarly, it can be a form of resource when the function is completed before a specific time or within a particular duration [23].

In the second step, the variability of outputs from a function is determined. Variability occurs as a result of the function itself, the impact of an upstream function, and the working environment. There is variability in "*time*" and "*precision*" [20].

In the third step, functional resonance is determined. The variability of the output from an upstream function might affect the downstream function. The FRAM model illustrates such interactions between functions [20].

When building the FRAM model, it is necessary not to get lost in details. Patriarca et al. [39] integrated three abstraction levels (i.e., functional purpose, generalized function, and physical function) into FRAM models to ease the modeling of complex systems.

In the fourth step, performance variability is managed with a focus on the aspects which lead to uncontrolled performance variability [20].

Results

In this study, FRAM has been applied in the Turkish coffee-making process. In total, seventeen functions were identified at the physical function level (Table 1).

All functions were described by considering six aspects, which are outlined in Table 2. However, it should be noted that not all functions were describable in these six aspects. The use of FRAM does not force its users to identify all six aspects [23].

The FRAM Model Visualiser (FMV) was used to draw the FRAM model. The FMV automatically links the functions if any of their six aspects are the same.

Table 1 The list of functions for three abstraction levels

Agency abstraction	Consumer
Functional purpose (FP)	Consumer's FP: to drink coffee at home
Generalized function (GF)	Buy ingredients
	Prepare all ingredients and kitchen appliances
	Make coffee
	Serve coffee
Physical function (PF)	Go to the supermarket
	Find coffee
	Buy ground coffee
	Bring coffee to home
	Take the coffee pod
	Put the coffee pod on the stovetop
	Add cold water
	Add Turkish coffee
	Turn on the stove
	Stir coffee
	Boil coffee
	Turn off the stove
	Transfer foam into the cup
	Put coffee into the cup
	Prepare the tray to be served
	Serve it with water
	Serve it with Turkish delight

Table 2 The description of the function <boil coffee>

Function <boil coffee>	
Description	Turn the heat to a low to a medium level to boil coffee.
Aspects	
Input	Turning on the stovetop
Output	Boiling coffee
Precondition	Putting the coffee pod on the small rack
Resource	Coffee, stove and coffee pod
Control	Check the color of the foam on the coffee
Time	Complete the activity just after boiling the coffee, after approximately 5 s

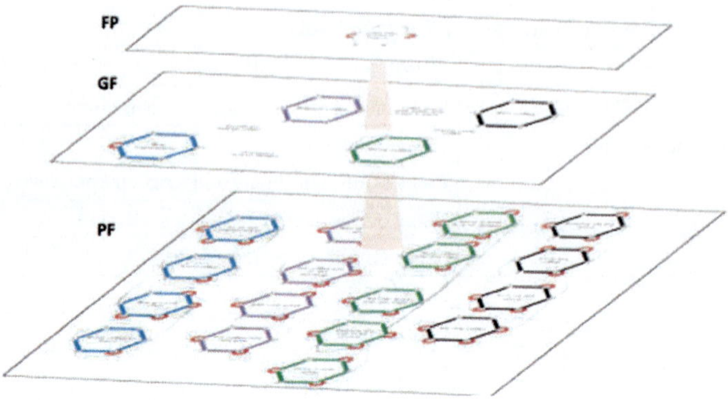

Fig. 2 The functional abstraction framework for the coffee-making process

The FRAM model does not aim to describe functions in sequence. However, the functions involved in the coffee-making process were described in sequence due to the selected process being rather linear and consisting of a series of instructions. Figure 2 represents the FRAM models for three abstraction levels.

The variability of the output of each function was determined concerning *Time* (i.e., too early, on time, too late and not at all) and *Precision* (precise, acceptable, imprecise, and wrong). For example, the function <boil coffee> can be imprecise when the stove was turned off too soon.

The FRAM models help determine aggregated variability. For instance, if the function <boil coffee> is too late, variability in this function will delay the occurrence of the functions <pour coffee into cups> and <turn off the stove>. This might have an impact on the variability of the remaining downstream functions. The downstream function might also affect the variability of the upstream function (Fig. 3).

After considering aggregated variability, scenarios were developed to understand when and how the performance variability becomes uncontrolled. Here, recommendations were provided to manage performance variability, and, in turn, to assess risks. For instance, a recommendation could be given to "*watch coffee at all times when boiling it.*" This is because timing is critical in the function <boil coffee> , and any delay or imprecision in this function might have an impact on the achievement of the goals (i.e., making tasty Turkish coffee). Similarly, another recommendation could be given regarding the quality of the coffee as "*checking the expiration date of the Turkish coffee when buying it.*" If the coffee is stale, the flavor notes of the coffee will not be strong, so the coffee will not produce adequate foam, and it will not taste good.

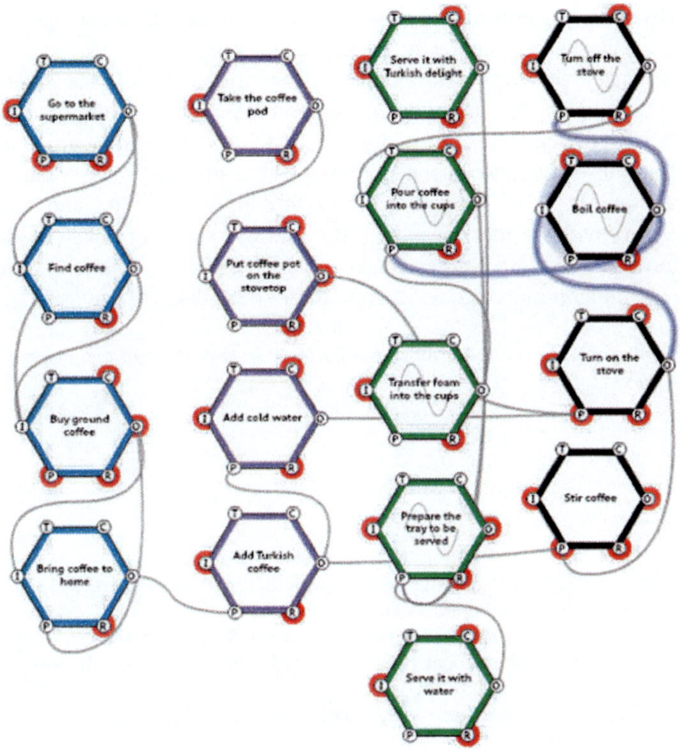

Fig. 3 The aggregated variability of the function <boil coffee>

Discussion and Conclusion

This study has used the coffee-making process as an example of FRAM. However, it must be noted that the FRAM application is suitable for complex systems, and it requires the involvement of a multidisciplinary team [23].

FRAM has predominantly been used to assess risks or investigate incidents [2, 21, 26, 40]. In this study, risk can be determined as "*uncontrolled performance variability*," and it might have both positive and negative consequences. This description might seem to be similar to the definition provided in the traditional approach. However, FRAM differs in terms of how risk is assessed. In FRAM, the risk is assessed by considering impacts without focusing on causes or probabilities. FRAM does not focus on causes; it instead focuses on sustaining success by accepting flexibility in the system and variability in performance [16, 20, 36]. Performance variability in socio-technical systems is normal and, sometimes, even required [2, 16].

Furthermore, FRAM differs from the point where it focuses on the actual work rather than the work described. Most of the traditional tools and techniques analyze the system by assessing how the actual work is differentiated from the work described. In practice, the work might not be the same as in the policies and procedures [20, 22].

Nevertheless, most safety risk management applications still use traditional risk assessment techniques. The safety risk management practice in complex systems can, therefore, be strengthened by combining FRAM with traditional techniques [40, 44].

Several studies have shown the potential value of FRAM to ensure safety in complex systems [12, 14, 18, 27, 45]. FRAM supports a better understanding of a system and provides a more in-depth analysis by revealing risks emerging from system interactions. Future studies could apply FRAM in comparison to traditional risk assessment techniques to demonstrate their strengths.

Acknowledgements This work was supported by the Research Fund of the İstanbul Medeniyet University. Project number: 1695.

References

1. Adriaensen A et al (2019) A socio-technical analysis of functional properties in a joint cognitive system: a case study in an aircraft cockpit. Ergonomics 1–19
2. Anvarifar F et al (2017) An application of the Functional Resonance Analysis Method (FRAM) to risk analysis of multifunctional flood defences in the Netherlands. Reliab Eng Syst Saf 158:130–141
3. Aven T (2008) Risk analysis—assessing uncertainties beyond expected values and probabilities. Wiley, New York
4. Aven T (2012) The risk concept-historical and recent development trends. Reliab Eng Syst Saf 99:33–44
5. Aven T (2016) Risk assessment and risk management: review of recent advances on their foundation. Eur J Oper Res 253(1):1–13
6. Baybutt P (2015) Designing risk matrices to avoid risk ranking reversal error. Process Saf Prog 35(1):41–46
7. Borghesi A, Gaudenzi B (2013) Risk management: how to assess, transfer and communicate critical risks. Springer, London
8. BSI (2001) BS IEC 61882: hazard and operability (HAZOP) studies. British Standards Institution, London. Available at https://cdn.auckland.ac.nz/assets/ecm/documents/Hazard-Operability-Studies.pdf
9. BSI (2006) BS EN-60812: Analysis techniques for system reliability: procedure for failure mode and effects analysis (FMEA). British Standards Institution, London
10. BSI (2009) BS ISO 31000: risk management: principles and guidelines. British Standards Institution, London
11. BSI (2010) BS EN 31010: risk management: risk assessment techniques. British Standards Institution, London
12. Cabrera Aguilera MV et al (2016) Modelling performance variabilities in oil spill response to improve system resilience. J Loss Prev Process Ind 41:18–30
13. Card A, Ward J, Clarkson P (2014) Generating options for active risk control (GO-ARC): introducing a novel technique. J Healthcare Qual 36(5):32–41
14. de Carvalho PVR (2011) The use of Functional Resonance Analysis Method (FRAM) in a mid-air collision to understand some characteristics of the air traffic management system resilience. Reliab Eng Syst Saf 96(11):1482–1498

15. Chemweno P et al (2015) Development of a risk assessment selection methodology for asset maintenance decision making: An analytic network process (ANP) approach. Int J Prod Econ 170:663–676
16. Clay-Williams R, Hounsgaard J, Hollnagel E (2015) Where the rubber meets the road: using FRAM to align work-as-imagined with work-as-done when implementing clinical guidelines. Implementation Sci 10:125
17. Cox LA (2008) What's wrong with risk matrices? Risk Anal 28(2):497–512
18. Furniss D, Curzon P, Blandford A (2016) Using FRAM beyond safety: a case study to explore how sociotechnical systems can flourish or stall. Theor Issues Ergon Sci 17(5–6):507–532
19. Gadd SA, Keeley DM, Balmforth HF (2004) Pitfalls in risk assessment: examples from the UK. Saf Sci 42:841–857
20. Hollnagel E (2012) FRAM: the Functional Resonance Analysis Method modelling complex socio-technical systems. Ashgate, Surrey
21. Hollnagel, E. (2013) An application of the Functional Resonance Analysis Method (FRAM) to risk assessment of organisational change. Stockholm
22. Hollnagel E (2014) Safety I and safety II. Ashgate, Surrey
23. Hollnagel E (2018) The functional resonance analysis method. Report
24. IEC (2006) IEC 61025 fault tree analysis (FTA). Geneva
25. Kaya GK, Canbaz HT (2019) The problem with traditional accident models to investigate patient safety incidents in health-care. In: Calisir F, Cevikcan E, Akdag CH (ed) Industrial engineering in the big data era. Lecture notes in management and industrial engineering. Springer, Cham, pp 481–488
26. Kaya GK, Hocaoglu M (2020) Semi-quantitative application to the Functional Resonance Analysis Method for supporting safety management in a complex health-care process. Reliab Eng Syst Saf 0(0):1–9
27. Kaya GK, Ovali HF, Ozturk F (2019) Using the functional resonance analysis method on the drug administration process to assess performance variability. Saf Sci 118:835–840
28. Kaya GK, Ward JR, Clarkson PJ (2019) A framework to support risk assessment in hospitals. Int J Qual Health Care 31(5):392–401
29. Kaya GK, Ward JR, Clarkson PJ (2019) A review of risk matrices used in acute hospitals in England. Risk Anal 39(5):1060–1070
30. Kazaras K, Kirytopoulos K, Rentizelas A (2012) Introducing the STAMP method in road tunnel safety assessment. Saf Sci 50:1806–1817
31. Leveson N (2011) Engineering a safer world: systems thinking applied to safety. The MIT Press, Massachusetts
32. Leveson N (2015) A systems approach to risk management through leading safety indicators. Reliab Eng Syst Saf 136:17–34
33. Li J, Bao C, Wu D (2018) How to design rating schemes of risk matrices: a sequential updating approach. Risk Anal 38(1):99–117
34. Mullai A (2006) Risk assessment frameworks and techniques. In: Ojala L (ed). DaGoB, Turku
35. Patriarca R, Bergström J et al (2018) Resilience engineering: current status of the research and future challenges. Saf Sci 102:79–100
36. Patriarca R, Falegnami A et al (2018) Resilience engineering for socio-technical risk analysis: application in neuro-surgery. Reliab Eng Syst Saf 180:321–335
37. Patriarca R et al (2019) Systemic safety management in anesthesiological practices. Saf Sci 120:850–864
38. Patriarca R, Bergström J (2017) Modelling complexity in everyday operations: functional resonance in maritime mooring at quay. Cogn Technol Work 19(4):711–729
39. Patriarca R, Bergström J, Di Gravio G (2017) Defining the functional resonance analysis space: Combining Abstraction Hierarchy and FRAM. Reliab Eng Syst Saf 165:34–46
40. Patriarca R, Di Gravio G, Costantino F (2017) A Monte Carlo evolution of the Functional Resonance Analysis Method (FRAM) to assess performance variability in complex systems. Saf Sci 91:49–60

41. Pitblado R, Weijand P (2014) Barrier diagram (bow-tie) quality issues for operating managers. Am Inst Chem Eng 33(4):355–361
42. de Rademaeker E et al (2014) A review of the past, present and future of the European loss prevention and safety promotion in the process industries. Process Saf Environ Prot Inst Chem Eng 92:280–291
43. Simsekler MCE et al (2019) Integration of multiple methods in identifying patient safety risks. Saf Sci 118:530–537
44. Sujan M, Felici M (2012) Combining failure mode and functional resonance analyses in healthcare settings. In: Ortmeier F, Daniel P (eds) International conference on computer safety, reliability and security (SAFECOMP). Springer, Berlin Heidelberg, pp 364–375
45. Tian J et al (2016) FRAMA: a safety assessment approach based on Functional Resonance Analysis Method. Saf Sci 85:41–52

The Problems of Project Management Companies in the Turkish Construction and the Real Estate Industries

Levent Sumer and Fatih Kiraz

Abstract Effective project management has always been important for the success of different industries. From scope, time, and cost management to quality, risk, and procurement management, all aspects of project management need an integrated approach by using skilled staff and providing efficient communication among all stakeholders. While some companies establish in-house teams for managing their projects, some companies hire professional companies to act on their behalf. It is critical to measure the success of these companies, yet there are major obstacles that project management companies face while they are performing their responsibilities, which may cause failures. In that context, this study, as far as being the first research in the literature, investigates the main problems of project management (PM) companies in Turkey by focusing on the PM services in the construction and real estate industries. A survey was conducted among PM professionals working for PM companies, contractors, employers, designers, vendors, and sub-contractors to find out these problems. The need for PM standards, the balance between the responsibility and the authority of the PM companies, the existence of the PM companies providing very low prices, excessive employer intervention and delayed payments by employers were found to be the most important factors that cause problems. Developing PM standards, issuing legal regulations to be a PM company, right pricing policy, and on-time payments by the project owners were recommended by the participants as solutions.

Keywords Project management · Authority · Standardization · Payment · Success

L. Sumer (✉) · F. Kiraz
Department of Business Administration, Faculty of Economics, Administrative and Social Sciences, Istinye University, Istanbul, Turkey
e-mail: levent.sumer@istinye.edu.tr

F. Kiraz
e-mail: fatih.kiraz@istinye.edu.tr

F. Calisir (ed.), *Industrial Engineering in the Internet-of-Things World*,
Lecture Notes in Management and Industrial Engineering,
https://doi.org/10.1007/978-3-030-76724-2_2

Introduction

The construction industry requires a professional project management approach due to the level of the difficulty of its nature. The uniqueness of each project and many potential obstacles and problems faced in a construction site such as adverse weather conditions, limited transportation, lack of labor or materials, etc., define the construction project management as chaos management. The level of complexity of the construction industry make the conditions less predictable, and that may cause inefficient results [14]. In that context, it is essential to assign professional teams who are experienced in project management services, including scope, time, design, cost, contracts, health, safety, quality, procurement, risk, and other management disciplines. While some companies establish in-house teams for managing their projects, some companies hire professional companies to act on their behalf. It is critical to measure the success of these companies, yet there are major obstacles that project management companies face while they are performing their responsibilities, which cause failures.

One of the earliest studies about the challenges of project management made by MacDonald [11] determined the problems of project management as poorly defined project management controls, insufficient work definition, unrealistic schedules, underestimated costs, and inadequate cost control and accounting practices. Wright [16] defined the project management challenges as undefined goals, scope changes, inadequately skilled personnel, lack of accountability, improper risk management, ambiguous contingency plans, poor communications, impossible deadlines, resource deprivation, and lack of stakeholder engagement. Arditi [2] pointed out the link between the company culture and the project delays, one of the major problems of construction project management. Kale and Arditi [8] investigated the effects of the strong relationship between the contractors and subcontractors on contractors' performance, and they figured out that a strong relationship among parties affects the performance positively. Sometimes the perception of different parties regarding the project management problems differs as [1] pointed out in their study in which they measured owners' and contractors' perception about the time delays and [12] stressed the project management consultancy perspective for time overruns and pointed out the cash problems, poor site management, insufficient experience and planning and scheduling inefficiency of the contractors. Chou and Yang [4] seek the effects of project management knowledge on construction project outcomes based on the Project Management Body of Knowledge Book (PMBOK) Guide published by Project Management Institute (PMI), one of the most well-known NGO focusing on project management and emphasized the importance of communication. Kashyap [9] also underlined the poorly defined goals and objectives, unrealistic deadlines, insufficient team skills, miscommunication, risk management, scope, and lack of accountability as the main challenging issues. Keeping teams on the same page, using the right project management software, and teamwork problems are the other points he stressed about the problems of project management.

Regarding the problems of project management companies, the studies in the literature are very limited [15], mainly underlining the relationship-based issues, analyzed the problems faced by project management companies in New Zealand and summarized the problems as,misunderstanding and lack of respect of other professionals to project management companies, employer-related problems such as the difficulty of obtaining or understanding the brief, coordination or communication problems and on-time payment. Another study was conducted by [10] by focusing on the constraints of the development of construction project management in China, and their study exhibited that the lack of qualified and experienced project managers, misunderstanding the roles and responsibilities, limited knowledge and skills, distorted competition in the market, the time of appointing the project management companies are the major limitations of the development of the project management companies in the Chinese construction industry. In that context, for sustainable success, companies should invest in intangible assets [7].

In the last decades, the Turkish construction and the real estate industry grew rapidly. Although the general contractors gained international experience especially in Russia, CIS countries, the Middle East, and North Africa, the real estate investment projects within the country were constructed by less qualified and less experienced construction companies. That situation caused a need for professional project management services, and while some international project management companies entered into the Turkish construction market and completed successful projects, new local project management companies were also established and started competing with the international firms. In the last few years, the real estate industry slowed down, and some multinational companies left the market, and the remaining companies tried to survive in the hard competition where the market contracted. These companies need to overcome major problems while performing their responsibilities.

In that context, this study, as far as being the first research in the literature, investigates the main problems of project management (PM) companies in Turkey by focusing on the PM services in the construction and real estate industries. A survey was conducted among project management professionals working for PM companies, contractors, employers (investors, developers), designers, vendors, and sub-contractors. The respondents were asked to scale the major problems faced by the PM companies as well as to provide their own opinions, and the results of the survey were analyzed by using R programming language and R core team's "stats" package [13]. The need for project management standards, the balance between the responsibility and the authority of the PM companies, the existence of the PM companies providing unrealistic prices, excessive employer intervention, and delayed payments by employers were found the most important factors that cause problems. Developing PM standards, issuing legal regulations to be a PM company, right pricing policy, and on-time payments by the project owners were recommended by the participants as solutions. Baes on the results, some important recommendations are made for enhancing the successes of project management companies in the Turkish construction and real estate industries.

Table 1 Organization type and level of education

Organization type	Frequency (%)	Level of education	Frequency (%)
Project management company	*26.61*	Ph.D.	11.01
General contractor	*19.27*	Masters	53.21
Employer (investor, developer)	*24.77*	Undergraduate	34.86
Others (designer, vendor, sub-contractor)	*29.36*	High School	0.92

Table 2 Occupation and position

Occupation	Frequency (%)	Position	Frequency (%)
Civil Engineer	*61.11*	General Manager	*24.77*
Architect	*11.11*	Deputy GM	*2.75*
Electrical Engineer	*4.63*	Director	*13.76*
Mechanical Engineer	*7.41*	Manager	*28.44*
Urban Planner	*2.78*	Others	*30.28*
Interior Designer	*0.92*		
Others	*12.04*		

Methodology

Data

A survey was conducted, the questions shown in Tables 1, 2, 3, 4 and 5 were asked, and 109 responses were collected to find out the perception of the project management professionals regarding the problems and the challenges they face while managing the construction projects. Likert scale was used while collecting the responses. 27% of the 109 respondents work for a PM company, which is followed by 25% with

Table 3 Experience and gross yearly income of the company

Experience (years)	Frequency (%)	Gross yearly income of the company (Million TL)	Frequency (%)
> 20	*37.61*	> 50	*48.57*
15–20	*34.86*	20–50	*7.62*
10–15	*17.43*	10–20	*3.81*
5–10	*8.26*	5–10	*11.43*
< 5	*1.83*	< 5	*28.57*

Table 4 The problems of the project management companies

The problems of the project management companies
Lack of balance between authority and responsibility
Excessive employer intervention
Inadequacy of project management standards
Low pricing policy due to fierce competition
Lack of qualified employee
Construction and real estate industries are in a crisis
Oppressive contract terms
Bank guarantee requirements
Delayed payments

Table 5 Recommendations for resolving the problems of the PM companies

Recommendations for resolving the problems of the PM companies
Providing fair contracts
Right pricing policy by the employers
No interference of the employers to the authority areas of the PM companies
On-time payment
No pressure regarding the timeline
Applying project management standards
Legal requirements for being a project management company

project owners (investors and developers) and 19% with general contractors, respectively. Almost all the respondents have a university degree, while 64% of them have a master or a Ph.D. degree. Civil engineering, with 61%, is the main occupation among the respondents, architecture is ranked in second place with 11%. 70% of the total responses are answers of the people who work as a manager or at an upper management position where every 1of 4 respondents are the general manager of the company they work for. 90% of the total respondents have at least ten years or more experience in the construction or real estate industry. The respondents with more than 20 years of experience have the highest rate among all responses, with 38%. Besides, almost half (49%) of the responses are from the employees working for companies that have more than 50 million TL annual gross income. The categorical questions and the frequencies of the responses are shown in Tables 1, 2 and 3.

Bredillet [3] underlines the existence of standards as a key issue in the economic growth and performance of firms. Crawford [6] emphasizes the collective opinions of experienced managers while developing project management standards. Clarke [5] puts standardization of project management, perception about the benefits of the project management such as being considered as a reporting tool, inadequate change control system, project overloads at the same time, cultural and individual issues on the top where the project management companies challenge to increase their effectiveness.

Results

The respondents were asked to scale the questions the potential problems obtained throughout the literature, the past real estate and construction management experiences of the authors, and the discussions with senior professionals. The results exhibited that; inadequacy of project management standards with a 69% response rate seems to be the most important problem of the PM firms. That is followed by the lack of balance between the authority and responsibility of PM companies (63%), parallel to the findings of Liu et al. [10] low pricing policy due to fierce competition among PM companies (61%) and excessive employer intervention (59%) are ranked third and fourth respectively. The delayed payment, an important variable from [15], is ranked as the fifth important problem in this study.

One of the striking results of this study is that all respondent groups share similar feelings about the top 4 major problems above (each higher than 50% score) and also their importance levels. There is no statistically significant difference between any possible pair. Chi-Squared test results are summarized in Table 6. On the other hand, Fig. 1 shows, for each variable, the percentages of respondents who agree that the given statement is a major problem of PM companies.

Unfair contract terms and bank guarantee requirements seem to be the least important problems, according to the participants.

The majority of the PM services (43%) are provided as Agency Construction Management, where the PM company acts as the agent of the Employer to manage the project. Design-bid-build and design-build methods had almost the same rate with 35% and 36%, respectively. Construction management at risk, the model where the PM firm carries all the risks of the project, is not a very common method in Turkey. Figure 2 shows the distribution of the responses in terms of the project delivery system used.

The respondents were also asked to rank the possible solutions to resolve the problems of the PM companies. Parallel to the problems, they ranked applying PM standards as the most important solution with a 70% response rate. Legally regulating the PM services, which may lead to constraining the non-qualified companies to compete in the market by providing unrealistic prices, were found the second and third important solutions. No interference of the employer was also ranked as the 4th recommended solution chosen by the participants of the survey. The results were also validated by the X-squared test. Figure 3 and Table 7 shows the results of the analysis.

Table 6 The results of Pearson's chi-squared tests

Inadequacy of project management standards	Not a Problem	Main Problem
Project management firm	7	24
General contractor	8	13
Employer (investor, developer)	9	20
Others	10	18
X-squared = 1.8045, df = 3, *p*-value = 0.614		
Lack of balance between authority and responsibility	Not a problem	Main problem
Project management firm	11	20
General contractor	5	16
Employer (investor, developer)	12	17
Others	13	15
X-squared = 2.8676, df = 3, *p*-value = 0.4125		
Excessive employer intervention	Not a problem	Main problem
Project management firm	10	21
General contractor	11	10
Employer (investor, developer)	14	15
Others	10	18
X-squared = 3.0518, df = 3, *p*-value = 0.3837		
Low pricing policy due to fierce competition	Not a problem	Main problem
Project management firm	11	20
General contractor	5	16
Employer (investor, developer)	13	16
Others	14	14
X-squared = 4.0105, df = 3, *p*-value = 0.2603		
Lack of qualified employee	Not a problem	Main problem
Project management firm	24	7
General contractor	10	11
Employer (investor, developer)	15	14
Others	17	11
X-squared = 6.1092, df = 3, *p*-value = 0.1064		
The industry is in a crisis	Not a problem	Main problem
Project management firm	22	9
General contractor	13	8
Employer (investor, developer)	16	13
Others	12	16
X-squared = 4.994, df = 3, *p*-value = 0.1722		

(continued)

Table 6 (continued)

Oppressive contract terms	Not a problem	Main problem
Project management firm	25	6
General contractor	16	5
Employer (investor, developer)	24	5
Other	25	3
X-squared = 1.5517, df = 3, p-value = 0.6704		
Bank guarantee requirements	Not a problem	Main problem
Project management firm	28	3
General contractor	16	5
Employer (investor, developer)	26	3
Others	22	6
X-squared = 3.2322, df = 3, p-value = 0.3572		
Delayed payments	Not a problem	Main problem
Project management firm	18	13
General contractor	11	10
Employer (investor, developer)	19	10
Others	9	19
X-squared = 7.004, df = 3, p-value = 0.07177		

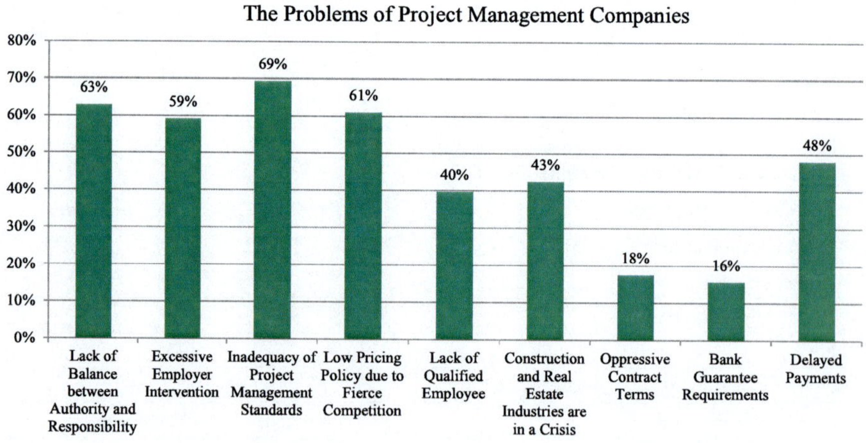

Fig. 1 Agreement levels on the major problems of PM companies

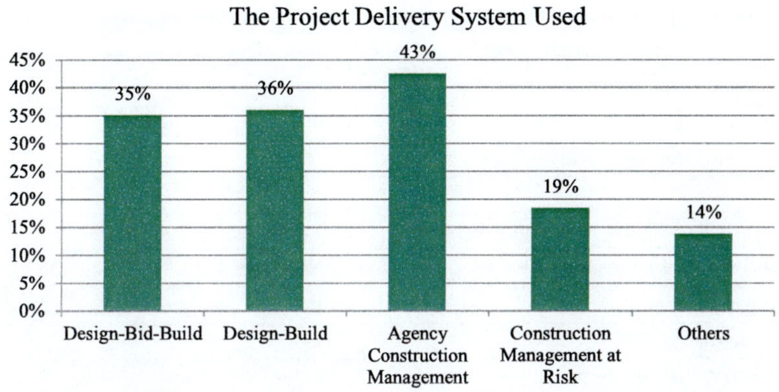

Fig. 2 The project delivery systems used

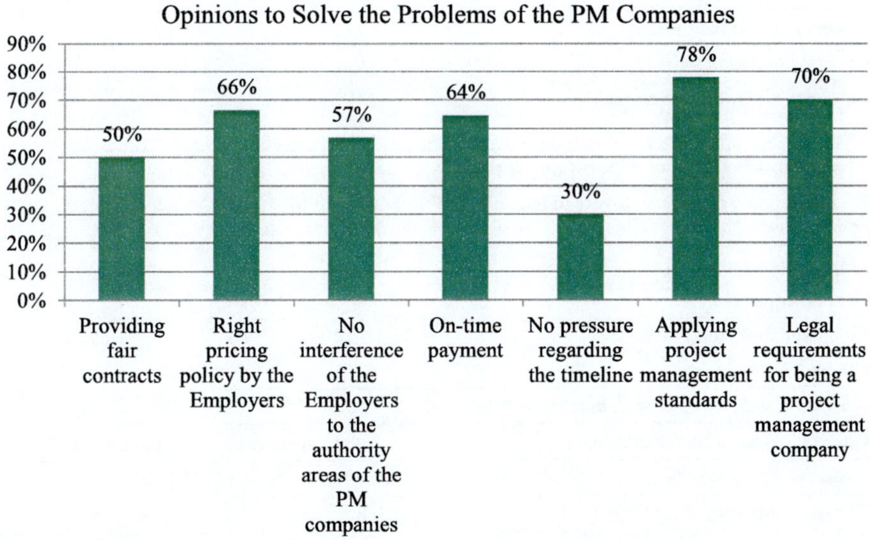

Fig. 3 Opinions of the respondents to solve the problems of the PM companies

Discussion and Conclusion

Professional project management service provided by PM companies is an important success criterion for construction projects. The results of this study exhibit that the project management companies in Turkey mostly suffers from the inadequacy of project management standards, the lack of balance between the responsibility and the authority of the PM companies while performing their tasks, the existence of

Table 7 Opinions of the respondents to solve the problems of the PM companies

Applying project management standards	Not suggested	Suggested
Project management firm	3	28
General contractor	4	17
Employer (investor, developer)	11	18
Others	10	18
X-squared = 8.3995, df = 3, *p*-value = 0.03844		
No pressure regarding the timeline	Not suggested	Suggested
Project management firm	18	13
General contractor	19	2
Employer (investor, developer)	21	8
Others	20	8
X-squared = 6.477, df = 3, *p*-value = 0.09057		
Right pricing policy by the employers	Not suggested	Suggested
Project management firm	11	20
General contractor	9	12
Employer (investor, developer)	10	19
Others	10	18
X-squared = 0.43552, df = 3, *p*-value = 0.9328		
No interference of the employers to the authority areas of the PM companies	Not suggested	Suggested
Project management firm	11	20
General contractor	14	7
Employer (investor, developer)	14	15
Others	11	17
X-squared = 5.5612, df = 3, *p*-value = 0.135		
On-time payment by the employer	Not suggested	Suggested
Project management firm	17	14
General contractor	8	13
Employer (investor, developer)	12	17
Others	5	23
X-squared = 8.6346, df = 3, *p*-value = 0.03457		
Providing fair contracts	Not suggested	Suggested
Project management firm	17	14
General contractor	10	11
Employer (investor, developer)	16	13
Others	14	14
X-squared = 0.41981, df = 3, *p*-value = 0.9361		

(continued)

Table 7 (continued)

Legal requirements for being a PM firm	Not suggested	Suggested
Project management firm	7	24
General contractor	6	15
Employer (investor, developer)	9	20
Others	14	14
X-squared = 5.4166, df = 3, p-value = 0.1437		

the PM companies providing unrealistic prices, excess employer's involvement and delayed payments by the employers.

In Turkey, some institutions are established to improve the PM knowledge in the country such as Turkey Chapter of Project Management Institute (PMI) and Istanbul Project Management Association. Besides, the International Project Management Institute (UPYE) has focused on the real estate and construction industries and has been working on developing construction project management standards. This study may help them to focus on certain issues while structuring the PM standards and presenting it to the government authorities.

In this study, some of the respondents, which are senior experts in the construction project management, also provided additional ideas and enhanced the value of the research. These senior professionals classified the problems of the project management companies as; the lack of project management approach in the country, the requests of the employers to complete the projects before deadlines, undervaluing of the services provided by the project management companies, considering the project management services as a cost item, inadequateness of project management systems, cost changes due to changes in the economic conditions, contractual conflicts occurred due to the insufficient knowledge of the employers about project management, the mispositioning the project management company by the employer and evaluating them as a third party which is trying to find their mistakes, underusing the project management company because of the lack of knowledge of what they provide, insisting on selecting the cheapest project management company offer, mismanagement of design processes which causes additional effort carried by all the parties. The lack of balance between the responsibility and authority which is ranked as the second important problem here is a major issue for the PM firms because although the employers transfer all the responsibilities regarding the works on-site, they still keep the last word about the decisions to be given such as extending the payment durations, applying or waiving the delay penalties of the contractors, approving or changing the design without complying with the durations stated in the contracts. That makes the PM company inefficient while managing the parties on site. This major issue can only be resolved if the employers are aware of the value of the services they get from professional PM companies. One of the senior professionals underlined the contribution of the project management companies to the project cost by making value engineering. Considering the project management company as a party that is responsible for quantity surveying only and the dependency of the project management company to the employer or contractor were listed as additional problems of the PM companies.

In addition to the findings of this study regarding the solutions, the respondents suggested providing a minimum PM fee standard by the chambers or the government agencies, selecting the project management company before the design process, training the employers about project management services by third parties, putting additional effort for right planning, determining the project management quality standards, paying more attention to planning and design phases and allocating more time and budget on these issues.

The number of PM companies in Turkey is not a high figure. This is one of the reasons this study also covered other related groups. Another reason is the curiosity about whether the results would change if we also look from different perspectives. A future study, especially when there are more PM companies, may focus largely on the employees of the PM companies, and then the results may be compared with the findings of this initial study in the relevant literature. Further researches may also consider using different international standards such as ISO 21500: 2012 Guidance on project management and Construction Management Association of America (CMAA) standards and compare their approach with the current market practice in Turkey.

References

1. Andi A, Lalitan D, Loanata VR (2010) Perceptions toward factors causing delays in structural and finishing works. Civ Eng Dimension 12(1):8–17
2. Arditi D, Nayak S, Damci A (2017) Effect of organizational culture on delay in construction. Int J Project Manage 35(2):136–147
3. Bredillet CN (2003) Genesis and role of standards: theoretical foundations and socio-economical model for the construction and use of standards. Int J Project Manage 21:463–470
4. Chou J, Yang J (2012) Project management knowledge and effects on construction project outcomes: an empirical study. Proj Manage J 43(5):47–67
5. Clarke A (1999) A practical use of key success factors to improve the effectiveness of project management. Int J Project Manage 17:139–145
6. Crawford L (2005) Senior management perceptions of project management competence. Int J Project Manage 23:7–16
7. Ertugrul M (2020) Direct and indirect value relevance of R&D capitalization. İstanbul Ticaret Üniversitesi Sosyal Bilimler Dergisi 19(37):781–802
8. Kale S, Arditi D (2001) General contractors' relationships with subcontractors: a strategic asset. Constr Manage Econ 19(5):541–549
9. Kashyap S (2019) 10 common challenges in project management and how to solve them. Available at https://www.proofhub.com/articles/project-management-challenges
10. Liu G, Shen Q, Li H, Shen L (2004) Factors constraining the development of professional project management in China's construction industry. Int J Project Manage 22:203–211
11. MacDonald RH (1974) Project management problems which can be avoided. Project Manage Q 5(2):34–36
12. Memon AH, Abdul RI, Abdullah MR, Abdulaziz AA (2011) Time overrun in construction projects from the perspective of project management consultant (PMC). J Surveying Constr Property 2(1):54–66

13. R Core Team (2019) R: a language and environment for statistical computing. R Foundation for Statistical Computing, Vienna, Austria. https://www.R-project.org/
14. Ribeiro P, Paiva A, Varajao J, Dominguez C (2013) Success evaluation factors in construction project management -some evidence from medium and large Portuguese companies. KSCE J Civ Eng 17(4):603–609
15. Wilkinson S (2001) An analysis of the problems faced by project management companies managing construction projects. Eng Constr Archit Manage 3:160–170
16. Wright S (2016) Project management challenges in the construction industry. Available at https://project-management.com/project-management-challenges-in-the-construction-industry/

Forecasting of Retail Produce Sales Based on XGBoost Algorithm

Yakup Turgut and Mustafa Erdem

Abstract Sales forecasting of vegetables and fruits imposes a challenging task for the retailers because the demand for them varies depending on several factors, such as temperature, season, holiday. Poor sales forecasting can cause too much cost for retailers since these products are unusable after deterioration. Also, people tend to consume these products freshly. This research aims to compare the forecasting performance of traditional statistical and new machine learning methods. We apply seasonal ARIMA to forecast daily sales of fruits and vegetables as a traditional method. As a machine learning algorithm, we apply LSTM and XGBoost algorithms. The results indicate that the XGBoost algorithm gives more accurate results than the other two methods.

Keywords Machine learning · Perishable foods · Sales forecasting · SARIMA · XGBoost · LSTM

Introduction

Determining the stock levels of perishable products is more complicated than nonperishable products due to their short shelf life and customer behavior toward them [2, 26]. Therefore, it is necessary to develop different stock policies (van Donselaar 2006) and supply chain (SC) strategies [7, 28] for these foods. Fruits and vegetables, which are largely studied in this study, are classified as perishable foods with particular storage characteristics. Managing the fruits and vegetables supply chain is complex and difficult because of their fluctuating demand pattern. Several factors,

Y. Turgut (✉)
Industrial Engineering Department, Management Faculty, Istanbul Technical University, Istanbul, Turkey
e-mail: turgut16@itu.edu.tr

M. Erdem
Mechatronics Engineering Department, ITU Institute of Science and Technology, Istanbul Technical University, Istanbul, Turkey
e-mail: erdemm@itu.edu.tr

F. Calisir (ed.), *Industrial Engineering in the Internet-of-Things World*, Lecture Notes in Management and Industrial Engineering,
https://doi.org/10.1007/978-3-030-76724-2_3

such as weather conditions (Agnew and Tornes 1995), price changes [15], seasonality [19], are identified as the causes of variation in the demand of these foods. For Arunraj and Ahrens [5], factors can be classified as controllable factors, partially controllable factors, or uncontrollable factors. The authors state that the first includes price and product characteristics, the second includes substitution and cannibalization, and third includes events, weather, seasonality, and the number of customers. If organizations do not take into account these factors in stock management, they may suffer financial losses due to the waste of food and loss of customers. Optimized replenishment of these foods at the retail level is key to reducing the waste of these foods and increasing the efficiency of the fruits and vegetables supply chain. Optimal replenishment orders depend upon the sales forecasting [22]. Sales forecasting do not only enable retail managers to cope with stochastic demand, but also helps to maintain a competitive advantage in SC management.

Although the replenishment process for these foods usually performed manually by managers based on experience and the point of system (POS system) the necessity of analytical techniques has been understood lately, and more recent research has occurred in the demand forecasting of the perishable foods based on traditional statistical methods (causal models, time series and econometric methods) and machine learning (ML) methods. Sankaran [23] uses the Seasonal ARIMA model to forecast the daily demand for onion at a wholesale market and conclude that forecasting performance is satisfactory with an erratic demand. Raju et al. [21] investigate the factors causing the stochastic demand for perishable foods and examines the forecasting performance of linear and nonlinear forecasting methods. It concludes that temperature is the predominant factor that influences the demand, and nonlinear methods generate more accurate results than linear methods. Yang and Sutrisno (2018) bring a new perspective to forecast the demand for the bakery at franchise stores. The idea is to use sales occurring in the early morning hour to forecast the sales of the rest of the day. They also compare the forecasting performance of Feed Forward Neural Network (FFNN) and Regression analysis. They conclude that this approach is very promising to generate online-forecasting, and FFNN gives better results than regression analysis. Sridama and Siribut [24] propose a decision support system for demand forecasting of perishable foods to improve the inventory management of these foods. They analyze the forecasting performance of the following time series methods: Single exponential smoothing, Adaptive-response-rate single exponential smoothing, and Holt's two parameters linear exponential smoothing. They conclude that the Single exponential smoothing method gives better results than others. Huber and Stuckenschmidt (2017) propose a decision support system (DSS) based on the hierarchical clustering approach to obtain demand forecasts of perishable food at different organizational levels. They implemented the proposed DSS in the bakery chain of a company. The authors use multivariate ARIMA as a forecasting method. They conclude that the proposed approach gives acceptable results to increase the efficiency of the supply chain, and also decreases the computational time. The approach enables us to develop replenishment strategies based on product categories exhibiting similar demand patterns. Yang and Hu (2008) apply an ARIMA model to forecast the demand for cabbage.

Chen and Ou [13] propose an extended neural network model to forecast the daily demand for milk in a convenience store, and they compare the proposed model with an ARIMA model. The results indicate that the proposed model generates better results than the ARIMA model. Du et al. [14] develop an algorithm based on the Support vector machine and fuzzy theory to forecast the demand for perishable farm foods. They conclude that the proposed algorithm gives more promising results than radial basis function neural networks.

Although time series forecasting is generally superior to judgemental and econometric forecasting techniques for forecasting retail sales [16], it still lacks capturing the sudden changes in demand due to characterized nature of perishable foods. The forecasting performance of these methods may be improved by using hybridized versions of them [5]. Chen and Ou [10] propose a model which combines gray relational analysis and multilayer functional network model to forecast the sales of perishable food in a convenience store.

While these hybrid methods have provided considerable improvement in forecasting accuracy, much of it was not focusing especially on forecasting the fruits and vegetable sales at the retail levels. Furthermore, the forecasting accuracy is still required to improve by applying new algorithms. To improve the forecasting accuracy, this study focuses on the application of a gradient boosting ML algorithm, Extreme Gradient Boosting algorithm (XGBoost) due to its capabilities to handle the sparse data (Chen and Guestrin 2016), and computational efficiency, and it is popularity in ML competitions (Chen and He 2015). The results of the model are encouraging and show that XGBoost outperforms than classical SARIMA and LSTM models.

The rest of the paper is organized as follows: Sect. 2 outlines our method, which is used in this study and presents the description of data. Section 3 discusses the results of the performance of applied methods. Finally, Sect. 4 gives the conclusion and future research of the study.

Methodology

Description of Data

This study uses daily sales data of vegetables and fruits from a supermarket in Istanbul, Turkey, as a case study from January 2014 to December 2017. It would be more appropriate to conduct the forecasting on the product level due to the product-specific nature of demand pattern. Unfortunately, aggregate daily sales of vegetables and fruits are considered due to a lack of sales data on the product level. Figure 1 shows the daily sales data of vegetables and fruits in the time series plot. This time series plot shows that there is no obvious trend in data. There is an increasing trend in the sales of vegetables and fruits on some dates, such as the last week of the year.

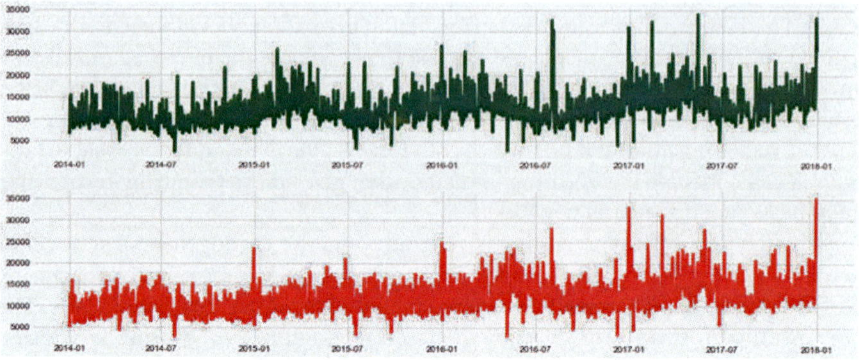

Fig. 1 The annual sales of vegetables and fruits (green indicates vegetables, red indicates fruits)

This cyclic pattern is repeated every year. This weekly or seasonal variation may be attributable to some causes such as the impact of weather or holidays.

Time Series

A time series is a sequential set of observations measured at successive times [8]. There are four key components of time series data, which should be analyzed before applying an algorithm: (1) Trend, (2) Seasonality, (3) Cyclical, and (4) Irregularity. Trend describes the general direction of observations over a long time. Seasonality explains the variation in the observations over a period of one year, usually caused by weather conditions, holidays, vacations, etc. Cyclical refers to the nonperiodic variations caused by circumstances, which occur in a repeating pattern. The duration of these variations lasts several years. Irregularity refers to the random variations in observations caused by unforeseeable reasons, such as earthquakes, floods, epidemic diseases, etc. Expectedly, these variations do not have a particular pattern. Time series plots may reveal these patterns or a combination of these patterns [3].

One of the most critical behaviors of a time series is stationarity. The stationarity of a time series data indicates its statistical behavior in time. When a time series exhibits this property, the statistical behavior of that series does not change in time. This means that it has a constant probability distribution. A time-series data must have this property because nonstationary data cannot be forecasted due to its unstable nature. If a time series data does not have stationarity behavior, it should be converted to a stationary form before performing any forecasting. There are two options to check the stationarity of a time series data: (1) Rolling statistics: plot the rolling average (moving average) and see how it varies with time, (2) ADCF (Augmented Dickey-Fuller Test) provides a formal statistical test to detect the stationarity property. The null hypothesis claims that the time series is nonstationary.

Seasonal ARIMA (SARIMA) Model

ARIMA stands for the autoregressive moving average, and it has three parameters: (p, d, q). AR component is referred to the use of past values in the regression equation for the series Y. The parameter p indicates the number of lags used in the model. MA component represents the error of the model as a combination of previous error terms. The parameter q specifies the number of terms to include in the model. The parameter d represents the degree of differencing in the integrated component. Differencing a series involves simply subtracting its current and previous values d times. It is used to stabilize the data to satisfy the stationarity assumption [9]. SARIMA (Seasonal ARIMA) is an extension to ARIMA, which allows the direct modeling of seasonal behavior of data. SARIMA model is represented by the following notation: ARIMA (p, d, q) (P, D, Q)$_s$. The lower-case (p, d, q) is the same as the nonseasonal ARIMA model. The upper-case (P, D, Q) represents the seasonal parameters of the model. The subscripted letter s indicates the length of the period in each season. For example, in monthly data, s = 12. Let d and D are nonnegative integers. A SARIMA model general form is given in Eq. (1)

$$\emptyset_P(B)\Phi_P\left(B^s\right)Y_t = \theta_q(B)\Theta_Q\left(B^S\right)\varepsilon_t \tag{1}$$

$$\emptyset_P(B) = 1 - \phi_1 B - \phi_2 B^2 - \ldots - \phi_P B^P \tag{2}$$

$$\Phi_P\left(B^S\right) = 1 - \Phi_S B^S - \Phi_{2S} B^{2S} - \ldots - \Phi_{PS} B^{PS} \tag{3}$$

$$\theta_q(B) = 1 - \theta_1 B - \theta_2 B^2 - \ldots - \theta_q B^q \tag{4}$$

$$\Theta_Q\left(B^S\right) = 1 - \Theta_S\left(B^S\right) - \Theta_{2S}\left(B^{2S}\right) - \ldots - \Theta_{QS}\left(B^{QS}\right) \tag{5}$$

where $\{X_t\}$ is the original series; $Y_t = (1 - B)^d(1 - B^s)^D X_t$ is differenced series to eliminate seasonality component; B is the lag operator; $\emptyset(B)$ and $\theta(B)$ are polynomials of order p and q, respectively; $\Phi\left(B^S\right)$ and $\Theta\left(B^S\right)$ are polynomial in B of degrees P and Q respectively; p is the order of nonseasonal autoregression; d is the number of regular differences; q is the order of nonseasonal moving average; P is the order of seasonal autoregression; D is the number of seasonal differences; Q is the order of seasonal moving average; and S is the length of the season [9]. The steps of the model are explained in the section of the application of the SARIMA model.

Long Short-Term Memory (LSTM)

Feedforward neural networks are not suitable for sequential data due to their fixed-size input/output. Therefore, they cannot be used to model memory. On the other hand, recurrent neural networks (RNN) are designed for capturing information from time-series data. In an RNN, thanks to the recurrence relation, each state is dependent on all previous computations. In theory, RNNs are capable of remembering information for long sequential data. However, in practice, this is not feasible due to the vanishing/exploding gradient problem. A similar problem is observed in deep feedforward networks. The source of this problem is the nature of RNN, which is using the same weight matrix to compute all the state updates. Even though the theory states that RNN can be used to learn long-term dependencies, due to vanishing/exploding gradient problems, they only seem to limit themselves to learn short-term dependencies.

LSTM solves the vanishing gradient problem and gives more accurate results compared to regular RNN. LSTM consists of three gates (forget, input, and output) and one cell state [17]. These are defined as follows:

$$f_t = \sigma\left(w_f h_{t-1} + w_f x_t\right) \tag{6}$$

$$i_t = \sigma(w_i h_{t-1} + w_i x_t) \tag{7}$$

$$o_t = \sigma(w_o h_{t-1} + w_o x_{t)} \tag{8}$$

$$\widetilde{c_t} = \tanh(w_c h_{t-1} + w_c x_t) \tag{9}$$

$$c_t = (i_t * \widetilde{c_t}) + (f_t * c_{t-1}) \tag{10}$$

$$h_t = o_t * \tanh(c_t) \tag{11}$$

Here f, i, o are forget, input and output gates respectively, c is cell state, h is a hidden state and x is the input. The complete structure of the LSTM is illustrated in Fig. 2.

LSTM can hold a combination of different information blocks at each time step. The main advantage of LSTM comes from the cell state. Cell state provides the possibility of explicitly information writing or removing. This cell state can only be altered by the gates which are responsible for letting the information pass through.

From previous studies, we know that the convolution operation works well to extract features as local input patches. This allows modular and efficient data representations. In our forecasting problem, we accept the time as a spatial dimension and process the data by applying 1D convolution operations to extract subsequences

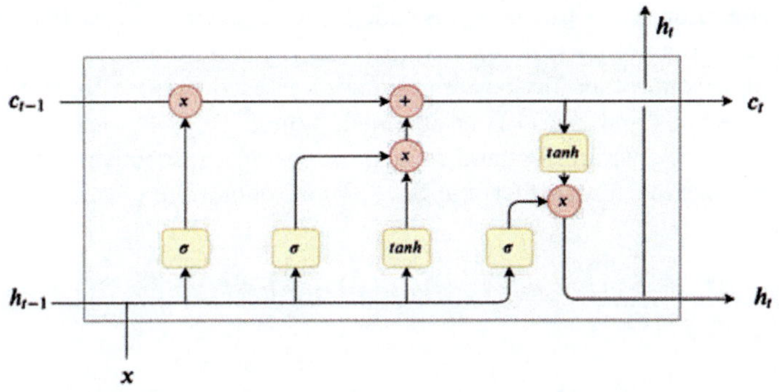

Fig. 2 LSTM structure

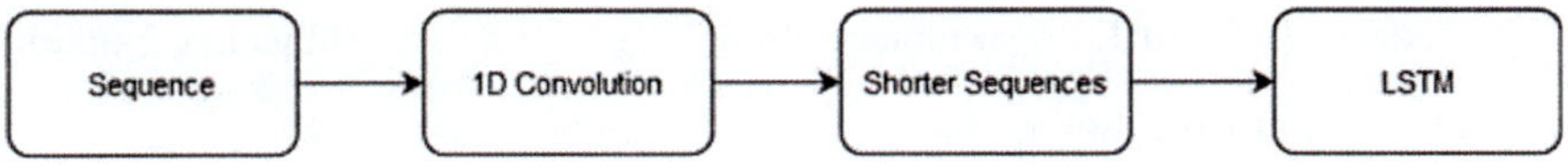

Fig. 3 Data feed processing

from the sequence. This allows us to recognize local patterns since the same transformation is applied to every patch. However, it is not possible to get reasonable results in forecasting problems just by using convolution operation. Since the convolution operation processes the input patches independently, it is not sensitive to the order of the timestep, unlike LSTM. One way to use the advantageous feature of convolution is to use it as a preprocessing step before LSTM. Figure 3 illustrates our data processing step combined with LSTM.

XGBoost (Extreme Gradient Boosting)

XGBoost is a scalable end-to-end novel machine learning algorithm based on tree learning gradient boosting that can handle sparse data in a highly efficient way. The mathematical formulation of the model is described by Chen and Guestrin (2016) as follows: Eq. (12) describes the objective function (also called loss function) which should be minimized at iteration t where $\hat{y}_i^{(t)}$ indicates the prediction of ith instance at ith iteration; $\Omega(f_t)$ describes the regularization term, which helps to model to avoid over-fitting results.

$$\mathcal{L}^{(t)} = \sum_{i=1}^{n} l\left(y_i, \hat{y}_i^{(t-1)} + f_t(x_i)\right) + \Omega(f_t) \tag{12}$$

At each iteration, a new function f_t is added, which provides the best improvement for the model. It is not possible to solve Eq. (12) by using traditional optimization methods. So second-order Taylor approximation is used to get a solvable form by traditional methods and Eq. (13) is obtained. Where $g_i = \partial_{\hat{y}^{(t-1)}} l\left(y_i, \hat{y}^{(t-1)}\right)$ and $h_i = \partial^2_{\hat{y}^{(t-1)}} l\left(y_i, \hat{y}^{(t-1)}\right)$ are first and second-order terms respectively at iteration t. When we remove the constant term in Eq. (13), we obtain Eq. (14).

$$\mathcal{L}^{(t)} = \sum_{i=1}^{n} \left[l(y_i, \hat{y}_i^{(t-1)} + g_i f_t(x_i) + \frac{1}{2} h_i f_t^2(x_i) \right] + \Omega(f_t) \tag{13}$$

$$\widetilde{\mathcal{L}}^{(t)} = \sum_{i=1}^{n} \left[g_i f_t(x_i) + \frac{1}{2} h_i f_t^2(x_i) \right] + \Omega(f_t) \tag{14}$$

If regularization term $\Omega(f_t)$ is replaced by $\gamma T + \frac{1}{2}\lambda \sum_{j=1}^{T} w_j^2$, then objective function takes its final form as follows:

$$\begin{aligned} \widetilde{\mathcal{L}}^{(t)} &= \sum_{i=1}^{n} \left[g_i f_t(x_i) + \frac{1}{2} h_i f_t^2(x_i) \right] + \gamma T + \frac{1}{2}\lambda \sum_{j=1}^{T} w_j^2 \\ &= \sum_{j=1}^{T} \left[\left(\sum_{i \in I_j} g_i \right) w_j + \frac{1}{2} \left(\sum_{i \in I_j} h_i + \lambda \right) w_j^2 \right] + \gamma T \end{aligned} \tag{15}$$

The optimal value of weights w_j at leaf j is obtained by using Eq. (16) and the optimal value is calculated by Eq. (17). This equation can be understood as a scoring function for a tree structure q. This score is similar to the impurity score for evaluating decision trees.

$$w_j^* = -\frac{\sum_{i \in I_j} g_i}{\sum_{i \in I_j} h_i + \lambda} \tag{16}$$

$$\widetilde{\mathcal{L}}^{(t)}(q) = -\frac{1}{2} \sum_{j=1}^{T} \frac{\sum_{i \in I_j} g_i}{\sum_{i \in I_j} h_i + \lambda} + \gamma T \tag{17}$$

It is a real challenge to enumerate all possible trees. The authors propose Eq. (18) to calculate the score of each tree structure for evaluating each one to select the best split where I_L and I_R denotes the instances set of left and right nodes after the split; $I = I_L \cup I_R$.

$$\mathcal{L}_{split} = \frac{1}{2} \left[\frac{\left(\sum_{i \in I_L} g_i\right)^2}{\sum_{i \in I_L} h_i + \lambda} + \frac{\left(\sum_{i \in I_R} g_i\right)^2}{\sum_{i \in I_R} h_i + \lambda} - \frac{\sum_{i \in I} g_i}{\sum_{i \in I_I} h_i + \lambda} \right] \tag{18}$$

Application of SARIMA Model

The steps involved in building a SARIMA model are as follows:

1. **Identification of Model Parameters**: the initial step of the SARIMA is to determine the values of parameters. The autocorrelation, partial autocorrelation, and Augmented Dickey-Fuller (ADF) test results for vegetables and fruits are shown in Fig. 4. The plots indicate that there is no regular or seasonal trend. ADF results also prove that both data satisfy stationarity property at 0.05. The

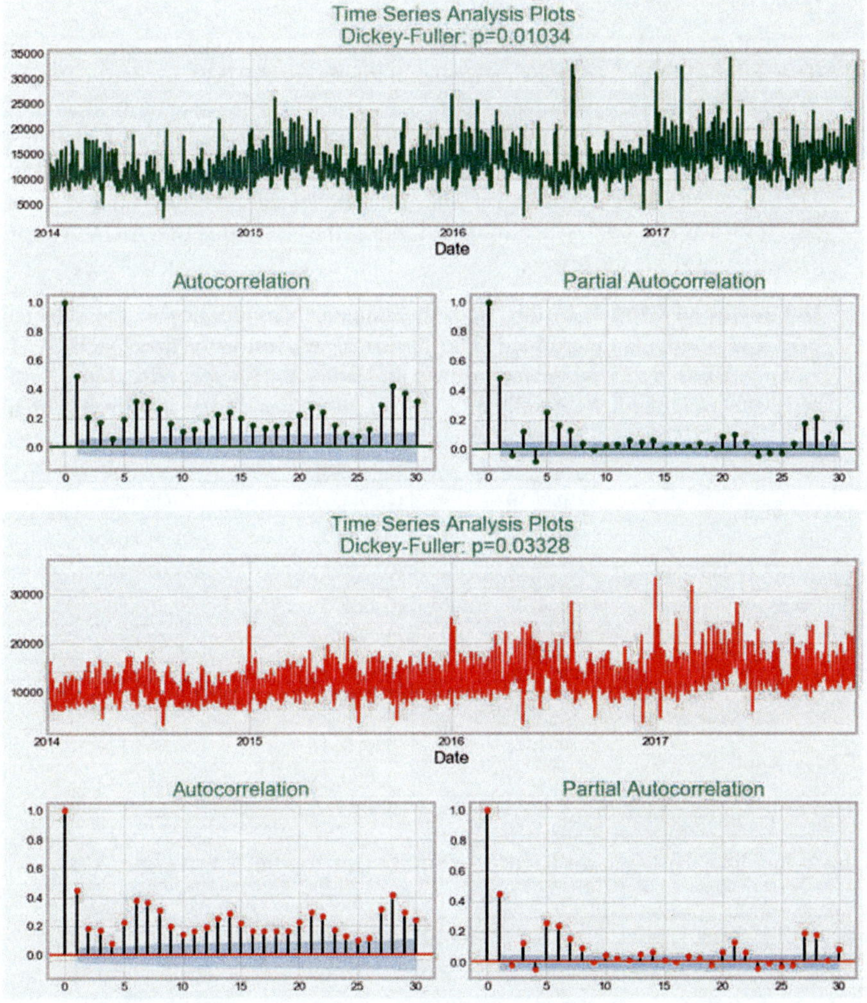

Fig. 4 Seasonal and stationary analysis of fruits and vegetables respectively

best values of parameters should be identified. To determine the parameters of the model, Akaike's Information Criterion (AIC) is commonly utilized. It is calculated as follows:

$$AIC(p) = n \ln\left(\frac{RSS}{n}\right) + 2K \tag{19}$$

where n is the number of observations, and RSS is the residual sums of squares. The parameters providing the minimum AIC value will be set as model parameters. Another approach for determining appropriate parameters of the model is to analyze (ACF) and (PACF) plots.

2. **Estimation of Model Parameters:** A grid search approach is employed for determining the best forecasting model in this study. ARIMA (p, d, q) (P, D, Q) m model requires six parameters: p, d, q, P, D, and Q. The value of m is set as 12 because used data are monthly with a period of 12. The AIC values of evaluated models are shown in Table 1. According to Table 1, SARIMA (1, 1, 1) × (1, 0, 1)12 shows the lowest AIC value. Thus, this model should be considered as the best forecasting model. According to Table 1, the AIC value of SARIMA (1, 1, 2) × (0, 0, 1)12 is the lowest.

3. **Diagnostic of Model**: In this step, the statistical importance of the selected SARIMA model is determined. The statistical test results of the SARIMA (1, 1, 1) × (0, 0, 1, 1) model are shown in Table 2. The second column indicates the weight of the coefficients. Since all values of P > |z| are less than

Table 1 AIC Values of SARIMA models

(p, d, q)	(P, D, Q, m)	AIC values
(0, 1, 0)	(0, 0, 0, 1)	21,115.035
(1, 1, 0)	(0, 0, 0, 1)	21,029.448
(0, 1, 1)	(0, 0, 0, 1)	20,763.500
(1, 1, 1)	(0, 0, 0, 1)	20,683.540
(1, 1, 2)	(0, 0, 0, 1)	**20,655.100**
(2, 1, 3)	(0, 0, 0, 1)	20,659.035
(0, 1, 2)	(0, 0, 0, 1)	20,663.134
(1, 1, 3)	(0, 0, 0, 1)	20,656.342

Table 2 Results of the diagnostics test of the SARIMA (1, 1, 2) × (0, 0, 1, 1) model

	coef	std err	z	P > |z|	[0.025	0.975]
ar.L1	−0.2861	0.071	−4.027	0.000	−0.425	−0.147
ma.L1	−0.3550	0.064	−5.550	0.000	−0.480	−0.230
ma.L2	−0.5916	0.060	−9.943	0.000	−0.708	−0.475
sigma2	9.581e+06	3.5e−05	2.74e+11	0.000	9.58e+06	9.58e+06

0.05 (significance level), we can conclude that all coefficients are statistically significant.

4. **Forecasting**: The model with estimated parameters is used to make forecasts. Data from January 2014 to December 2016 are used as a training set, and the remaining data is used as test data. We forecast the periods between January and December 2017 daily. Figure 5 shows the model forecast values and the actual value curve. Mean absolute percentage error (MAPE) on the test data set is 24.57 and 24.43% for fruits and vegetables respectively.

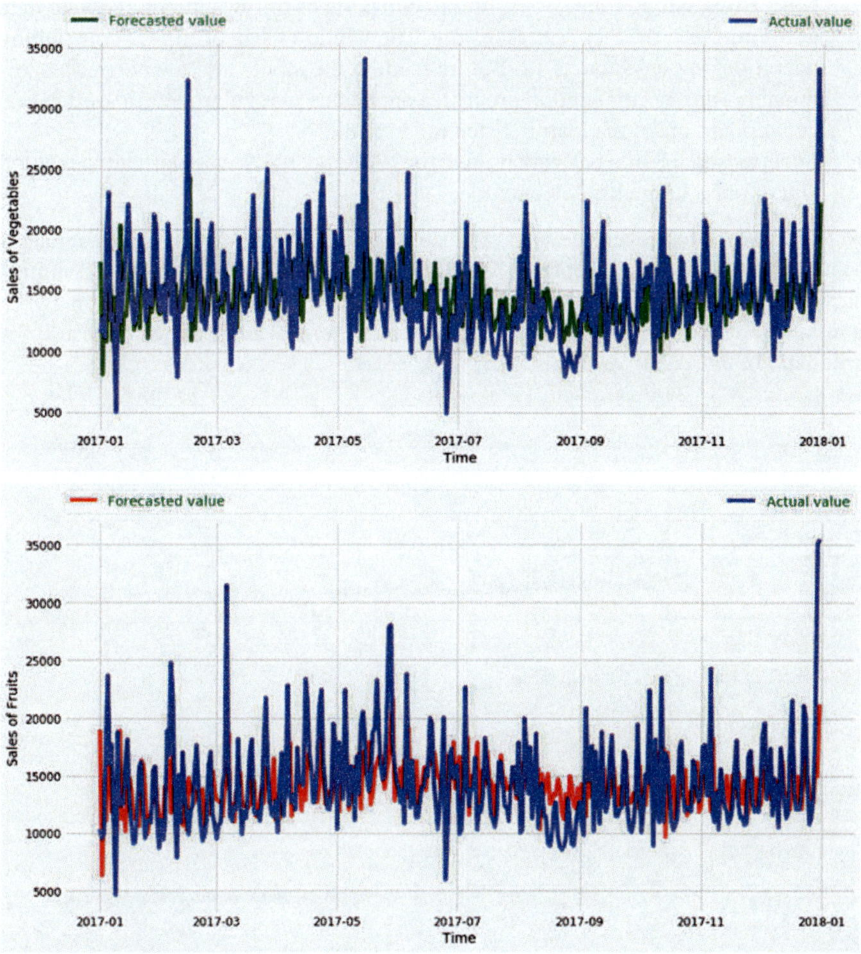

Fig. 5 Forecast value and actual value fitting curve

Application of LSTM Model

The algorithm is implemented in Python with the Pytorch library.

1. **Model Parameters**: Our network consists of 1 hidden layer with 9 neurons. The output activation function is linear. The input to the network is our 17 handcrafted features. Mean squared error loss is selected for loss calculation. A popular ADAM optimization algorithm was selected to optimize network weight values. Hyperparameters of the model and optimization algorithm are selected based on trial-and-error.
2. **Data Processing**: Total data split into two parts as follows: first three-year data as training and the last year as testing. The windows size of the 1D convolution operation was selected as 12. Before feeding the inputs into the network, additional scaling/normalization processes applied as in regular feedforward neural networks to make the learning step more stable.
3. **Forecasting**: Neural network trained for 2000 epochs. No regularization method was applied to the implementation.

Training loss for vegetable sales data is illustrated in Fig. 6. MAPE on test data is 24.6, which is inside an acceptable boundary compared to previous studies. Figure 8 shows the prediction of vegetable data. To forecast fruit data, the same LSTM network is retrained by using the same hyperparameters. Figures 7 and 8 show the results for fruit data. In this case, MAPE is 22.5.

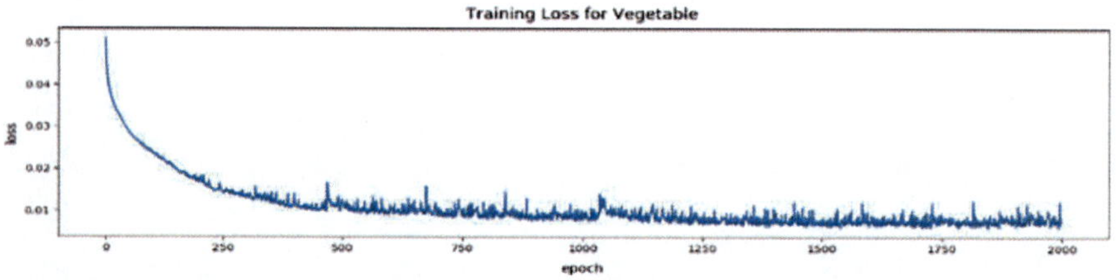

Fig. 6 LSTM traning loss for vegetable

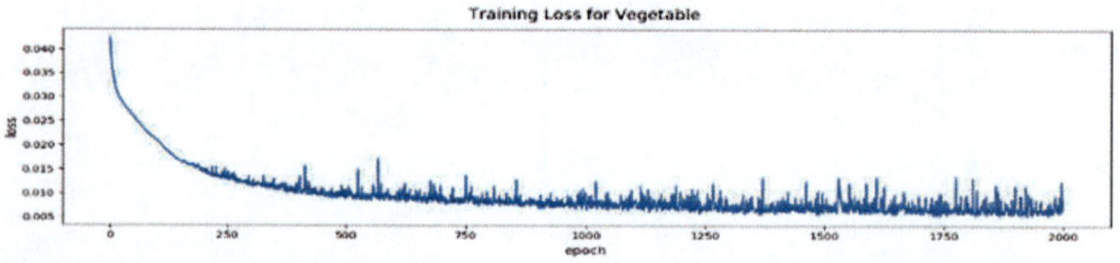

Fig. 7 LSTM traning loss for fruit

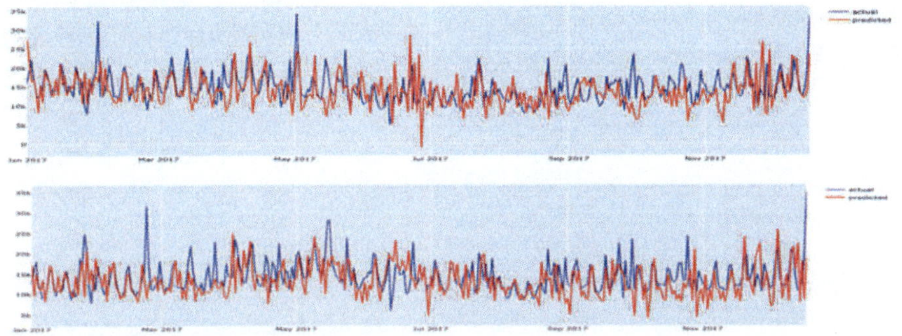

Fig. 8 Forecast value and actual value fitting curve for vegetable (top) and fruit (bottom)

Application of XGBoost (Extreme Gradient Boosting)

The algorithm is implemented in Python using open-source XGBoost libraries. An open-source framework provides a fast and easy implementation of the algorithm. The following steps are involved in the implementation of the algorithm.

1. **Feature extraction and selection**: We extract 17 features after reading the papers in the area and obtain the relative importance of these features, as shown in Fig. 9. The temperature and dollar rate are the most important features. The categorical variables are converted into the numerical form by using one-hot encoding.
2. **Parameter Tuning**: The XGBoost algorithm has too many parameters, and the values of these parameters highly affect the prediction performance of the model. So it is required to perform hyper-parameter tuning operation to obtain the more appropriate model. It is the most time-consuming part of the implementation of the algorithm. The custom grid search approach is used to find the parameter values. The booster parameters related to the learning task and model complexity

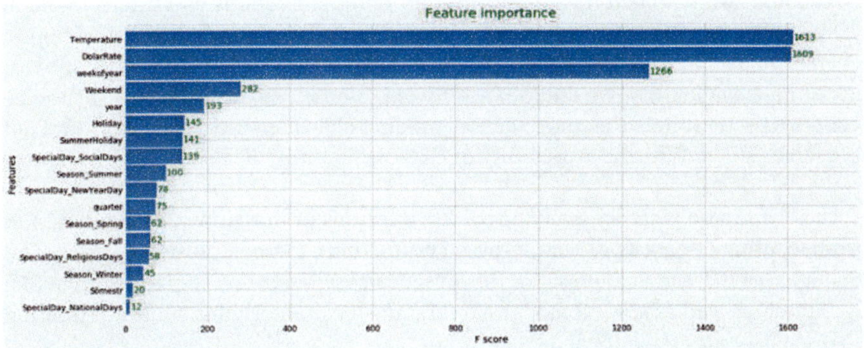

Fig. 9 Feature importance plot

Table 3 Hyper-parameter tuning values

Parameter	Parameter Values	Best Value
max_depth	[0, 10]	1
n_estimators	{100, 200}	100
subsample	[0, 1]	0.16
Colsample_bytree	[0, 1]	1
gamma	**{0, 1}**	**0**
reg_alpha	[0, 1]	0
learning_rate	[0.01, 0.1]	0.1
min_child_weight	{1, 3, 5}	1

are tuned by using the given values in Table 3. These values are determined based on the expert's suggestions. Column 2 indicates the used values in the grid search process, and column 3 represents the best value of tuned parameters. Default values are used for the remaining parameters.

3. **Forecasting**: After selecting the best model parameters, we can use the model to make forecasting. Figure 10 shows the model forecast values and the actual value curve. The forecasting results indicate that XGBoost can make better predictions than SARIMA and LSTM (see Table 4).

Summary and Outlook

In this study, we focused on the applicability of the XGBoost algorithm to forecast the daily sales of perishable foods. A specific focus of the study was directed toward two special perishable food categories: vegetables and fruits. The test cases were performed in the retail market. The results show that XGBoost yields better predictions compared to SARIMA and LSTM. The outcomes of this study can give several useful insights for managers, such as the development of stock policy, investments in SC.

Although we obtained meaningful results, some limitations and future research avenues may be emphasized. First, the advantage of using ML techniques highly depends on data availability. Since all factors are not included in the model due to a lack of data availability, we could not fully exploit the advantages of ML methods. Second, the controllable factors, such as product characteristic, promotions are not used in this study. With the availability of all these features, the most relevant features can be identified.

These limitations imply many possible extensions of this study in the future. For example, product-based or location-based estimation can be employed to understand the impact of these factors on sales. Besides, it is possible to investigate different ML algorithms, as well as new methods of combining these algorithms while considering the respective accuracy of each.

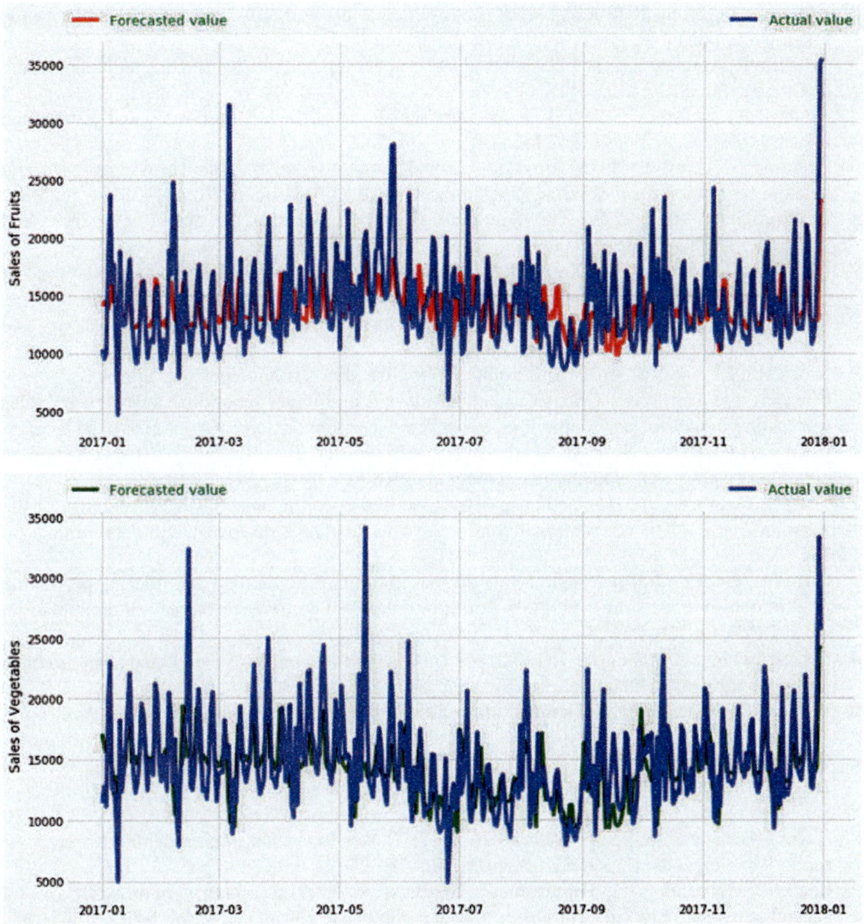

Fig. 10 Forecast value and actual value fitting curve

Table 4 Performance comparison of forecasting methods based on MAPE

Model	MAPE for fruits (%)	MAPE for vegetables (%)
SARIMA	24.43	24.57
LSTM	22.5	24.6
XGBoost	16.57	16.43

References

1. Aburto L, Weber R (2007) Improved supply chain management based on hybrid demand forecasts. Appl Soft Comput 7(1):136–144

2. Adebanjo D, Mann R (2000) Identifying problems in forecasting consumer demand in the fast-moving consumer goods sector. Benchmarking Int J
3. Adhikari R, Agrawal RK (2013) An introductory study on time series modeling and forecasting. *arXiv preprint* arXiv:1302.6613
4. Agnew MD, Thornes JE (1995) The weather sensitivity of the UK food retail and distribution industry. Meteorol Appl 2(2):137–147
5. Arunraj NS, Ahrens D (2015) A hybrid seasonal autoregressive integrated moving average and quantile regression for daily food sales forecasting. Int J Prod Econ 170:321–335
6. Ashagidigbi WM, Adebayo AS, Salau SA (2019) Analysis of the demand for fruits and vegetables among households in Nigeria. Sci Lett 7(2):45–51
7. Blackburn J, Scudder G (2009) Supply chain strategies for perishable products: the case of fresh produce. Prod Oper Manag 18(2):129–137
8. Brockwell PJ, Davis RA, Fienberg SE (1991) Time series: theory and methods: theory and methods. Springer Science & Business Media
9. Brockwell PJ, Davis RA (2016) Introduction to time series and forecasting. Springer
10. Chen FL, Ou TY (2009) Gray relation analysis and multilayer functional link network sales forecasting model for perishable food in convenience store. Exp Syst Appl 36(3):7054–7063
11. Chen T, He T (2015) Higgs boson discovery with boosted trees. In: NIPS 2014 workshop on high-energy physics and machine learning, pp 69–80
12. Chen T, Guestrin C (2016) Xgboost: a scalable tree boosting system. In Proceedings of the 22nd ACM SIGKDD international conference on knowledge discovery and data mining, pp 785–794
13. Chen FL, Ou TY (2008) A neural-network-based forecasting method for ordering perishable food in convenience stores. In: 2008 fourth international conference on natural computation, vol 2. IEEE, pp 250–254
14. Du XF, Leung SC, Zhang JL, Lai KK (2013) Demand forecasting of perishable farm products using support vector machine. Int J Syst Sci 44(3):556–567
15. Durham C, Eales J (2010) Demand elasticities for fresh fruit at the retail level. Appl Econ 42(11):1345–1354
16. Geurts MD, Kelly JP (1986) Forecasting retail sales using alternative models. Int J Forecast 2(3):261–272
17. Hochreiter S, Schmidhuber J (1997) Long short-term memory. Neural Comput 9(8):1735–1780
18. Huber J, Gossmann A, Stuckenschmidt H (2017) Cluster-based hierarchical demand forecasting for perishable goods. Expert Syst Appl 76:140–151
19. Locke E, Coronado GD, Thompson B, Kuniyuki A (2009) Seasonal variation in fruit and vegetable consumption in a rural agricultural community. J Am Diet Assoc 109(1):45–51
20. Paam P, Berretta R, Heydar M, Middleton RH, García-Flores R, Juliano P (2016) Planning models to optimize the agri-fresh food supply chain for loss minimization: a review. Ref Mod Food Sci 19–54
21. Raju Y, Kang PS, Moroz A, Clement R, Hopwell A, Duffy AP (2016) Investigating the demand for short-shelf-life food products for SME wholesalers
22. Reddy AS, Chakradhar P, Santosh T (2018) Demand forecasting and demand supply management of vegetables in India: a review and prospect. Int J Comput Technol 17(1):7170–7178
23. Sankaran S (2014) Demand forecasting of fresh vegetable product by seasonal ARIMA model. Int J Oper Res 20(3):315–330
24. Sridama P, SiributｃC (2018) Decision support system for customer demand forecasting and inventory management of perishable goods. J Adv Manag Sci 6(1)
25. van Donselaar K, Van Woensel T, Broekmeulen RACM, Fransoo J (2006) Inventory control of perishables in supermarkets. Int J Prod Econ 104(2):462–472
26. Van Woensel T, Van Donselaar K, Broekmeulen R, Fransoo J (2007). Consumer responses to shelf out-of-stocks of perishable products. Int J Phys Distrib Logistics Manag
27. Yang CL, Sutrisno H (2018) Short-term sales forecast of perishable goods for franchise business. In 2018 10th international conference on knowledge and smart technology (KST). IEEE, pp 101–105

28. Yang S, Xiao Y, Kuo YH (2017) The supply chain design for perishable food with stochastic demand. Sustainability 9(7):1195
29. Yang HX, Hu J (2013) Forecast of fresh agricultural products demand based on the ARIMA model. Guangdong Agric Sci 11:52

Future of Engineering Education: Cyber-Physical Systems Engineering

Banu Yetkin Ekren and Vikas Kumar

Abstract Recent Industry 4.0 developments have created disruptive changes in industries. It has also changed the way people work and the required skills of the workforce for the future. This new technological advancement has created a requirement of workforce combining information technology (IT) and production knowledge. Unfortunately, many schools and universities have still been training students behind Industry 4.0 developments and requirements. Therefore, it is important to understand what characteristics and knowledge, as well as skills, are required for the future of jobs and engineering profiles to shape the new education requirements. This paper aims to address the future of engineering education, specifically by focusing on a promising engineering department that is the cyber-physical systems (CPS) engineering. We also aim to discuss how the curriculum of that novel discipline, CPS engineering, should be for the future of engineering requirements.

Keywords CPS engineering · Engineering education · Future of education · Cyber-physical systems engineering · Industry 4.0

Introduction

The fourth industrial revolution includes an integration of advanced automation in factories to create a smart process environment. To carry out an efficient production, all machinery and equipment in the production environment are coordinated through the Internet with the help of sensors and all required data with the cloud storage. Kagermann et al. [6] define Industry 4.0 as the technological evolution from embedded systems to cyber-physical systems (CPSs). With the help of embedded

B. Y. Ekren (✉)
Department of Industrial Engineering, Yasar University, Bornova, Izmir, Turkey
e-mail: banu.ekren@yasar.edu.tr

V. Kumar
Bristol Business School, University of the West of England, Bristol, UK
e-mail: Vikas.Kumar@uwe.ac.uk

F. Calisir (ed.), *Industrial Engineering in the Internet-of-Things World*,
Lecture Notes in Management and Industrial Engineering,
https://doi.org/10.1007/978-3-030-76724-2_4

systems, IoT, MtoM communication, and CPS technologies, it is possible to integrate the virtual environment with the physical world. As a result of that, smart factories are emerging to overcome the complexity of the production environment [14]. Researches also illustrate the importance of Information and Communication Technologies (ICTs) as well as embedded systems, IoT, CPS, Industrial Integration and Industrial Information Integration in the infrastructure of tomorrow's smart factories.

The first three industrial revolutions are the result of mechanization, electricity, and IT. Industry 4.0 is a result of the introduction of the IoT into the manufacturing environment. In the future, industries are estimated to have global networks incorporating CPS oriented machinery and production facilities. With the help of CPSs, smart machines and production inputs will be able to exchange information, make decisions, and control systems, intelligently. Hence, significant improvements in industrial processes, material usage, and supply chain management, as well as life cycle management, are expected results of Industry 4.0. By the creation of smart factories, there will be smart products spreading all over the facilities, running all the time and, keeping their own history, tracking data and current status, and alternative routes to achieve a pre-defined target. By also vertically and horizontally networked processes in manufacturing environments and facilities, these smart products will enable the intelligently managing of the industries.

Evolution towards a smart environment in industries has changed the future of the required workforce structure [13]. According to the 2019 World Development Report [12], due to the recent technological developments, while some kinds of jobs are being eliminated, some others are created, and some others are altered. It is suggested for companies that in order to remain competitive and to keep pace with it, they should invest capital more in humans and adapt their employees by training them to gain the required skills. For instance, in the 2019 World Development Report [12] a new trend in employment strategy in companies is declared to be a flexible employment strategy aligning with companies' needs.

For that future of the changing workforce requirements and the CPSs jobs of the future, some of the hot topics have emerged. These are artificial intelligence (AI), programming, software engineering, autonomy, statistics, computation, automation, complex and integrated systems. These topics are also to diffuse in the engineering higher education curricula. Thus, engineering faculties should rethink and renovate their education curricula by integrating novel hot trends towards Industry 4.0 developments. In this paper, our aim is to shed light on novel engineering discipline, CPS engineering curricula promising to grow in the near future. We researched on the subject and summarized our findings in the following sections.

Literature Review

We summarize the literature works on how jobs change and new requirements emerge in workforce structure. According to the Rockefeller Foundation's one of the surveys

[10], almost half of recent college graduates do not use the skills they gained during their college education at work. Eighty-six percent declare that they learn new skills outside of college.

According to Long et al. [8], engineering education requires a different design concept than science education. For instance, engineering education includes designing and building things. Therefore, it must be up-to-date with the latest techniques, technologies, and trends. They also declare that a four-year education period is a very short time period for engineering education. Therefore, the curricula of engineering education should be designed carefully with the relevant topics for the twenty-first century.

Susskind and Susskind [11] declare an estimation that within 20 years, 50% of US jobs could be automated. Frey and Osborne [5] also estimate a similar automation replacement prediction for the future. According to them, some professional jobs such as doctors, lawyers, teachers, etc. will dramatically change. However, jobs that cannot be automated will mostly involve skills such as complex communications, social intelligence, creativity, and perception and adaptability, etc.

It is well known that U.S. universities have a good reputation in engineering education. However, according to a recent report by News and World Report (2020), Tsinghua University is ranked number one among six of the top 10 Asian universities. For instance, in the Computer Science discipline, Tsinghua is ranked as number one, and two other Chinese universities are ranked in the top ten. This ranking of Tsinghua University is above well-known universities such as Carnegie-Mellon, Princeton, Georgia Tech, Cambridge, Oxford, Illinois, Michigan, and Caltech, etc. That ranking work includes 28 subjects in the sorting process. According to the McKinsey Global Institute's latest report [9], it is evaluated that jobs such as computer scientists, engineers, and IT administrators, which are related to developing and deploying new technologies, are expected to grow by 2030.

According to Linkedin's 2018 U.S. Emerging Jobs Report [7], 2018's top five emerging jobs are blockchain developer, machine learning engineer, assurance staff, and sales development representative.

Brynjolfsson and McAfee [2] state that: "Not only are the new technologies exponential, digital, and combinatorial, but most of the gains are still ahead of us. In the next twenty-four months, the planet will add more computer power than it did in all previous history. Over the next twenty-four years, the increase will likely be over a thousand-fold."

Another future of job alert is given by the White House, USA. In the AI Report [1], the need for Americans to be educated and trained towards software and AI-related subjects is declared.

Zivi [15] presents a work on one of the unique practical CPS education on the U.S. Naval Academy's Cyber Science education curriculum. He gives information on how a CPS education should be in terms of the contents.

Ekren and Kumar [4] have studied the future of engineering education by focusing on the Industrial 4.0 revolution. They summarize the findings on what the future of higher education would look like in the near future.

Information technology is changing at an exponential rate. Hence, there will not be enough time to retrain students after they graduate. If students leave universities unprepared for the twenty-first century, they may never catch up due to the rapidly changing world. Hence, engineering faculties should focus on renovating the engineering curricula towards novel requirements and developments. In this paper, we propose a trending engineering education towards Industry 4.0 developments that is CPS engineering by investigating how the curricula should be configured to meet the future of work requirements. We investigated the universities' related curricula and current projects progressing on this related subject and summarized our findings.

Areas Related with CPS Engineering

In the USA, a multiyear NSF funded project pursued by the National Academies of Sciences, Engineering, and Medicine focuses on developing and addressing effective CPS education [3]. In that report, it is declared that the future of the CPS workforce is likely to include a combination of engineers in the fields of computer science, mechanical engineering, systems engineering, and electrical and computing engineering. The report summarizes the related areas of the CPSs as:

- ***Robotics***: It is an interdisciplinary branch of engineering and science, including mechanical engineering, electrical and electronics engineering, information engineering, computer science, and other disciplines. This area studies on systems with sensors and actuators operating autonomously or semi-autonomously in cooperation with humans. Robotics include the topics of kinematics, dynamics, robot hardware, and control software, perception, sensing, and state estimation, as well as control of manipulators and vehicles. Although many robots are often considered to be CPS, many CPS are not considered as robots. In the robotics program, particularly some of the subjects particularly cover that field.
- ***Systems engineering***: It is an interdisciplinary field of engineering and engineering management, focusing on how to design and manage complex systems over their life cycles. This area contributes significantly to CPS due to the complexity of the environment with the recent highly connected systems. Systems engineers typically focus on the organization, management, and integration required for large complex systems. However, they do not necessarily address the detailed technological requirements within the cyber aspects of systems.
- ***The Internet of Things (IoT)***: This subject is a system component composed of computing, mechanical and digital devices, objects, people, etc. It has the ability to transfer data over a network associated with users and their environments. With the recent advancements, the IoT applications require CPS characteristics such as real-time control, real-time response, etc. For instance, smart city implementations

require IoT applications and are becoming more sophisticated and dependent on the CPS environment.

- ***The Industrial Internet***: It includes interconnected sensors, devices, and other instruments which are also connected with computers' industrial applications, including manufacturing and energy management. This connectivity facilitates data collection, data exchange, and data analysis. By that, improvements in productivity and efficiency, as well as other economic benefits, can be accomplished.

Mainly by focusing on the above four main CPS related subjects, we draw the future of the CPS Engineering Education curriculum below, also with the help of current curricula at some universities as well as reports of projects [3].

A CPS Engineering Education

In the era of Industry 4.0, "smart things," industries are eager to implement CPS. To be able to correctly design, invent, build, and deploy these systems, interdisciplinary skillsets are declared to be significantly necessary than ever before. In the CPS Education Report [3], it is declared that all computing and engineering-related disciplines will make widespread use of CPS. It is also mentioned that the requirement of expert knowledge on CPS is significant. We summarize our findings on CPS Engineering Education by also considering that report, in this paper.

In the CPS Education Report [3], it is also declared that designing a CPS course or degree program would involve a careful balancing of physical and cyber aspects and application knowledge. To develop the CPS curricula, the report includes model curricula from multiple perspectives [3]. The report identifies six overarching foundations. They are described in the following sections.

Foundation 1: Basic concepts for computing

It is declared that expertise in CPS cannot be achieved by solely one or two programming courses. It also requires solid training in computing such as data structures and algorithms; programming; models of computation, embedded hardware; discrete event systems; software engineering; model-based design, real-time operating systems, and programming for networks.

Foundation 2: Physical world computing

For CPS education, the requirement of computing foundations with physical-world properties and constraints should be emphasized. Real-world comprises of complex systems. It is hard to anticipate the complexity of the system by the system designers and by software. Therefore, the first step for students is to thoroughly understand the concepts of sensors, actuators and their programming, analysis of signals, realtime control of systems, embedded systems, resource management, constraint management for time, memory size, and power, as well as techniques for managing unreliability in physical systems.

Foundation 3: Discrete and continuous mathematics

Since it is declared that both discrete and continuous mathematics are foundational skills for all CPS engineers, it is critical for students to learn how to deal with the integration of these techniques with the systems. Besides mathematics, students will also require to understand graph theory, combinatorics, probability and statistics, stochastic processes, logic, linear algebra, calculus and differential equations, etc.

Foundation 4: Cross-cutting application of sensing, control, communication, actuation, and computing

The report declares that Foundation 4 is significant due to the cross-cutting nature of CPS, and communication networks and sensing, signal processing, and actuation with real-time constraints. Hence, the curriculum should be designed with an inter-disciplinary concept. At the core of this foundation, there is knowledge of control, signal processing, embedded system design, and their implementations. To detail, the below issues are suggested to be covered in the CPS curricula [3]:

- Control algorithms for both linear and nonlinear systems, stochastic processes, adaptive control, system identification, and hybrid control;
- Optimization as well as optimal control of dynamic systems;
- Real-time analysis, time-triggered or event-triggered control, and decision-making with noisy data (data-driven models);
- Wireless communications, synchronous and asynchronous communications, and ad hoc networking;
- Effect of physical properties on software requirements;
- Control-based signal processing, network management, computation, and communication models;
- Safety, reliability, and dependability;
- Security and privacy;
- Human factors related to humans-in-the-loop as well as behavioral aspects;

Foundation 5: Control, computing, and communication of heterogeneous and dynamic systems

This foundation requires control, communications, and computing subjects. Linear and nonlinear models, stochastic models, and discrete-event and hybrid models, optimization, probability theory, and dynamic programming as relevant topics. Key concepts also include uncertainty and risk; properties of the physical world and computational devices; communication of systems (e.g., wireless communications); error detection and correction; merge of physical and computational models; and analysis and integration of signals and systems and finite-state automata.

One Model for 4-Year, 40-Course Undergraduate Degree in CPS

Math and Natural Science (10 courses)

- Calculus I and II
- Differential Equations
- Linear Algebra
- Probability and Statistics
- Logic
- Physics I (Mechanics and Dynamics)
- Physics II (Electrical Circuits)
- Chemistry or Biology
- Discrete Math

CPS Core (12 courses)

- Introduction to CPS (Freshman Laboratory Course)
- Computer Programming
- Data Structures and Algorithms
- Programming Physical Systems
- Software Engineering
- Model-Based System Design
- Heterogeneous Models of Computation
- Formal Methods and Synthesis
- Resource-Aware Real-Time Computing
- Control Systems
- Optimization
- Digital Signal Processing

Fig. 1 A 4-yer undergraduate degree curriculum for CPS engineering. *Source* CPS Education Report [3]

Foundation 6: System development for CPS

For the CPS development, initial requirements, certification on safety–critical systems, high confidence, and resiliency, etc. are considered. Through it, students are suggested mastering these key concepts: safety, resilience, security, privacy, assurance cases, hazard analysis, verification and validation, model-based design and tools, system design, Internet of Things, cloud computing, testing the developed CPS.

In the report, a sample of a bachelor's degree CPS engineering curriculum is also provided (see Fig. 1). From that curriculum, it can be observed that the CPS programs include both computer science and electrical engineering programs related courses. It is also mentioned that the difference in CPS curricula would be due to whether the program mostly focuses on computer science or an electric and computer engineering perspective.

Current Bachelor Degree Curricula in CPS Engineering

We searched the current bachelor's degree cyber-physical systems engineering related departments at colleges and universities in the world. We found that there exist mostly graduate levels in this discipline. However, two schools, U.S. Naval Academy and Depaul University in the US, provide a related bachelor degree program at their schools. The curriculum details are given for the Naval Academy and Depaul University in Figs. 2 and 3, respectively. In Fig. 2, only the cyber science-related courses except the other core courses are summarized. In Fig. 3, the whole curriculum for the CPS Engineering program is provided.

U.S. Naval Academy
Cyber Science Courses

Cyber Security I
Cyber Fundamentals I
Cyber Systems Engineering
Systems Programming & OS Fundamentals
Data Structures for Cyber Operations
Applied Cyber Systems Architecture
Information Operations, Social Engineering, and Hacktivism
Web & Database Cyber Operations
Security: Fundamental Principles
Networking & Mobile Computing
Cyber Operations I
Cyber Operations II
Cyber Planning & Policy
Cyber Law & Ethics

Fig. 2 Department of cyber-science curriculum of U.S. Naval Academy

DePaul University
Cyber Physical Systems Engineering

Introduction to Computer Science I	University Physics II
Introduction to Computer Science II	University Physics III
or	Probability and Statistics I
Python for Programmers *and* 1 Major Elective	Networking for Cyber-Physical Systems
Data Structures I	Cyber-Physical Systems Engineering I
Discrete Mathematics I	Cyber-Physical Systems Engineering II
Discrete Mathematics II	Cyber-Physical Systems Engineering III
Calculus I	Embedded Systems I
Calculus III	Embedded Systems II
Calculus III	Cyber-Physical System Security
Data Structures II	1 Major elective
Design and Analysis of Algorithms	Linear Systems
Computer Systems I	Foundations of Cyber-Physical Computing
Computer Systems II	Cyber-Physical Systems Engineering Practicum I (Experiential Learning)
Multivariable Calculus I	Cyber-Physical Systems Engineering Practicum II (Capstone)
Linear Algebra	Technical Writing
Differential Equations	3 Major electives
University Physics I	1 open elective

Fig. 3 Department of cyber-physical systems engineering curriculum of DePaul University

Although currently, we could only find two bachelor degree programs in CPS engineering, due to the recent Industry 4.0 developments, it would be promising to grow in the near future. Since the programs could be shaped towards either more electrical engineering, or software engineering, or systems engineering, students could select the program and the universities depending on their choices. Therefore, while shaping the program curriculum, the university first should decide what qualified graduates it is aiming to give after its education.

Conclusion

This paper aims to study the future of engineering education, specifically by focusing on a promising engineering department, cyber-physical systems (CPS) engineering. We also discuss how the curriculum of this novel discipline, CPS engineering, should be in order to meet the future of engineering requirements. After providing the required information for understanding the CPS discipline properly, we provide two current curricula form this program from different universities. Besides, we also present an NSF funded project's work by providing a developed curriculum for the program.

As future work, this study can be extended in many directions. For instance, one may focus on a specific curriculum in CPS engineering by considering some expertise classes under the program. Namely, first, those expertise classes (e.g., software, electrical, systems engineering-oriented, etc.) could be defined under the CPS engineering, and then, specific curricula for those classes can be proposed followingly. Besides, the other current engineering disciplines, such as industrial engineering, computer science, etc. could be investigated under a CPS environment. How those current departments' curricula should be renovated towards the CPS environment would also be a promising work as a future study.

Acknowledgements We would like to express our sincere thanks to Newton Katip Celebi, TUBITAK, and Royal Academy of Engineering (Industry-Academia Partnership Programme 2018/2019, No. 4180046) for funding this project.

References

1. Artificial Intelligence, Automation and the Economy (2016) Executive office of the president. White House, Washington, DC. https://obamawhitehouse.archives.gov/sites/whitehouse.gov/files/documents/Artificial-Intelligence-Automation-Economy.PDF
2. Brynjolfsson E, McAfee A (2016) The second machine age: work, progress, and prosperity in a time of brilliant technologies. W. W. Norton
3. CPS Education Report (2016) A 21st century cyber-physical systems education, report, committee on 21st century cyber-physical systems education. National Academies Press. www.nap.edu/catalog/23686/a-21st-century-cyber-physical-systems-education

4. Ekren BY, Kumar V (2020) Engineering education towards industry 4.0. In: 10th IEOM international conference proceeding, Scopus-index, Dubai, March 2020, ISSN: 2169-8767, ISBN: 978-1-5323-5952-1
5. Frey CB, Osborne M (2013) The future of employment. Oxford Martin Programme on Technology and Employment, University of Oxford, UK
6. Kagermann H, Helbig J, Hellinger A, Wahlster W (2013). Recommendations for implementing the strategic initiative INDUSTRIE 4.0: securing the future of German manufacturing industry; final report of the industrie 4.0 working group. Forschungsunion
7. Linkedn's 2018 U.S. Emerging Jobs Report (2018) https://economicgraph.linkedin.com/research/linkedin-2018-emerging-jobs-report
8. Long LN, Blanchete S, Kelley T, Hohnka M (2019) The crucial need to modernize engineering education. Presented at IEEE aerospace conference, Mar 2019
9. McKinsey Global Institute Report (2017) Jobs lost, jobs gained: workforce transitions in a time of automation. https://www.mckinsey.com/featured-insights/future-of-work/jobs-lost-jobs-gained-what-the-future-of-work-will-mean-for-jobs-skills-and-wages
10. Rockefeller Foundation March 2017 Survey (2017). https://assets.rockefellerfoundation.org/app/uploads/20170320171306/Impact-Hiring-Survey-Results.pdf. Accessed Feb 2020
11. Susskind R, Susskind D (2016) The future of the professions: how technology will transform the work of human experts. Oxford University Press, Oxford
12. World Development Report (2019) The changing nature of work. http://documents.worldbank.org/curated/en/816281518818814423/pdf/2019-WDR-Report.pdf. Accessed Feb 2020
13. World Economic Forum (2018) The future of jobs report. Global challenge insight report. https://www.weforum.org/reports/the-future-of-jobs-report-2018. Accessed Feb 2020
14. Xu LD, Eric L, Xu L, Ling L (2018) Industry 4.0: state of the art and future trends. Int J Prod Res 56:2941–2962
15. Zivi E (2017) Teaching cyber-physical systems. ASME Mech Eng 139(03):S3–S8. https://doi.org/10.1115/1.2017-Mar-4

Industry 4.0 and Components in Production Enterprises

Fatih Ozturk

Abstract The globalizing world economy is constantly challenging producers with difficult competitive conditions. Manufacturer companies determine various methods to overcome these difficulties. Industry 4.0 and IoT applications, which have been popular in recent years, are the main ones of these methods. With these methods, solutions covering almost every area of our lives started to be produced. This approach, which we also call digitalization, has spread rapidly all over the world. In order for manufacturers to have more advantageous status, they need to reduce production costs in order to achieve more production flexibility and more efficient processes. Digitalization provides great advantages for efficiency increase. This method, which covers the product life cycle in all aspects and provides instant answers, seems to be developed for many years and will remain in the focus of the manufacturers.

Keywords Industry 4.0 · Production enterprises · Internet of things · Industrial automation · Smart factories

Introduction

Industry 4.0 and IoT applications in production enterprises have been one of the most demanded issues in recent years [16]. While rapidly developing digitalization applications produce solutions for almost every area of our lives, the production sector is also affected. The production sector is facing serious competitive conditions within the globalizing world economy [3]. Production companies have to reduce production costs, provide more production flexibility, and more efficient processes in order to become more advantageous than the firms they compete with. The best way to take advantage of competitive advantages is through integrated and innovative electrification, automation, and digitization solutions [15].

F. Ozturk (✉)
Department of Industrial Engineering, Faculty of Engineering and Natural Sciences, Istanbul Medeniyet University, Istanbul, Turkey
e-mail: fatih.ozturk@medeniyet.edu.tr

F. Calisir (ed.), *Industrial Engineering in the Internet-of-Things World*, Lecture Notes in Management and Industrial Engineering,
https://doi.org/10.1007/978-3-030-76724-2_5

We see that Digital Transformation applications, which are developing rapidly all over the world, are now being used in production lines [18]. It is seen that IoT, big data, autonomous robots, simulation, system integration, cybersecurity, and Augmented Reality (AR) applications, which are accepted as the basic components of Industry 4.0, are being preferred for different purposes in different processes in production lines. Digital transformation applications play a very important role, especially in production lines, where there is continuous production. In such production lines, it is one of the demands that the process is required to continue the production continuously [17].

Companies around the world face major challenges due to environmental, social, economic, and technological developments. The most important way to overcome these challenges is to monitor all production processes and have the ability to manage all value chains quickly and responsibly. Companies will need virtual and physical structures that allow close collaboration and rapid adaptation throughout the entire lifecycle, from innovation to production and distribution [12].

Industrial automation systems and digitalization are the most important elements for enterprises aiming to increase efficiency in engineering and operations, reduce operating costs, and improve product quality [23].

By using process control and automation systems, production systems can be managed at a lower cost, process information can be optimized, and energy efficiency can be increased [6]. Digital services play an increasingly important role in decisions such as analyzing or not analyzing process and plant data, applying system components, or simply improving processes.

The fourth industrial revolution, known as Industry 4.0, paves the way for the systematic implementation of a modernized energy network to manage ever-growing energy demand by integrating renewable energy sources [7]. Internet of Things (IoT), industrial internet, cloud-based production [9], and new concepts such as intelligent production partly address these requirements and are often classified by the visionary concept of a fourth industrial revolution.

Industry 4.0 refers to the latest technological advances that the Internet and supporting technologies serve as a backbone that integrates physical objects, human actors, intelligent machines and production lines [27].

By using real-time and high-value support systems, intelligent production has enabled a coordinated, and performance-oriented production initiative that responds quickly to customer demands, minimizes the use of energy and materials, radically improves sustainability, productivity, innovation, and economic competitiveness [32]. Businesses aiming to increase efficiency in engineering and operations, reduce operating costs, and improve product quality want to invest in industrial automation systems and digitalization.

Digitalization Concept and Industry 4.0

In order to talk about Digitalization and Industry 4.0, first of all, it is necessary to make sure that the industrial automation and control systems are fully and accurately designed and operating. Some of the tools used in industrial automation systems can be listed as follows.

PLC (Programmable Logic Controller)

In industrial applications, PLC devices are mostly used as control systems. PLC devices are long-lasting and reliable devices designed for industrial environments. However, hardware prices are high. There is an extra charge for software licenses used to program PLC devices [13].

For each process used in different sectors, the performance requirements of industrial automation and control systems are different. For 'Process Automation' applications, the powerful and reliable PLC controller is of great importance.

Distributed Input and Output Modules (I/O)

Distributed I/O modules have been used in industrial automation and control systems in recent years. Similarly, the use of distributed I/O modules in Process Automation applications has shown a rapid increase. Particularly in 'Process Automation' applications, devices, and sensors within the plant are distributed over a wider area. The electrical connections of the signals from these regions to the I/O modules on the PLC cause both serious labor and high costs. By keeping the PLC control system in the central control room and distributing the I/O modules close to the devices and sensors, it was observed that the electrical wiring to be made was minimized, and the time and cable cost were minimized. Figure 1 shows the deployment of distributed I/O modules in four different locations.

The distributed I/O modules are fully adaptable to the plant structure with a modular, flexible, and integrated structure. It also provides significant savings in many areas such as wiring, assembly, engineering, commissioning, maintenance. The communication protocols of the distributed I/O modules are shown in Table 1 [28].

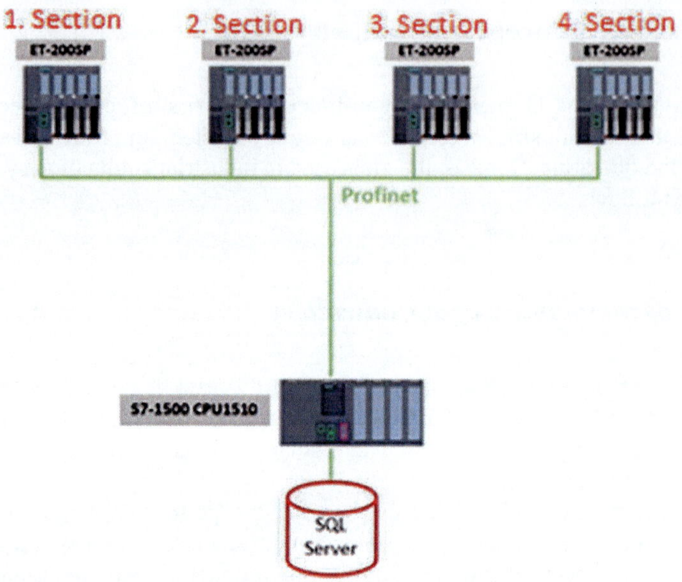

Fig. 1 Application example of distributed I/O modules

Table 1 Distributed I/O modules communication protocols

Pos	Communication protocols
1	PROFINET IO
2	PROFIBUS DP V0/V1
3	IO-Link V1.1
4	AS-Interface
5	Modbus TCP
6	Point to Point (RS 232, RS 485, RS 422)
7	Freeport
8	3964 (R)
9	USS
10	Modbus RTU (master/slave)

HMI (Human Machine Interface)

HMI devices are one of the most used devices after PLC in industrial applications. It is used to monitor process and input information to the PLC program. HMI devices have different sizes and specifications.

Interfaces for audit control in HMI engineering process applications can be regarded as an information integration task. This is especially true for configuration screens, which are the core elements of the high-performance HMI [31].

Industrial Network

Planning, implementing, and linking industrial networks to the enterprise network require a lot of talent and accessible expert knowledge. The same applies when a communication network needs to be secured, diagnosed, or optimized. One of the most important components in Industry 4.0 and digitalization applications is the error-free installation of the industrial network system with the right devices. With the recent use of Industry 4.0 and digitalization applications, there has been an increase in the use of industrial network systems. Thus, the hardware and software costs of industrial network systems used to provide monitoring, control, and diagnostics in industrial systems are reduced. However, there are some application challenges when these new technologies are applied in industrial environments. Given that many of these systems are made up of different distributed systems, management, and coordination are a major challenge [10].

PROFINET (PN) is an industrial Ethernet supported high-level open network communication protocol that uses TCP/P and IT standards in industrial control systems. PN communicates separately with the industrial controllers (PLCs, HMIs, servo drives, RFID (Radio-frequency identification) readers, etc.) scattered throughout the system, ensuring horizontal and vertical integration, ensuring safety, alarm and data transmission faster. According to the timing characteristics, PROFINET defines three types of traffic, as shown in Table 2.

Non Real-Time (NRT) traffic is a typical example of PROFINET CBA networks, real-time (RT) Class 1 can be processed by both PROFINET CBA and IO. RT class 2 performances are only possible with PROFINET IO. Such a class, also known as synchronous, relies on a highly precise and synchronized loop that requires the use of advanced switches capable of predicting delays in which they identify and correct them.

PROFINET IO refers to four types of devices that can be found in the network:

1. IO controller,
2. IO device,
3. IO supervisor,
4. IO parameter server.

    1. The IO controller is "smart" devices, such as personal computers or programmable controllers that perform automation tasks.

Table 2 Profinet timing features [8]

Traffic type	Periodicity/reaction times	Jitter (%)
Non real time (NRT)	$\geq$ 100 ms	$\geq$ 100
Real time class 1	$\geq$ 5 ms	$\geq$ 15
Real time class 2	$\geq$ 250 μs	$\leq$ 0.4 (1 μs)

2. IO devices are devices that perform the interface between IO controllers and field. Examples of these devices include sensors, actuators, valve batteries, electronic terminators, etc.
3. IO supervisors, IO controllers and/or IO devices and devices that change both configuration and diagnostic data.
4. IO parameter servers are devices used to change configuration data for applications with IO devices. Data traffic generated by both IO supervisors and IO parameter servers usually occurs during off-line stages and does not imply real-time performances. Conversely, the exchange of data between IO controllers and IO devices is often related to automation tasks carried out and therefore has critical timing requirements. [8].

After addressing industrial automation and control systems, we can move on to Industry 4.0 and its components.

Industry 4.0

Industry 4.0 is an initiative that combines a range of technologies that help achieve high-throughput production processes. An important feature of industrial production in Industry 4.0 is that physical elements such as sensors, devices, and corporate assets are connected to each other and to the internet. In this environment, devices and sensors generate an increasing amount of data.

An important consideration is that the execution of industrial processes depends not only on the internal situation and user interactions, but also on the context of execution, and provides value-added information to create contextual awareness and improve monitoring [11].

Components of Industry 4.0

When talking about Industry 4.0 and digitalization, it will be useful to give a brief overview of its main components. There are different technologies in the Industry 4.0 concept. These technologies require individual expertise and competence.

Autonomous Robots

One of the first components that come to mind when we say industrial automation is robots or robotic technologies. Robots, which have been used extensively in almost all sectors, have become an important component of Industry 4.0.

Since robots can be applied to all sectors, and their costs are low, industrial robots have been widely used in production areas in recent years [4].

Simulation

Nowadays, while developing technological products, different alternatives are encountered in design studies. It is important to predict how these alternatives affect the performance of the product during the actual application. Computer-aided engineering simulation helps engineers anticipate the effects of a design change by predicting the outcome of any design change on the real-world performance of their products.

Facility design is an important factor affecting the overall performance of production systems. Facility layout design deals with the allocation of machinery/departments in a facility and has a major impact on the efficiency and efficiency of production activities. An effective layout can reduce production costs and improve system performance. Discrete event simulation is an appropriate tool to evaluate the current order, to assess potential alternative areas for improvement by evaluating different settlement alternatives. Therefore, various researchers have applied simulation to different plant layout problems [20]. Simulations will be used more extensively in plant operations to use real-time data to mirror the physical world into a virtual model that can include machines, products, and people, thereby reducing machine setup times and improving quality [26].

Vertical and Horizontal Integration

Vertical and horizontal integration is the integration of all stages of production processes into the system in Industry 4.0 applications. What is essential at Industry 4.0 is that all interconnected systems must constantly communicate with each other and monitor and control each other as necessary.

Two of the most important factors for production in Smart Factories are horizontal and vertical integration. Horizontal integration means creating a network between individual machines, pieces of equipment, or production units, ensuring an uninterrupted flow between each step in the production and planning process, as well as between the steps in the production and planning processes of different enterprises. Vertical integration networks beyond traditional production hierarchy levels mean uninterrupted communication and flow in technological infrastructure. In this context, data obtained from sensors, valves, motors, and enterprise resource planning software and business intelligence applications are handled within this scope.

Internet of Things (IoT)

Businesses who want to increase their production and reduce their production costs need to professionalize their production methods [29]. IIoT (Industrial Internet of Things) can make a definition like this. It is the use of certain IoT technologies in combination with some smart devices in cyber-physical systems in an industrial environment. It can also be expressed as follows, The IIoT vision of the world is a place where interconnected smart things, a larger system or activity as part of the systems that make up intelligent production [5].

With the inclusion of information technologies in production processes, production power can be increased, and processes can be accelerated. With intelligent production processes, intelligent products and remote-control systems can provide a high level of flexibility to production processes. At the same time, it is possible to produce products that can be customized at low costs in line with customer requests by allowing the customers to be integrated into the system.

It is mentioned that the devices make life easier by communicating with the internet of objects, that is, devices communicating with other devices. The data produced by the interaction between the objects enables the industrial processes to be conducted in a more controlled way, to make more detailed analyzes, and to make more dynamic and effective decisions [22]. With this structure, which is also called the Internet of Industrial Objects, factories become smart. Thus, many different and complex structures can be produced in less time and optimum quality. When the devices are connected to each other, the generated data can be transferred rapidly with the support of high-speed internet, and the results obtained from the data obtained can be made faster, and the most effective decisions can be made. These decisions can be transferred to the personnel at work, the manager who follows the work, and all the devices at the same time and can work synchronously. In short, digitalization is not only by transferring the data in the field to digital environments, but also ensuring that these data can be used managerially. Communication systems, which create a vast communication network with the Internet of Things and thus aim to lift the boundary between real and virtual worlds, constitute one of the fundamental forces of Industry 4.0. Industry 4.0-based production processes are based on the fact that systems connect to different networks through different interfaces and communicate with different services. Communication of all devices operating in industrial environments is only possible within a well-designed and configured healthy network. With the developing wired and wireless communication technologies, communication can be made on any platform as well as open-source software and standards developed to provide easier, safer, and manageable communication both in local and wide area networks.

Industrial identification systems are part of industrial communication systems and are one of the main components of Industry 4.0 and IoT systems. Identification systems, also known as RFID (Radio Frequency Identification) systems, digitize organizations with new generation communication technologies and transform them into more efficient and effective systems.

Cyber Security

As a result of the development of information technologies, it became an inseparable part of our lives, digital transformations of our private and public institutions and the targeted and coordinated development of cyber threats caused cyberspace to expand, and cyber threats became a threat to our national security. Today, information and communication technologies (ICT) support most of the industrial production processes. The ICT revolution brought about important transformations in the first and second industrial revolutions, similar to the high effects of mechanics and electricity. This development supported the emergence of cloud-based systems, the Internet of Things (IoT), big data, and the emergence of the Industry 4.0 concept. However, new technological solutions always carry security vulnerabilities that pose unexpected risks [24].

With Industry 4.0 and digitalization applications, cybersecurity has started to pose a major threat to industrial enterprises. The fact that every point in production can communicate with each other in a secure way, the different production facilities can communicate with each other, and the production information is recorded locally or in the cloud makes the data security of production companies important.

Cloud Computing

Cloud computing promises to demand endless and inexpensive resources to provide appropriate infrastructure support for user applications. For this reason, it is increasingly recognized by many businesses that try to move their applications to the cloud to reduce costs and automate reconfiguration. These businesses are supported by free or private cloud platforms that support application deployment and restructuring in the cloud [19].

Cloud security is one of the most important factors preventing cloud usage today. To overcome this reluctance, the cloud application development process must address potential security issues from the outset and adapt to the flexibility offered by the cloud paradigm, as well as the security constraints posed by developers and cloud customers [2].

With the help of Cloud Computing, users can use the applications required for the business in any area over the internet through the computers in the service provider instead of keeping them in on-site computers or data centers. Thus, more economical, flexible, and faster data management can be achieved.

Cyber-Physical Management Systems

Cyber-Physical Systems (CPS) connect the physical world with the virtual computing world with the help of sensors and actuators. CPS's, which are composed of different constituent components, create global behaviors in collaboration. These components often include software systems, communication technologies, sensors/actuators, including embedded technologies, to interact with the real world.

In Smart Factories, production processes mean that devices and systems communicate with each other instantly and identify and regulate all the needs within production processes. For example, if there is a need for material at any stage of production, the required material order is automatically placed, failures can be detected instantly, and on-site and solutions can be offered. Thus, productivity is increased while errors are minimized. CPS are systems that can effectively integrate cyber and physical components using the modern sensor, computing, and networking Technologies [33].

CPS uses data from various sensors for complete information. There is a difference between a malfunction of a sensor during operation and a malfunction of a faulty sensor, and therefore the user can receive incorrect information [30].

Smart production, smart buildings and infrastructures, smart city, smart cars, smart systems and so on. CPS in development is a basic concept [1].

Augmented Reality

Augmented Reality (AR) is a new form of human–machine interaction that encompasses computer-generated information in a real-world environment. The information provided by AR is derived from the actual environment and is, therefore, context-sensitive. The AR can therefore improve the user's appearance with virtual information sensitive to the current state of the surrounding true thickness [25]. With the introduction of the Industry 4.0 concept, AR applications were introduced in digitalized smart factories.

Application examples are used in different scenarios ranging from field service operations to maintenance operations, training, and quality control applications. Within the scope of Industry 4.0, automotive is one of the sectors that benefited the most from AR. AR is used not only in production and maintenance but also in sales. Customers can have driving experience with AR simulation applications if they wish before buying their vehicles. If they wish, they can see the color or accessory options of their favorite vehicles in the AR application.

Big Data and Data Analytics

With the development of technology, devices that communicate with each other, electronic transactions, data traffic, mobile applications, e-mails, videos, pictures, scientific data, etc. The size of the data produced is increasing day by day. In the last few years, especially since 2013, all data sources have increased in terms of the number of records. Trends for scientific publications and funding archives followed a similar path [14]. Keeping such large amounts of data on secure systems and analyzing them and converting them into meaningful information is especially important for production companies.

Particularly in production enterprises, tens of information about production processes can be read and converted into meaningful information, and production can be analyzed, and a serious productivity increase can be achieved.

Result

The transition to a digital transformation to Industry 4.0 naturally has difficulties. In production sectors where technological development and large investments are required, serious costs may initially arise. The promise that manufacturers have an economical size helps them to take the chance of the entire advanced automation and digitization process. Digital transformation and Industry 4.0 bring us convenience in our daily lives and continue to be a part of our lives. It is seen that our quality of life has increased with this transformation, which makes the connections of objects and systems with each other effective and efficient.

In order to show how important and effective the quality management system is, the systematic put forward by Industry 4.0 is important for businesses.

With increasing digitalization, the errors caused by the human factor are reduced, and the system becomes automated, so productivity in production increases, the number of faulty products decreases, and costs are minimized. High efficiency, cost, and minimum errors are a result desired by the quality management system [21]. With Industry 4.0, the adoption of a structure that will ensure the maximum level of customer satisfaction in production and service encourages digitalization for businesses.

References

1. Alguliyev R, Imamverdiyev Y, Sukhostat L (2018) Cyber-physical systems and their security issues. Comput Ind 100:212–223
2. Casola V, De Benedictis A, Rak M, Rios E (2016) Security-by-design in clouds: a security-SLA driven methodology to build secure cloud applications. CLOUD FORWARD: from distributed to complete computing. Proc Comput Sci 97:18–20

3. Celik N, Ozturk F (2017) The Upcoming issues of industry 4.0 on occupational health and safety specialized on turkey example. Int J Econ Bus Manage Res 1(5):236–256
4. Chen Y, Dong F (2013) Robot machining: recent development and future research issues. Int J Adv Prod Technol 66:1489–1497
5. Conway J 2015 The industrial internet of things: an evolution to a smart production enterprise. Schneider Electric Whitepaper, pp 2
6. Edgar TF, Pistikopoulos EN (2018) Smart production and energy systems. Comput Chem Eng 114:130–144
7. Faheem M, Shah SBH, Butt RA, Raza B, Anwar M, Ashraf MW, Ngadi MdA, Gungor VC (2018) Smart grid communication and information technologies in the perspective of Industry 4.0: Opportunities and challenges. Comput Sci Rev 30:1–30
8. Ferrari P, Flammini A, Vitturi S (2006) Performance analysis of PROFINET networks. Comput Stand Interf 28(4):369–385
9. Gao R, Wang L, Teti R, Dornfeld D, Kumara S, Mori M, Helu M (2015) Cloud-enabled prognosis for production. CIRP Ann-Prod Technol 64(2):749–772
10. Gholami M, Taboun MS, Brennan RW (2017) A wireless intelligent network for industrial control. 27th Int Conf Flexible Autom Intell Prod 11:878–888
11. Giustozzi F, Saunier J, Zanni-Merk C (2018) Context modeling for industry 4.0: an ontology-based proposal. Proc Comput Sci 126:675–684
12. Gligor DM, Holcom MC (2012) Understanding the role of logistics capabilities in achieving supply chain agility: a systematic literature review. Surg Endosc Other Interv Tech 17(4):438–453
13. Hoxha V, Bula I, Shala M, Hajrizi E (2016) Cost-oriented open source automation potential application in industrial control applications. IFAC-PapersOnLine 49(29):212–214
14. Huang Y, Porter AL, Cunningham SW, Robinson DKR, Liu J, Zhu D (2018) A technology delivery system for characterizing the supply side of technology emergence: Illustrated for big data and analytics. Technol Forecast Soc Chang 130:165–176
15. Kayar A, Ayvaz B, Öztürk F (2018) Akıllı Fabrikalar, Akıllı Üretim: Endüstri 4.0'a Genel Bakış". In: International Eurasian conference on science, engineering and technology (EurasianSciEnTech 2018), Nov 22–23, Ankara, Turkey, pp 1661–1668
16. Kayar A, Ozturk F (2019) Predictive maintenance in Industry 4.0 applications. Int Conf Life and Eng Sci ICOLES 27(29):16
17. Kayar A, Ozturk F (2019) Machine data collection and analysis in industry 4.0 applications. Int Conf Life Eng Sci ICOLES 27(29):317–322
18. Kayar A, Ozturk F, Kayacan O (2019) Fast fault solving methods in smart manufacturing lines with augmented reality applications. Recent Adv Data Sci Bus Anal 25(28):182–187
19. Kritikos K, Massonet P (2016) An integrated meta-model for cloud application security modelling, CLOUD FORWARD: from distributed to complete computing. CF, Madrid, Spain, pp 18–20
20. Negahban A, Smith JS (2014) Simulation for production system design and operation: literature review and analysis. J Prod Syst 33:241–261
21. Ozturk F (2014) Qualität, Effizienzsteigerung und integrierte Managementsystemen im türkischen Eisenbahnsektor. Soc Nat Sci J 8(2):14–19
22. Ozturk F, Kayar A (2019) Product lifecycle management in smart factories: industry 4.0 applications. In: Proceedings of 10th international symposium on intelligent manufacturing and service systems. Istanbul/Turkey, pp 1420–1427
23. Ozturk F, Kayar A, Vatansever A (2019) Advanced manufacturing with industry 4.0 applications. In: 5th international conference on advances in mechanicalengineering. Istanbul/Turkey, pp 17–19
24. Pereira T, Barreto L, Amaral A (2017) Network and information security challenges within Industry 4.0 paradigm. In: Production engineering society international conference. Vigo (Pontevedra), Spain, pp 28–30
25. Reinhart G, Patron C (2003) Integrating augmented reality in the assembly domain—fundamentals benefits and applications. CIRP Ann 52(1):5–8

26. Rüßmann M, Lorenz M, Gerbert P, Waldner M (2015) Industry 4.0: the future of productivity and growth in production industries, 09 Apr 2015, pp 1–14
27. Schumacher A, Erol S, Sihn W (2016) A maturity model for assessing Industry 4.0 readiness and maturity of production enterprises. Proc CIRP 52:161–166
28. Siemens Industry Mall Web Site (2019) https://w3.siemens.com/mcms/distributed-io/en/ip20-systems/et-200sp/et200sp-communication-modules/pages/default.aspx
29. Singh D, Tripathi G, Jara AJ (2014) A survey of internet-of-things: future vision, architecture, challenges and services. In: IEEE world forum on internet of things 2014. Seoul, pp 287–292
30. Tang L-A, Yu X, Kim S, Gu Q, Han J, Leung A, La Porta T (2013) Trustworthiness analysis of sensor data in cyber-physical systems. J Comput Syst Sci 79:383–401
31. Urbas L, Obst M, Stöss M (2012) Formal models for high performance HMI engineering. IFAC Proc 45(2):854–859
32. Yuan Z, Qin W, Zhao J (2017) Smart production for the oil refining and petrochemical industry. Engineering 3(2):179–182
33. Zeadally S, Jabeur N (2016) Cyber-physical system design with sensor networking technologies. The Institution of Engineering and Technology, London UK

s, S Inventory Control Optimization Under Inventory Sharing Policy for Omni-Channel Network

Damla Izmirli, Banu Yetkin Ekren, and Ecem Eroglu

Abstract Rapidly developing technology and multi-channel integrations in marketing have led to the emergence of omnichannel retailing. In omnichannel, consumers can take advantage of all shopping channels interactively. It overcomes the drawbacks of traditional supply chain applications, but uncertain stock demands cause serious difficulties and costs. In this work, an inventory sharing policy study presented for an omnichannel network in order to reduce the risks caused by stock and demand uncertainty. In literature, this policy also referred to as lateral transshipment. Recently, due to the increase in e-commerce, it has become significant to respond to customer demands shortly for customer satisfaction. Inventory sharing allows the exchange of items between the same echelon of the supply chain network enabling high service levels with low costs. Since, in an omnichannel network, physical transshipment may not take place due to the electronic order case, we prefer to state this policy as inventory sharing instead of lateral transshipment. We develop simulation models for three different omnichannel scenarios and optimize the total network cost by considering those three sharing policies. We compare each scenario's optimal cost result. For the modeling and optimization purpose, we use the ARENA 16.0 commercial software and OptQuest to optimize the (*s*, *S*) inventory levels.

Keywords Omni-channel · Lateral transshipment · (*s, S*) inventory · Inventory control · E-commrce

Introduction

The acceleration of globalization and economic developments have increased service expectations in customer demands. Hence, competition among enterprises has been established on how quick response is given for customer demand. Therefore, for

D. Izmirli · B. Y. Ekren · E. Eroglu (✉)
Department of Industrial Engineering, Faculty of Engineering, Yasar University, İzmir, Turkey

B. Y. Ekren
e-mail: banu.ekren@yasar.edu.tr

F. Calisir (ed.), *Industrial Engineering in the Internet-of-Things World*,
Lecture Notes in Management and Industrial Engineering,
https://doi.org/10.1007/978-3-030-76724-2_6

businesses, in order to succeed and survive in an intensely competitive environment, they should manage their activities well in all aspects of the supply chain such as logistics, supply, production, and marketing. Logistics is an important issue for the business to compete and, at the same time, ensure customer satisfaction [4]. Supply chain management has become one of the most important management issues in increasing the global competitiveness of businesses today. An effective supply chain management application is required to produce better quality products by companies that want to increase their profitability to respond to customer demand faster [20].

The recent widespread use of the Internet has provided numerous opportunities, such as cost reduction and service improvements for businesses and the supply chain process [14]. Along with technological developments, digitalization in marketing and retailing started to dominate the businesses [15]. Retailing companies also took a role in e-commerce. They are keen on the trial of different marketing strategies in customer satisfaction. For instance, both online and offline (physical) channels are to be combined in marketing strategies.

Recent technological developments and changing consumer habits compel retailers to use all channels and provide uninterrupted service for customers. Before today's technological developments became widespread, customers were served by retailers through physical stores, that is, through a single channel. Today, with the spread of internet access, the online store has become one of the most popular channels, making it easy and fast to reach customers. Offline channels have started to turn into the online and complex structure [2]. The shopping experience offered to customers is transformed from single-channel to multi-channel, cross-channel, and omnichannel trade. For instance, the main purpose of the omnichannel is to enable customers the use of all channels in one operation. Companies aim to enable customers to access online stores, mobile devices, and physical stores simultaneously through omnichannel. Therefore, retailers combine digital and traditional retail applications to meet the demands of their customers and enable customers to move freely between mobile devices, offline and online stores [19]. Advantages of both physical and online store operations and the increased cost incurred in a single channel customer have also pushed the new way of marketing strategies in businesses. Retailers are eager to implement the omnichannel approach for this purpose. However, a transition from the traditional supply chain to omnichannel applications requires redesigning supply chains and integrating channels, but these changes should create some difficulties [3]. The differences between physical and online store operations create some financial difficulties for businesses [12]. In this work, we also aim to provide different scenarios for omnichannel integrations of retailers.

Inventory management and logistics have become important issues in the supply chain. The high competitive environment forces firms to hold the excess amount of stock in order to respond to customer requests promptly. However, it may increase costs due to excess stock holding costs and inadequate sales. To prevent these, it is aimed to improve customer services by keeping the cost at the lowest level by companies. Technological applications in inventory systems provide communication between inventory points, reduce the risks that may arise, and ensure the implementation of the concept of lateral transshipment, thus reducing the shortage of stock [11].

There are several studies dealing with lateral transshipment with different supply chain models. Lateral transshipment approaches are divided into two groups, namely reactive and proactive, that they differ according to the time of sharing by [18]. In reactive approaches, sharing can be done at moments when ignoring situations [16]. In proactive approaches, sharing can be done at predetermined times before demands are observed [5].

In the supply chain, it is critical to define the optimal level of order quantity by determining the level of re-order point and security stock [10]. Arrow et al. [1] developed a simple model to determine the best order level and the best maximum stock point as a function of demand distribution, installation, and stocking costs. Ozdemir et al. [17] conducted a study to minimize the cost of a model with different cost and demand parameters, a single supplier, and multiple dealers. In his study, he demonstrated that the base stock policy is the optimum stock replacement policy to minimize the cost. They used an optimization procedure to find out the optimal *s*, *S* levels.

Ekren and Heragu [6] studied the multi-location transshipment problem by considering lateral inventory share in the system. They complete simulation optimization to solve the problem. The soonest work of Ekren and Arslan [9] studies *s*, *S* inventory control model to compare the different lateral transshipment policies in a single-echelon network. In that study, they model a single item to minimize cost. In their work, it was concluded by them that lateral transshipment scenarios are better than lateral transshipment. Ekren and Ornek [7, 8] studied *s*, *S* inventory control problem by using simulation-based optimization.

The aim of this study is to adopt the versatile channel approach (i.e., omnichannel) and apply a reactive inventory sharing policy to let the retailers share their items in the backorder condition to minimize the total network cost. We assume two companies that are having connected offline and online marketing channels. We seek the best inventory sharing policy (also known as pooling policy in the literature), optimizing the total network cost. The decision variables are considered to be the safety-stock and up-to levels of stocking locations optimized by the OptQuest tool in ARENA 16.0 simulation software. Three scenarios are considered for inventory share policies based on the connected network design. The results are compared and analyzed.

Methodology

Problem Definition

As mentioned previously, omnichannel marketing is gaining popularity with recent technological developments. This is also due to the result of more connected and more integrated enterprises than ever. By the horizontal and vertical integration in enterprises, management issue has become more complex and more challenging in the network. Hence, companies seek for true approaches to manage their complex supply network while reducing their total cost. Determining the true inventory control policy is one of the emerging approaches in the efficient management of a supply chain network. For instance, lateral transshipment applications have become one of the efficient implementations in real-time visible and connected network designs in efficiently meeting customer demands. However, in this work of omnichannel marketing having e-marketing, since a physical lateral transshipment may not take place, we preferred stating this policy as an inventory sharing policy. In other words, because the shipment takes place to the customer directly from the inventory shared company, no physical lateral transshipment takes place.

In this work, we assume that the stocking locations are covered by the main depot with infinite capacity. The considered networks are omnichannel companies having both online and offline demand distributions separately. The inventory sharing takes place when demand cannot be met by a stocking location. The main purpose of this study is to determine which pre-specified policy works better for companies under these assumptions. The network designs, along with the sharing policy assumptions, are detailed in the following sections.

Simulation Modelling and Assumptions

In this study, the considered designs are simulated by Arena 16.0, and the s, S inventory levels are optimized by using the OptQuest optimization tool in this simulation software. This optimization tool is developed on heuristic-based algorithms. It combines the meta-heuristics of tabu search, neural networks, and scatter search into a single search heuristic [13]. The tool allows the user to define integer and linear constraints for the simulation inputs. Initially, it requires the user to specify the lower, suggested, and the upper values for the decision variables (i.e., s, S levels) to be optimized. The suggested values are for the starting points in the search procedure.

As mentioned, inventory sharing is also known as inventory pooling policy in the supply network, referring to a firm's ability to serve multiple demanding points, each with its own uncertain demand. In the non pooling case, all demand is served from a single stock of inventory. Therefore, the pooling policy is more suitable in a connected supply network management. The inventory control model is developed based on the safety stock level (i.e., s, S). We defined three different control policies and compared each scenario's performance based on the optimized total cost. Thus, the aim of this study can be summarized as the minimization of total network cost under omnichannel commerce by considering that there is an inventory sharing policy in the system. The inventory sharing policies are defined under three scenarios summarized in the following sections, along with their simulation assumptions.

Simulation Model Assumptions

The common assumptions of all simulated models are as follows:

- There are two stocking locations at the same echelon level with both offline and online stores. The stocking locations have separate demand distributions where their online stores receive more demands than offline stores.
- Demand arrives at the beginning of each day at each stocking location separately by a normal distribution.
- The normal distribution for demand amounts of stocking locations is set to Online1 Normal (35, 10), Online 2 Normal (30, 10), Offline1 Normal (25, 10), Offline2 Normal (28, 10).
- Inventory share between stocking locations takes place in this order: first, the demand arriving stocking location of the company checks its online/offline stocking location's inventory level. Then, the other connected stocking locations are checked according to the policies explained in the below section.
- The main depot has an infinite capacity, and lead time from the main depot to the stocking locations is considered to be constant and two days for each.
- When there is an order on the road for a stocking location, that storage location also considers it in its replenishment calculation.
- The capacity of a truck is considered to be 20 items per truck that are from the main depot to the stocking locations. The fixed transportation cost per truck is $80.
- Total network cost calculations include holding, ordering, transportation, backorder, and inventory share costs.
- The simulation is run for one year with ten independent replications. A steady-state analysis is conducted, and the warm-up period is considered to be one month.
- The inventory level at each stocking location is checked at the end of each day.
- If total demand cannot be met by the whole network, the total demand amount becomes a lost sale, and the regarding backorder cost is incurred in the total cost.
- Verification is done by debugging the models, and validation is done by an expert's model check.

The notations that are used for calculations are summarized below:

Tc	Fixed transportation cost of a single truck from the main warehouse to any stocking location (i.e., $80/truck)
d_{it}	demand amount arriving at online/offline store i, $i = \{1, 2, 3, 4\}$, at the beginning of day t
I_{it}	inventory level at stocking location i, at the end of day t
AI_i	annual average inventory amount carried at stocking location i
h	annual holding cost per item
C_T	truck capacity from the main depot (i.e., 20 item/truck)
Oc	ordering cost per item from the main depot

(continued)

(continued)

LC_{ji}	inventory share cost per item from stocking location *j* to *i*.
L_{jit}	total inventory share amount from stocking location *j* to *i* at the end of day *t*
Q_{it}	order amount given from stocking location *i*, at the end of day *t*
s_i	safety stock level of stocking locations *i*, $i = \{1, 2, 3, 4\}$
S_i	up-to level of stocking location *i*, $i = \{1, 2, 3, 4\}$
L	lead time from the main depot to any stocking location (i.e., 2 days)
n_{it}	number of trucks shipped from the main depot to the stocking location *i* at day *t*, (i.e., an integer)
b_{it}	lost sale amount at stocking location *i*, at the end of day *t*
C_b	back order cost per item at any stocking location ($10/item)

h and C_b are set to $1/item—year and $10/item, respectively. LC_{ij} is set for $1/item if it takes place between the same company depots (i.e., online to offline or vice versa), and it is set for $1.5/item if it takes place between different company depots. *Oc* is set for $1/item. Note that the inventory level at each stocking location is checked at the end of each day. If the inventory level of stocking location *i* is less than s_i, then Q_{it} amount of order is sent to the main depot to the stocking location *i* where Q_{it} is computed by (1).

$$Q_{it} = \begin{cases} S_i - I_{it}\ if\ I_{it} \leq s_i \\ 0, \qquad otherwise \end{cases} \tag{1}$$

Total network cost (*TC*) is calculated by the (2) where *T* is the total simulation run time (i.e. 365 days) and *k* is the total number of stocking locations (i.e. $k = 4$):

$$TC = \sum_{t=1}^{T} \sum_{i=1}^{k} \sum_{j \neq i}^{k} [(AI_i \times h) + (b_{it} \times C_b) + (n_{it} \times O_C) + (n_{it} \times T_C) + \left(L_{jit} \times LC_{ji}\right)] \tag{2}$$

where n_{it} and L_{jit} are computed by (3) and (4), respectively. Note that n_{it} is rounded up to the next integer.

$$n_{it} = Q_{it}/C_T \tag{3}$$

$$L_{jit} = I_{it} - d_{it} \tag{4}$$

As mentioned previously, three pooling policies are considered based on network designs. These scenarios are explained in the following section.

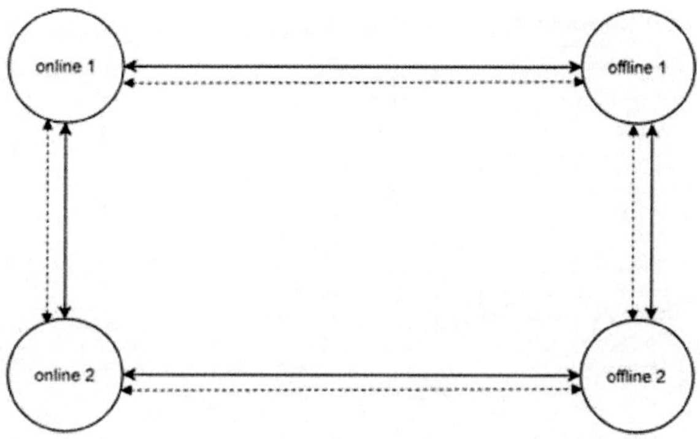

Fig. 1 Network design for policy 1

Inventory Share Policies

Three different inventory share policies are defined based on the level connection of the network. Remember that there are two companies in the same echelon. These companies have both its online and offline stocking locations. In each scenario, it is always assumed that a specific company's online and offline stores are connected, and they always share their inventories. The determined policies are summarized below:

Policy 1

Figure 1 shows the network design of Policy 1. Note that in that figure, dashed and straight lines show information and physical product flows between stocking locations, respectively. According to that figure, Company 1's online and offline stores are connected. The same is also correct for Company 2. Two-way information and product flow can take place between those connected stores. Besides, online and offline stores of different companies are also connected on their own. The flow chart of the simulation model of this policy is presented in Fig. 2 as an example. Policy 1 assumptions are summarized below.

- Demand arrives at each stocking location at the beginning of each day, whose mean amount follows a normal distribution with the parameters described in the general system assumption's part.
- Online/offline store first checks whether or not the current inventory could meet the arriving demand amount. If the current inventory amount is sufficient, then, the stocking location meets the whole arriving demand immediately. Otherwise, it checks whether or not it can meet the remaining amount from another stocking location at a time. Namely, the remaining amount is not split among different stocking locations. If there is no stocking location that can meet the whole remaining amount, then, the whole demand becomes a lost sale. And, backorder cost is incurred in the total cost.

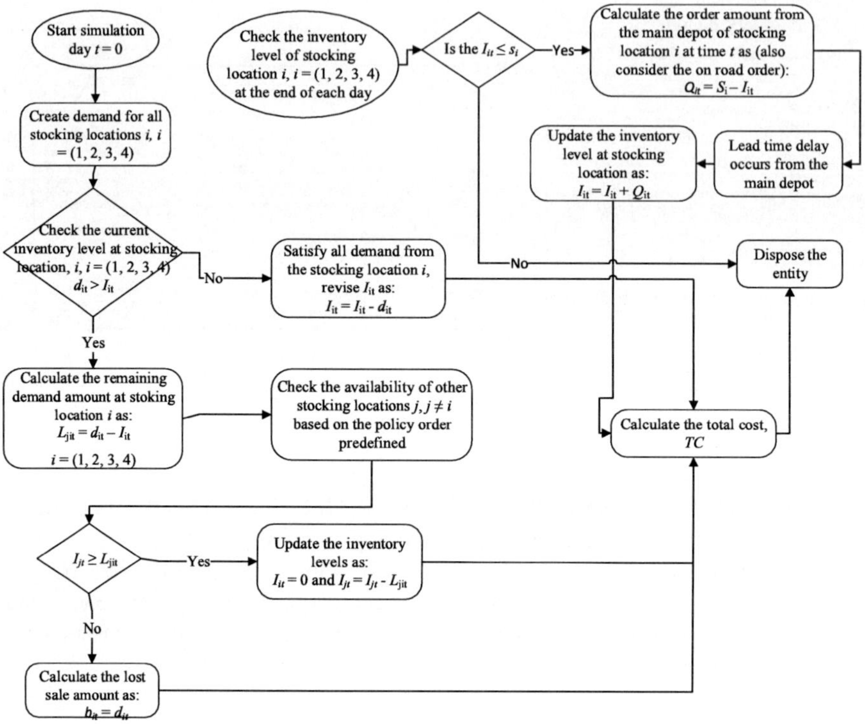

Fig. 2 Simulation flow chart of policy 1

- For the remaining amount of order, the availability of inventories in stocking locations are checked in this order: First, its own company's online/offline store is checked. Then, the others are checked.
- In different companies, inventory share only takes place between online to online or offline to offline stores.

Policy 2

Figure 3 shows the network design for Policy 2. According to that policy, Company 1's online and offline stores are connected, and Company 2's online and offline stores are also connected. Besides, the online store of Company 1 and Company 2 are also connected. The assumptions of this policy are similar to that of Policy 1. The main difference is the connectedness of the stores. Since there is no connection between the two offline stores, they cannot make any inventory share between them.

Policy 3

Figure 4 shows the network design for Policy 3. This policy is the policy having the highest level of connectedness among the stores. Namely, every stocking location is connected with the other one. In this case, our aim is to compare how the inventory

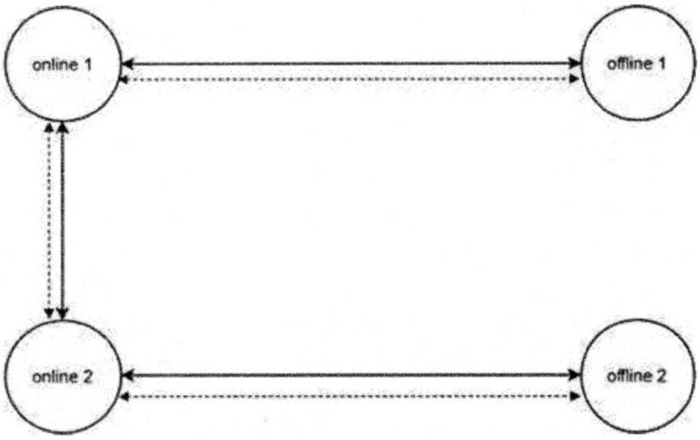

Fig. 3 Network design for policy 2

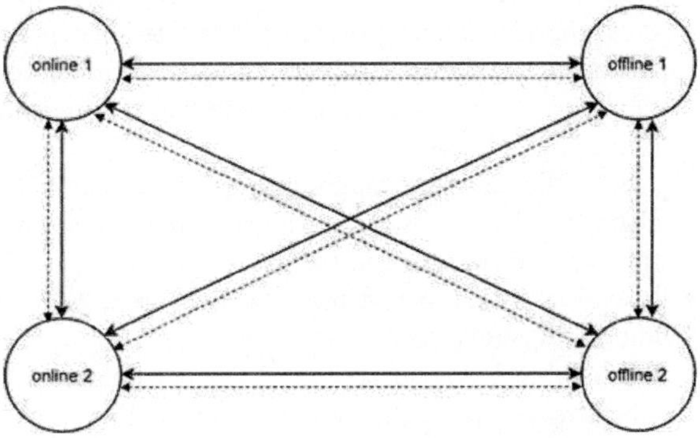

Fig. 4 Network design for policy 3

share condition at the highest level affects system performance. This policy's assumptions are the same as in Policy 1. However, due to the difference in connectedness, the available inventory seeking policies are as follows:

- When there is not sufficient inventory in the Offline 1 store, then, the remaining demand amount availability is checked in this order: Online 1, Offline 2, and Online 2.
- When there is not sufficient inventory in Offline 2 store, then, the remaining demand amount availability is checked in this order: Online 2, Offline 1, and Online 1.
- When there is not sufficient inventory in the Online 1 store, then, the remaining demand amount availability is checked in this order: Offline 1, Offline 2, and Online 2.
- When there is not sufficient inventory in the Online 2 store, then, the remaining demand amount availability is checked in this order: Offline 2, Offline 1, and Online 1.

Simulation Results

Table 1 shows the average optimal cost results obtained by the OptQuest tool. Regarding s_i, S_i levels based on the OptQuest results are summarized in Table 2.

In Table 1, we also provide the total costs separately incurred in the total cost calculations. For instance, it is observed that the total inventory share cost is the highest in Policy 3. This is probably due to that in that design; the network connection is at the highest level. It is also observed that the least backorder cost is obtained again by Policy 3 where the connection is at the highest level in that design.

According to Tables 1 and 2, the findings are summarized below:

- The lowest optimal cost is obtained in Policy 3, where the connectedness is the highest one among the network designs.
- Note that the connectedness level of the design policies can be ordered from the most connected one to the least connected one as Policy 3, Policy 1, and Policy 2. According to Table 1, it is observed that higher connectedness level reduces the total cost. Hence, we can conclude that the inventory sharing policy works well in the studied networks.
- Also note that, when total inventory share cost increases, the overall network cost also (*TC*) decreases. It seems that there is a negative correlation between the total inventory share cost and the total network cost.
- It is also observed that the total backorder cost decreases with the increase of lateral inventory cost.
- From Table 2, it is observed that when the network connectedness is high, a single stocking location behaves as if a distribution depot for inventory sharing. In this

Table 1 Cost results obtained by the OptQuest

Policy	TC ($)	Total backorder cost ($)	Total holding cost ($)	Total transportation cost ($)	Total inventory share cost ($)	Total ordering cost ($)
1	255,795	23,841	30,545	158,890	3239	39,280
2	260,350	33,631	36,957	150,385	1695	37,682
3	248,900	6457	25,877	165,220	9669	41,677

Table 2 *s*, *S* value results

Policy	Online store		Offline store		Online store		Offline store	
	Company		Company		Company		Company	
	1	2	1	2	1	2	1	2
	s_1	s_2	s_3	s_4	S_1	S_2	S_3	S_4
1	60	30	65	185	80	70	150	300
2	85	20	55	150	100	85	90	275
3	45	30	40	265	85	60	105	385

case, Company 2's offline store takes such a role. This is because its *s* level is pretty small, and *S* level is high, letting this store keep more inventory.

Discussion and Conclusion

This paper studies inventory sharing policy in omnichannel networks in terms of total network cost. We study three different integrated network designs under inventory sharing policies. We optimize the *s*, *S* inventory levels in each design by minimizing the total cost of the network. The simulation results are compared based on ten independent replications. It is observed that, when the connection level increases in the network, the optimal cost decreases. Hence, inventory sharing works well to minimize the total network cost.

To test the sensitivity of the results on the utilized parameter values, we aim to complete a sensitivity analysis in the studied network designs as future work. Besides, different sharing policies would also worth trying to test the effectiveness of the network designs.

Acknowledgements We would like to express our sincere thanks to Newton Katip Celebi, TUBITAK, and Royal Academy of Engineering (Industry-Academia Partnership Programme 2018/2019, No. 4180046) for funding this project.

References

1. Arrow K, Harris T, Marschak J (1951) Optimal inventory policy. Econometrica 19:250–272
2. Beck N, Rygl D (2015) Categorization of multiple channel retailing in multi-, cross-, and omni-channel retailing for retailers and retailing. J Retail Consum Serv 27:170–178
3. Cao L, Li L (2015) The impact of cross-channel integration on retailers' sales growth. J Retail 91:198–216
4. Carvalho JLG, Campomar MC (2014) Multichannel at retail and omni-channel: challenges for marketing and logistics. Bus Manag Rev 4:103–113
5. Diks E, de Kok A (1998) Optimal control of a divergent multi-echelon inventory system. Eur J Oper Res 111:75–97
6. Ekren BY, Heragu SS (2008) Simulation based optimization of multi-location transshipment problem with capacitated transportation. In: Mason SJ et al (ed) Proceedings of the 2008 winter simulation conference. IEEE, Piscataway, New Jersey, pp 2632–2638
7. Ekren BY, Ornek MA (2015) Determining optimum (s, S) levels of floor stock items in a paint production environment. Simul Model Pract Theory 57:133–141
8. Ekren BY, Ornek MA (2015b) Simulation-Based Inventory Control in a Chemical Industry. In: Ihsan Sabuncuoglu, Bahar Y. Kara, Bopaya Bidanda (eds) Industrial engineering applications in emerging countries, CRC Press—Taylor & Francis Group, pp. 67–79. ISBN 978-1-4822-2689-8
9. Ekren BY, Arslan B (2019) Simulation-Based lateral transshipment policy optimization for s, S inventory control problem in a single-echelon supply chain network. Int J Optim Control Theor Appl (IJOCTA) 10:9–16
10. Freeman R (1957) Ss inventory policy with variable delivery time. Manage Sci 3:431–434

11. Glazebrook K, Paterson C, Rauscher S (2015) Benefits of hybrid lateral transshipments in multi-item inventory systems under periodic replenishment. Prod Oper Manage 24:311–324
12. Ishfaq R, Defee CC, Gibson BJ (2016) Realignment of the physical distribution process in omni-channel fulfillment. Int J Phys Distrib Logistics Manage 46:543–561
13. Kleijnen JPC, Wan J (2007) Optimization of simulated systems: OptQuest and alternatives. Simul Model Pract Theory 15:354–362
14. Lancioni R, Smith M, Oliva T (2000) The role of the internet in supply chain management. Ind Mark Manage 29:45–56
15. Larsen TS, Kotsab H, Grieger M (2003) Electronic marketplaces and supply chain relationships. Ind Mark Manage 32:199–210
16. Olsson F (2010) An inventory model with unidirectional lateral transshipments. Eur J Oper Res 200:725–732
17. Ozdemir D, Yucesan E, Herer YT (2006) Multi-location transshipment problem with capacitated transportation. Eur J Oper Res 175:602–621
18. Paterson C, Kiesmüller G, Teunter R, Glazebrook K (2011) Inventory models with lateral transshipments: a review. Eur J Oper Res 210:125–136
19. Piotrowicz W, Cuthbertson R (2014) Introduction to the special issue information technology in retail: toward omnichannel retailing. Int J Electron Commer 18:5–16
20. Presutti W (2003) Supply management and e-procurement: creating value added in the supply chain. Ind Mark Manage 32:219–226

An Application Using Stability Increasing for the Grinding Machine Performance Improvement in the Automobile Industry

Mehmet Cakmakci, Ece Sonmez, and Melis Kucukyasar

Abstract At the beginning of the third millennium, the use of the Industry 4.0 process has been accelerated and spread in different parts of production. With the rapid development of technology in production, the workload also increases due to the role of the human being. The workload is a concept in which the human processing system is expressed in terms of its ability to process information and produce responses within the framework of its physical and mental characteristics. Especially in the manufacturing sector, the contribution of the human being to the production increases with the technological function while physically decreasing with the developing machinery technologies. With this study, it aims to increase the production capacity in line with meeting the increasing demands in these production units, and improve the production plans of these units with the help of solution models to be developed by using appropriate analysis techniques of injector nozzles model measurements especially by using SMED approach within the lean manufacturing.

Keywords Performance improvement · Grinding machine · Improvement potentials · Injector nozzles model · Single minute exchange of Dies-SMED · Workload analysis

Introduction

At the beginning of the third millennium, the steps of change worldwide are accelerating. The most important of these changes is the globalization of production. It is prominent for companies operating internationally to spread their activities to other countries and continents. Market demands contain more product variants in parallel to customization. This evolution is not limited to certain types of industry,

M. Cakmakci (✉) · E. Sonmez · M. Kucukyasar
Industrial Engineering Department, Engineering Faculty, Dokuz Eylul University, Izmir, Turkey
e-mail: mehmet.cakmakci@deu.edu.tr

M. Kucukyasar
The Graduate School of Natural and Applied Sciences, Yasar University, Izmir, Turkey

F. Calisir (ed.), *Industrial Engineering in the Internet-of-Things World*,
Lecture Notes in Management and Industrial Engineering,
https://doi.org/10.1007/978-3-030-76724-2_7

rather it is a general phenomenon. As a result, customer demands and diversity are also increasing, so in order to respond to these demands, industries need qualified personnel to meet the required workforce. In order to adapt to globalization, businesses aim to reduce their annual production costs by using their resources more effectively and have a larger share in the market. Due to globalization, customer demands and demand diversity increase in industrialized countries. We see continuous improvement and lean production in key roles in the rational use of resources. Therefore, customer-oriented production becomes more important.

At this point, the terms "continuous process improvement" and "lean manufacturing," namely Toyota Production System comes into play. Lean manufacturing systems must have the ability to achieve responsive, small-batch manufacture so that they can meet rapidly changing market demands [1–5]. In fact, the lean manufacturing system is a part of corporate culture, such as tools and approaches.

The Bosch Group's business activities with the largest area in Turkey is the Bosch Automotive. It was established on an area of 92,000 square meters in Bursa. It manufactures products for diesel injection systems, namely injector nozzles, complete injectors, and Common Rail and Unit Injectors, with a team of 3000 people. The spray nozzle consists of three main parts: hard body, needle, and spring plate. These three parts are produced in three different product lines in the Spray Nozzle department and then combined in the assembly line in the same department.

In this study, the effective use of the labor force that exists in the production process and therefore in the injector nozzle production processes is examined in order to meet the need of qualified personnel, which is one of the factors required for the industries to be at the targeted level or rate in production and sales in the globalized world. With this study, it aims to increase the production capacity in line with meeting the increasing demands in these production units, and improve the production plans of these units with the help of solution models to be developed by using appropriate analysis techniques of injector nozzles model measurements (see Fig. 1).

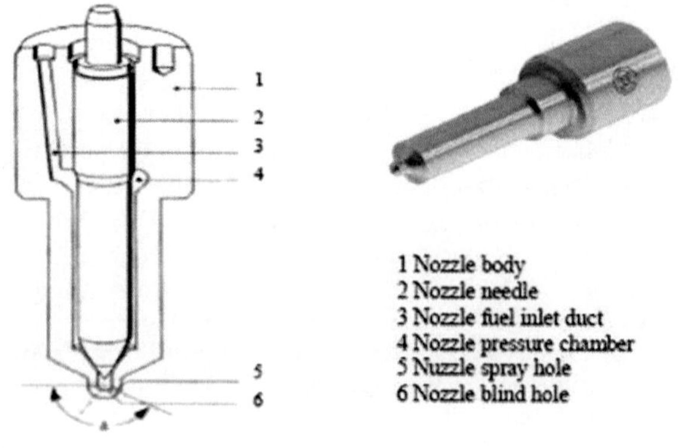

Fig. 1 Sectional view of an injector nozzles model [8, 9]

Grinding Process

An abrasive machining process using a grinding wheel as a cutting tool, the grinding machine can make precise cuts and allow very fine surfaces to be formed. In the grinding machine, the grinding head can be controlled to move along a fixed working area. Otherwise, the workpiece can be moved while the grinding head remains in a fixed position. The most important component of a precision grinding machine is the power-driven grinding head that runs at speed it should be. There is also a bearing with a fixture to guide and hold the workpiece. Grinding machines remove the material from the workpiece by abrading the material with the process where a significant amount of heat emerges due to friction. Cooling oil is used to prevent this heat from reaching levels that will change the crystal structure of the material. In high precision cylindrical and surface grinding machines, the final grinding steps are usually set to be removed approximately 2/10000 mm per pass.

Workload Analysis of Measurement Process of Injector Nozzles Model

Performance is defined as the effectiveness of a roughly specific task and is influenced by two main factors [6, 7]. These are defined by the degree of task, environment and work environment, operator duties, and features of the human factor, i.e., operator. Operators who take part in production can be affected during their duties due to their work or working environment. This interaction determines both operator workload and operator performance (hence system and task performance).

The workload is a general term used to describe the cost of fulfilling task requirements for the human element of human-machine systems (Sandra and Wickens). In this study, it is aimed to increase and optimize the performance; hence, the workload of the operator in injector nozzles model measurements. Measurement activities and actions that support and supplement these activities are grouped as follows (see Fig. 2).

A. Seat angle and cone angle measurement of injector nozzles model:

- Six pieces are taken for the frame consisting of 150 pieces
- Go to the measuring table with measuring apparatus
- By holding compressed air, oil and particles in 6 pieces are cleaned
- With the technoscope, six pieces are subjected to eye control, and errors such as scratches, wrong grinding that can be found in the piece are checked
- Taking six pieces by the operator and walk to Mahr machine
- Three pieces are connected to Mahr machine
- The machine makes its measurements automatically. Whether the computer and data are within the quality control limits are checked by the operator

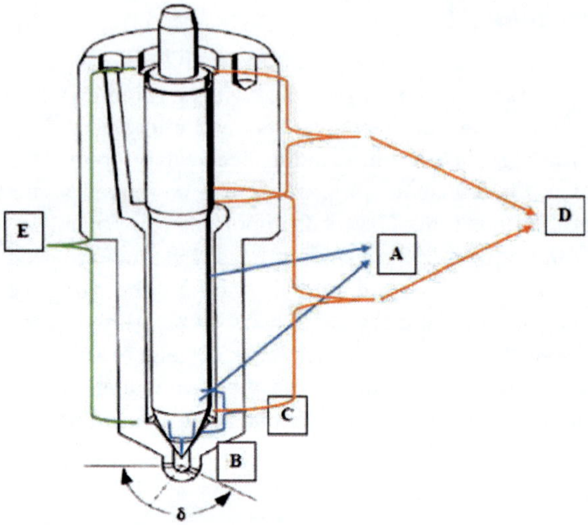

Fig. 2 Measurement process groups and activities of an injector nozzle

- According to the data received, if necessary, the machine is given the correct setting by the operator, and the correct operation of the machine is ensured

B. Blind Hole Diameter Measurements injector nozzles model:

- It takes 6 pieces for the operator shift
- Go to Squat Diameter measuring device
- In the device, the cord diameter is measured. The part is connected to the machine. It measures the diameter values with the program on the computer
- The values taken to save on the computer are noted by the operator
- The data is saved by walking to the computer

C. Blind Hole Depth Measurements injector nozzles model:

- Six pieces are taken for the operator shift
- Walking to the measuring apparatus table
- Parts are placed in the blind hole depth measuring apparatus
- With the help of the apparatus, the blind hole depth of the piece is measured

D. Shoulder and Belt Length Measurements of injector nozzles model:

- Taking six pieces by the operator from the machine
- Walking to the hommel machine
- With the air holding apparatus located next to the machine, the air is drawn into them
- One piece is connected to the machine by the operator, and the result is observed on the computer
- Then it repeats the same process for the other for five parts

E. Plan Length Measurements (plan length includes half of the breast body from the end of the initial seating line):

- Taking six pieces by the operator from the machine
- Walking to the measuring apparatus table
- With the help of measuring devices, the length of the injector nozzles body is measured.

Using Single Minute Exchange of Dies—SMED Approach for the Measurement Process of the Injector Nozzles Model

Short set-up times are a must in any industry today. In current situations, there is a good methodology to reduce these set-up times. In practice, even for very modern and high-tech equipment, there is much that can be improved about the technical concept of the equipment to make it more installation-friendly. Just in time production or JIT is one of Lean's main pillars. It is all about doing whatever the customer wants, when he wants, wherever he wants, and in the desired amounts without keeping anything in the inventory. As can be seen from the applications and literature, SMED adapts to the production approach just in time. First of all, it allows you to adapt to an increasing market, where customers are increasingly demanding diversity and responsiveness. The degree of flexibility will need to be increased (takt time) to better adapt to customer demand. This requires a reduction in inventory and a move towards working with small batch sizes. In addition, the flow of processes needs to be improved and balanced (see Table 1).

The ideal process is the process of having One Piece Flow, creating only one product when searched. Generally, One Minute Change (SMED) techniques will be used to shorten set-up times to reduce batch size. In this context, the Spaghetti diagram, from which SMED techniques are used, in which the movements of the operator to the machines, the travel distances are measured, the time and purpose of each operation performed by the operator is measured, the vehicles and devices used are determined. Therefore, all the data related to the operator are observed, analyzed, and analyzed.

Table 1 The portion of basic set-up steps before SMED implementation [3]

Steps in set-up	The proportion of setup time before SMED (%)
Preparation, after-process adjustments, checking of materials and tools	30
Mounting and removing blades, tools, and parts	5
Measurements, settings, and calibrations	15
Trial runs and adjustments	50

By observing the movement of the operator with the spaghetti diagram, it will be seen which movement has occurred within a limited time by constantly monitoring its movement in the floor plan of the area in which it operates. A few hours will be enough to draw the motion model. The result can help identify the movement that can be reduced by rearranging the workplace. Although a rough outline of the floor plan along with the basic equipment is sufficient, the drawing should be done on an architectural layout. A graph that shows the actual movement that takes place when a particular transaction is executing or doing a job in a particular area. It can help identify unnecessary movements and opportunities for improvement. This finding may mean changes in the location of the equipment, reassigning job responsibilities, or changing the layout of the facility.

Relationship Between the Set-up Time Reduction—SMED and Spaghetti Diagram

Short set-up times in production, set-up, is a must in every industry today. In current cases, there is a good methodology to reduce these set-up times. In practice, even for very modern and high-tech equipment, there is much that can be improved about the technical concept of the equipment to make it more installation-friendly. Just in time production or JIT is one of the lean's main pillars. It is all about doing whatever the customer wants, wherever he wants, and in the desired quantities without keeping anything in the inventory. As can be seen from the applications and literature, a single minute exchange of dies (SMED) adapts to the timely production approach. First of all, it ensures adaptation to an increasing market, where customers are increasingly demanding diversity and responsiveness. To better adapt to customer demand, it is necessary to increase the degree of flexibility (takt time).

This results in a reduction in inventory and working with small batch sizes. In addition, the flow of processes needs to be improved and balanced. The ideal process is the process of having one piece flow; it is to create only one product when it is searched. Generally, a single minute exchange of dies (SMED) techniques will be used to shorten set-up times to reduce the batch size (see Table 1). In this context, the Spaghetti diagram will be used within the framework of the SMED technique, in which the movements of the operator to the machines, the distance of movement, the time and purpose of each operation performed by the operator are measured, the tools and devices used are determined. Therefore, all data related to the operator are observed and analyzed (see Fig. 7).

By observing the movement of the operator with the spaghetti diagram, it will be seen which movement has occurred within a limited time by constantly monitoring its movement in the floor plan of the area where it works. A few hours will be enough to draw the motion model. The result can help to identify the movement that can be reduced by rearranging the workplace. Although a rough outline of the floor plan along with the basic equipment is sufficient, the drawing must be made on an architectural layout. This chart shows the actual movement that takes place while a particular transaction is being carried out or doing business in a particular area. It can help identify unnecessary movements and opportunities for improvement. This finding may mean changes in the location of the equipment, reassigning job responsibilities, or changing the layout of the facility.

Measurement Process Improvement of Injector Nozzles Model

The aim of improving the process is to increase the number of machines used by the operator for measurement operations from two to three while minimizing the operation times; in other words, optimization. Within the framework of the lean manufacturing approach, when using the spaghetti diagram technique in the context of the SMED technique and assuming that 3MMB is improved, the total working load rate was reduced to 63.07% (see Fig. 5). During the improvement process, the basic actions required are determined when analyzing any manual operation by the operator. Each of these basic movements required a predetermined standard time based on the factors affecting it. These operations will be carried out using the Method Time Measurement MTM (Methods-Time Measurement) technique.

Problem Definition

As can be seen in this study, the purpose of the study is to increase the number of machines used by the operator for measurement operations from two to three. In other words, the machines that the operator manages operations are increased from two to three. Before improvement, when the operator-controlled two machines, the current working load, i.e., the total occupancy rate, was 60.29% (see Fig. 3). In the improvement process, when the operator's workload increased from two machines to three machines, the operator's workload, i.e., the total occupancy rate, increased to 87.10% (see Figs. 4, 6 and 7).

An Application

In the industry, which is an automotive supplier industry and maintains its leadership in the market of injector production, the problem about the effective use of the workforce has been addressed to be realized in the inside grinding process (see Fig. 5). Inside grinding is a very precise process in the production of the injector nozzle body. The inside of the nozzle body is grinded from three main points and

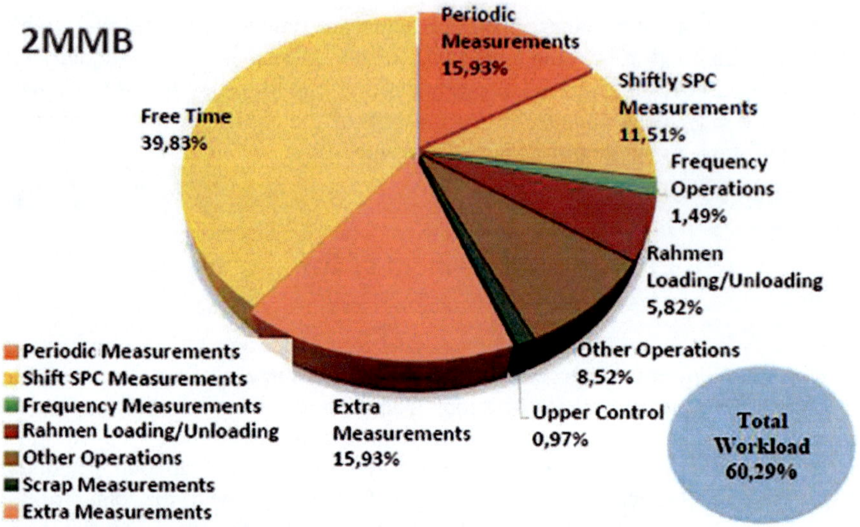

Fig. 3 Measurement process groups and activities of an injector nozzle

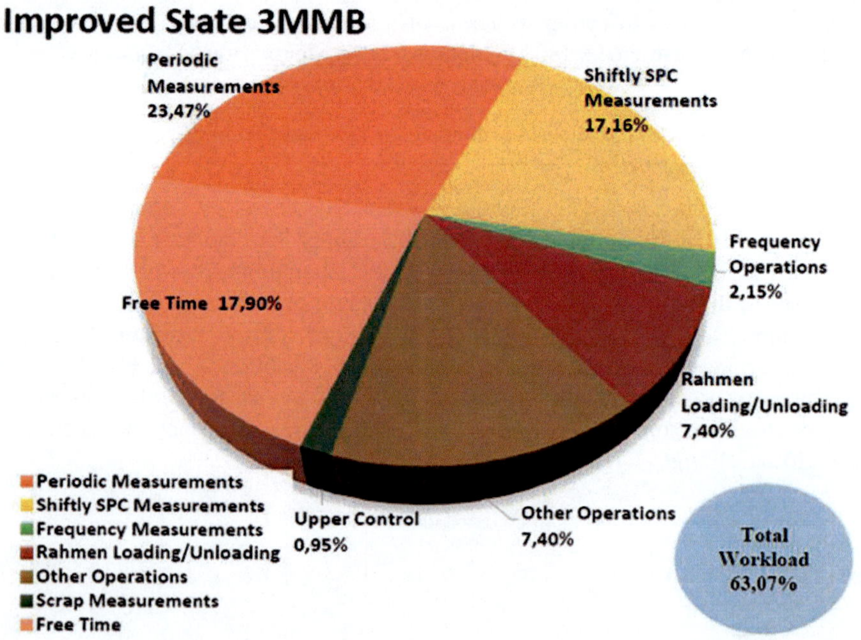

Fig. 4 Measurement process groups and activities of an injector nozzle after process improvement

Fig. 5 Inner grinding 3MMB model nozzle body process [8]

observed with the control cards in order to keep it at the desired level. Control of machine stability is provided by various measurements. Thus, measurements are made whether the part is at the desired level or not.

In the initial situation in the problem under the lens, the operator controls and manages two machines. The main task of the operator is to perform standard works and keep the machine under control within the specified time. At this point, it is very important to provide this data from the average operators at the production site to understand and analyze how full the operator is during the working period. In the targeted situation, the operator is intended to look at three machines. In order to understand and analyze the data just mentioned (see Figs. 6 and 7);

- Single Minute Exchange of Dies (SMED)
- Spaghetti Diagrams
- MTM Analysis
- Layout location selection with AutoCad
- Workload Analysis
- Shift Machine Tracking.

work will be done.

Conclusion and Future Works

Future studies will be aimed at increasing the performance of the operator, making measurements on multiple machines in the production of injector nozzles model using different techniques determined in this study. As a continuation of this study, data will be collected by the operator from the ongoing measurement processes. The obtained data will be evaluated by using the techniques determined in this study. The operator load will be analyzed primarily, and the improvement will be provided to the results to be obtained. Secondly, it will be discussed whether different techniques determined to improve the measurement process of the injector nozzles model, such as SPC techniques, simulation, SMED technique, spaghetti diagram, and so on techniques are accurate. By using the techniques listed below, the bottlenecks that may occur during the measurement process will be recognized, thereby preventing performance losses from bottlenecks.

- Sensitivity analysis of performance measures
- Analytic Hierarchy Process (AHP)
- Comparison of Discrete Event Simulation Modelling with current and future situation
- Identification of Improvement Potentials.

The use of simulation techniques in the production of the injector nozzle module is applied as a successful technique in increasing operator performance in measurement processes. Within the framework of the simulation technique, product, process, and system design can be made, and the configuration can be tested and approved. The system can be modeled and analyzed with the simulation program, and forecasting alternative models to the system can be provided.

In real life, it can be impossible or too costly to observe any system or sequence of operations. The system observed can be so complex that it may be impossible to define this system with mathematical equations and to obtain predictive analytical solutions for system operation. Even if the mathematical model of the system under consideration can be established, the analytical techniques required to solve the model may be insufficient. Experiments to verify mathematical models that define the system can be either impossible or too costly.

References

1. Cakmakci M (2009) Process improvement: performance analysis of the set-up time reduction-SMED in the automobile industry. Int J Adv Manuf Technol 41(1–2):168–179
2. McIntosh RI, Owen GW, Culley SJ, Mileham T (2007) Changeover improvement: reinterpreting Shingo's SMED methodology. IEEE Trans Eng Manage 54(1):98–111
3. Shingo S (1985) A revolution in manufacturing, the SMED system. Productivity Press, University Park, IL
4. Van Goubergen D, Van Landeghem H (2002) Rules for integrating fast changeover capabili ties into new equipment design. Rob Comput-Integr Manuf 18(3–4):205–214
5. Womack JP, Jones DT (1997) Lean thinking. Simon and Schuster, London
6. Miller S (2001) "Workload Measures" literature review, Document ID: N01-006
7. Paas F, Vanmerrienboer JJG (1993) The efficiency of ınstructional conditions an approach to combine mental effort and performance-measures. Hum Factors 35(4):737–743
8. Bosch Nozzles (2019) https://www.boschautoparts.com/en/auto/diesel-parts/nozzles. Accessed 22 Oct 2019
9. Dogru B (2007) Enjektör parametrelerinin motor performansına etkisi. Master thesis
10. Sandra GH, Wickens CD, Workload assessment and prediction, manprint, pp 257–296

Two Scheduling Rule Comparisons for Operations of Shuttles in Tier-to-Tier SBS/RS

Melis Kucukyasar and Banu Yetkin Ekren

Abstract The aim of this paper is to study and compare two different scheduling rules for operations of transactions in shuttle-based storage and retrieval system (SBS/RS) to observe how scheduling rule affects the system performance. SBS/RS is one of the emerging automated warehousing technologies mostly utilized by large distribution centers due to its advantages of fast transaction process rate. Although the increased transaction rate is because of the system (warehouse) design of having multiple numbers of autonomous shuttles in the system, operational rules such as scheduling of waiting transactions in queues may also play a significant role in increasing the system performance. For this purpose, we implement first-in-first-out (FIFO) and shortest processes (i.e., travel) time scheduling rules for selecting waiting transactions in shuttle queues to investigate how these rules affect the performance of the system. We simulate the system under different racking designs and arrival rate scenarios and observe the system performance metrics in terms of average cycle time (e.g., throughput rate) and average energy consumption per transaction, separately. The results suggest that the SPT scheduling rule outperforms FIFO rule, especially for the throughput rate performance metric.

Keywords Shuttle-based storage and retrieval system · Tier-to-tier SBS/RS · Automated warehousing · SBS/RS · Warehousing

Introduction

Recent increased competition and customer-oriented supply chain management have created the requirement of quick response time for demand in companies. Recent technological developments allow autonomous operations in industries that could

M. Kucukyasar (✉) · B. Yetkin Ekren
Department of Industrial Engineering, Faculty of Engineering, Yasar University, Bornova, Izmir, Turkey

B. Yetkin Ekren
e-mail: banu.ekren@yasar.edu.tr

F. Calisir (ed.), *Industrial Engineering in the Internet-of-Things World*, Lecture Notes in Management and Industrial Engineering,
https://doi.org/10.1007/978-3-030-76724-2_8

result in increased efficiency and productivity, as well as decreased error rates and lead times.

In automated warehousing technologies, there are several studies [1–5]. Shuttle based storage and retrieval systems (SBS/SR) is one of the extensively utilized technologies preferred as an alternative design to mini-load storage and retrieval system in e-commerce. SBS/RS is capable of efficiently meeting the recent demand profile with high variability and low volume requesting for short response time. Lehrer et al. [6] showed that SBS/RS works with a high throughput rate by reduced cycle time compared to its simulants. They focus on a traditional tier-captive SBS/RS design in which each tier of the racking system has a dedicated shuttle. Differently, in this paper, we study a tier-to-tier SBS/RS design where shuttles are not tier-captive and can travel between tiers. The advantage of this tier-to-tier design is the decreased investment cost due to the decreased total number of shuttles in the system.

In literature, mostly tier-captive SBS/RS design is studied. One of the earliest studies has been completed by Carlo and Vis [7]. They represented a heuristic solution approach for scheduling of lifts in an SBS/RS warehouse consisting of a conveyor system, non-passing lifts and multiple shuttles that are dedicated to specific single tiers in an aisle. They develop two functions for the sequencing of the requests. Wang et al. [8] conducted a time-based algorithm to sequence tasks by using a mathematical modeling approach for tier-captive design. A non-dominated genetic algorithm was used to determine the optimal solution. Lehrer et al. [9] introduced an analytical travel time model by considering characteristics of shuttles and lifting mechanisms by including acceleration, deceleration, and maximum velocity profiles. They develop the models for both single and dual command scheduling rules in the system. Ekren et al. [10] studied an analytical-based tool that can estimate the mean and variance of travel time of lifts and shuttles per transaction for tier-captive system design. That tool can also estimate the mean energy consumption and mean energy regeneration per transaction based on the velocity profiles of shuttles and lifts. Recent studies are carried out by Ekren [12]. In Ekren [11], the author performs a statistical experimental design analysis to determine the significant factors affecting some critical system performance metrics from the studied SBS/RS. The results show that the most significant factor is the number of aisles design factor in the warehouse. In Ekren [12], the author studies warehouse design by a multi-objective optimization procedure by considering both energy consumption and cycle time performance metrics in the system.

There are few studies in the literature studying tier-to-tier SBS/RS. Ha and Chae [13] studies a tier-to-tier SBS/RS and compare two operation approaches of shuttles called free and non-free balancing. They are developed to prevent collisions of shuttles. Then, they proposed a decision model to determine the number of shuttles in the tier-to-tier system design [14]. Finally, Zhao et al. studied an integer programming scheduling model to minimize the idle time of the lifting mechanism and shuttles in tier-to-tier system design by using simulation-based optimization software [15]. This study is the lack because it solely considers retrieval transactions in the system. Different from the existing studies, we study scheduling rule in tier-to-tier SBS/RS to compare how the two performance metrics, average cycle time, and energy consumption per transaction, are affected. We test the performance of the system under different racking designs.

Methodology

In a tier-to-tier SBS/RS system design, there is less number of shuttles than the number of tiers that exist in an aisle. Shuttles are aisle-captive, meaning that they cannot travel between aisles. However, they are allowed to travel between tiers in their dedicated aisle by using a lifting mechanism located at the end of each aisle. There are two lifting mechanisms in the system, one of which performs vertical travel for loads, and the other performs vertical travel for shuttles. The system details, along with the simulation assumptions, are given in the below section.

System Definition

Figure 1 shows the side and the top view of an aisle for the studied tier-to-tier SBS/RS design. According to that, the Lift 1 mechanism performs travel for loads (i.e., totes), and the Lift 2 mechanism performs travel for shuttles. Namely, shuttles use Lift 2 in Fig. 1, to travel to other tiers. Note that there are two storage sides in each tier located at the left and right sides. Two types of transactions arrive at the system, storage, and retrieval transactions. Storage refers to the storage of incoming items in racks. Retrieval refers, retrieving the items from storage racks to ship them for customers.

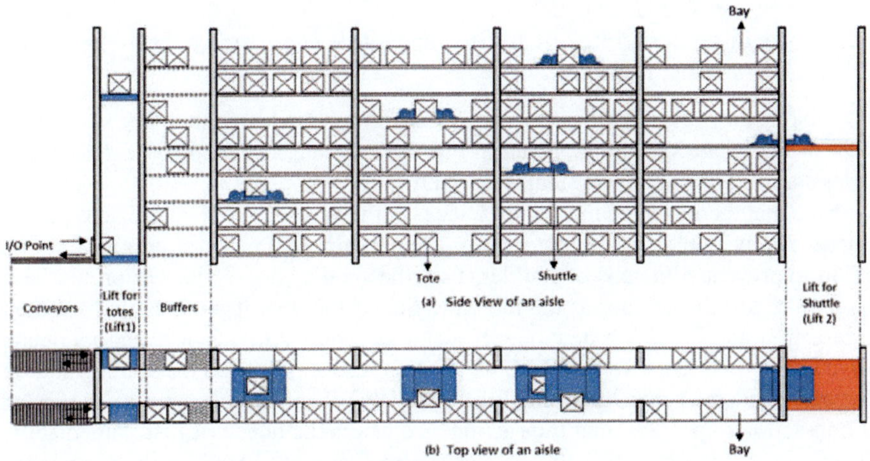

Fig. 1 The physical configuration of a tier-to-tier SBS/RS

In the racking system, it is assumed that each bay can hold a single tote. Lift 1 is assumed to have two separate lifting tables that can travel independently. Lift 2 and shuttles can carry a single shuttle and tote, respectively.

The other assumptions are considered in the simulation model are as follows:

- Incoming transactions arrive at the Input/Output (I/O) points at the first tier of each aisle.
- To process the storage transactions, first, totes are dropped-off by Lift 1 at the buffers. Then, they are picked up by shuttles to be stored in their bay addresses.
- In retrieval transactions, totes are dropped-off by shuttles at the buffers. Then, they are picked up by Lift 1 to be released at the I/O point.
- If a transaction is in the first tier, Lift 1 is not utilized.
- The buffers at each tier can hold three totes at a time.
- Dwell point policy of lifts and shuttles is assumed to be the point where they complete their last process.
- When a shuttle is assigned for a transaction located at a tier different than the shuttle's current tier, the shuttle travels to the Lift 2 location immediately while it also enters the Lift 2 queue to seize it.
- To prevent the collisions of shuttles at the same tier, a single shuttle is allowed to travel within a tier.
- Storage and retrieval transactions are created with equal probability distributions. The mean arrival rate follows a Poisson distribution.
- Lift and shuttle acceleration and deceleration delays are considered to be 2 m/sec^2. The maximum velocity that they can reach are also determined to be 2 m/sec.
- A pure random storage policy is considered.
- Loading and unloading time delays for totes are ignored.
- Distance between two adjacent bays and distance between the buffer area and the first bay is equal and 0.5 m in each tier. Besides, the distance between two adjacent tiers is equal and 0.35 m.
- The weights of a shuttle, Lift 1, Lift 2 and tote is considered to be 40 kg, 60 kg, 60 kg, and 20 kg, respectively.

Table 1 shows the notations that are used in the system description. The simulation flow charts of the system are given in Fig. 2. Note that, FIFO and SPT scheduling rules are run for the same system, separately.

Note that there is three number of queues in the system: Lift 1, Lift 2, and shuttle queue. An available shuttle first selects a transaction by the scheduling rule. In an effort to prevent collisions of shuttles, first, the shuttle checks whether or not there is another shuttle running at the targeting tier. If there is, it releases that selected transaction and process a new one. If a shuttle requires Lift 2, it enters its queue immediately while it also travels to its location. Lift 2 processes the waiting shuttle by the considered scheduling rule, FIFO, or SPT. When a shuttle arrives at a tier for a transaction to process, that transaction also enters the queue of Lift immediately, if it is not at the first tier. Lift 1 queue is also processed based on the considered scheduling rule.

Table 1 Notations for performance metrics and parameters

Notation	Description	Unit
T	Average cycle time	sec/transaction
E_C	Average energy consumption	kWh/transaction
E_R	Average energy regeneration	kWh/transaction
E	Average net energy consumption	kWh/transaction
λ	Mean arrival rate	Number of transactions/month
U_S	Average shuttle utilization	%
U_{L1}	Average Lift 1 utilization	%
U_{L2}	Average Lift 2 utilization	%
t_i	i-th transaction at the transaction queue	
s_j	j-th shuttle	
A_i	0, if there is no shuttle at the i-th transaction tier address; 1, otherwise	

Note that awaiting transaction does not enter the Lift 1 queue until a shuttle arrives at that transaction's tier address. The simulation model is run for forty days with ten days warm-up period. The warm-up period is determined by the eye-ball technique by checking the average cycle time of the transaction performance metric. Verification is done by debugging the codes as well as animating the model. The validation is done by checking the simulation model results by an expert. Performance metrics are calculated for the averages of five independent replications, and they are given in the following sections with their 95% confidence intervals.

Performance Metrics of the System

The performance of the system is evaluated in terms of average cycle time, average energy consumption, and average energy regeneration per transaction. Since shuttles are mostly the bottleneck, we run each scenario under 90%, 95% and 98% average utilization values of shuttles by adapting the arrival rates accordingly. Namely, based on the utilization value scenarios, we increase or decrease the arrival rate so that we could obtain the desired average utilization levels for the shuttles. By that, we observe how high throughput rates can be obtained by the regarding scheduling rules. Under the same utilization levels, we compare the throughput rates of the scheduling rules. In other words, the cycle time performance metric is treated as the throughput rate in the evaluation procedure by fixing the utilization levels in the system. Energy regeneration is the reproduced energy by the decelerated shuttles and lifts. In this work, we assume that there is a regeneration mechanism in the system [10].

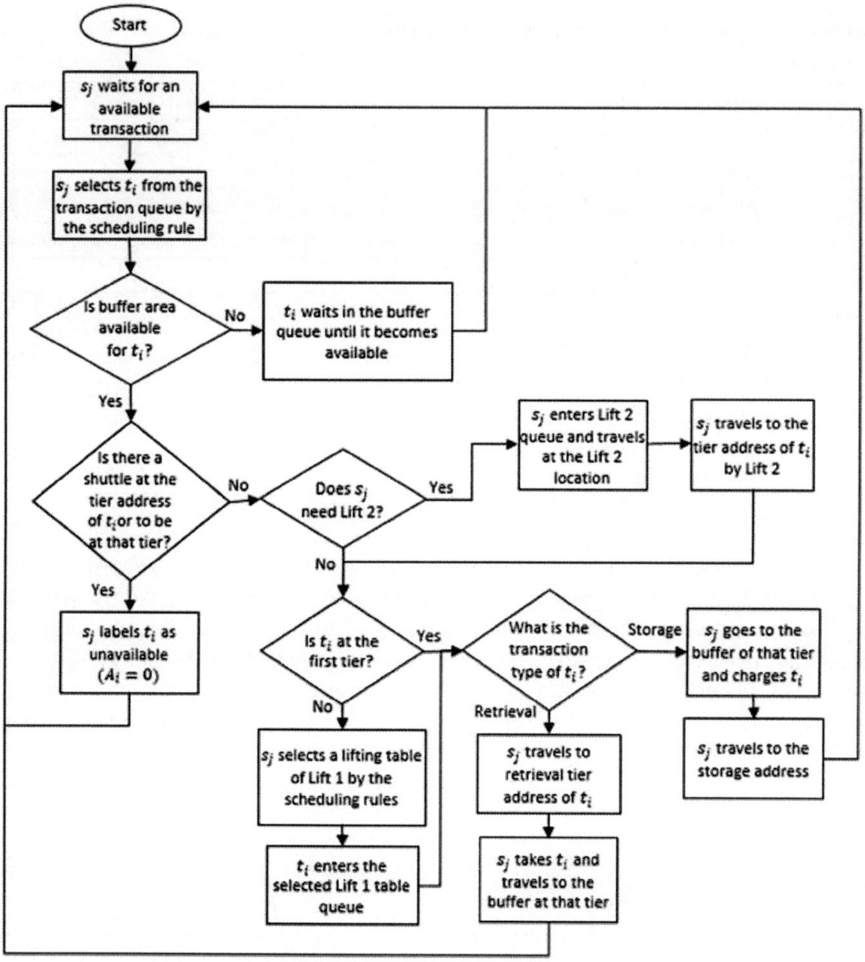

Fig. 2 Flow chat of tier-to-tier SBS/RS

As mentioned previously, to reach the objective average utilization for shuttles (i.e., 90, 95 and 98%) arrival rates are adjusted. Experiments are done for two warehouse design scenarios: 15 tiers, 50 bays: and 25 tiers of and 30 bays. A single-aisle is simulated by assuming that the mean arrival rate is the same for each aisle by the random storage policy. The number of the shuttle is considered to be five shuttles for both warehouse designs. We consider the average net energy consumption as a performance metric by decreasing the energy regeneration amount from the average consumed energy, as shown by (1).

$$E = E_C - E_R \tag{1}$$

Scheduling Rules, FIFO and SPT

Remember that two scheduling policies are tested for each queue. These are FIFO and SPT scheduling rules. FIFO processes the transactions according to the first-come, first-serve rule. For SPT, the shuttle queue evaluates the waiting transactions according to their tier addresses. Namely, the shuttle selects the transaction whose tier address is closest to its current tier. However, if there is currently a shuttle running at that tier, it considers the following closest transaction. The lift queues also process the transactions based on the shortest distance travel.

Design Scenarios

In this study, two warehouse designs with 15 tiers and 50 bays, 25 tiers, and 30 bays with 5 shuttles are considered to analyze the effect of scheduling rules on different physical configurations. Hence, the total tote capacity for each aisle is fixed as 750 bays. In addition to this, scheduling rules are applied for both physical configurations with three different arrival rate scenarios where we could obtain average utilization values for shuttles to be 90, 95 and 98%, separately. Table 2 shows a summary of the conducted experiments.

Table 2 Conducted experiments

Design No.	Number of tiers	Number of bays	Number of shuttles	Scheduling rule	Shuttle utilization (%)
1	15	50	5	FIFO	90
					95
					98
				SPT	90
					95
					98
2	25	30	5	FIFO	90
					95
					98
				SPT	90
					95
					98

Results

Remember that Table 2 experiments are run for 40-days with 10-days warm-up period for 5 independent replications. Table 3 summarizes the results at 95% confidence intervals for Design 1 and Design 2, separately. Note that the λ values represent the value for a single aisle.

Table 3 results are also summarized by Figs. 3 and 4 graphes. From the results, it is observed that in terms of the throughput rate performance metric, SPT always works better (Fig. 3). Moreover, energy consumption per transaction has decreased dramatically in the SPT rule against the FIFO rule when the throughput rate is high. In addition, in terms of the energy consumption performance metric, Design 1 always

Table 3 Simulation results for the conducted experiments

Design No.	Scheduling rule	U_S (%)	λ (transaction/month)	T (sec)	E (kWh/transaction)
1	FIFO	90	720,000	37.89 ± 0.33	2.31E-03 ± 2.05E-06
		95	751,304	54.17 ± 1.00	2.31E-03 ± 1.37E-06
		98	773,731	103.99 ± 3.16	2.31E-03 ± 1.71E-06
	SPT	90	773,731	29.39 ± 0.04	1.98E-03 ± 1.72E-06
		95	836,129	34.23 ± 0.09	1.86E-03 ± 2.12E-06
		98	893,793	41.02 ± 0.05	1.72E-03 ± 1.40E-06
2	FIFO	90	674,122	48.04 ± 1.14	3.38E-03 ± 1.60E-06
		95	696,774	76.7 ± 2.69	3.38E-03 ± 1.34E-06
		98	710,137	137.1 ± 10.96	3.38E-03 ± 2.04E-06
	SPT	90	740,571	30.67 ± 0.12	2.98E-03 ± 2.85E-06
		95	810,000	38.16 ± 0.19	2.77E-03 ± 4.82E-06
		98	864,000	49.03 ± 0.25	2.56E-03 ± 4.07E-06

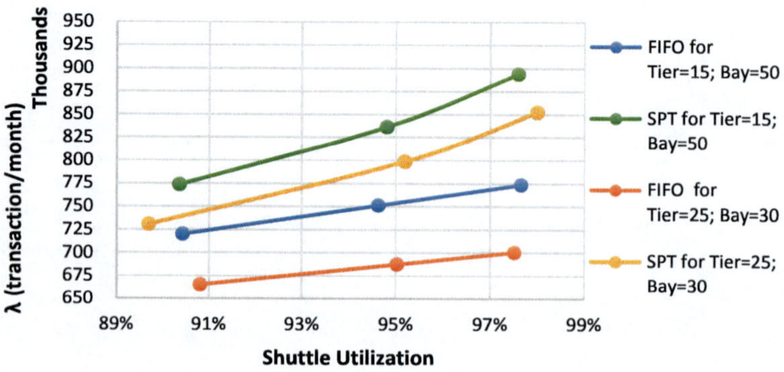

Fig. 3 Shuttle utilization versus throughput rate per month (λ) results

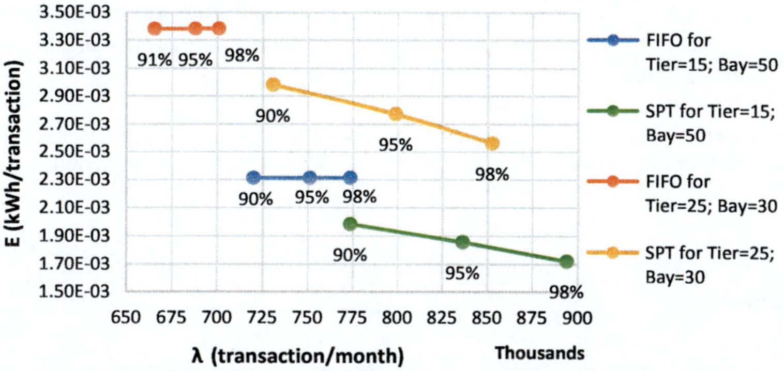

Fig. 4 Average net energy consumption per transaction (E) versus throughput rate per month (λ) results

works better (Fig. 4). Note that Design 1 has less number of tiers in the system than Design 2. Hence, it can be concluded that the increased number of tiers increases energy consumption in the system.

Discussion and Conclusion

In this paper, we study two different scheduling rules for operations of transactions in tier-to-tier shuttle-based storage and retrieval system (SBS/RS) to observe how scheduling rule affects the system performance. We implement first-in-first-out (FIFO) and shortest processes (i.e., travel) time scheduling rules for the process of waiting transactions in shuttle queues. We simulate the systems under different racking designs and observe the system performance in terms of average cycle time (or throughput rate) and average energy consumption per transaction, separately. We conduct 12 different experiments based on scheduling rules, warehouse designs, and utilization levels of shuttles. The results suggest that the SPT scheduling rule outperforms FIFO rule, especially in average cycle time performance metrics (i.e., throughput rate). The results show that there are no remarkable differences in energy consumption per transaction by the change of shuttle utilization in FIFO rule. However, in SPT rule, there is a descending trend by the increased shuttle utilization. Therefore, SPT rule provides an advantage for reduced energy consumption on high shuttle utilization.

There is a trade-off between cycle time and energy consumption per transaction since high utilization also causes high cycle time. Moreover, it is shown that a low number of tiers warehouse design options work well from the energy consumption point of view against the cycle time per transaction point of view.

As future work, this study can be extended by considering more scheduling rule policies as well as dynamic operating rules by considering the agent-based modeling approach in the system.

Acknowledgements This work was supported by The Scientific and Technological Research Council of Turkey and Slovenian Research Agency: ARRS [grant number: 118M180].

References

1. Ekren BY (2017) Graph-based solution for performance evaluation of shuttle-based storage and retrieval system. Int J Prod Res 55:6516–6526
2. Ekren BY, Heragu SS (2012) Performance comparison of two material handling systems: AVS/RS and CBAS/RS. Int J Prod Res 50(15):4061–4074
3. Ekren BY, Heragu SS (2011) Simulation-based performance analysis of autonomous vehicle storage and retrieval system. Simul Model Pract Theory 19:1640–1650
4. Ekren BY, Sari Z, Lerher T (2015) Warehouse design under class-based storage policy of shuttle-based storage and retrieval system. IFAC-PapersOnLine 48(3):1152–1154
5. Lerher T, Ekren BY, Sari Z, Rosi B (2016) Method for evaluating the throughput performance of shuttle based storage and retrieval systems. Tehnički vjesnik 23(3):715–723
6. Lehrer T, Ekren YB, Sari Z, Rosi B (2015) Simulation analysis of shuttle based storage and retrieval systems. Int J Simul Modell 14(1):48–59
7. Carlo HJ, Vis IF (2012) Sequencing dynamic storage systems with multiple lifts and shuttles. Int J Prod Econ 140(2):844–853
8. Wang Y, Mou S, Wu Y (2015) Task scheduling for multi-tier shuttle warehousing systems. Int J Prod Res 53(19):5884–5895
9. Lerher T, Ekren BY, Dukic G, Rosi B (2015) Travel time model for shuttle-based storage and retrieval systems. Int J Adv Manuf Technol 78(9–12):1705–1725
10. Ekren BY, Akpunar A, Sari Z, Lerher T (2018) A tool for time, variance and energy related performance estimations in a shuttle-based storage and retrieval system. Appl Math Modell 63:109–127
11. Ekren BY (2020a) A simulation-based experimental design for SBS/RS warehouse design by considering energy related performance metrics. Simul Modell Pract Theory. (in press). https://doi.org/10.1016/j.simpat.2019.101991
12. Ekren BY (2020b) A multi-objective optimisation study for the design of an AVS/RS warehouse. Int J Prod Res. (in press). https://doi.org/10.1080/00207543.2020.1720927
13. Ha Y, Chae J (2018) Free balancing for a shuttle-based storage and retrieval system. Simul Modell Practice Theory 82:12–31
14. Ha Y, Chae J (2018) A decision model to determine the number of shuttles in a tier-to-tier SBS/RS. Int J Prod Res 57(4):963–984
15. Zhao X, Wang Y, Wang Y, Huang K (2019) Integer programming scheduling model for tier-to-tier shuttle-based storage and retrieval systems. Processes 7(4):223

Urban Transportation Planning Model for Long Term Refugee Camps Development: Case of Naher El Bared Camp in Lebanon

Nabil Nehme and Hani AlNaghi

Abstract This paper investigates the urban transportation planning framework related to long term refugee camps and their integration with the surrounding areas. The research considers the case of Naher El Bared Camp (NBC) in North Lebanon. First, the transport study for NBC before 2007 is analyzed via implementing the four steps model and performing a micro analysis using Emme regional transportation model linked with ArcGIS and Synchro software for signalized intersection traffic analysis. Then, a new transport model for the year 2030 is developed subject to several to social, economic, land use, and political constraints. This paper is the first documented study that investigates urban transportation planning for long term refugee camps in the Middle East region countries.

Keywords Four steps model · EMME · Urban transportation planning · Transport model camp · Long term refugee camp

Introduction

The Naher El Bared Camp (NBC) is one of the 58 recognized Palestinian refugee camps spread between Lebanon, Jordan, Syria, the West Bank, and the Gaza Strip. The area is known as the "Old Camp," has an area of approximately 0.2 km^2, and was first created in December 1949 by non-profit humanitarian organizations, such as the Red Cross, to host more than 5000 Palestinian refugees. However, the NBC has expanded over the years and transformed from a temporary 'tent-site' into a hyper-congested mass of multi-story buildings, characterized by concentrations of poverty and extreme overcrowding [1]. In the summer of 2007, a battle started between the Lebanese Armed Forces and an Islamist militia group called Fateh al-Islam and lasted

N. Nehme (✉)
School of Engineering, Lebanese American University, Byblos, Lebanon
e-mail: nabil.nehme01@lau.edu.lb

H. AlNaghi
Department of Civil and Environmental Engineering, American University of Beirut, Beirut, Lebanon

F. Calisir (ed.), *Industrial Engineering in the Internet-of-Things World*, Lecture Notes in Management and Industrial Engineering,
https://doi.org/10.1007/978-3-030-76724-2_9

for three-and-a-half months inside NBC. During this battle, the majority of the camp was destroyed, and the adjacent municipalities were also affected. Before the battle, NBC was the second most populous Palestinian refugee camp in Lebanon, with a population of more than 30,000 persons [2–4].

In 2008, funded by the United Nations Development Programme (UNDP) and the United Nations Reliefs and Works Agency for Palestine in the Near East (UNRWA), an assessment and planning unit for the camp's reconstruction was established and has successfully guided the camp residents to develop a reconstruction plan [5–7]. This traffic impact assessment was required to assist the planning unit and to establish a satisfactory level of service in the proposed master plan for the horizon year of 2030. This assessment followed the procedure that is recommended by the Directorate General of Urban Planning in Lebanon. In some instances, information beyond the guidelines is provided in this study.

The NBC is located 16 km away from the city of Tripoli along the northern coast of Lebanon, in the Mhammara cadastral region of the Akkar Caza. It is bounded by the Bared River from the South, the Tripoli-Syria highway from the East, El Aabdeh junction from the North, and the Mediterranean Sea from the West [6, 8]. Figure 1 illustrates the general location map of Naher el-Bared and highlights the location of the new camp compared to the new camp. The NBC "Old Camp" is referred to as the geographical area before 2007, and the "New Camp" is composed of the "Old Camp" in addition to nine traffic zone sectors added after the expansion of the camp in 2007. The study area and the road network are represented in Fig. 2.

Prior to the 2007 war, the NBC site was a mixed-use development along a commercial spine, the main road extending from El-Minieh Sea Road to El-Aabdeh Roundabout. Two-third of the population were living in the dense "Old Camp" (population of 21,587 persons in a 0.2 km^2) while the "New Camp" included large areas of agricultural fields with a lower population density (population of 12,155 persons in a 1.13 km^2). In addition, the site included a total of 2741 non-residential units (commercial and institutional). Figure 3 illustrates sector divisions of the camp area, and Fig. 4 illustrates the spatial distribution of the non-residential units throughout the camp area. The total population reported was estimated to be 33,729 persons, with a total of 2,741 non-residential units. The detailed breakdown of residential populations and non-residential units for each of the sectors is presented in Table 1.

After the end of military operations, a preliminary master plan was developed to reconstruct NBC between the UNRWA and the community-based Nahr el-Bared Reconstruction Commission for Civil Action and Studies [5]. This preliminary master plan has six objectives. The first objective is to replicate the camp's landmarks. The second objective is to improve the camp's environment. The third objective is to enhance the urban infrastructure. The fourth objective is to improve the access by (i) establishing new pedestrian walkways, (ii) increasing the parking space outside the camp, and (iii) prohibiting vehicular access to residential blocks. The fifth objective is to expand the road and pedestrian networks by (i) reconstructing the road networks, (ii) widening the main road to 12 m and secondary roads to 6 m, (iv) widening alleyways to 4.5 m for pedestrian access to the residential building blocks, and (v) increasing parking capacity. The sixth objective is to enhance emergency

Fig. 1 General location of Naher El Bared camp

services by restricting specific roads to emergency services such as ambulances and fire trucks.

The reconstruction of the completely destroyed "Old Camp" consists of 4,705 residential units and 901 non-residential units (communal and commercial) in addition to the "New Camp" infrastructure, and related infrastructure works in the adjacent areas, including utility connections. The "Old Camp" reconstruction work was divided into eight packages. Figure 5 illustrates the "Old Camp" reconstruction plan layout and the corresponding blocks for each package.

The paper is structured as follows. Section 2 presents a brief review of the relevant literature. In Sect. 3, the current traffic conditions are generated and analyzed using the four steps model. In Sect. 4, the future traffic condition is forecasted, based on the four steps model generated in Sect. 3, using Emme/2 software. Section 5 concludes the paper and proposes future recommendations.

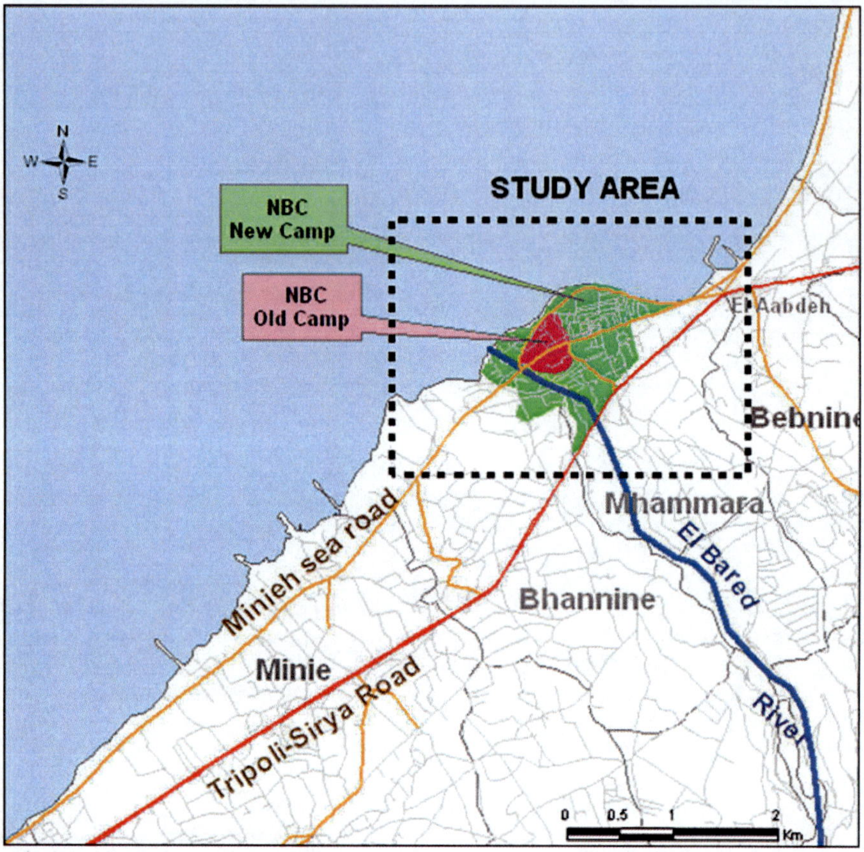

Fig. 2 Road network and the study area

Literature Review

Damme [9] discussed, for the first time, the transformation of refugee camps to permanent or long term camps, more than thirty years, by considering the cases of Goma and Guinea camps in Africa. Damme argued that the refugee, in this case, becomes completely dependent on outside help, and the cost per refugee is more than the gross national product per head of the host country. Damme differentiated between giant camps and small camps and discussed the high cost of operations and maintenance of long terms refugee camps. It is worth noting that in the literature, there is no documentation of any transportation planning study for refugee camps, especially for small permanent refugee camps such as Naher el-Bared refugee camp (NBC), where the population is less than 50,000. Therefore, this literature review is limited to transportation plans and mobility inside 'small cities.'

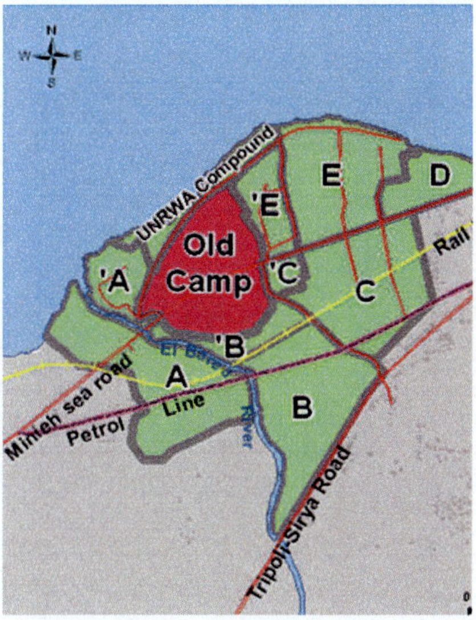

Fig. 3 NBC sectors

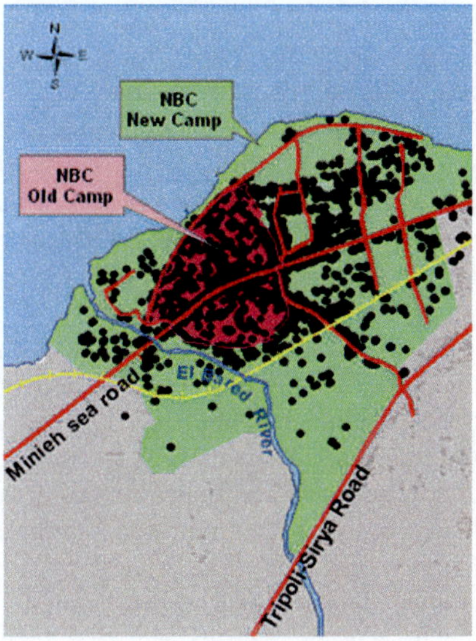

Fig. 4 Distribution of commercial units

Anderson et al. [10] developed a new methodology to predict external trip circulations in small cities. The authors identified the impact of roadway infrastructure and nearby cities on a small community traffic flow. They developed a statistical regression model using SPSS by taking as parameters: the origin and destination stations, the average daily traffic, the number of lanes, the nearby major city, the percent of trucks, the route continuity, and internal-external factors. First, they used a stepwise

Table 1 Breakdown of population and non-residential units

Sector		Population	Non-residential units
Old camp		21,578	898
New camp	*A'*	*820*	55
	B'	*1037*	280
	C'	*245*	103
	E'	*2523*	254
	A	*1273*	125
	B	*1101*	81
	C	*2441*	276
	D	*293*	122
	E	*2418*	514
	UNRWA compound	–	33
	Sub-total	12,151	1,843
GRAND total		33,729	2,741

regression model that included four variables: average daily traffic at the destination station, the nearby major city at the destination station, the route continuity between origin and destination stations, and internal-external factor for a condition when origin station is the same as the destination station. Then, a multiple regression analysis was performed using the four variables to determine the percent trip between every station. The proposed model was able to forecast 56% of the external trip data with an error of less than 20% compared to the National Cooperative Highway Research Program (NCHRP) model that predicted only 50% of the external trip data with the same error margin.

Moreover, Autunes et al. [11] discussed a new accessibility maximization approach to the inter-urban road network for long term planning. The authors developed a nonlinear combinatorial optimization model to minimize the travel time across the facilities by considering no action for the current connection, construction of new connection, and link improvement as decision variables. The traffic flows were estimated based on the four steps model, and the model considered accessibility and budget constraints. The authors used heuristic solutions, such as local search and simulated annealing to solve the model.

Drezner and Wesolowshy [12] analyzed the network design problem by formulating a multi-objective function model to minimize construction and transportation costs. The transportation cost was based on traffic and demand service generated. The authors did not consider the volume of traffic on each link. They used a descent algorithm, simulated annealing, tabu search, and a genetic algorithm as a solution methodology. In addition, Bigotte et al. [13] maximized the accessibility to facilities via promoting center and links to a higher hierarchy. They formulated an optimization model to minimize the overall summation for the movement of people in terms

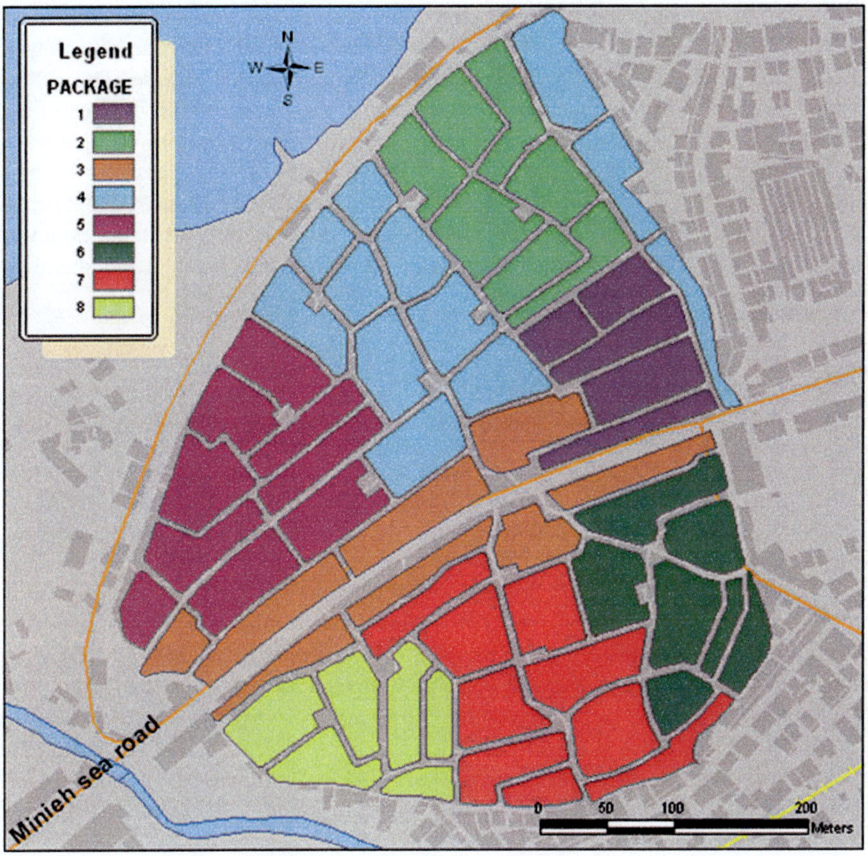

Fig. 5 Packages of the reconstruction of the old camp

of the total travel time. But the maximum capacity or the peak of the flow was not considered in the proposed model. The authors used a three-steps heuristic solution methodology similar to Nested Partition Algorithm.

Bose [14] studied the mobility for refugees settling in a non-traditional immigrant destination in Vermont, a state located in the northeastern side of the United States of America. The methodology used is based on a multi-year qualitative study of travel behavior preferences. The results of this study supported the hypothesis that lack of accessibility for refugees decreases their economic growth, deteriorates their medical health, and creates barriers to self-empowerment, acculturation, and integration within the new society.

Cetinkaya et al. [15] developed a Geographic Information System (GIS) based on a multi-criteria decision analysis model for refugee camps in southeastern Turkey to accommodate Syrian refugees. The authors used a fuzzy analytic hierarchy process to rank indicators generated from their developed GIS model. Fifteen new refugee camp locations are presented to replace the current camps allocated to Syrian refugees in southeastern Turkey. Gemenetzi [16] studied the evolution and the planning of Thessaloniki city in Greece after the Syrian refugee crisis in Europe. The author discussed the role of spatial planning in changing the geography of the city. According to Gemenetzi [16], Thessaloniki city has to adopt one of two options to respond to the Syrian refugee inflow in Greece. The first option is to follow passive pre-crisis strategies, which will transform Thessaloniki into an isolated city from geopolitical and economic perspectives. The second option is to adopt a radical resilience policy to frame the city's geography and ensures discursive narratives of strategic adaptability.

Current Traffic Conditions

Current Road Network

The Naher El Bared Camp (NBC) is adjacent to the Tripoli-Syrian Border highway, which is an international road with dual carriageway, two lanes per direction, and can be accessed through six access points as illustrated in Fig. 6. The six access points are (1) El-Aabdeh main entrance, (2) El-Mhammara entrance, (3) El-Minieh Sea Road entrance (bridge), (4) El-Mhammara minor and limited entrance, (5) Pedestrian Bridge on the Petrol Line and (6) Pedestrian Bridge on the Rail Line. The NBC internal road network is mainly the commercial spine (Camp Main Road connecting the Minieh Sea Road with El-Aabdeh interchange), which intersects several minor transverse roads and a narrow Sea Front Road that is poorly paved. In addition, the abandoned rail line and petrol line right of ways are paved with base coarse and are used as parking and internal access roads.

Traffic Counts

In order to estimate traffic flows in and around the project site, previous traffic counts data that were collected at various locations for the years 1998, 2001, and 2005 were adopted in this study [8]. The traffic counts locations are illustrated in Fig. 7 and includes: (i) TMC (Turning Movement Counts) over 14 h at two intersections: El-Aabdeh Roundabout, and El Mhammara intersection, (ii) ATC (Automatic Traffic Counts) over a whole week on the highway adjacent to the CAMP, and (iii) MCC (Manual Classified Counts) at all TMC and ATC locations.

The automatic traffic counts ATC1 were collected during the years 1998, 2001, and 2005, using electronic counters that were installed for 7 days and provide hourly traffic variation for every day of the week. The average week's daily traffic (AWDT) was estimated to be 31,386 vehicles divided equally between the South Bound and

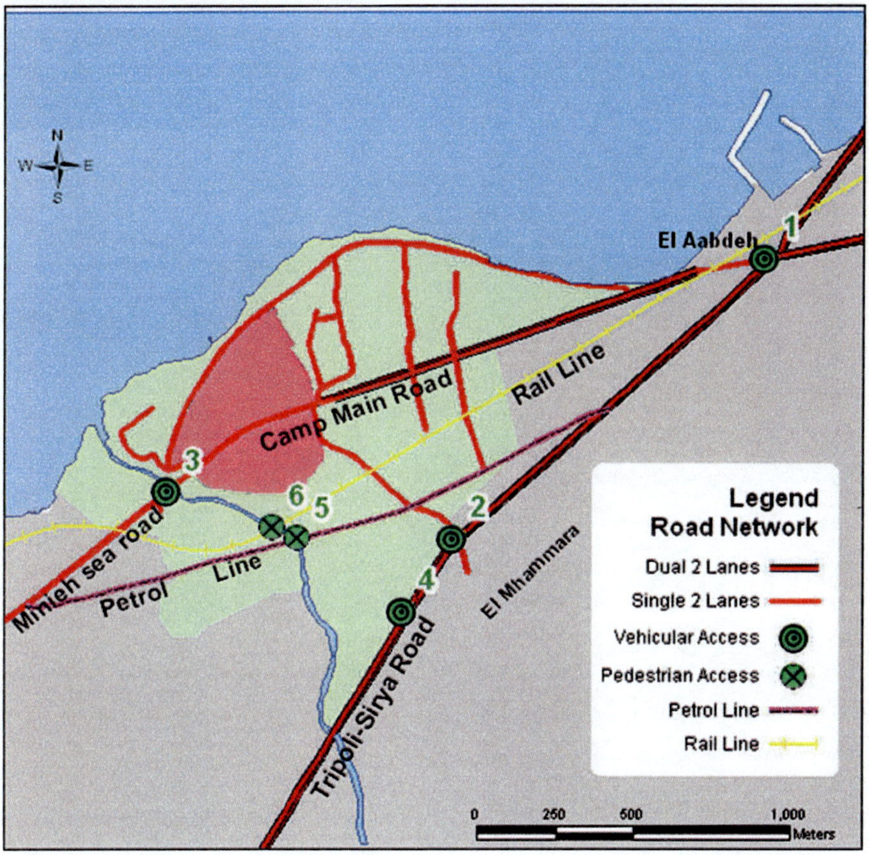

Fig. 6 Road network and site accessibility

the North Bound. The summary of the average daily traffic counts of the various years is presented in Table 2.

It can be noted that between the years 1998 and 2001, the yearly growth (around 9%) was higher than the years between 2001 and 2005 (around 2%). In fact, the Council of Development and Reconstruction report [17] has adopted a yearly growth factor of 2.5% for the forecast of the years 2010, 2015, and 2020. Therefore, the yearly growth factor of 2.5% was adopted in this study. Hourly traffic variation of the Tripoli-Syria Highway from the ATC1 counts for the year 2006 are presented in Fig. 8, showing the morning peak to be the highest (Total both directions of 2300 vehicles per hour being 10% of the total daily traffic). In addition, Fig. 9 represents hourly traffic variation of the camp-related traffic (from TMC1 and TMC2) indicates a noon peak hour of (approximately 8% of the total daily traffic).

Turning movement counts at TMC1, located at El- Aabdeh Intersection, are reported in Table 3. A total of 31,796 turning movements were reported at this TMC

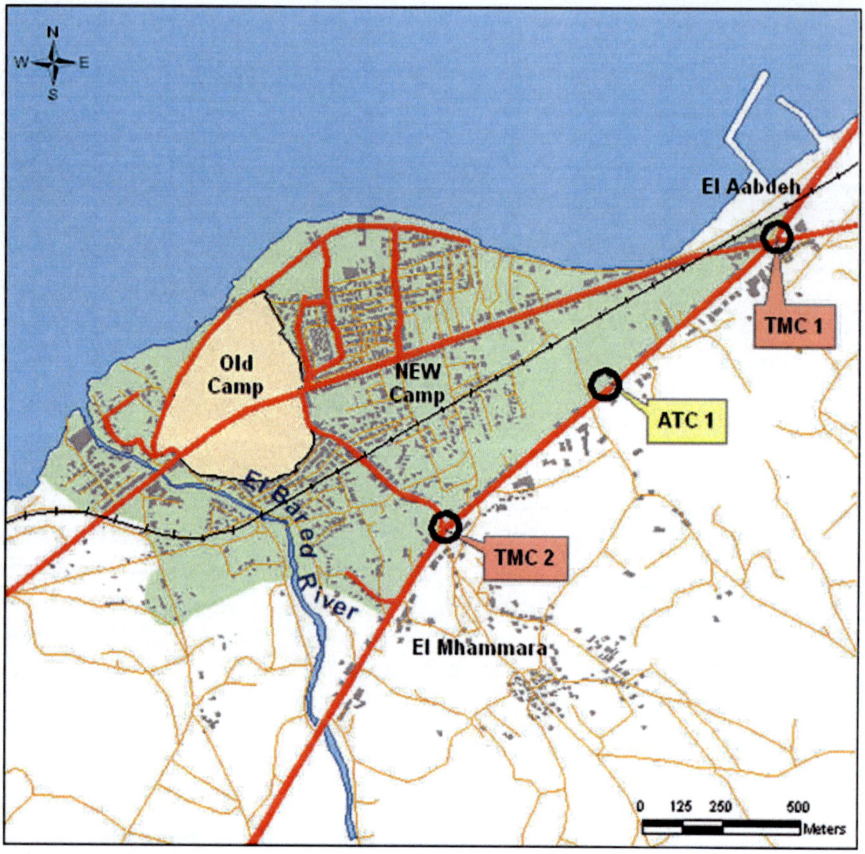

Fig. 7 Locations of traffic counts

Table 2 ATC1-average week daily traffic

Year	SouthBound	NorthBound	TOTAL
1998	*11,410*	*11,053*	22,463
2001	*14,347*	*14,848*	29,195
2005	*15,693*	*15,693*	31,386

over the four different bounds (East, South, West and North) divided as follows. 18,552 for passenger car, 5,437 for Taxi, 4,128 for Van and Minibus, 162 for Bus, 2,517 for Light Truck, 233 for Medium Truck, 418 for Heavy Truck and 348 for other types of vehicles.

Turning movement counts at TMC2, located at El-Mhammara Intersection, are reported in Table 4. A total of 26,621 turning movements were reported at this TMC over the four different bounds (East, South, West, and North) divided as follows. 14,182 for passenger car, 5387 for Taxi, 3780 for Van and Minibus, 195 for Bus,

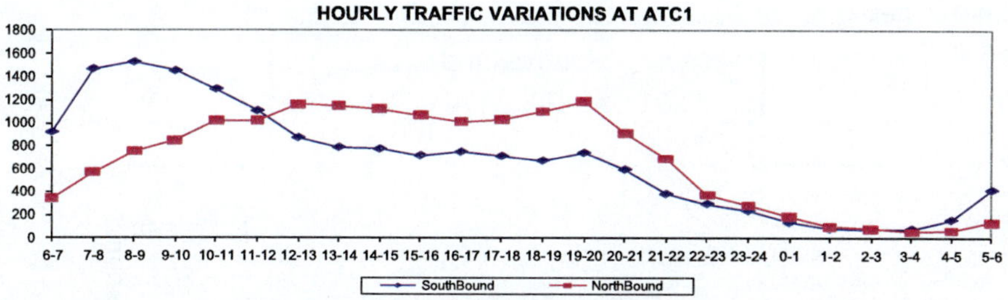

Fig. 8 Hourly variation of traffic flows on both the tripoli-sirya road

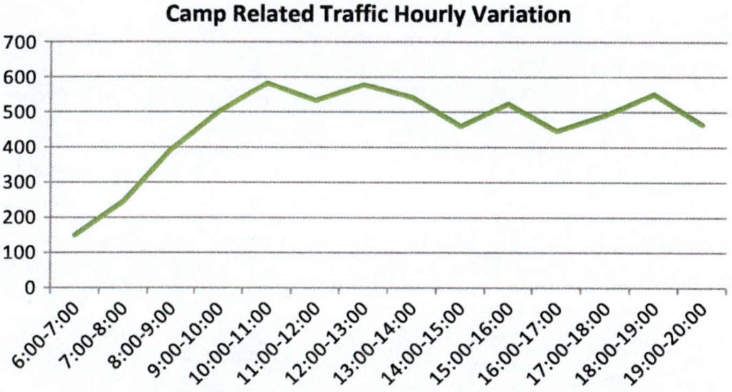

Fig. 9 Hourly variation of the camp-related traffic

Table 3 Turning movement counts at El-Aabdeh intersection (TMC1)

		SouthBound				
	886	2,713	1,283	0		
	↵	↓	↳	↺		
EastBound	0 ⮌				↥ 1,126	WestBound
	376 ↰				← 2,678	
	1,974 →				↧ 8,033	
	1,162 ↴				↻ 0	
	↶	↰	↑	↱		
	0	908	1,912	8,746		
		NorthBound				

Table 4 Turning movement counts at El-Mhammara intersection (TMC2)

Approach				
SouthBound	191	11,835	174	16
EastBound	0	206	89	374
WestBound	266	101	274	0
NorthBound	128	375	12,228	365

2264 for Light Truck, 185 for Medium Truck, 482 for Heavy Truck and 146 for other types of vehicles.

Trip Generations

External traffic accessing the site from El-Aabdeh and El-Mhamara intersections (access points 1 and 2, respectively), as illustrated in Fig. 6 are available from the turning movement counts and are shown in Table 5. The Minieh Sea Road (access point 3) traffic was conservatively assumed to be equivalent to 20% of the total of both access points 1 and 2. The internal trips are mostly pedestrian trips due to low car ownership, parking constraints, and most destinations are within walking distance (schools and high density residential/commercial developments). Therefore, internal vehicular trips were conservatively estimated to be equivalent to 20% of the total external trips.

Table 5 Study area daily trip generations

Approach	Total	Ext. Split (%)
El Aabdeh approach	*9505*	*78*
El Mhammara approach	*1590*	*13*
Sea road approach	*1110*	*9*
Total external trips	12,205	100
Internal trips	2,441	
Total trip generation	14,645	

The total daily vehicular trip generation for the new and old camp area for the year 2006 is estimated to be 14,645 trips, assuming no additional through traffic due to the configuration of the road network. The population of this area is estimated to be around 33,000 persons and includes more than 2300 non-residential units (commercial and institutional). A daily vehicular person trip generation/attraction rate is estimated to be 0.44 trips/person/day for the NBC.

The total daily vehicular trip generation for the new and old camp area for the year 2006 is estimated to be 14,645 trips, assuming no additional through traffic due to the configuration of the road network. The population of this area is estimated to be around 33,000 persons and includes more than 2300 non-residential units (commercial and institutional). A daily vehicular person trip generation/attraction rate is estimated to be 0.44 trips/person/day for the NBC.

Trip Distributions

External trip distribution was generated using the results of the turning movement counts, while the internal trip distribution was estimated proportionally based on population densities of each sector and assuming negligible friction factor inside the camp for a gravity model. The peak hour trip distributions are illustrated in Fig. 10. The study area is then represented in 19 Traffic Analysis Zones (TAZ), being 14 internal TAZs for NBC and 5 external TAZs as illustrated in Fig. 11. The distribution of the 3,774 peak hour trip generations distributed as follows: 2346 trips external to external trips, and 1428 trips related to NBC internal as external as indicated in Fig. 1. Then, trip generations were expanded into a full Origin-Destination (OD) matrix based on the 19 TAZs, as shown in Table 6.

Mode Choice

Results generated from the manual classified counts on the roads entering/exiting the camp area from TMC1 and TMC2 are summarized in Table 7. The percentage of total vehicles traffic distributions were deduced as follows (1) Passenger Car: 61.8%, (2) Taxi and Service (Jetni): 14.7%, (3) Van and Minibus: 11.7%, (4) Bus: 0.3%, (5) Pickup and Light Truck: 8.5%, (5) Medium Truck: 1.1%, (6) Heavy Truck: 0.6% and (7) Other type of vehicles: 1.2%.

Trip Assignment

The traffic assignment model developed in this study takes into account the urban specifications of the Naher El-Bared Camp, including trip generation, internal road

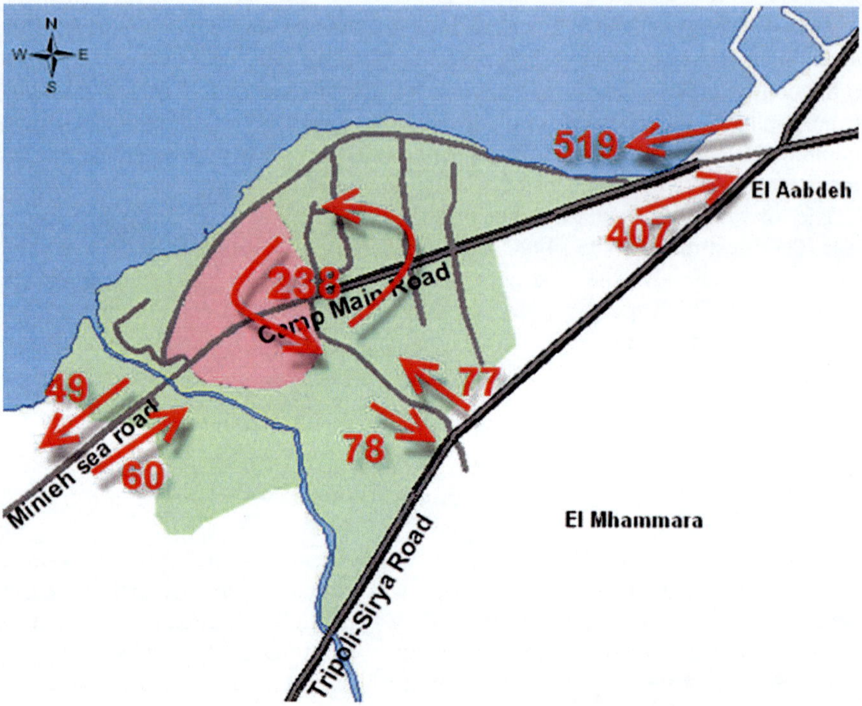

Fig. 10 NBC peak hour trip distribution

	ID	SECTOR
INTERNAL	1	Old Camp A
	2	Old Camp B
	3	Old Camp C
	4	Old Camp D
	5	New Camp A'
	6	New Camp B'
	7	New Camp C'
	8	New Camp E'
	9	New Camp A
	10	New Camp B
	11	New Camp C
	12	New Camp D
	13	New Camp E
	14	UNRWA Compound
EXTERNAL	15	Sea Road
	16	Tripoli Road
	17	El Mhammara
	18	Halba Road
	19	Syrian Road

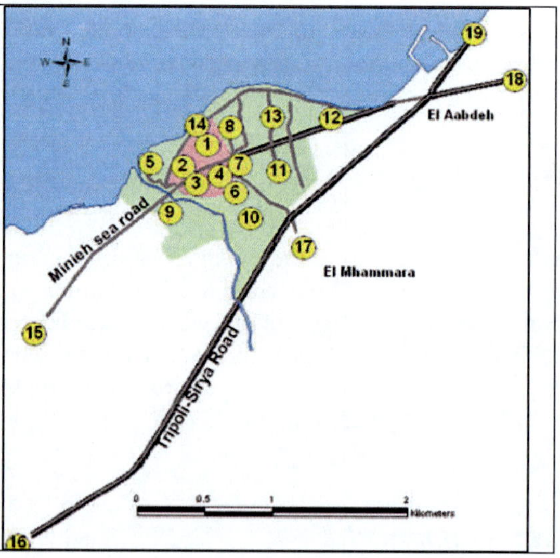

Fig. 11 Traffic analysis zones

Table 6 Expanded OD matrix

O/D	1	2	3	4	5	6	7	8	9	10	11	12	13	14	15	16	17	18	19	TOTAL
1		7.38	4.56	5.32	0.98	1.25	0.27	3.04	1.52	1.30	2.93	0.33	2.88	1.19	3.26	68.44	3.42	65.13	18.50	192
2	7.38		3.95	4.61	0.85	1.08	0.23	2.63	1.32	1.13	2.54	0.28	2.49	1.03	2.82	59.28	2.96	56.42	16.04	167
3	4.56	3.95		2.85	0.52	0.67	0.14	1.63	0.81	0.70	1.57	0.17	1.54	0.64	1.74	36.61	1.83	34.85	9.90	105
4	5.32	4.61	2.85		0.61	0.78	0.17	1.90	0.95	0.81	1.83	0.20	1.80	0.75	2.03	42.73	2.13	40.65	11.55	122
5	0.98	0.85	0.52	0.61		0.14	0.03	0.35	0.17	0.15	0.34	0.04	0.33	0.14	0.37	7.85	0.39	7.47	2.12	23
6	1.25	1.08	0.67	0.78	0.14		0.04	0.44	0.22	0.19	0.43	0.05	0.42	0.17	0.48	10.03	0.50	9.54	2.71	29
7	0.27	0.23	0.14	0.17	0.03	0.04		0.10	0.05	0.04	0.09	0.01	0.09	0.04	0.10	2.18	0.11	2.07	0.59	6
8	3.04	2.63	1.63	1.90	0.35	0.44	0.10		0.54	0.46	1.05	0.12	1.03	0.43	1.16	24.41	1.22	23.23	6.60	70
9	1.52	1.32	0.81	0.95	0.17	0.22	0.05	0.54		0.23	0.52	0.06	0.51	0.21	0.58	12.21	0.61	11.62	3.30	35
10	1.30	1.13	0.70	0.81	0.15	0.19	0.04	0.46	0.23		0.45	0.05	0.44	0.18	0.50	10.46	0.52	9.96	2.83	30
11	2.93	2.54	1.57	1.83	0.34	0.43	0.09	1.05	0.52	0.45		0.11	0.99	0.41	1.12	23.53	1.18	22.40	6.37	68
12	0.33	0.28	0.17	0.20	0.04	0.05	0.01	0.12	0.06	0.05	0.11		0.11	0.05	0.12	2.62	0.13	2.49	0.71	8
13	2.88	2.49	1.54	1.80	0.33	0.42	0.09	1.03	0.51	0.44	0.99	0.11		0.40	1.10	23.10	1.15	21.98	6.25	67
14	1.19	1.03	0.64	0.75	0.14	0.17	0.04	0.43	0.21	0.18	0.41	0.05	0.40		0.46	9.59	0.48	9.13	2.59	28
15	2.66	2.30	1.42	1.66	0.31	0.39	0.08	0.95	0.47	0.41	0.92	0.10	0.90	0.37		21.36	1.07	20.33	5.78	61
16	71.04	61.54	38.01	44.34	8.14	10.41	2.26	25.35	12.67	10.86	24.43	2.71	23.98	9.95	27.15		28.51	543.04	154.24	1,099
17	3.47	3.01	1.86	2.17	0.40	0.51	0.11	1.24	0.62	0.53	1.19	0.13	1.17	0.49	1.33	27.90		26.55	7.54	80
18	64.26	55.66	34.38	40.11	7.37	9.41	2.05	22.92	11.46	9.82	22.10	2.46	21.70	9.00	24.56	516.18	25.79		139.51	1,019
19	26.48	22.95	14.17	16.53	3.04	3.88	0.84	9.45	4.72	4.05	9.11	1.01	8.94	3.71	10.12	212.78	10.63	202.48		565
TOTAL	201	175	110	127	24	30	7	74	37	32	71	8	70	29	79	1,111	83	1,109	397	3,774

Table 7 Results of the manual classified counts

Vehicle type	%
Passenger Car	61.8
Taxi/Service	14.7
Van/Minibus	11.7
Bus	0.3
Pickup/light truck	
Medium truck	1.1
Heavy truck	0.6
Other vehicles	1.2
Total	100

network, and connections with the external area-wide road network. The modeling framework and software used are Emme/2 regional transportation model linked with ArcGIS and Synchro7 for signalized intersection traffic analysis. The analysis was performed for the year 2006, before the war, noon peak hour period, and the Emme/2 assignment output is presented in Fig. 12.

Traffic Analysis

Results of the 2006 assignment revealed that road links were operating with a low to moderate Volume-to-Capacity (V/C) ratio. The highest ratio reported was 500 vehicles per direction on the Camp Main Road, and congestions were observed on junctions, namely the major junction, as shown in Fig. 14. This junction is a four-way unsignalized intersection with the north-south approaches staggered at a distance. Then, Synchro software was used to analyze this junction, and the output is presented

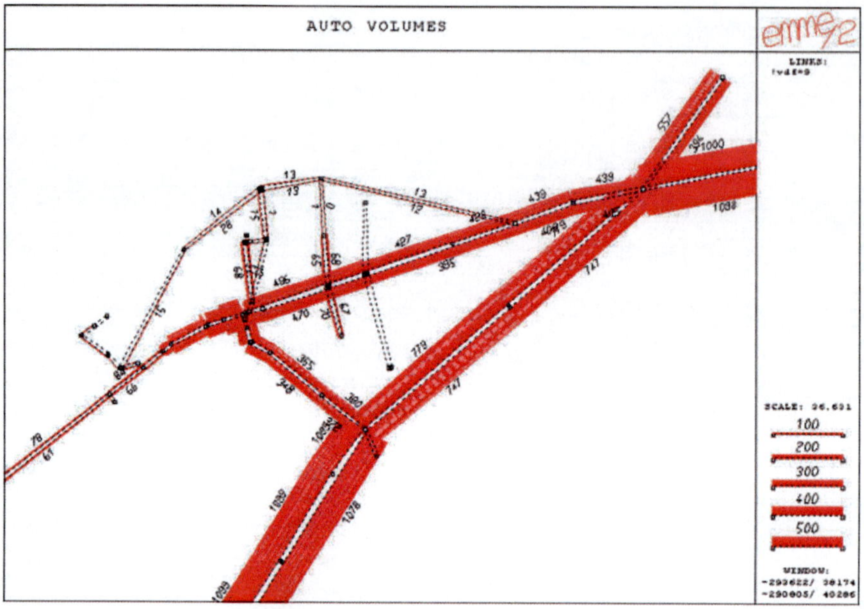

Fig. 12 Emme/2 assignment output

in Fig. 13. This junction was modeled as two 3-way intersections, and the LOS (level of service) was deduced to be F.

Fig. 13 Synchro output for the main camp junction

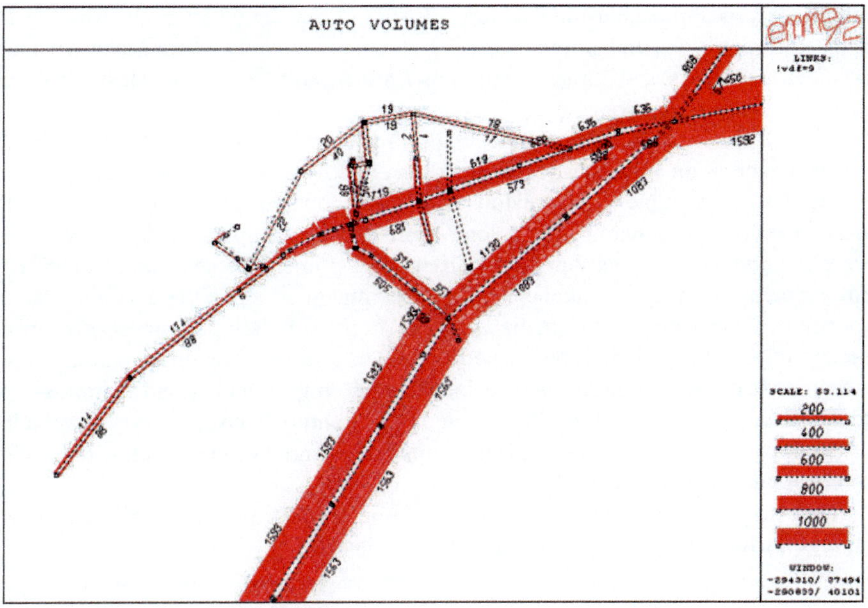

Fig. 14 Emme/2 assignment output (year 2030)

Future Traffic Condition

Currently, the old camp is closed by the Lebanese army for security reasons; only construction workers with special permits are allowed into the site. With more than 70% of the camp residents being displaced, traffic operations in the year 2010 are considered temporary. Therefore, the population of the camp for the year 2015, at the full reconstruction of the old camp, was expected to be equal to the year 2006. Therefore, a future scenario for the year 2030 (forecasted at 2.5% yearly growth) was evaluated in Emme/2. Figure 14 showed the results, where the Volume-to-Capacity (V/C) ratios are still acceptable (the highest ratio on the camp main road reached 700 vehicles per direction with a capacity of 1200 vehicles per lane, V/C is equal to 0.58). The main junction continues to operate at a level of service F and is recommended to be converted into a roundabout which can accommodate higher traffic volumes with the minimal conflicting points.

Conclusion and Recommendations

This paper analyzed the urban transportation planning framework related to long term refugee camps, specifically the case of Naher El Bared Camp in North Lebanon. A transport study was conducted using a combination of the four steps model and a micro-analysis approach using Synchro software. A new transport model was developed to forecast demand in the year 2030, taking into consideration land use and political constraints, and implemented using Emme/2 software. Several insights were deduced from the developed model.

First, despite the high-density population inside the long term refugee camp, a low trip generation was recorded, approximately 0.44 trips/person/day. To the best of our knowledge, none of the previous research conducted in the literature recorded any value for trip generating in a long term refugee camp. Second, congestions were mainly on junctions due to the absence of proper geometry and traffic law enforcement. Therefore, roundabouts are recommended since the accommodated capacities are higher for unsignalized intersections, simulating the level of service of any future intersection, as performed by [18], is a worthwhile direction for future research. It is also recommended to increase parking widening and sidewalks to facilitate and enhance mobility inside the refugee camp. Thus, Analysis of vehicle ownership and operating costs inside the camp is a potential venue for future research, as illustrated by Mansour et al. [19].

This study is a first step toward documenting and formulating urban transport studies for long term refugee camps in the Middle East.

References

1. United Nations High Commissioner for Refugees (UNHCR) (2016) The situation of palestinian refugees in lebanon
2. Ramadan A (2009) Destroying Naher el-bared: sovereignty and urbicide in the space of exception. Political Geography 28:153–163
3. Ramadan A (2010) In the ruins of Naher al-Barid: understanding the meaning of the camp. J Palestine Stud 40:49–62
4. Ramadan A (2013) Spatialising the refugee camp. Trans Inst Br Geogr 38:65–77
5. Naher el-Bared reconstruction commission for civil action and studies (2008) Preliminary Master Plan and Guidelines for the Reconstruction of Naher el-Bared Palestine Refugee Camp. United Nations Relief and Works Agency, Lebanon
6. United Nations relief and works agency for palestine refugees in the near east (UNRWA) (2012) Reconstruction of Naher El-bared camp and UNRWA compound
7. United Nations development programme (UNDP) (2008) The contribution of UNDP to the early recovery process of Palestinian and Lebanese populations affected by Naher El Bared Crisis. United Nations Development Programme, Lebanon
8. Government of Lebanon (2008) A common challenge: a shared responsibility. Report of the international donor conference for the recovery and reconstruction of the Naher el-Bared palestinian refugee camp and conflict-affected areas of North Lebanon. Vienna
9. Damme WV (1995) Do refugees belong in camps? experiences from Goma and Guinea. The Lancet 346:360–362
10. Anderson MD, Abdullah YM, Gholston SE, Jones SL (2006) Development of a methodology to predict through-trip rates for small communities. J Urban Planning Dev 132:112–114
11. Antunes A, Seco A, Pinto N (2003) An accessibility-maximization approach to road network planning. Comput-Aided Civil Infrastruct Eng 18:224–240
12. Drezner Z, Wesolowsky G (2003) Network design: selection and design of links and facility location. Transp Res Part A Policy Pract 37:241–256

13. Bigotte JF, Krass D, Antunes A, Berman O (2010) Integrated modeling of urban hierarchy and transportation network planning. Transp Res Part A Policy Pract 44(3):506–522
14. Bose P (2014) Refugees in Vermont: mobility and acculturation in a new immigrant destination. J Transp Geogr 36:151–159
15. Çetinkaya C, Özceylan E, Erbaş Kabak M (2016) GIS-based fuzzy MCDA approach for siting refugee camp: a case study for southeastern Turkey. Int J Disaster Risk Reduction 18:218–231
16. Gemenetzi G (2017) Thessaloniki: the changing geography of the city and the role of spatial planning. Cities 64:88–97
17. Council of Development and Reconstruction (2001) Traffic study for the feasibility of the deir-ammar-el aabdeh corridor. http://www.cdr.gov.lb
18. Fawaz W, El Khoury J (2016) An exact modelling of the uniform control traffic delay in undersaturated signalized intersections. J Adv Transp 50:918–932
19. Mansour C, Haddad M, Zgheib E (2018) Assessing consumption, emissions and costs of electrified vehicles under real driving conditions in a developing country with an inadequate road transport system. Trans Res Part D Transp Environ 63:498–513

POLCA Approach on Make to Order Production System: An Application from the Hydraulic Industry

Yildiz Kose, Emre Cevikcan, and Samet Resul Celik

Abstract Applications of production control systems such as pull systems, which are also called Kanban control systems have attracted attention not only in the field of production but also in the academic world. Suri (1998) has introduced a different production control system to ensure attractiveness in production systems with high product variety: Paired-cell Overlapping Loops of Cards with Authorisation (POLCA) production control system. The POLCA system aims to ensure the effective use of the pull system in a wide variety of products and in a low volume. In the literature, the POLCA system has been examined for make-to-order systems (MTO), but its advantages have not been seen clearly. In MTO systems, the simulations are not fully understood, and the comparisons may not yield clear results. In this paper, a more general perspective on the design issues in the POLCA system has been introduced, and the tools and methods that can be understood by the people who are not licensed and who manage the production company have been summarized. Within this paper, the definition of POLCA is mentioned first. Then the industrial application area and conditions of use have touched on. The current and proposed examining the advantages and disadvantages of the systems, the basic calculations, and researches on the shaping of the production line according to the proposed POLCA system have been presented.

Keywords POLCA system · Make-to-order · Simulation · Hydraulic industry

---

Y. Kose (✉) · E. Cevikcan · S. R. Celik
Industrial Engineering Department, Management Faculty, Istanbul Technical University, Istanbul, Turkey
e-mail: kose18@itu.edu.tr

E. Cevikcan
e-mail: cevikcan@itu.edu.tr

S. R. Celik
Industrial Engineering Department, Faculty of Engineering, Bogazici University, Istanbul, Turkey

F. Calisir (ed.), *Industrial Engineering in the Internet-of-Things World*,
Lecture Notes in Management and Industrial Engineering,
https://doi.org/10.1007/978-3-030-76724-2_10

Introduction

Reducing delivery times, reducing costs, eliminating waste and increasing flexibility are the main objectives of any manufacturing company. Traditional push production systems cannot fully satisfy or facilitate these objectives. Therefore, the transformation of the production system from the push system to the pull system becomes critical. Today's new developments have given companies the ability to custom-engineer and then manufacture products for individual clients without incurring the high additional cost. These developments can occur in the field of production capacity control, lot sizing, and material control approach [11].

Quick Response Manufacturing (QRM) was developed by [18] in order to increase customer service level by reducing the lead time in a low volume high mix manufacturing environment. To achieve this goal, a control strategy tool called POLCA (Paired-cell Overlapping Loops of Cards with Authorization) is designed and implemented to regulate the workflow at various stations of the manufacturing system [2]. A material control system designed for make-to-order or engineer-to-order companies such as POLCA provides short throughput times to these companies. Thus, companies can cope with a wide range of customized products and strong pressure [13]. POLCA uses overlapping card loops between successive pairs of cells in routing a job and applies a work-in-process (WIP) cap at each loop [4].

In this study, the POLCA system has been systematically addressed. POLCA model has been adapted to a company with low-volume, high-variety products. As a result of the arrangements made with POLCA in the company that is currently implementing a push system, the changes in some performance criteria such as value-added time, waiting time in queue, amount of WIP have been analyzed.

Most of the information about the POLCA material control system is given in the Suri [18] book and at the QRM conferences by Suri [17, 20]. Other articles draw attention to POLCA when comparing various control systems [8, 10, 14, 16].

Because the POLCA system shows several similar features with the Kanban system, reorganization into a cellular structure is needed [15]. Therefore, the economic analysis of cellular production systems has also attracted great attention [1, 6, 9].

Although there are a few case studies of the researchers, the number of published case studies can be called insufficient [3, 15]. Several successful applications of POLCA have been dealt with in the QRM workshops, according to the best knowledge of authors, these applications are limited to parts of the production system and not across the company. Therefore, there is still a lack of insight into the application problems and the barriers encountered in the design and implementation.

It is clear that there is a lack of literature on well-prepared case studies that explain and analyze the application of the POLCA system. In QRM workshops, a few successful applications of POLCA have been addressed, but according to the best knowledge of the authors, these applications are limited to parts of the production system, not the whole company. In this context, this study contributes to the relevant literature by means of developing a proposed model based on an application of the POLCA simulation model in the valve industry.

The rest of the paper is organized as follows. In Sect. 2, a material control system is given. The POLCA system is explained in Sect. 3. Section 4 includes an industry application. A comparison of two systems is analyzed in Sect. 5, and finally, conclusions are provided in Sect. 6.

Material Control System

Material planning and control strategies can be examined in three classes; push, pull, and hybrid systems. The difference between the push and pull systems can be understood by looking at how the products to be produced are transmitted to the stations. It is determined according to the expiry time of the order in push systems, but it is expected to withdraw from the finished products for production in pull systems. Push systems are related to Material Requirements Planning (MRP) systems, and pull systems are also included in the literature as Kanban Control Systems (KCS). Companies must first analyze their production systems so that they can select the control system. These can be examined in four classes; make-to-stock (MTS), make-to-order (MTO), assembly-to-order (ATO), and engineering-to-order (ETO) [12]. QRM can be defined as the fast design and production of a customized product according to customer requirements [19]. Traditional control systems such as Kanban and CONWIP have their own disadvantages in low volume and in various production environments [5]. If we apply Kanban to a wide variety of environments, it results in a very high WIP at each phase.

POLCA System

POLCA is a material control technique that is a hybrid push-pull system that combines the best features of push/MRP systems and pull/Kanban control and designed for make-to-order or engineer-to-order companies [13]. These companies produce a wide range of products to meet customer demands. It tries to attract the interest of the customers by showing the number of varieties and offering them different options. A new material control strategy called POLCA is introduced for this situation. It mainly focuses on the execution part of the QRM philosophy.

Principles of POLCA

The basic principle in this strategy is that it uses the card as an existing capacity signal between two cells, unlike Kanban. In POLCA, if the material flows from two

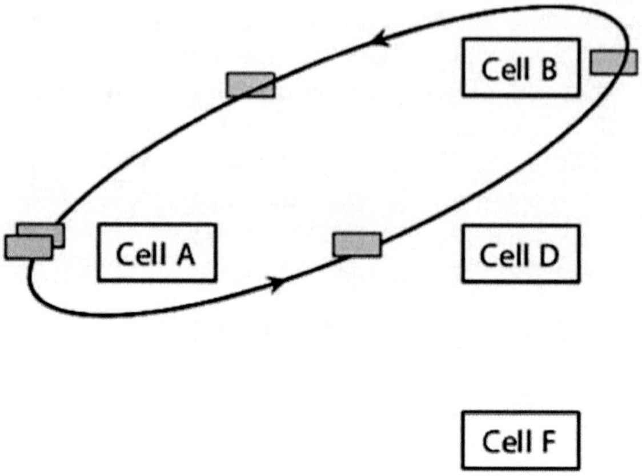

Fig. 1 POLCA card and cell relation

cells (from Cell A to Cell B), these cells are bound by a POLCA ring (see Fig. 1). This cycle is provided by POLCA cards, which are unique to that loop only. The card is labeled according to the source, and the target cell, e.g., the card used in the cycle shown in the figure, is called the A/B card. If Cell A has to start a job that is directed to Cell B, it must have an A/B card to start the job. When Cell A completes the job, it sends the job to the Cell B with the A/B card. This is one of the basic features of the POLCA system. The card from cell B sends the message, "We finished the work. You can send us another one." In other words, the return of POLCA cards means that there is a capacity gap for this resource in future cells.

Kanban works with the inventory signal while POLCA works with capacity signal. This difference shows why the POLCA works for low volume and special parts and that the Kanban system is suitable for product flow.

After making the MRP system calculations, the system creates a dispatch list that lists all jobs that are sorted by the authorization date and that are not yet started by the cell. This list is created for each cell in Table 1. Products in this list are divided into two parts. As shown in the table, the day shows June 23rd. Orders for June 23 and prior orders are authorized. Orders after June 23 are not authorized. They cannot go into production even if the machines are idle. This makes it easier for teams in the cell to decide the job that first to start. Whichever job is above, the list starts first.

Table 1 Dispatch list example

Cell name: Cell A Date: 23 June 2019			
Job ID	Authorization date	Next cell	Additional job data
MM01	June 21	D	…
MM02	June 23	B	…
MM03	June 23	D	…
MM04	June 24	F	…
MM05	June 25	B	…

POLCA cards enable the system to use capacity in the most effective way. If the number of cards is low, the efficiency of the machines will be reduced. For example, an order that should start on June 21 has not started for 2 days. The cell we are in is Cell A. The target cell is the Cell D. Then, the POLCA card A/D card is required. If there is a card in the cell, it will be added to the job and started working. But if this card cannot be found, it cannot be started. Because the absence of this card indicates that the Cell D is busy. The cards serve as a capacity signal and work to reduce the number of WIP.

The performance of POLCA can be measured by operational parameters and financial parameters. Flow time, work in process, throughput, delivery time, and effort in expediting orders for the customer have been seen operational parameters to measure the performance of the POLCA system [7].

Advantages and Disadvantages

The advantages of POLCA are as follows; card number stability, a clear objective for cells, ease of resource planning, reduced staff cost, overall use efficiency, reducing WIP, and intercellular control. The workflow between cells is synchronized, and waiting times are reduced via POLCA cards. Thus, while speeding up the work transfer, it also provides a balance between the machines and cells in the system [5]. Decisions on when and what to leave to the workshop have a great impact on both the workshop output time and delivery performance. POLCA can provide a stable delivery time by making the machine workers' time and machine controls. POLCA does not need to be limited to production. Companies may also include office cells and warehouses in the POLCA system. This system does not require administrators to allocate a great time and for tracking and editing. POLCA cards ensure the best use of capacity; they also help prevent congestion in the workplace due to the fact that jobs are being started early or are stacked in bottlenecks. Since POLCA cards are not connected to products as in Kanban, there is no inventory spread when you have a wide variety of products. It also allows avoiding overloading with the capacity signal. The most important thing in the POLCA system is to be able to understand which production system cannot be applied. If some conditions are not met, POLCA also has great disadvantages.

Formulation

Firstly, the intercell traffic must be analyzed. This means the possible routines of the product family should be found. After analyzing the routines, collected data is put in a table. This table is called from/to table. The from/to table is a table that displays the number of orders. Thanks to this table, it can be seen which cells are flowing.

Data analysis and table placement will be shown in the application. After the table is created, the frequencies between the cells are seen.

The number of POLCA cards is calculated according to these frequencies. The formula for calculation is [19]:

$$The\,number\ of\ A/B\ POLCA\ cards = \left[[LT_A + LT_B] * NUM_{(A,B)}\right] * \alpha$$

LT_A, LT_B, $NUM_{(A,B)}$ and α represent a lead time for cell A, lead time for cell B, the number of jobs go from Cell A to Cell B and safety factor, respectively. The logic of the formula is to create enough cards to support the frequency between loops. If more cards are generated, this increases the number of WIPs. If the number of cards is missing, the required flow cannot be achieved.

Industrial Application

To demonstrate the effectiveness of POLCA, it has been employed to evaluate a hydraulic company located in Istanbul. In the company, which is adopted the push control approach as production control, valves are produced according to the order. The problem is that the company produces a wide variety of products with low volume. The product variety of this company has the qualification required for the POLCA approach. After simulation models, a system including the application of the POLCA production control technique as an alternative to the production control system present in the company is modeled, and simulation is performed on the Arena 11.0 program. The results of this study are compared with the results of the current situation, and the benefits of the proposed POLCA system are revealed.

Analysis of Current System

The company selected for the application produces hydraulic valves with make-to-order. Valve castings will be included in the production line after they come from the foundry as raw materials. At this stage, it is useful to examine the company's production line. The company manufactures 12 types of hydraulic valves which have a similar production process. The production process of the valves in the company is shown in Fig. 2. These valves are varied in properties such as the volume of fluid, the number of eyes, which it allows to pass through. For example, 45 cc fluid volume

Fig. 2 Production process of the valves

(a) (b)

Fig. 3 **a** GMS 452; **b** GMS 905

and 2-arm types are called GMS 452 (see Fig. 3a), 90 cc fluid volume, and 5-arm types are called GMS 905 (see Fig. 3b).

The units defined as vertical machining and horizontal machining centers in the production system have 2 and 4 CNC machines, respectively, with the same functionality. If there is no empty machine in the queue, the machine which has at least a queue is assigned. All of the vertical machinings in the system can process all products, and there are no machine-related differences in production times. After the bodies processed in the vertical machining center, if there is no empty machine in the queue, the machine is assigned as the machine with at least a queue in a horizontal process. After this assignment, the bodies belonging to any order from the vertical process center are transferred to the cleaning section. Just as in the vertical process center, the machines in the horizontal process center can process all product types with the same processing time. In the cleaning section, the deburring of the machined parts is carried out, and they are prepared for honing. This process is performed by a single cleaner, and the finished bodies are sent to the honing section. The honing section is the biggest problem in the system. For honing, 150 pieces must come to the honing queue. This increases the waiting time per piece. Long queues and intermediate stocks are formed in front of the source as the bodies of all orders have to pass over the single source in this section. The produced parts are sent to the assembly and passed through the testing phase. Thus, the production process is completed. One of the features that should be taken into consideration is that the valve bodies that circulate in the system move according to orders. For example, when the company receives 110 orders from the GMS 453 product, the body of these 110 valves is brought from the casting as raw material and is placed in the queue of a vertical processing machine for vertical operations. After this stage, these 110 GMS 453 valve bodies never separate in the system and move together until they are removed from the assembly. In this case, the company causes high intermediate stock levels. This is perceived as a problem by the company and is being solved.

Current System Simulation

Before moving on to the modeling of the existing system, it is useful to transfer the data of the existing system and the assumptions to the Arena 11.0 program. The order quantities that resemble real data are used to protect privacy. Firstly, the company's net operating time, which is 10 h, is transferred to the model. A 127-day order list is created for the 12 types of products produced by the company. A virtual day counter is defined in the system for the day-to-day exchange of incoming orders, and orders were entered from the "Expression" section. Orders can move together if the "Batch Module" is used. These products are separated after entering the system and are combined once again for batch processing before the honing process. The entrance of the products is made by the "Create" module. After that, the features of each product are assigned with the "Assign" module, and the products are combined with the "Batch" module in order for the products to be entered into the system.

The processing times of 12 types of products in different production units are given in Table 2. All of these processing times are defined in the "Expression" section of the Arena program. It is also assumed that the processing times in these 5 units are constant and do not vary.

Another issue that is reflected in the model related to the processing times is the setup time in vertical and horizontal processing machines. Each of the 6 machines in vertical and horizontal machining centers has the setup times when switching from one product type to another. These times are 1 h for each of the 2 machines in the vertical machining center and 1.5 h for each of the 4 machines in the horizontal machining center. These times are defined in the model as in Fig. 4a, b as well as the processing times of the machines. After the machines understand the difference between the products they process, and the next product is different, an "Assign" module is used to the machines and the time is conditionally extended.

Table 2 Processes times of valves (Second)

Product code	Vertical	Horizontal	Honing	Cleaning	Assembly	Test
451	204	528	180	60	90	40
452	228	417	360	60	180	40
453	213	603	540	60	270	40
454	270	748	720	120	360	80
455	232	1425	900	180	450	100
456	265	1680	1080	180	540	100
901	285	823	180	180	90	100
902	288	962	360	60	180	40
903	294	1212	540	60	270	40
904	305	1872	900	120	450	80
905	291	3123	1080	120	540	80
906	115	240	180	180	90	100

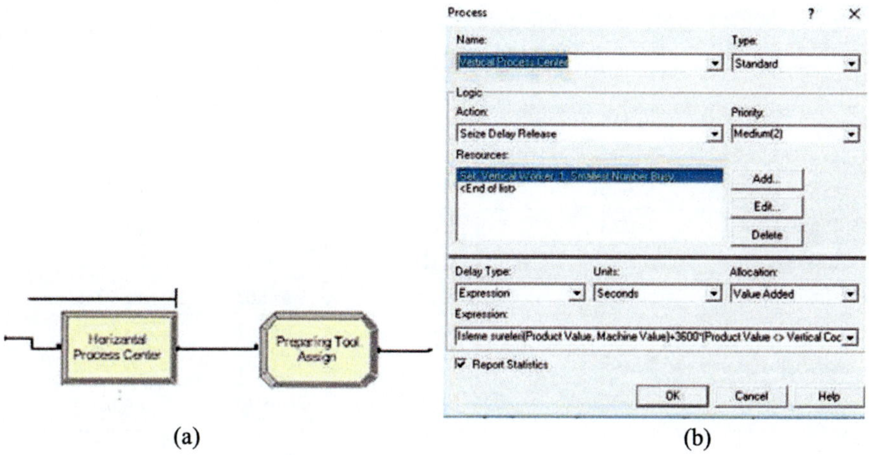

Fig. 4 **a** Preparing assign module; **b** setup time expression

For vertical and horizontal machines, the "Set" module is created for the resources, and the product is sent to the machine in which the queue contains fewer products. After the vertical and horizontal machining centers are modeled, cleaning, honing, assembly, and test sections are added to the model, and the products that are ready for shipment are removed from the system with the "Dispose" module. The current system modeled on the arena is shown in Fig. 5.

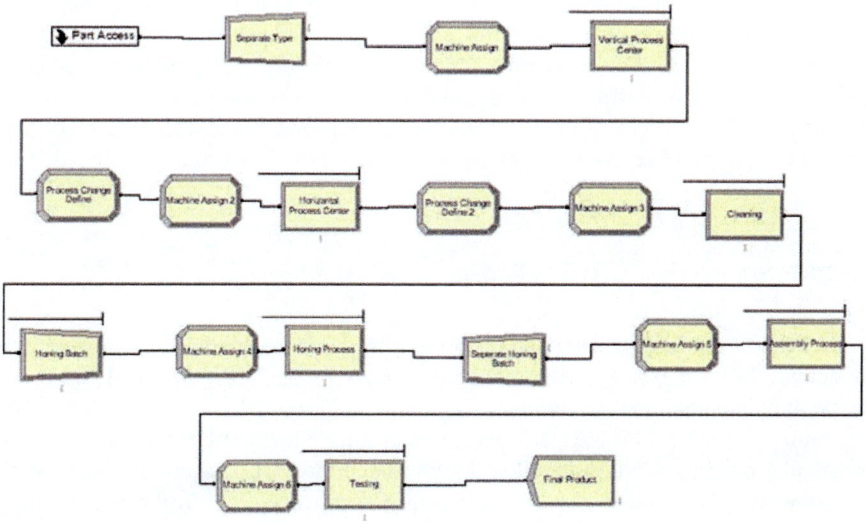

Fig. 5 Current system simulation

Table 3 Current system ARENA results

Number In	Number in	VAT	WT	WIP
Entity 1	128	0	0	0
GMS451 Model	435	2.9335	34.6993	8.4924
GMS452 Model	1795	3.3579	63.4185	82.8636
GMS453 Model	940	7.8424	50.7558	35.2933
GMS454 Model	738	5.4892	41.8742	21.3820
GMS455 Model	637	8.1268	66.0455	27.9012
GMS456 Model	336	8.4011	97.4863	15.9342
GMS901 Model	536	8.1793	31.6401	13.8705
GMS902 Model	1493	3.9791	74.4841	79.6638
GMS903 Model	1139	3.9303	58.1613	46.6948
GMS904 Model	671	5.7099	138.55	57.8293
GMS905 Model	285	7.4879	132.76	15.8216
GMS906 Model	235	4.3354	55.5996	4.5120
Total	9240	69.7728	845.47	410.2587

Current System Simulation Results

After realizing the current situation model under the assumptions and conditions stated in the previous section, the total number of products released from the system is 9240, and these values according to the product types are shown in Table 3. Meanwhile, this table shows the value-added time (VAT) according to the types of products. The system is repeated 30 times for the obtained values. The waiting times (WT) of products are shown in the fourth column of the table. One of the reasons for long periods of time is that the honing process is due to bulk production rather than the part. The WIP values of the current system are shown in the fifth column of the table. The Entity 1 value is a virtual input created for the day count.

Simulation of POLCA Application

In the POLCA model, all order and process information used in the current situation model will be used exactly. The innovation brought by the POLCA model will be at the point of leaving the orders within the system. In this model, incoming orders must be combined with a POLCA card before entering the queue of one of the vertical processing machines. No orders that do not match the POLCA card are permitted. In this way, it is aimed to prevent excess in-process inventory. In the new model, POLCA cards are provided with a "Hold" module. A constraint is defined between each binary process.

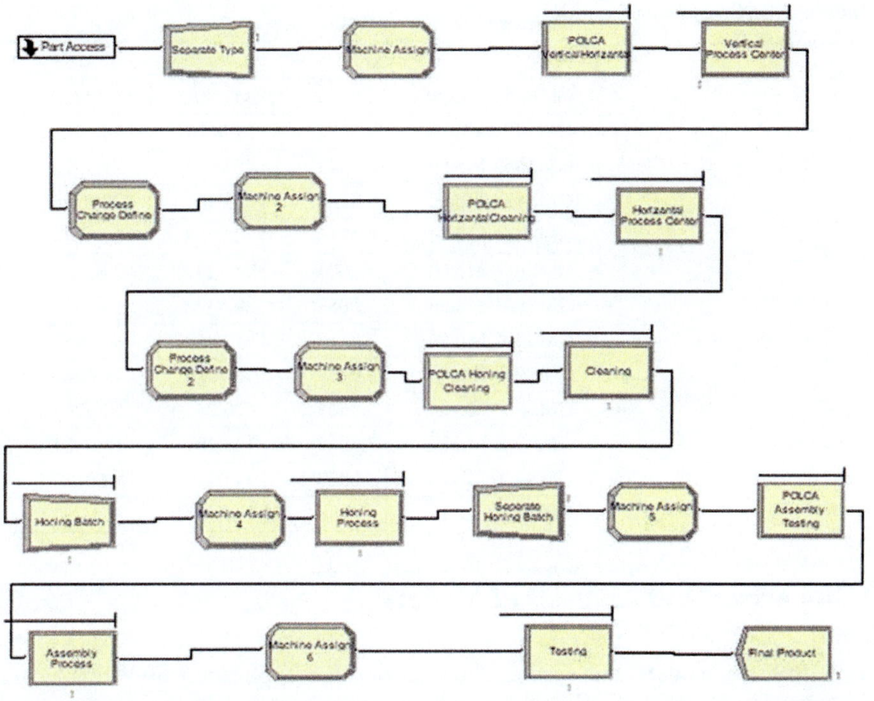

Fig. 6 The proposed POLCA system

Thus, it is aimed to reduce the amount of WIP in the system. The system was re-run with previous assumptions. The general picture of the system is shown in Fig. 6.

POLCA Application Simulation Results

After running the POLCA application model under the assumptions and conditions mentioned in the previous section, the total number of products exiting the system is 9157. This number is very close to the number in the current system. Table 4 shows value-added time, wait time of products, and the number of works in process details for application POLCA. All of the analyses are calculated hourly based. In the virtual day counter, there is not any process. So, it does not wait at queue. If honing batch is improved, the waiting time can be decreased.

Table 4 POLCA system ARENA results

Number In	VAT	WT	WIP
Entity 1	0	0	0
GMS451 Model	2.9179	34.6701	8.4823
GMS452 Model	3.9931	57.6677	76.5208
GMS453 Model	4.9949	40.0692	21.1222
GMS454 Model	5.5186	45.5901	29.0316
GMS455 Model	8.1313	70.7788	29.6824
GMS456 Model	5.2005	98.8257	7.8272
GMS901 Model	8.2087	28.6328	14.5319
GMS902 Model	3.2123	61.8869	66.0951
GMS903 Model	3.9467	56.7989	45.6846
GMS904 Model	5.6505	120.25	50.4702
GMS905 Model	7.3872	123.28	14.7401
GMS906 Model	4.0867	57.0419	4.6018
Total	63.2484	795.4921	368.7902

Comparison and Analysis of Models

Two different models, the current situation and the application of POLCA, are performed under the same conditions, and the results are taken. By comparing these results, the benefits of POLCA application in a company that implements traditional push production control and makes production according to order are examined. The most significant improvement is for the queue waiting times and numbers in the queue. There is a significant improvement in waiting in the queue compared to the current system in Table 5. The number of waiting in both models queues are not shown in Table 5 because the numbers of waitings are zero.

First, there is no difference between the two models running in the same period in terms of the number of outputs. Secondly, it is seen that the POLCA system provides improvement by decreasing WIP on the 8 product types. Moreover, POLCA provides an improvement of approximately 6% of the total amount of WIP. Thirdly, considering the average wait times in the processes per order, it is seen that the POLCA application provides a decrease of 4, 95%. Finally, it is observed that the number of queues decreases by about 99% in the balanced system in assembly, cleaning, honing, horizontal, test, and vertical processes. In the honing process, the waiting in the queue has completely disappeared.

Table 5 Number of queues waiting in the current and proposed system

Name of queue	Number waiting in the current system	Name of queue	Number waiting in the proposed system
Assembly, Queue	68.138	Assembly, Queue	0.0882477
Cleaning, Queue	95.8785	Cleaning, Queue	0.79
Honing Batch, Queue	79.9804	Honing Batch, Queue	83.332
Honing, Queue	0.01528404	Honing, Queue	0
Horizontal, Queue	0.3546	Horizontal, Queue	0.02243254
Testing, Queue	0.1254	POLCA assembly testing, Queue	6.6983
Vertical, Queue	195.86	POLCA honing celaning, Queue	71.7327
Total	440.352	POLCA horizontal cleaning, Queue	0.269
		POLCA vertical cleaning, Queue	183.71
		Testing, Queue	0.00030785
		Vertical process.Queue	1.4922
		Total	348.125

Discussion and Conclusion

POLCA application is a very simple but effective production control method with the cards defined on pairs in production units. It is possible to apply the POLCA technique in systems that produce demand according to the order in which the demand is unstable, and the product variety is wide. In order to ensure the successful implementation of the POLCA strategy, it seems necessary to have an effective system that brings together all parts of the QRM strategy. Due to the lack of a systematic approach, wrong POLCA applications can be seen in industrial-sized systems.

With this study, a real-life application suitable for the POLCA system is selected, and the proposed system with the same assumptions and parameters is performed by making arrangements on the current system. The results are compared between the current and proposed systems according to certain performance measures. It is possible to figure out that POLCA provides improvements by decreasing the values such as waiting times, machine queues, WIP in a company that produces according to the order. Therefore, it is an ideal alternative to solve the existing problems of systems by using the POLCA pull system.

It is foreseen that the application will be further elaborated and larger developments will be achieved if new trends such as artificial intelligence and big data are used.

References

1. Baysan S, Kabadurmus O, Durmusoglu MB (2007) Economic analysis of polca system design via simulation: a multi-cell manufacturing case study. In: 37th International conference on computers and industrial engineering 2007
2. Bong CS, Chong KE, How WC (2018) Job shop material control based on the principles of quick response manufacturing. J Adv Manuf Technol 12(1–2):15–30
3. Farnoush A, Wiktorsson M (2013) POLCA and CONWIP performance in a divergent production line: An automotive case study. J Manage Control 24:159–186
4. Fernandes NO, Carmo-Silva S (2017) Load-based generic polca: performance assessement using simulation. Int J Mechatron Appl Mechan 1:198–201
5. Germs R, Riezebos J (2010) Workload balancing capability of pull systems in MTO production. Int J Prod Res 48(8):2345–2360
6. Irizarry MDLA, Wilson JR, Trevino J (2001) A flexible simulation tool for manufacturing-cell design, II: response surface analysis and case study. IIE Trans Inst Ind Eng 33(10):837–846
7. Krishnamurthy A, Suri R (2009) Planning and implementing POLCA: A card-based control system for high variety or custom engineered products. Prod Planning Control 20(7):596–610
8. Krishnamurthy A, Suri R, Vernon M (2004) Re-examining the performance of MRP and kanban material control strategies for multi-product flexible manufacturing systems. Int J Flex Manuf Syst 16:123–150
9. Leep HR (1995) Planning, design, and analysis of cellular manufacturing systems. Manuf Res Technol
10. Lödding H, Yu KW, Wiendahl HP (2003) Decentralized WIP-oriented manufacturing control (DEWIP). Prod Planning Control 14(1):42–54
11. Luan S, Jia G, Kong J (2013) A model approach to POLCA system for quick response manufacturing. J Comput Inf Syst 9(3):1167–1174
12. Maganha I, Silva C, Ferreira LMDF (2019) An analysis of reconfigurability in different business production strategies. IFAC-PapersOnLine 52(13):1028–1033
13. Riezebos J (2010) Design of POLCA material control systems. Int J Prod Res 48(5):1455–1477
14. Ryan SM, Choobineh FF (2003) Total WIP and WIP mix for a CONWIP controlled job shop. IIE Trans Inst Ind Eng 35(5):405–418
15. Severino MR, Godinho Filho M (2019) POLCA system for supply chain management: simulation in the automotive industry. J Intell Manuf 30:1271–1289
16. Stevenson M, Hendry LC, Kingsman BG (2005) A review of production planning and control: The applicability of key concepts to the make-to-order industry. Int J Prod Res 43(1):869–898
17. Suri (2003) QRM and POLCA: a winning combination for manufacturing enterprises in the 21st century 21st century markets are here. In: Technical report, center for quick response manufacturing, May
18. Suri R (1998) Quick response manufacturing: a companywide approach to reducing lead times. Productivity Press, Portland, OR
19. Suri R (2010) It's about time—the competitive advantage of quick response manufacturing. In: CRC Press
20. Suri R, Krishnamurthy A (2003) How to plan and implement POLCA–A material control system for high variety or custom-engineered products. Center for quick response management technical report

An Agent-Based Simulation Model for Deadlock Prevention in an Aisle-to-Aisle SBS/RS

Ecem Eroglu and Banu Yetkin Ekren

Abstract Flexibility and high throughput rate in storage and retrieval systems are essential criteria in today's competitive marketing. Recent developments in information technology enable the intelligent design of systems. This study aims to propose a tier-captive aisle-to-aisle shuttle-based storage and retrieval system (SBS/RS) where shuttles can make autonomous decisions to prevent deadlocks and collisions as well as the efficient process of transactions. Deadlock prevention algorithms are one of the primary concerns in today's autonomous vehicle environment. In the considered tier-captive aisle-to-aisle SBS/RS, multiple shuttles can travel between aisles in a dedicated tier. The advantage of this design is that there may be the fewer total number of shuttles running in the system compared to a traditional tier-captive SBS/RS. Due to the complexity of the proposed system and autonomous shuttle-based decision-making target, we utilized the advantage of an agent-based modeling approach by simulating the system. Agent definitions, roles, and behaviors are specified to ensure that no collision and blockage take place in the system. Thanks to the intelligent abilities of agents so that the system can run effectively by using real-time information.

Keywords Agent-based simulation · Deadlock prevention · SBS · RS · Tier-captive · Aisle-to-aisle SBS · RS · Smart decision

Introduction

Warehouses are critical for supply chains. Autonomous vehicle-based warehousing technologies are emerging and providing challenging advantages for the efficient operation of warehouses. Shuttle-based storage and retrieval system (SBS/RS) is

E. Eroglu (✉) · B. Yetkin Ekren
Department of Industrial Engineering, Faculty of Engineering, Yasar University, Bornova, Izmir, Turkey
e-mail: ecem.eroglu@outlook.com

B. Yetkin Ekren
e-mail: banu.ekren@yasar.edu.tr

F. Calisir (ed.), *Industrial Engineering in the Internet-of-Things World*, Lecture Notes in Management and Industrial Engineering,
https://doi.org/10.1007/978-3-030-76724-2_11

one of such technologies mostly utilized for large distribution centers due to its high capacity of transaction rate. It is an alternative design for other traditional automated storage and retrieval systems [7, 9]. An SBS/RS is composed of storage racks, shuttles, and lifts. Shuttles perform horizontal travel for storage and retrieval transactions. There is a single lifting mechanism at each aisle installed at an endpoint of the aisle carrying loads (i.e., totes) between tiers. Figure 1 shows a typical SBS/RS design where there is a dedicated shuttle in each tier (i.e., multiple aisles of a tier).

Note that in a tier-captive SBS/RS there is a tier-captive shuttle in a tier of an aisle [2]. It is known to be a traditional SBS/RS design in the literature. In that design, the average utilization of shuttles is usually very low compared to the average utilization of lifts. Namely, lifts are mostly bottlenecks in those designs [1, 3–5, 8, 11, 12]. In an effort to increase the utilization of shuttles and decrease the initial investment cost of SBS/RS, we propose an alternative design for SBS/RS in which shuttles are tier-captive and aisle-to-aisle. In that design, shuttles can travel between aisles within a tier. The advantage of this design is that it may have a relatively lower total number of shuttles and decreased investment cost compared to a traditional design. However, the disadvantage of this design might be the complexity of operational management due to collision and deadlock possibilities of shuttles while traveling in the same area. Thus, developing efficient operation rules for shuttles resulting in efficiently processing becomes a significant issue in this case. In this paper, we study to develop smart operational rules in order to prevent collisions and deadlocks of shuttles by utilizing agent-based simulation modeling.

The first study of SBS/RS was carried out by Carlo and Vis [2]. They focused on scheduling of two non-passing lifts in traditional SBS/RS. They introduced two functions to evaluate candidate solutions and developed heuristic solutions for the problem. Marchet et al. [15] presented an analytical model by using an open queuing network modeling approach for tier-captive SBS/RS to estimate some performance metrics from the system. To validate the models, they compared the results with their simulation results. Later, Marchet et al. [16] studied a simulation-based work

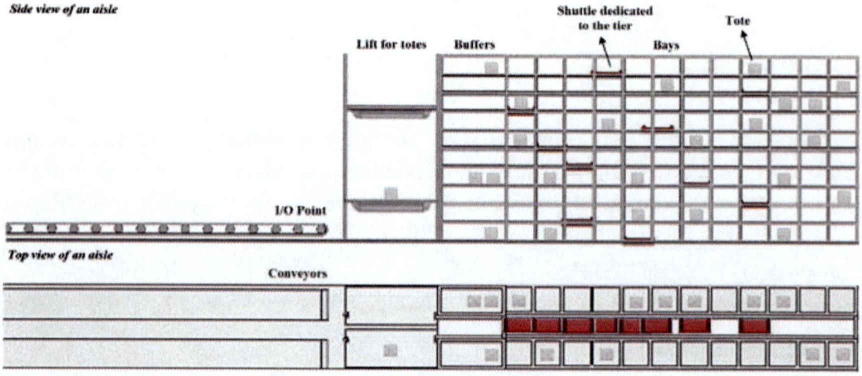

Fig. 1 A typical SBS/RS warehouse with dedicated shuttles

presenting trade-offs in tier-captive SBS/RS designs. Ekren et al. [8] consider a class-based storage policy in the process of SBS/RS, resulting in decreased cycle time.

Lerher [10] developed analytical travel time models for SBS/RS with aisle changing shuttle carriers as one of the most related work to this proposed study. However, there is a single dedicated shuttle for a tier of multiple aisles to prevent any collision or deadlock in the system. Differently, we consider multiple shuttles that can travel between tiers by a separate lifting mechanism in the system and propose collision and deadlock prevention algorithms in the model.

Ekren [3] studied a simulation-based approach for the design of a traditional tier-captive SBS/RS. In order to evaluate the performance metrics based on designs promptly, she draws several graphs under various design concepts. Ekren et al. [6] proposed an open queuing network-based model that can estimate the mean and variance of travel time of lifts and shuttles per transaction in a tier-captive SBS/RS. This tool can also estimate the energy-related performance metrics based on several design parameters. Recently, Ekren [4, 5] has studied an experimental design and multi-objective optimization procedure for the design of tier-captive SBS/RS by considering the optimization of average cycle time per transaction and average energy consumption per transaction performance metrics simultaneously.

The agent-based simulation is an effective tool to evaluate the behavior of complex systems, as we showed in the proposed paper. An agent can be described as anything that can be regarded as perceiving its environment through sensors and taking action upon that environment through effectors [18]. The decision processes of agents can be described by the developers clearly at the micro-level in an agent-based simulation model. The macro-level structure of the whole system emerges as a result of the actions of the agents and the interaction between agents and the environment [19].

As mentioned previously, deadlock prevention is one of the primary concerns in this paper. Deadlock prevention includes defining some rules beforehand to prevent deadlocks. Deadlock avoidance investigates the system state and bypasses deadlocks in real-time. Deadlock prediction is a previous step of deadlock avoidance to learn the location of deadlocks in advance [20]. Lienert and Fottner [13] presented a model applying the time window routing method to move shuttles safely. They focused on tier-to-tier and aisle-to-aisle system configurations. Roy et al. [17] developed protocols for three types of vehicle blocking. Their numerical studies indicated that delays caused by blocking increases transaction cycle time significantly (10–20%).

By the advancement of information technologies enabled the smart design of systems, agent-based simulation modeling is found to be an appropriate approach to analyze such complex systems correctly. We use the ARENA 16.0 commercial software for this purpose. Since the modeling approach focuses on real-time system control and requires real-time information and communication, we utilize a higher level of modeling approach that is agent-based modeling. After developing the proper agent behaviors and decision rules, to test their robustness, we try the models under different scenarios in terms of the number of shuttles in the system.

Methodology

The System Description

The physical warehouse design of the studied tier-captive aisle-to-aisle SBS/RS differs from the traditional SBS/RS warehouse design. Figure 2 shows the physical configuration of the studied system.

Unlike a classical design, there are transition points considered for the travel of shuttles between aisles. We develop a generic simulation model such that the physical design (e.g., the number of aisles and transition points) can be changed. Except for the transition points, to prevent the deadlocks, we consider escape points attached to the waiting points. Note that after a shuttle completes a process and it becomes idle, it travels to the closest decision point not to cause a deadlock. When a busy shuttle tends to pass through a decision point, and there is an idle shuttle waiting in that point, then that idle shuttle moves to the closest waiting point and then to the escape point attached to that waiting point. The intersection points of aisles are the points where shuttles make decisions for where to travel. Namely, a shuttle first stops at the intersection point and then navigates to the target address or a new decision point. In

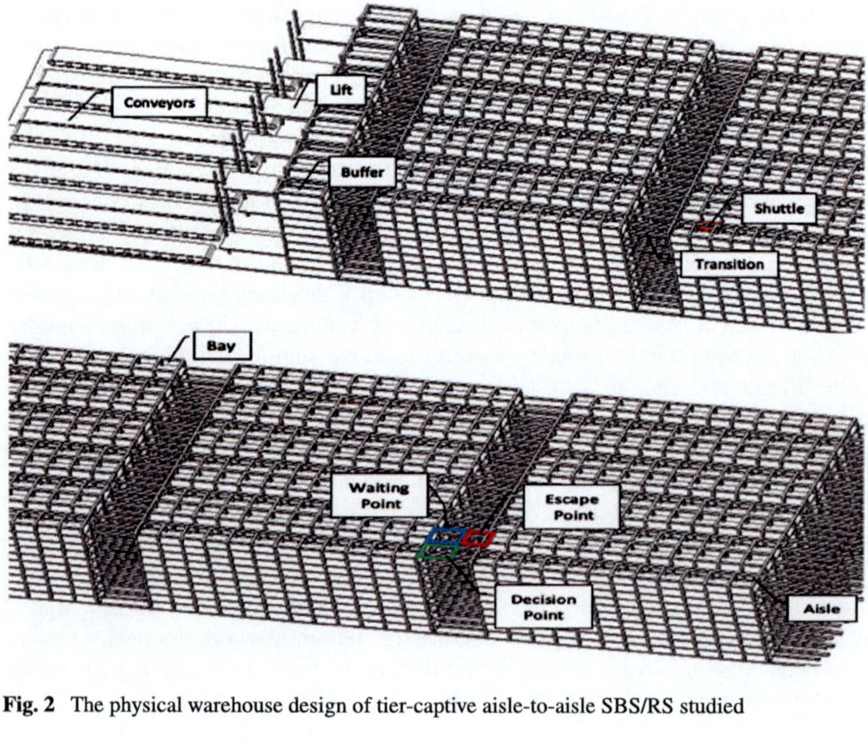

Fig. 2 The physical warehouse design of tier-captive aisle-to-aisle SBS/RS studied

this paper, three different number of shuttle scenarios in a tier are tested. These are one, two, and three number of shuttles.

To detail, there are two types of transactions arriving at the system, storage and retrieval. For storage transactions, the shuttle picks up the load from a buffer location to transfer it to its storage address (i.e., bay). For retrieval transactions, the shuttle carries the load from a bay address to a buffer location.

Agent Definitions and Roles in the Simulation Model

A typical agent-based model has three elements. First is the set of agents, their attributes, and their behaviors. Second is the set of agent relationships and methods of interactions. The third is the environment of the agents. Agents interact with their environment as well as other agents [14]. In the proposed model, three types of agents are defined:

1. Demand agent,
2. Shuttle agent, and
3. Deadlock control agent.

Each agent is modeled such that it can make an independent decision. The agent interactions, i.e., communication of agents and the environment, are shown in Fig. 3.

All the agents interact with the environment. Shuttle agents making decisions as a result of communication are in bidirectional communication with the other agents. Real-time information on system status is provided by all agents, and all can evaluate those pieces of information. The usage of this communication in decision-making is called a bidding strategy that is essential in agent-based simulation. A description of

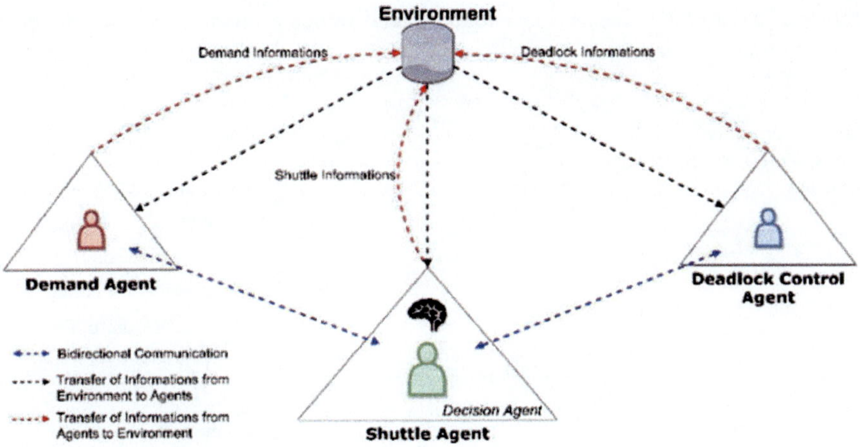

Fig. 3 Agent interactions

the rules for these behaviors is provided below. Agents in the system act under those predefined rules.

Agent Behaviors

Decisions about the system can be divided into two as design and control. Agent behaviors corresponding to the control level decisions are presented in Table 1.

Demand Agent

It tracks the arriving transaction type information as storage and retrieval requests in the system.

Distance Calculation

All, either idle or busy shuttles in the system, are candidates for any waiting demand (i.e., transaction) to be allocated. When a shuttle becomes available to process a transaction, the demand agent calculates the total travel distances based on all waiting transactions' addresses and shuttles' points. Namely, it calculates the total travel distance of a transaction when it is paired with all possible shuttle options. If the calculated shuttle is busy at that time, in travel distance calculation, the destination address of that shuttle is considered. Based on the distance results, the assignment of a transaction to the idle shuttle is decided. These decisions are taken by the shuttles. For that, first, the calculated distance pairs are sorted in increasing order, and the regarding the transaction is assigned to that idle shuttle. For instance, if there is a transaction waiting in queue whose total travel distance is the minimum one for that idle shuttle, however when it is paired with a busy shuttle its total travel distance

Table 1 Agent behaviors corresponding to control level decisions

Agent	The behavior of the agent
1. Demand Agent	1.1 Distance Calculation
2. Shuttle Agent	2.1 Demand Assignment
	2.2 Dwell Point Policy
	2.3 Decision Point Policy
	2.4 Deadlock Control Policy
	2.5 Direction Decision
	2.6 Triggering Policy
	2.7 Alternative Way Policy
3. Deadlock Control Agent	3.1 Deadlock Case–Control

is much smaller than that idle one, then this transaction is not assigned to that idle shuttle. The idle shuttle selects the next option for it.

Shuttle Agent

The action (i.e., transaction selection) decisions in the system are taken by the shuttle agents. The rule is explained below.

Demand Assignment
By considering the total travel distance values calculated by the demand agent, the available shuttle selects the shortest possible transaction to process. Namely, it does not always select the shortest travel distanced transaction if this transaction's total travel distance is less when it is paired with a busy shuttle.

Dwell Point Policy
An idle shuttle always waits at a decision point located at the upper level of its current condition. If this point is full, a triggering policy is applied.

Decision Point Policy
Shuttle always travels at a decision point through its direction. While a shuttle is traveling, if that target decision point is occupied by another shuttle then, a triggering policy is applied.

Deadlock Control Policy
A deadlock control policy is activated when a shuttle notices a collision possibility. Accordingly, a triggering policy or direction policy is applied.

Direction Decision

Depending on the direction of the demand, this decision selects a proper decision point to proceed.

Triggering Policy

Trigger to go to Waiting Points:

An active shuttle triggers the waiting shuttle at the decision point that is on its way to let it go to a waiting point. The active shuttle waits until it reaches to the waiting station.

Triggering to Escape Station Policy:

When an idle shuttle is at a waiting point, an active shuttle may trigger it to the escape point if it is on its way.

Alternative Way Policy:

While a shuttle tends to go to a wait point, if the decision point on its way is full then, this shuttle creates a new route towards its target wait point.

Deadlock Control Agent:

This agent exists to control and prevent deadlock situations.

Deadlock Case-Control:

If any deadlock case is shown in Fig. 4 (Case 1, 2, or 3) takes place, then a deadlock prevention policy is applied, also shown in the same figure. For instance, Case 1 is the case where two shuttles are to collide through their route. For the solution of this, 2.6.1 policy "Trigger to go to Waiting Points" is applied. For Case 2 problem, where a decision-making point is full while another shuttle heads to there, policy 2.6.2 "Triggering Policy" is applied, so on.

Simulation Assumptions

The system is simulated by using the Arena 16.0 commercial software. The simulation model assumptions are summarized as follows:

- The mean arrival rate for storage and retrieval transactions follow a Poisson distribution with equal mean.
- Mean arrival rate values are adjusted such that we obtain 95% average shuttle utilization in the system design (see Table 2).
- Arriving storage or retrieval addresses are specified randomly.
- The required time to load and unload the totes onto/from the shuttle is ignored.
- The maximum velocity that shuttles can reach is assumed to be 2 m/s. The acceleration and deceleration values for velocity are 2 m/s^2.
- The distance between all bays and points (i.e., buffers, decision, waiting, escape points) is assumed to be 0.5 m.
- It is considered that there are 10 aisles and 50 bays with a double side. Therefore, the warehouse capacity is 1,000 bays for each tier.
- The simulation run length is two months with a one-day warm-up period that is decided by the eye-ball technique.
- The model is run for five independent replications.
- The system performance metrics are considered to be the average flow time per transaction, the ratio of waiting time to flow time, and the number of transactions processed during the simulation run.
- Shuttles do not breakdown during the simulation.
- Verification and validation are done by debugging and animating the models.

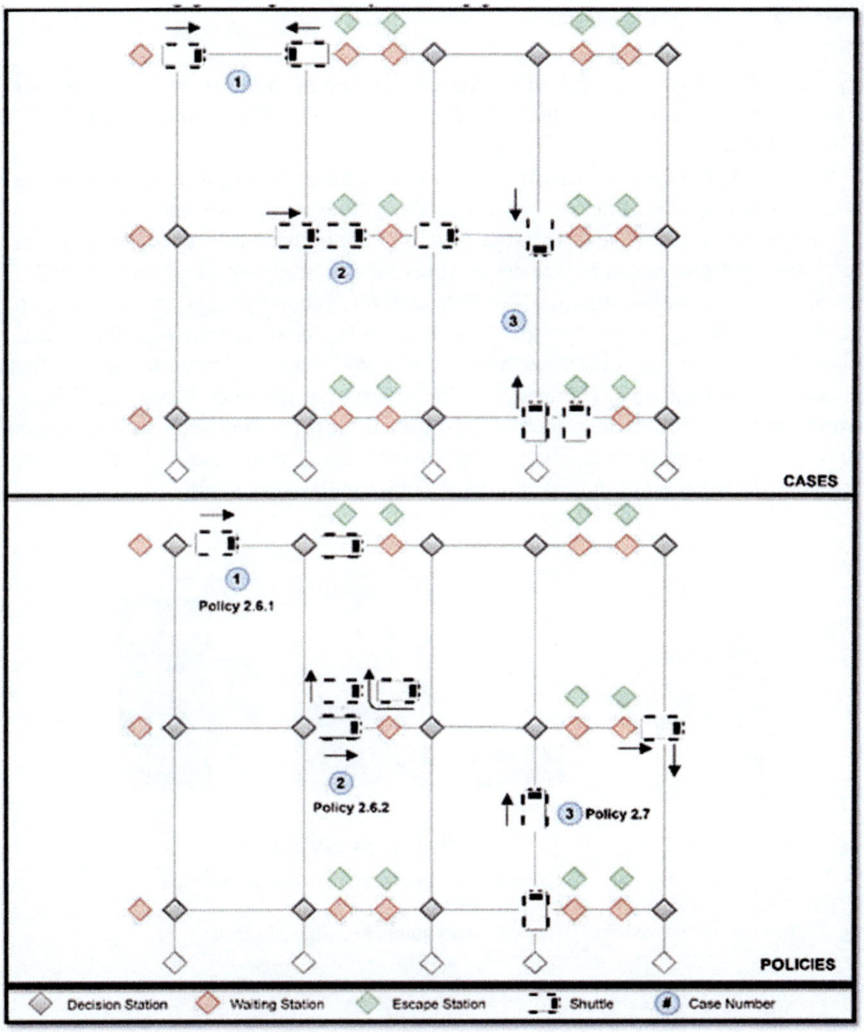

Fig. 4 Deadlock cases and applied policies to solve them

Table 2 Simulation results for 95% shuttle utilization

Number of shuttle	Average flow time per transaction (s)	The ratio of waiting time within flow time	Number of transaction processed (for 2 months)
1	78.94 ± 1.22	80.2%	314,890 ± 748
2	72.24 ± 0.96	65.0%	388,910 ± 983
3	58.16 ± 0.55	54.0%	547,040 ± 760

Results

The performance metrics for the different number of shuttles with 95% average shuttle utilization values are shown in Table 2. The results are summarized for 95% confidence intervals.

In Table 2, the system performance is evaluated in terms of three performance metrics: average flow time, the ratio of waiting time within average flow time, and the total number of transactions processed in two months. Table 2 scenarios are also tested for different transaction selection rules such as first-come-first-served (FCFS), and shortest processing time (SPT). Note that, the Table 2 results are for the agent-based (i.e., bidding-based) decision-making results. By also experimenting with the FCFS and SPT, our aim is to test how the bidding-based assignment policy affects the system performance. Figures 5 and 6 summarize the overall results. Figure 5 shows the results for the total number of transactions processed versus the number of shuttles for all experiments. Figure 6 shows the ratio of waiting time within average flow time versus the number of shuttles results for all experiments.

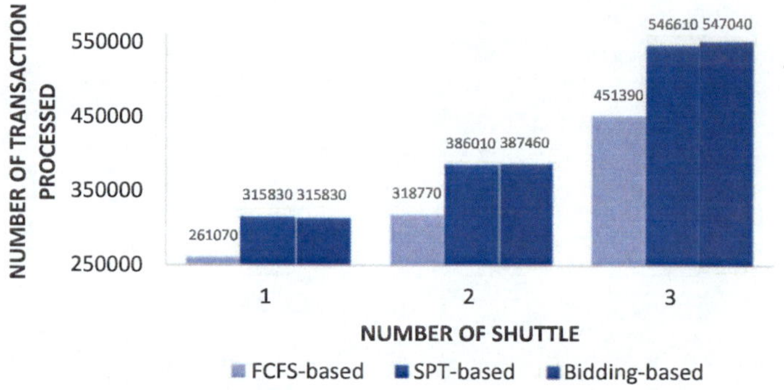

Fig. 5 Number of transactions processed versus number of shuttles results

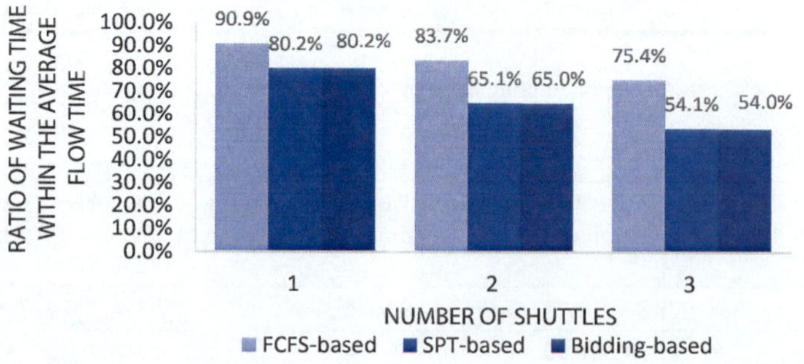

Fig. 6 Ratio of waiting time within the average flow time versus number of shuttles results

The results show that the worst performance metrics are always obtained at the single shuttle case as well as in the FCFS scheduling scenarios. From Figs. 5 and 6 it is also observed that both bidding and SPT scheduling algorithms produce close results. Bidding is relatively a little bit better than the SPT rule in each number of shuttle scenario. However, by improving the bidding rules as a future work could improve the system performance.

Discussion and Conclusion

In this paper, in an effort to reduce initial investment cost and increase the average utilization of shuttles in SBS/RS, we propose a novel tier-captive aisle-to-aisle SBS/RS design in which multiple shuttles can run within multiple aisles and a dedicated tier. The proposed system is designed for the use of industrial warehouses requiring an increased throughput rate with decreased investment cost compared to traditional SBS/RS. Since this system considers travel of shuttles between aisles within a single tier, the management of collision and deadlock of shuttles may become a significant issue. In order to prevent collisions and deadlock of shuttles, we study agent-based modeling to find out a good control policy. We define the agent's behaviors and rules and try them for three different number of shuttle scenarios by simulating the system. To be able to compare the effectiveness of the proposed agent-based working system, we also compare its results with two static alternative transaction selection procedures: FCFS and SPT.

The results are evaluated in terms of the average flow time of a transaction, the ratio of waiting time within average flow time, and the total number of transactions processed in two months. It is observed that the proposed bidding procedure works better; however, it could be improved more.

As a future work, it would worth studying more intelligent agent-based control policies for the proposed system. Also, it might be beneficial to compare the proposed tier-captive aisle-to-aisle design with alternative designs of SBS/RS (e.g., traditional tier-captive designs, solely tier-to-tier designs, etc.).

Acknowledgements This work was supported by The Scientific and Technological Research Council of Turkey and Slovenian Research Agency: ARRS [grant number: 118M180].

References

1. Borovinšek M, Ekren BY, Burinskienė A, Lerher T (2017) Multi-objective optimisation model of shuttle-based storage and retrieval system. Transport 32(2):120–137
2. Carlo HJ, Vis IF (2012) Sequencing dynamic storage systems with multiple lifts and shuttles. Int J Prod Econ 140:844–853
3. Ekren BY (2017) Graph-based solution for performance evaluation of shuttle-based storage and retrieval system. Int J Prod Res 55:6516–6526

4. Ekren BY (2020a) A Simulation-based experimental design for SBS/RS warehouse design by considering energy related performance metrics. Simulation Modeling Pract Theor (in press). https://doi.org/10.1016/j.simpat.2019.101991
5. Ekren BY (2020b) A multi-objective optimisation study for the design of an AVS/RS warehouse. Int J Prod Res (in press). https://doi.org/10.1080/00207543.2020.1720927
6. Ekren BY, Akpunar A, Sari Z, Lerher T (2018) A tool for time, variance and energy related performance estimations in a shuttle-based storage and retrieval system. Appl Math Modeling 63:109–127
7. Ekren BY, Heragu SS (2012) Performance comparison of two material handling systems: AVS/RS and CBAS/RS. Int J Prod Res 50(15):4061–4074
8. Ekren BY, Sari Z, Lerher T (2015) Warehouse design under class-based storage policy of shuttle-based storage and retrieval system. IFAC-PapersOnLine 48(3):1152–1154
9. Ekren BY, Heragu SS (2011) Simulation-based performance analysis of autonomous vehicle storage and retrieval system. Simulation Modeling Pract Theor 19:1640–1650
10. Lerher T (2018) Aisle changing shuttle carriers in autonomous vehicle storage and retrieval systems. Int J Prod Res 56:3859–3879
11. Lerher T, Ekren BY, Sari Z, Rosi B (2015) Simulation analysis of shuttle based storage and retrieval systems. Int J Simulation Modelling 14:48–59
12. Lerher T, Ekren BY, Sari Z, Rosi B (2016) Method for evaluating the throughput performance of shuttle based storage and retrieval systems. Tehnički vjesnik 23(3):715–723
13. Lienert T, Fottner J (2017) No more deadlocks—applying the time window routing method to shuttle systems. ECMS 169–175
14. Macal CM, North MJ (2010) Tutorial on agent-based modeling and simulation. J Simulation 4:151–162
15. Marchet G, Melacini M, Perotti S, Tappia E (2012) Analytical model to estimate performances of autonomous vehicle storage and retrieval systems for product totes. Int J Prod Res 50:7134–7148
16. Marchet G, Melacini M, Perotti S, Tappia E (2013) Development of a framework for the design of autonomous vehicle storage and retrieval systems. Int J Prod Res 51:4365–4387
17. Roy D, Krishnamurthy A, Heragu SS, Malmborg CJ (2013) Blocking effects in warehouse systems with autonomous vehicles. IEEE Trans Autom Sci Eng 11:439–451
18. Russell SJ, Norvig P (2003) Artificial intelligence: a modern approach. Prentice Hall, Hoboken
19. Siebers PO, Aickelin U (2008) Introduction to multi-agent simulation. In: Encyclopedia of decision-making and decision support technologies. IGI Global, pp 554–564
20. Yeh M-S, Yeh W-C (1998) Deadlock prediction and avoidance for zone-control AGVS. Int J Prod Res 36:2879–2889

A Comparison of the Multi-criteria Decision-Making Methods for the Selection of Researchers

Gulsum Kubra Kaya and Fatih Ozturk

Abstract Multi-criteria decision-making methods (MCDM) have been introduced to make effective decisions under conflicting criteria. This study used AHP-based VIKOR, TOPSIS, and MOORA methods to select two researchers among the twenty-six alternative candidates and to compare the findings of the different MCDM methods. The results showed that the AHP-based VIKOR and TOPSIS methods suggested the selection of the same candidates. However, different methods sorted the candidates in a significantly different order. This study reveals that MCDM methods might not always propose the same solution, although they are still useful in effective decision-making and easy to apply.

Keywords Decision-making · VIKOR · TOPSIS · MOORA · Researcher selection

Introduction

Decision-making is a situation that every individual frequently encounters in both daily and business life. A typical decision-making process involves three stages: the definition of the decision-making problem, the development and use of a decision-making model, and the creation of action plans [24]. Although the decision-making process is completed with the creation of action plans, the adverse effects of the inefficient decision-making process are inevitable to continue [8].

In the literature, several methods have been developed by using different algorithms [12, 30, 31]. As part of these methods, Multi-Criteria Decision-Making (MCDM) methods have been developed to make decisions under conflicting criteria

G. K. Kaya (✉) · F. Ozturk
Industrial Engineering Department, Faculty of Engineering and Natural Sciences, Istanbul Medeniyet University, Istanbul, Turkey
e-mail: kubra.kaya@medeniyet.edu.tr

F. Ozturk
e-mail: fatih.ozturk@medeniyet.edu.tr

F. Calisir (ed.), *Industrial Engineering in the Internet-of-Things World*, Lecture Notes in Management and Industrial Engineering,
https://doi.org/10.1007/978-3-030-76724-2_12

[1, 32]. MCDM methods combine many disciplines, predominantly mathematics, and they provide a systematic way of making decisions [14, 18, 24].

Each MCDM method was developed on different algorithms; therefore, they might reach different conclusions. In other words, different methods can suggest the selection of different alternatives. It is, at precisely this point, that the reliability of MCDM methods has been criticized by several researchers [4, 6].

However, despite all criticisms, the MCDM methods have been used in classification, selection, and ranking problems concerning various processes in many different industries. For example, Antmen and Mic [2] used fuzzy TOPSIS and Analytical Hierarchy Process (AHP) methods to select a ventilator in the pediatric intensive care unit. Ozturk and Kaya [23] used fuzzy VIKOR to select personnel in the automotive industry. Bedir and Yalcin et al. [5] used Analytical Network Process (ANP) and PROMETHEE methods to select subcontractors. Soba and Simsek et al. [29] used AHP based VIKOR to select doctoral students. Brauers and Edmundas et al. [8] used the MOORA method to select a contractor. In addition, several other MCDM methods have been used in decision-making. However, among all MCDM methods, TOPSIS, AHP and VIKOR methods were frequently used, and the MOORA method was promoted due to its ease of use and low-time requirement [13, 16, 17, 21, 34].

This study aims to use AHP-based TOPSIS, VIKOR, and MOORA methods for the selection of two researchers to an engineering faculty and to compare the findings of these MCDM methods.

Methodology

Study Design

The design of this study consists of three stages: determining criteria, estimating the criteria weights, and ranking alternatives (Fig. 1).

The determination of the criteria was made by reviewing similar studies in the decision-making literature [15, 17, 19, 22, 26, 28, 29] and conducting meetings with four academic members in the related faculty.

The AHP method was used to calculate the weights for each criterion. Initially, a questionnaire was designed to assess the relative importance of each criterion. In this questionnaire, a scale of 1 to 9 was used to make pairwise comparison [27]. Four faculty members completed the questionnaire, and they reached a consensus on the conflicted responses.

The VIKOR, TOPSIS, and MOORA (MOORA-rate system and MOORA-reference point theory) methods were used to rank the twenty-six candidates based on seven criteria. The application of MCDM methods was carried out using MS Office Excel.

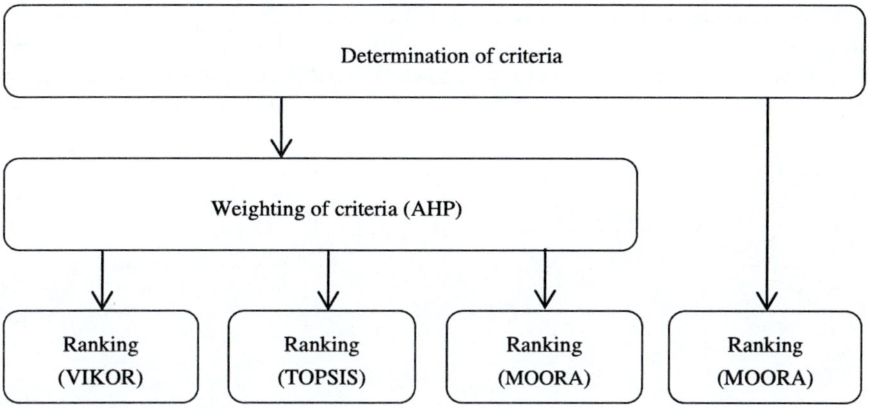

Fig. 1 The research design for the selection of two researchers

The AHP Method

AHP is based on the general measurement theory, and it aims to solve problems identified for a specific purpose. It is possible to describe the AHP method in four stages [27].

Step 1 Creating the hierarchical structure of the decision problem: It starts from the top level. Level 1 represents the goal; level 2 represents the criteria; level 3 shows the sub-criteria, and the lowest level shows the alternatives.

Step 2 Creating the binary comparison matrix: Binary comparison matrices for each level of the hierarchical structure are created by Eq. (1). Here, n criteria $(a_1, a_2, \ldots a_n)$ are compared by using the 1–9 scale of Saaty.

$$\begin{bmatrix} 1 & a_{12} & \ldots & a_{1n} \\ a_{21} & 1 & \ldots & a_{2n} \\ \vdots & \vdots & \ldots & \vdots \\ a_{n1} & a_{n2} & \cdots & 1 \end{bmatrix} \tag{1}$$

Step 3 Determination of criterion weights: The weight values of each criterion are calculated. For this, the matrix is normalized using Eq. (2) and, then, the weights are calculated by Eq. (3).

$$a_{ij}^* = \frac{a_{ij}}{\sum_{i=1}^{n} a_{ij}} \tag{2}$$

$$w_i = \frac{\sum_{j=1}^{n} a_{ij}^*}{n} \tag{3}$$

Step 4 Making consistency calculations: Consistency is calculated to obtain reliable results. The Consistency Rate (CR) is expected to be less than 0.10. For this, λ_{max} is calculated in Eq. (4), the Consistency Index (CI) by Eq. (5), and CR value by Eq. (6). Random Value Index (RI) in Eq. (6) is the value corresponding to n from the RI table.

$$\lambda_{\max} = \frac{\sum_{i=1}^{n}\left(\frac{d_i}{w_i}\right)}{n}, \quad [d_i]_{n\times 1} = \left[a_{ij}\right]_{n\times n} \times [w_i]_{n\times 1} \tag{4}$$

$$CI = \frac{\lambda_{\max} - n}{n - 1} \tag{5}$$

$$CR = \frac{CI}{RI} \tag{6}$$

The VIKOR Method

The VIKOR method is developed to calculate the closeness of the alternatives to the ideal solution, and, thus, it provides a compromise solution to the problem [20]. It is possible to describe the VIKOR method in five stages:

Step 1 The best f_i^* and the worst f_i^- values are determined: The decision matrix is created with the scores of the alternatives for each criterion (i = 1, 2,.. n), and the values of f_i^* and f_i^- are calculated based on the criterion features. Here, Eq. (7a) is for the criterion with the benefit feature and Eq. (7b) for cost.

$$\begin{aligned} f_i^* &= max_j x_{ij} \\ f_i^- &= min_j x_{ij} \end{aligned} \tag{7a}$$

$$\begin{aligned} f_i^* &= min_j x_{ij} \\ f_i^- &= max_j x_{ij} \end{aligned} \tag{7b}$$

Step 2 Calculation of S_j and R_j values: S_j (average group score) the score is calculated in Eq. (8) and R_j (worst group score) score in Eq. (9) for each alternative $(j = 1, 2, ..J)$.

$$S_j = \sum_{i=1}^{n} w_i \frac{f_i^* - x_{ij}}{f_i^* - f_i^-} \tag{8}$$

$$R_j = max_i \left[w_i \frac{f_i^* - x_{ij}}{f_i^* - f_i^-} \right] \tag{9}$$

Step 3 Calculation of Q_j value: For alternatives ($j = 1, 2, \ldots J$), the maximum group benefit (Q_j) is calculated by Eq. (10). The parameters S^*, S^-, R^*, R^- required to calculate Q_j are shown by Eq. (11). The v value in Eq. (11) represents the maximum group utility.

$$Q_j = \frac{v(S_j - S^*)}{(S^- - S^*)} + \frac{(1-v)(R_j - R^*)}{(R^- - R^*)} \tag{10}$$

$$\begin{aligned} S^* &= min_j S_j;\ R^* = min_j R_j \\ S^- &= max_j S_j;\ R^- = max_j R_j \end{aligned} \tag{11}$$

Step 4 Sorting S_j, R_j, Q_j values: These three values obtained by each alternative are sorted from lowest to highest.

Step 5 Checking the conditions: The reliability of the ranking ordering of alternatives is controlled by two conditions: acceptable advantage condition and acceptable stability condition.

Under the condition of acceptable advantage in Eq. (12), A^1 is the first (the lowest value) alternative ($j = 1, 2, \ldots J$) that ranks from lowest to highest and A^2 is the second.

$$Q_{A^2} - Q_{A^1} \geq \frac{1}{j-1} \tag{12}$$

Under the acceptable stability condition, A^1 is ranked the best by S_j and/or R_j.

When the first (acceptable advantage) of these conditions is met, but the second condition (acceptable stability) is not met, A^1 and A^2 are considered together as a compromised solution.

If the first condition is not met: all of the alternatives from A^1, A^2 $A^3 \ldots A^m$ are considered as compromised solutions. The value of m is determined according to Eq. (13).

$$Q_{A^m} - Q_{A^1} < \frac{1}{j-1} \text{ for maximum } m \tag{13}$$

The TOPSIS Method

The TOPSIS method chooses the alternative that is closest to the ideal solution, but the farthest to the negative ideal solution [11]. It is possible to apply the TOPSIS method in six steps.

Step 1 Calculate the normalized decision matrix: The normalization of the decision matrix is calculated by finding r_{ij} (normalized values) as in Eq. (14). Here, the criteria are specified with i ($i = 1, 2, ..n$)and alternatives with j ($j = 1, 2, ..J$).

$$r_{ij} = \frac{x_{ij}}{\sqrt{\sum_{j=1}^{J} x_{ij}^2}} \tag{14}$$

Step 2 Creating the weighted normalized decision matrix: The weights ($w_1, w_2, \ldots w_n$) of each criterion ($i = 1, 2, \ldots n$) are determined by the decision-maker. The weighted normalized value v_{ij} is calculated, as shown by Eq. (15).

$$v_{ij} = w_i r_{ij} \tag{15}$$

Step 3 Determination of the ideal and negative ideal solutions: The ideal solution (A^*) takes the maximum value when associated with benefit criterion (I'), and the minimum value when associated with cost criterion (I'') (Eq. 16a). The negative ideal solution $\left(A^-\right)$ applies the opposite (Eq. 16b).

$$\begin{aligned} A^* &= \left\{v_1^*, v_2^*, \ldots v_n^*\right\} \\ &= \left\{\left(max_j v_{ij} \| \mathrm{i} \in I'\right), \left(min_j v_{ij} \| \mathrm{i} \in I''\right)\right\} \end{aligned} \tag{16a}$$

$$\begin{aligned} A^- &= \left\{v_1^-, v_2^-, \ldots v_n^-\right\} \\ &= \left\{\left(min_j v_{ij} \| \mathrm{i} \in I'\right), \left(max_j v_{ij} \| \mathrm{i} \in I''\right)\right\} \end{aligned} \tag{16b}$$

Step 4 Calculate distance values: The distance from the ideal solution (D_j^*) is calculated by Eq. (17a) by using the Euclidean distance, and the distance from the negative ideal solution (D_j^-) is calculated by Eq. (17b).

$$D_j^* = \sqrt{\sum_{i=1}^{n} \left(v_{ij} - v_i^*\right)^2}, \; j = 1, 2, \ldots, J \tag{17a}$$

$$D_j^- = \sqrt{\sum_{i=1}^{n} \left(v_{ij} - v_i^-\right)^2}, \; j = 1, 2, \ldots, J \tag{17b}$$

Step 5 Calculation of the relative proximity to the ideal solution: Relative proximity (C_j^*) of alternative a_j to A^* is calculated by Eq. (18).

$$C_j^* = \frac{D_j^-}{\left(D_j^* + D_j^-\right)}, \; j = 1, 2, \ldots, J \tag{18}$$

Step 6 Rank the preference order: The ranking is made from the alternative having the largest C_j^* values (the best alternative) to the lowest.

The MOORA Method

The MOORA method is for multi-objective optimization with discrete alternatives. It has two approaches: the ratio system and reference point theory [7, 8].

The ratio system is carried out in two steps. In the first step, the normalization process is applied by Eq. (19). Here, i represents the objective ($i = 1, 2, \ldots, n$) and j alternative (j = 1, 2, ..., J).

$$x_{ij}^{*} = \frac{x_{ij}}{\sqrt{\sum_{j=1}^{J} x_{ij}^{2}}} \tag{19}$$

In the second step of the ratio system, the evaluation of the degree of the alternative (j) meeting the objective (i) is found by the optimization of the normalized values. Equation (20) is applied depending on the objectives of the criteria (i.e., maximization or minimization).

$$y_{j}^{*} = \sum_{i=1}^{i=g} x_{ij}^{*} - \sum_{i=g+1}^{i=n} x_{ij}^{*} \tag{20}$$

In Eq. (20), $i = 1, 2, \ldots, g$ represents objectives (of the criteria) to be maximized and $i = g + 1, g + 2, \ldots, n$ objectives to be minimized. The ranking order of each alternative is obtained by sorting the values of y_j^* from highest to lowest. The largest y_j^* values will be the best alternative.

Reference point theory measures the distances between alternatives (x_{ij}^*) and a reference point (r_i) by Eq. (21).

$$min_j max_i \left(\left|r_i - x_{ij}^{*}\right|\right) \tag{21}$$

In cases where criteria weights (s_i) are known, x_{ij}^* values in Eq. (20) and r_i and x_{ij}^* values in Eq. (21) are multiplied by the coefficient s_i, as in Eqs. (22) and (23).

$$y_{j}^{*} = \sum_{i=1}^{i=g} s_i x_{ij}^{*} - \sum_{i=g+1}^{i=n} s_i x_{ij}^{*} \tag{22}$$

$$min_j max_i \left(\left|s_i r_i - s_i x_{ij}^{*}\right|\right) \tag{23}$$

Results

In this study, seven criteria were identified, as shown in Table 1. These criteria were used to evaluate twenty-six candidates for the selection of two researchers.

Table 2 represents the evaluation findings of the candidates based on seven criteria. The relevant data for criteria C1 to C4 were directly obtained from the candidates; C5 was from the faculty; C6 was from a university ranking list; C7 were from faculty members.

In this study, AHP was used to calculate the weight of each criterion. Table 3 shows the criterion comparison matrix. Four faculty members participated in a questionnaire to generate the comparison matrix.

The comparison matrix has been normalized by the operations in Eq. (2). Then, the criterion significance weight in Table 4 has been calculated by using Eq. (3). Following that, the consistency ratio was calculated, as CR = 0.03. By reason of $CR < 0.1$, the values obtained are considered to be consistent.

After the calculation of the weights of the criteria, VIKOR was applied to calculate Q_j values; TOPSIS was used to calculate C_j^* values; the MOORA-rate system method was used to calculate y_j^* values; the MOORA-reference point theory method was used to calculate $\max_i\left(\left|r_i - x_{ij}^*\right|\right)$ and $\max_i\left(\left|s_i r_i - s_i x_{ij}^*\right|\right)$ values. Table 5 shows the ranking order of the twenty-six alternatives by using each method.

Table 1 Candidate selection criteria and their explanations

Criterion (C)	Explanation
ALES exam score (C1)	A general exam, including linguistic and mathematics tests. Maximization of this criterion is aimed.
Foreign language score (C2)	Candidates take YDS, YOKDİL, or similar language exams. Maximization of the exam score is aimed.
GPA average (C3)	Undergraduate grade average, in a 4-point system. Maximization of this criterion is aimed.
Work experience (C4)	It shows the years of work experience of candidates. Maximization of this criterion is aimed.
Written Exam (C5)	The candidates take a written exam prepared by the faculty. Maximization of this criterion is aimed.
University ranking (C6)	It represents the candidate's undergraduate or graduate degree university ranking order. Minimisation of this criterion is aimed.
Job fit (C7)	It represents the closeness of the candidate's fit to the applied program by assigning a score between 1 (the worst) and 5 (the best). Maximization of this criterion is aimed.

Table 2 Evaluation of candidates in terms of criteria

Alternative	C1	C2	C3	C4	C5	C6	C7
A1	90	70	3.2	6	75	12	4
A2	85	60	3.5	3,5	80	45	4
A3	96	80	2.7	5	85	34	1
A4	75	67	2.8	6	75	56	2
A5	76	70	2.9	7	70	89	3
A6	68	65	3	1	85	34	4
A7	89	75	3.1	2	70	5	5
A8	67	55	3.2	2	65	17	4
A9	75	60	2.88	3	60	34	3
A10	72	55	2.9	3	75	54	3
A11	70	57	3.05	4	70	23	3
A12	65	62	3.1	8	55	78	4
A13	65	75	3.15	10	50	32	4
A14	70	70	3.2	5	45	41	4
A15	75	65	3.4	4	56	53	4
A16	80	60	3.6	3.5	70	23	5
A17	75	60	2.7	12	65	61	3
A18	94	72	2.6	10	50	19	1
A19	78	70	2.5	8	45	10	1
A20	67	68	2.4	15	65	5	2
A21	85	65	2.8	5	55	12	2
A22	70	64	2.9	2	50	15	1
A23	75	62	3	1	60	16	5
A24	80	60	2.65	1	65	5	3
A25	80	58	2.75	1.5	70	45	2
A26	75	60	3.25	2	65	33	2

Table 3 Criterion comparison matrix

Criterion	C1	C2	C3	C4	C5	C6	C7
C1	1	0.5	3	2	0.25	0.5	0.2
C2	2	1	3	2	0.5	2	0.3
C3	0.33	0.33	1	0.5	0.25	0.33	0.2
C4	0.5	0.5	2	1	0.33	0.5	0.2
C5	4	2	4	3	1	2	0.5
C6	2	0.5	3	2	0.5	1	0.5
C7	5	3	5	4	2	2	1

Table 4 The weight of importance of each criterion

Criterion (i)	C1	C2	C3	C4	C5	C6	C7
Weight (w_i)	0.09	0.14	0.04	0.07	0.22	0.12	0.32

Table 5 The ranking order of alternatives with the use of VIKOR, TOPSIS, and MOORA methods

Alternatives	VIKOR Qj, v = 0.5	TOPSIS Cj	MOORA-rate	MOORA-reference	MOORA-weighted reference
A1	1	1	1	17	1
A2	3	5	4	15	1
A3	4	7	6	10	22
A4	12	15	16	8	14
A5	11	19	20	7	1
A6	17	6	7	24	1
A7	2	2	13	19	1
A8	21	11	12	19	1
A9	15	18	19	17	12
A10	18	14	17	17	1
A11	19	10	11	13	1
A12	26	25	24	4	19
A13	24	13	2	2	21
A14	23	24	23	10	25
A15	14	20	18	13	18
A16	5	3	3	15	1
A17	13	9	9	1	1
A18	10	16	14	5	22
A19	22	22	25	5	25
A20	20	4	5	3	14
A21	8	17	15	10	19
A22	25	26	26	19	22
A23	9	8	8	24	12
A24	6	12	10	24	1
A25	7	21	21	23	14
A26	16	23	22	19	14

Discussion and Conclusion

This study was conducted to select two researchers to an engineering faculty and to compare the findings of the different methods. The findings revealed that the VIKOR

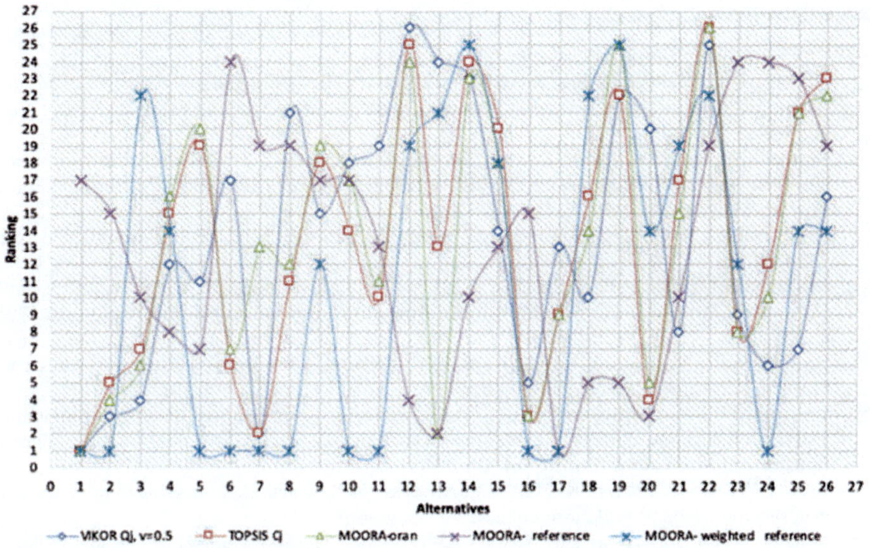

Fig. 2 Comparative ranking results of VIKOR, TOPSIS and MOORA methods

and TOPSIS methods suggest the selection of the same candidates. However, the ranking order of the rest of the candidates was considerably different. Moreover, the MOORA-reference method suggests the selection of entirely different candidates. Figure 2 illustrates the comparative ranking orders for twenty-six alternatives.

In this respect, although MCDM methods are useful to assist in decision-making, they might not always be reliable. It should be taken into account that MCDM methods might not always give the best results. Several researchers have criticized the use of MCDM methods from this perspective [4, 9, 32]. Here, two points should be mentioned. Firstly, MCDM methods have been built on linear mathematical algorithms. Decision-making might depend on the conditions, and thus, the use of linear methods might not fit well in real-life decisions. Individuals might use a non-linear algorithm when making decisions. Indeed, in real life, the decision-making process is rather complex. Secondly, the selection of the criteria and the evaluation of the candidates were based on expert judgments, which can be subjective. Researchers suggested using fuzzy logic to reduce the subjectivity and deal with uncertainty [3, 15, 23]. However, the use of fuzzy logic would still not solve the problem with the dynamic and non-linear features of the decision-making process. At this point, non-linear decision-making methods might provide more reliable results [10, 25, 33].

However, despite their limitations, MCDM methods have still been used to support decision-making due to their simplicity. Future studies might investigate the ideal selection of MCDM methods for specific conditions.

References

1. Altunok T, Özpeynirci Ö, Kazançoğlu Y, Yilmaz R (2010) Comparative analysis of multi-criteria decision making methods for postgraduate student selection. Eurasian J Educ Res 40:1–15
2. Antmen ZF, Miç P (2018) Çocuk yoğun bakım ünitesinde Çok Kriterli Karar Verme ile mekanik ventilatör seçimi ve bir uygulama örneği. Çukurova Üniversitesi Mühendislik ve Mimar. Fakültesi Derg 33:17–30
3. Asadabadi MR (2018) The stratified multi-criteria decision-making method. Knowl-Based Syst 162:115–123
4. Asadabadi MR, Chang E, Saberi M (2019) Are MCDM methods useful? A critical review of Analytic Hierarchy Process (AHP) and Analytic Network Process (ANP). Cogent Eng 6(1):1–11
5. Bedir N, Yalçın H, Özder EH, Eren T (2018) Çok Kriterli Karar Verme yöntemleriyle taşeron firma seçimi: Kırıkkale ilinde bir uygulama. Acad Platf J Eng Sci 6:25–33
6. Bernroider EWN, Schmöllerl P (2013) A Technological, organisational, and environmental analysis of decision making methodologies and satisfaction in the context of IT induced business transformations. Eur J Oper Res 224(1):141–153
7. Brauers WKM, Zavadskas EK (2006) The MOORA method and its application to privatization in a transition economy. Control Cybern 35(2):445–469
8. Brauers WKM, Zavadskas EK, Turskis Z, Vilutiene T (2008) Multi-objective contractor's ranking by applying the Moora method. J Bus Econ Manag 9(4):245–255
9. Chitsaz N, Banihabib ME (2015) Comparison of different multi criteria decision-making models in prioritizing flood management alternatives. Water Resour Manage 29(8):2503–2525
10. Ganzach Y (1998) Nonlinear models in decision making: the diagnosis of psychosis versus neurosis from the MMPI. Organ Behav Hum Decis Process 74(1):53–61
11. Hwang C-L, Yoon K (1981) Multiple attribute decision making: methods and applications. Lecture Notes in Economics and Mathematical Systems. Springer-Verlag
12. Ivlev I, Vacek J, Kneppo P (2015) Multi-criteria decision analysis for supporting the selection of medical devices under uncertainty. Eur J Oper Res 247:216–228
13. Kara İ, Ecer F (2016) AHP- VIKOR entegre yöntemi ile tedarikçi seçimi: tekstil sektörü uygulaması. DEU J GSSS 18(2):255–272
14. Kenger MD, Organ A (2017) Banka personel seçiminin çok kriterli karar verme yöntemlerinden Entropi temelli ARAS yöntemi ile değerlendirilmesi. Adnan Menderes Üniversitesi Sosyal Bilimler Enstitüsü Dergisi 4(4):152–170
15. Korkmaz O (2019) Personnel selection method based on topsis multi-criteria decision making method. Uluslararası İktisadi ve İdari İncelemeler Dergisi 23:1–16
16. Korkusuz AY, İnan UH, Özdemir Y, Başlıgil H (2020) Occupational health and safety performance measurement in healthcare sector using integrated multi criteria decision making methods. J Faculty Eng Archit Gazi Univ 35:81–96
17. Kumar D, Sameer SR, Suman KNS (2013) MADM methods for finding the right personnel in academic institutions. Int J U- and e-Service Sci Technol 6(5):133–144
18. Kuşakçı AO, Ayvaz B, Öztürk F, Sofu F (2019) Bulanik Multimoora ile personel seçimi: havacılık sektöründe bir uygulama. Ömer Halisdemir Üniversitesi Mühendislik Bilimleri Dergisi 8(1):96–110
19. Md Saad R, Ahmad MZ, Abu MS, Jusoh MS (2014) Hamming distance method with subjective and objective weights for personnel selection. Sci World J 1–9
20. Opricovic S, Tzeng GH (2004) Compromise solution by MCDM methods: a comparative analysis of VIKOR and TOPSIS. Eur J Oper Res 156:445–455
21. Özbek A, Erol E (2016) Analitik Hiyerarşi Süreci ve VIKOR yöntemleriyle işgören seçimi: tekstil sektöründe bir uygulama. Kırıkkale Üniversitesi Sosyal Bilimler Dergisi 6(1):93–108
22. Ozder EH, Bedir N, Eren T (2016) Academic staff selection with ANP and PROMETHEE method: a case study in Turkey. In: Proceedings of Academics World International Conference, 1–5

23. Ozturk F, Kaya GK (2020) Personnel selection with fuzzy VIKOR: an application in automotive supply industry. Gazi Univ Sci J Part C Design Technol 8(1):94–108
24. Pavan M, Todeschini R (2009) Multicriteria decision-making methods. In: Brown SD, Tauler R, Walczak B (eds) Comprehensive chemometrics: chemical and biochemical data analysis. Elsevier, Oxford, pp 591–625
25. Piegat A, Sałabun W (2014) Identification of a multicriteria decision-making model using the characteristic objects method. Appl Comput Intelligence Soft Comput 2014:1–14
26. Rouyendegh B, Erkan T (2013) An application of the Fuzzy ELECTRE method for academic staff selection. Hum Factors Ergon Manuf 23:107–115
27. Saaty RW (1987) The analytic hierarchy process-what it is and how it is used. Math Modelling 9(3–5):161–176
28. Saghafian S, Hejazi SR (2005) Multi-criteria group decision making using a modified fuzzy TOPSIS procedure. In :International Conference on Computational Intelligence for Modelling, Control and Automation and International Conference on Intelligent Agents, Web Technologies and Internet (IEEE)
29. Soba M, Şimşek A, Erdin E, Can A (2016) AHP temelli VIKOR yöntemi ile doktora öğrenci seçimi. Dumlupınar Üniversitesi Sosyal Bilimler Dergisi 50:109–132
30. Stanujkic D, Djordjevic B, Karabasevic D (2015) Selection of candidates in the process of recruitment and selection of personnel based on the SWARA and ARAS Methods. QUAESTUS Multidisc Res J 7:53–64
31. Supçiller AA, Deligöz K (2018) Tedarikçi Seçimi Probleminin Çok Kriterli Karar Verme Yöntemleriyle Uzlaşık Çözümü. Uluslararası İktisadi ve İdari İncelemeler Dergisi 18:355–368
32. Triantaphyllou E (2000) Multi-criteria decision making methods: a comparative study. Springer Science
33. Wu B, Cheng T, Yip TL, Wang Y (2020) Fuzzy logic based dynamic decision-making system for intelligent navigation strategy within inland traffic separation schemes. Ocean Eng 197:106909
34. Yıldırım B, Onder B (2018) Operasyonel, yönetsel ve stratejik problemlerin çözümünde çok kriterli karar verme yöntemleri. Bursa: Dora Yayıncılık

An Overview of Warehouse Operations for Cold Chain

Cansu Yurtseven, Banu Yetkin Ekren, and Ayhan Ozgur Toy

Abstract Warehouses play a significant role in cold chains as they do for regular supply chains. Although their goals are the same for both cold chains and regular supply chains, the operations of cold warehouses are more sophisticated since the cost of operation is considerably higher due to energy consumption and obsolesce of products in substandard conditions. Recently, there has been an enormous interest in the cold food supply chain to reduce food waste occurring along the chain. Hence, efficient management of cold warehouses becomes an important issue in this direction. Design and operation requirements in a cold warehouse may be different from a traditional non-cold warehouse. In this paper, we aim to provide an overview of cold chain operations, mostly by focusing on cold warehouse operations. We provide some statistics from a cold chain, design, and technology requirements for cold warehouses as well as warehouse operations shaped according to that warehouse features. It is observed that there are quite different design parameters in cold storage.

Keywords Warehouse operations · Cold supply chain · Cold warehouse · Technologies for cold chain

Introduction

Warehouses play a critical role in supply chains. The main purposes of warehouses are: (i) to store items in order to reduce demand variability for the upper echelon, and (ii) to decrease transportation lead times for customers. Storing and retrieving

C. Yurtseven · B. Y. Ekren (✉) · A. O. Toy
Department of Industrial Engineering, Faculty of Engineering, Yasar University, Bornova, Izmir, Turkey
e-mail: banu.ekren@yasar.edu.tr

C. Yurtseven
e-mail: cansu.yurtseven@yasar.edu.tr

A. O. Toy
e-mail: ozgur.toy@yasar.edu.tr

F. Calisir (ed.), *Industrial Engineering in the Internet-of-Things World*,
Lecture Notes in Management and Industrial Engineering,
https://doi.org/10.1007/978-3-030-76724-2_13

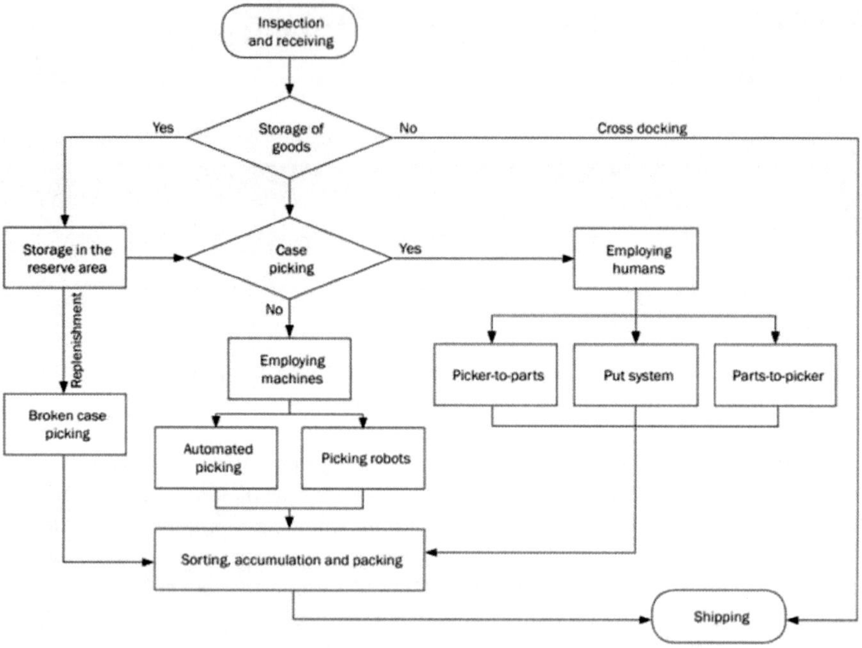

Fig. 1 Standard warehouse processes [15]

products promptly without any errors is desired to avoid any mistakes when satisfying customer demands. The main operations in warehouses such as shipment planning, order picking, assigning orders to vehicles with limited capacity, and routing of these vehicles are complex operations and have to be performed effectively. In the planning process, all of these operations should be considered simultaneously and optimized. Figure 1 shows a typical flow of operations/processes in a warehouse.

In warehouses, typically first, products arrive at the receiving docks, and they are unloaded from the trucks in that area. Second, products are sent to the storage area for their temporary storage. However, sometimes storage of products does not take place; in this case, products are sent directly from the receiving docks to the shipping docks that are called cross-docking. Those products stored in the reserve area wait until an order is received from a customer for their dispatch. The stocking policy of products may vary according to product type, size, and special requirements of those products. When a customer order arrives, interior vehicles or automated machines pick those products from their storage addresses and then send them to the shipment area where they will be loaded in the transporters.

Warehouses are designed based on the storage requirements of the products they will be used for. Storage requirements may change for raw materials, semi-products, or final products. Besides, based on the product types to store, the warehouse design requirements might vary. For instance, for frozen foods, pharmaceutical products, cold products, etc. refrigerated warehouses are required to keep the products in the right condition. It is challenging in cold warehouses to maintain the right cold storage temperature for the product inside while keeping personnel and equipment warm enough to perform at their optimal capacity.

The cold chain industry is growing and improving fast. As in recent trends in order to profile in the cold chain, customers value convenience, wholesome food, and just-in-time delivery. Moving away from highly processed packaged foods with a long shelf life to temperature-sensitive perishable food products requires an adjustment in the food supply chain. According to Lempert [17], the year 2017 goes down in history as the most important ever in grocery. He declares that the food world is changing at an incredible speed, and the industry must evolve to adapt to this change. He explains that for instance, groceries are now "cool," and they become industries attracting talents from the best schools and companies who would not even have thought about a career in grocery or food industry a few years ago [17].

Recently, it has been observed that the cold storage warehouse capacity has increased drastically. For instance, it has been reported that in 2018, India had the world's largest refrigerated warehouse capacity, weighing 150 million cubic meters following the United States with 131 million cubic meters. Refrigerated warehouses worldwide had a total capacity of 616 million cubic meters [23]. In an effort to attract readers on cold warehouses, in this paper, we aim to present a cold warehouse framework study by focusing on what operations, design metrics as well as critical issues to consider for operating them.

As mentioned, in this paper, we aim to shed light on cold chain operations, specifically by focusing on cold warehouses. We present an overview of operations in cold warehouses along with the points that should be focused on and improved in order to decrease the food loss.

Cold Chain

The cold chain includes the production of refrigerated products as well as the storage and distribution operations of those products throughout the chain. Cold chains are used in two main sectors: food and bio-pharmaceutical. Both sectors have considerably high revenues. More than $18 billion of the global third-party logistics revenue was received from the food and grocery market in 2017, whereas the healthcare market produced more than $17 billion in revenue [22].

Food chain is a recent significant consideration in sustainability due to high amount of food loss along with the distribution network. According to the Food and Agriculture Organization of the United Nations, almost 1/3 of the world's food was lost or wasted each year [12]. This means that roughly 1.3 billion tons of food is wasted every year. Chabada et al. [7] claim that two main causes of food waste at the wholesaler are due to substandard quality and short remaining shelf life of the products. Effective cold chain management would yield a decrease in food waste, therefore becoming a field of scientific studies.

Table 1 Products with required temperature and humidity ranges in cold storage

Product	Temperature (°C)	Humidity (%)
Chocolate	15–18	50–60
Flower	8–16	70–75
Vegetables	6–8	80–90
Fruits	4–6	80–90
Mushrooms	0	90–95
Fresh meat	−2–0	–
Frozen fruits and vegetables	−10–0	–
Frozen meat	−20	–
Frozen tuna	−40–60	–

Management of a food product chain is a challenging issue, especially due to the perishability property of food, their requirement of variable shelf life, certain storage temperatures, time for cooling, and specific packaging and logistics requirements. These products are quite vulnerable to storage conditions. For instance, storing under favorable temperature conditions and complying with the suitable temperature conditions in transportation operations are significant steps to avoid any food waste. Hence, refrigerated (cold) warehouses play a critical role in cold supply chains. Cold warehouse operations start with receiving of the refrigerated products and continue with storing them within ideal temperature conditions. Each product type may require a different storage temperature and condition. Table 1 depicts the ideal storage temperatures and humidity ranges for the storage of different product types as an example [19]. Obviously, when the product variety increases, cold warehouse management becomes more complex due to different storage requirements of each product type.

As mentioned, cold chain logistics strategies are commonly used for storing and transporting fresh foods, vegetables, meat, frozen foods, medicines, and drugs, etc. The main issue in cold chain logistics is the management of temperature-controlled operations (monitoring, etc.) along the supply chain. Monitoring requires devices and systems able to track the condition of the cold chain, such as temperature and humidity, throughout the stages. Cold chain products also require specific equipment and/or methods for fabrications. For example, vacuum packing is often used for the packing of meat to extend its shelf life. Unlike any other products, cold chain products are rarely available for immediate consumption. Hence, the analyses and requirements might be non-traditional in cold warehouses. Transportation of cold products requires specific operations as well as special physical environments, such as transport terminals, ports, and airports dedicated to cold chain logistics. For instance, a container in port terminals might be dedicated to refrigerated products. Terminal facilities have on-dock refrigerated warehouses. A range of transport technologies is available, and it has been improving for the transportation of cold chain goods. Other common technologies that are utilized in transportation are reefer vehicles and containers (maritime containers and unit load devices). They usually require

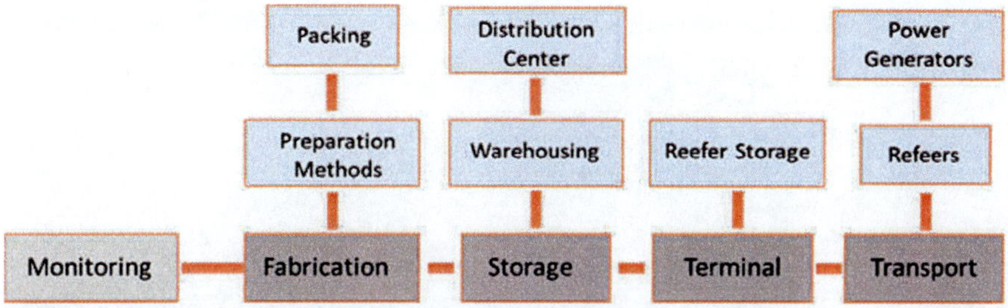

Fig. 2 Cold chain process

refrigeration plants with a power generator. Figure 2 shows the cold chain processes that are closely interacting in a sequential manner [25].

Cold Keeping Areas in Cold Chain

There are a variety of cold storage places in cold chains to keep the products cold for a period of time. These are designed from individual units to entire dedicated facilities. Refrigerated containers are the most basic and often the most cost-effective option for cold storage of small quantities of temperature-sensitive products. They can also be mobile, providing an extra advantage in flexibility. Blast freezers and chillers are ideal for companies requiring quickly cooling and storing foods before they reach the end consumer. In Figs. 3 and 4 show refrigerated container and blast freezer views, respectively.

Cold chain transportation operations should be designed to keep the ambient temperature of those cold areas constant. Shipment operations should be managed such that those temperature conditions are met properly. Otherwise, cold chain efficiency would decrease, and energy consumption, as well as cost, would increase. For example, route selection is an important issue in the shipment process, in that case, depending or in winter and summer [20]. In extreme temperatures, companies should select routes minimizing the number of times the doors should be opened. For example, there are specialty doors for cold warehouses for both on-site facilities and transport vehicles helping to maintain temperatures. The automated high-speed rapid door is a design for a cold chain that can open when it requires and automatically closes once the person/forklift/vehicle has passed through.

In the following section, we summarize the technologies that are frequently utilized in cold chain and cold facilities.

Fig. 3 Refrigerated container [21]

Fig. 4 Blast freezers [24]

Technologies Used in Cold Chain

Energy consumption and energy-saving issues have become one of the main considerations in almost all industries. However, because cold chain facilities would require the highest energy usage due to the refrigerated environment requirements, the factors eliminating energy loss should be considered carefully. Companies operating within a cold chain are investing in technologies both for minimization of energy consumption and waste. Some of those technologies are summarized below.

High-Speed Rapid Door

Increased adoption of newer technologies, including cascade refrigeration systems, high-speed doors, energy-efficient walls, and automated cranes, etc. have helped the increased efficiency and decreased operating costs in cold facilities. Every second that the door remains open longer than it is needed causes losing valuable temperature conditioned air. There are also a number of specialty doors available for both on-site facilities and transport vehicles, helping to conserve temperatures. Those automated high-speed rapid doors are only opening when required, and automatically closing once the person/forklift/vehicle passes through are magnificent technologies for cold facilities. A view of a freezer high-speed rapid door for a cold warehouse is shown in Fig. 5.

Refrigerated Trucks

Temperature monitoring during transportation is crucial in cold chains. Refrigerated trucks, vans, vehicles, and trailers are used for food, pharmaceutical, and agricultural transportation. Using temperature monitoring technologies in transportation reduce losses, save costs, and improve efficiency. Temperature sensors are installed into the trailer to track the real-time temperature data along each route of each load transported. These sensors are usually connected to GPS tracking devices. The software allows us to receive real-time alerts on temperature, whether it falls below or rises above a certain temperature threshold that the user selects.

However, for some refrigerated vehicles carrying chilled, frozen, or perishable goods, a reliable temperature monitoring system gives the warning to take immediate action to protect the refrigerated cargo and reduce losses.

For instance, Fig. 6 shows a truck monitoring system from the Guardmagic Company. Various data are collected from digital temperature sensors, fuel level sensors. These are sent to monitoring stations by a network. In monitoring stations, the received data is analyzed, and necessary reports and graphs are generated.

Fig. 5 Freezer high-speed doors [4]

Warehouse Management System (WMS)

Today, most warehouses, distribution centers, third-party logistics (3PLs), and fulfillment centers rely on warehouse management system (WMS) technology as the brains of their operations. Data from operational processes flow into the warehouse management software through automated data capture methods such as barcode scanners, mobile computers, and radio frequency identification (RFID) enabled devices to track temperature and condition of inventory. The data is then used throughout the system to track inventory through the progression of warehouse processes. Cold storage warehouse providers have more complex conditions when handling refrigerated and frozen inventory. Combined with the complications of handling food and pharmaceuticals, this creates the need for a warehouse management system that is specifically developed for cold storage warehouse management.

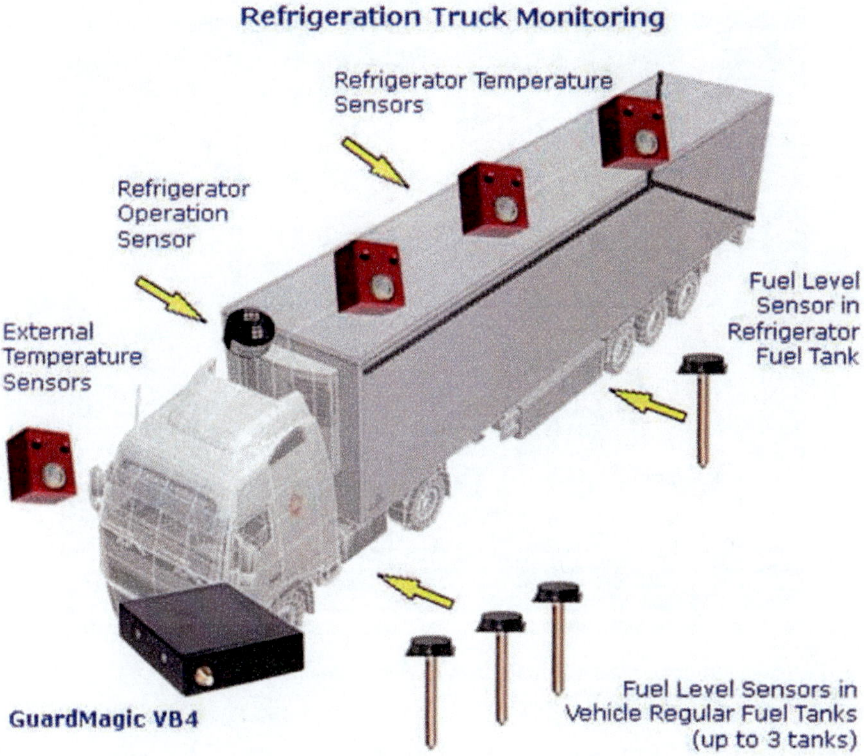

Fig. 6 Refrigerator truck monitoring system [14]

Automated Storage and Retrieval System (AS/RS)

AS/RSs are robotic systems designed to automate storage and retrieval operations. These systems contribute to cold storage in terms of size requirements, service capacities, and temperature conservation. High-density and rack assisted storage type allows deep and long designs that maximize a facility's cube capacity and minimize the footprint of a facility [10, 11]. High-density storage provides not only a smaller area for cooling but also an area that minimizes heat loss. The amount of warmer air entering the temperature-controlled zone is minimized by automatic storage [18]. This automation system provides energy saving in storage areas. Advanced robotics can take over the majority of picking operations, filling in the gaps created by insufficient staffing. Thanks to automation systems, it is possible to prevent the difficulties of using the labor force in the cold environment. Since staff no longer need to brave the cold, it significantly improves overall working conditions. Automated solutions also offer greater speed and flexibility, which can help warehouses get through upticks in order volume and optimize material flow from day to day. Figure 7 shows a long deep AS/RS technology.

Fig. 7 Automated storage and retrieval system in a warehouse [13]

Temperature Monitoring Technologies

As monitoring technologies advanced over time, many application possibilities opened up the tracking of the temperature of perishable products throughout storage and transportation in the cold chain. RFID Wireless devices can be installed in cold storage facilities to control temperature and humidity levels to help meet the quality standards. In order to provide a comprehensive monitoring solution, this method consists of wireless temperature and humidity sensors. If any threshold is exceeded, the related wireless devices can send e-mail or text message alerts. The responsible operator can examine temperature trends and variations over time with the aid of wireless cold unit tracking and historical data recording, and make adjustments when necessary [3]. Furthermore, temperature data loggers are used for temperature and humidity monitoring.

Cold Warehouse Operations

Cold storage warehouse management is a highly specialized discipline that requires extensive experience in the industry. Operations and logistics managers of cold storage facilities are responsible for both ensuring the quality and safety of their important products and for the protection and well-being of their employees who operate under these specific environmental conditions [16]. We summarize the main warehouse operations specifically on the cold warehouse case in the following sections.

Receiving

Products arrive at a warehouse, and they are unloaded from the trucks at the receiving docks. However, in a cold warehouse, receiving operations must be quick to avoid any decline in a product's shelf-life in cold storage. If the load remains exposed during the unloading process, or the reefer truck door stays open for a long time, this poses a serious issue. This problem may directly lead to food waste for the perishability of products. Also, it can shorten the shelf life of the products. Therefore, high-speed rapid doors have an important effect on finding a solution to this type of problem.

Storage

Storage is a significant operation in a warehouse that brings many problems and questions to be answered. The main issues are where the items should be stored, the types of items should be stored together, what storage policy should be implemented, how many SKUs should be stored in how much warehouse space, etc. [8]. Since the storage assignment also impacts the order picking process, its role becomes much more important. In the literature, possible storage policies are generally summarized as random storage, dedicated storage, class-based storage, cube-per-order index, and correlation-based storage [2, 5]. These storage policies should be reconsidered for the cold warehouses by taking in to account the fixed shelf life of products and different storage requirements of products.

In cold warehouses, single-deep and multi-deep AS/RS are mostly used. Multi-deep storage is beneficial for space utilization; hence energy saving takes place. In single-deep storage, products can be picked from both sides of the aisles with forklifts. Figure 8 shows a single-deep racking system.

Compared to the single-deep system, a considerable aisle floor space can be saved in a multi-deep system. Thus, cold warehouse owners usually prefer a multi-deep system in order to save energy and space in their warehouses. In that system, multi-products can be stored as lined up in a row by its storage compartment design. Figure 9 shows a multi-deep racking system.

A pushback racking system is utilized well in a multi-deep storage environment. In that system, pallet storage can be done from either side of an aisle through a deep storage compartment. It provides a higher storage capacity than other types of racking systems [1].

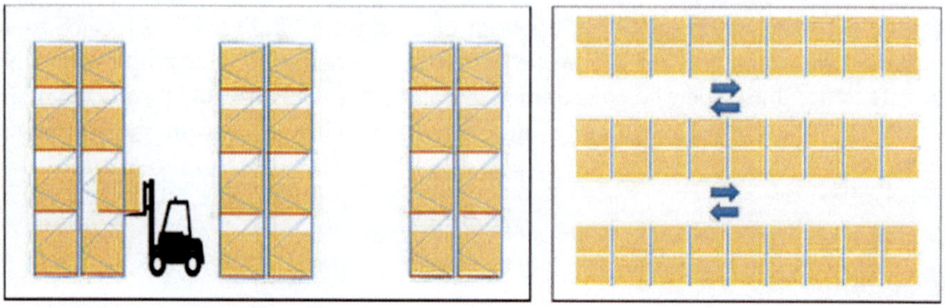

(a) side view (b) top view

Fig. 8 Single-deep racking system

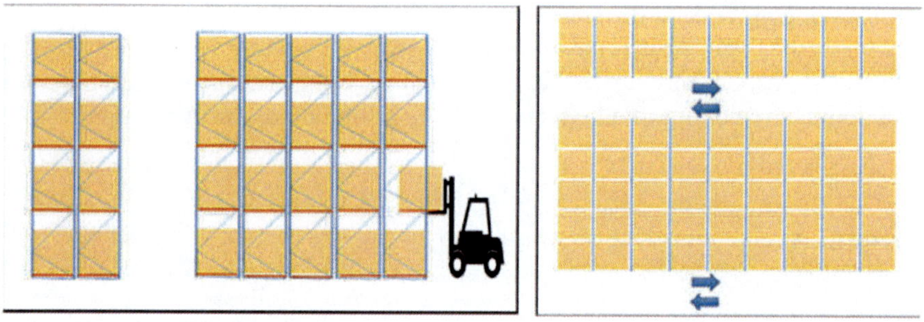

(a) side view (b) top view

Fig. 9 Multi-deep racking system

When a multi-deep storage system is used in a cold warehouse, the order picking policy becomes critical. Since there are only one entry and exit point of the pallets, the correct collection policy should be followed to minimize re-storage movements. Also, the storage policy should be considered well such that in the shipment, the product with the earliest expiration date should be retrieved easily.

Order Picking

Order picking system design and control in warehouses are complicated due to its many critical factors. Since the order picking system is highly dependent on the storage strategy of the products, the development of integrated optimal storage strategies is very important. Order picking is the most costly activity for almost every warehouses. Of all the warehouse costs, 55% are estimated to be belonging to order picking operations [9].

In the order picking operations, the decision of which strategy and sequence to pick up the product to decrease their preparation time and operation cost for shipment is critical. The order picking strategy could be order-based, priority-based, expiration date-based, distance-based, and vehicle-based. Order-based is a strategy in which products are collected from the warehouse in case of the orders receiving from the customer. The priority-based strategy is to prioritize the orders when picking the products. In the expiration date-based strategy, the expiration dates of the products are considered primarily for order picking. This strategy is may be useful for food products having expiration dates. Because traveling distance and time are also crucial for order picking, distance-based strategies, including expiration date collection policies, might be beneficial for food warehouses. Finally, the vehicle-based strategy considers the routes and travelled distance of the exterior vehicles while making the shipment.

Shipment

A shipment process comprises all processes of transporting products from warehouses to customers. If a shipment is not adequately prepared and conditioned accordingly, then the food and energy loss may increase. For example, product assignment policy for capacitated trucks and route selection are significant operations in shipment operations. In extreme temperatures, companies may prefer selecting routes that minimize the number of times truck doors must be opened. In road transportation, refrigerated trucks are mostly used while in ocean transportation, refrigerated containers and ships are preferred. In cold chain transportation, several methods are used for refrigerating trucks or containers. These are mostly water ice, dry ice, liquid nitrogen, and mechanical refrigeration systems, etc. [6].

Discussion and Conclusion

In this work, we aim to present an overview study for cold warehouses in the cold supply chain. Recently, there is an increasing interest in the efficient operation of cold chains to reduce food waste as well as decrease energy and operational costs along with the network. Due to the perishability of foods in cold food chain operations, temperature monitoring and effective temperature control management are significant problems that should be applied correctly throughout the network. Otherwise, the result would be increased food loss, energy consumption, as well as cost.

With the recent technological development, real-time visibility and tracking of a chain are possible. These technologies also enable the remote control of operations. Warehouse design should be reconsidered for the cold chain necessities as well. In an effort to reduce energy consumption and decrease utilized floor space, cold warehouses are mostly designed based on multi-deep storage racks. If storage operations in multi-deep storage rack environments are not designed well, then this may cause non-value-added operations to complete in order picking. Namely, for instance, if a product that is to be retrieved is stored at the very backside of a multi-deep rack, then, while order picking, to be able to reach that product, all the other products sorted in front of it should be taken out. To eliminate such non-value added operations, optimization of storage processes in such cold warehouses becomes important. As future work, we aim to develop an integrated optimization procedure considering both storage and retrieval processes simultaneously.

References

1. 3D Storage Systems (2020) https://www.3dstoragesystems.com/pushback-racking-basics
2. Accorsi R, Manzini R, Bortolini M (2012) A hierarchical procedure for storage allocation and assignment within an order-picking system. A case study. Int J Logistics Res Appl 15(6): 351–364
3. Aranet White Paper (2020) https://d1l0iodieynis3.cloudfront.net/wp-content/uploads/2018/01/01145614/COLD-CHAIN-whitepaper-small.pdf
4. Arrow-Industrial (2020) https://www.arrow-industrial.co.uk/industrial-doors/freezer-high-speed-doors/
5. Bindi F, Manzini R, Pareschi A, Regattieri A (2009) Similarity-based storage allocation rules in an order picking system: an application to the food service industry. Int J Logistics Res Appl 12(4): 233–247
6. Brecht JK, Sargent SA, Brecht PE, Saenz J, Rodowick L (2019) Protecting perishable foods during transport by truck and rail. UF, Ifas Extension https://edis.ifas.ufl.edu/pdffiles/HS/HS132800.pdf
7. Chabada L, Damgaard CM, Dreyer HC, Hvolby HH, Dukovska-Popovska I (2014) Logistical causes of food waste: a case study of a Norwegian distribution chain of chilled food products. In: IFIP international conference on advances in production management systems. Springer, Heidelberg, pp 273–280
8. Davarzani H, Norrman A (2015) Toward a relevant agenda for warehousing research: literature review and practitioners' input. Logistics Res 8(1): 1
9. De Koster R, Le-Duc T, Roodbergen KJ (2007) Design and control of warehouse order picking: a literature review. Eur J Operational Res 182(2): 481–501
10. Ekren BY, Sari Z, Lerher T (2015) Warehouse design under class-based storage policy of shuttle-based storage and retrieval system. IFAC-PapersOnLine 48(3): 1152–1154
11. Ekren BY, Heragu SS (2012) Performance comparison of two material handling systems: AVS/RS and CBAS/RS. Int J Prod Res 50(15): 4061–4074
12. Food and Agriculture Organization of the United Nations (2020) http://www.fao.org/food-loss-and-food-waste/en/
13. Gharehgozli AH, Yu Y, Zhang X, Koster RD (2017) Polynomial time algorithms to minimize total travel time in a two-depot automated storage/retrieval system. Transp Sci 51(1): 19–33
14. Guardmagic (2020) https://www.guardmagic.com/01-engl/11e-solution/14e-refrigerator-online/00e-refrigerator-online.htm
15. Habazin J, Glasnović A, Bajor I (2017) Order picking process in warehouse: case study of dairy industry in Croatia. Promet-Traffic Transp 29(1): 57–65
16. Interlake Mecalux (2018) https://www.interlakemecalux.com/blog/challenges-cold-storage-warehouses
17. Lempert P (2018) 10 Food Trends That Will Shape 2018. https://www.forbes.com/sites/phillempert/2017/12/13/10-food-trends-that-will-shape-2018/#117dbe564104

18. Logistics Management (2012) https://www.logisticsmgmt.com/article/best_practices_for_managing_a_cold_storage_warehouse
19. PT Capricorn Indonesia Consult (2019) A cold chain study of Indonesia. In: Kusano E (ed) The cold chain for agri-food products in ASEAN. ERIA Research Project Report FY2018 no.11, Jakarta: ERIA, pp 101–147
20. Remax Doors (2020) https://www.remaxdoors.com/warehouse-management-blog/best-practices-for-managing-a-cold-storage-warehouse
21. SeaNews (2015) https://www.seanews.com.tr/nyk-orders-5-500-technologically-advanced-refrigerated-containers/156361/
22. Statista (2018) https://www.statista.com/topics/4321/cold-chain-logistics/
23. Statista (2020) https://www.statista.com/statistics/915897/refrigerated-warehouses-capacity-country/
24. TeknoTek (2020) https://www.teknoteksogutma.com/blast-freezer/
25. The Geography of Transport Systems (2020) https://transportgeography.org/?page_id=6603

Design and Optimization of Automated Storage and Retrieval Systems: A Review

Zeki Murat Cinar and Qasim Zeeshan

Abstract Automated Storage and Retrieval Systems (AS/RS) are warehouses, specifically designed for material handling in modern manufacturing systems and are extensively utilized as distribution centers, due to the advantages of improved inventory control, cost-effective utilization of time, space and equipment. With the advent of smart manufacturing in the Industry 4.0 era, the significance of adopting automation technology in warehousing becomes increasingly imminent. Techniques and tools are being developed to manage, control, design, and optimize the AS/RS(s). Over the years, many researchers have focused on the design, analysis, and optimization of various AS/RS configurations. This paper aims to present a systematic literature review on the research in Configuration design and Optimization of AS/RS, by classifying the research according to the design objectives, configurations and optimization algorithms used, in order to highlight and expand the awareness on the current best practices, progress, and future research directions.

Keywords Automated storage and retrieval systems · Configuration design · Optimization · Warehouse

Introduction

Material handling systems (MHS) have been of interest to researchers over the past three decades. Material handling constitutes 15–70% of the total manufacturing cost of a product [39]. MHS integrates equipment and other technologies of manufacturing systems. The Automated Storage & Retrieval System (AS/RS) is a typical application of automation technology in modern MHS. AS/RS is specifically designed for the material handling process and is extensively utilized in modern MHS within the

Z. M. Cinar (✉) · Q. Zeeshan
Mechanical Engineering Department, Eastern Mediterranean University (EMU), via Mersin 10, Famagusta, Northern Cyprus, Turkey

Q. Zeeshan
e-mail: qasim.zeeshan@emu.edu.tr

F. Calisir (ed.), *Industrial Engineering in the Internet-of-Things World*,
Lecture Notes in Management and Industrial Engineering,
https://doi.org/10.1007/978-3-030-76724-2_14

production, pharmacy, libraries, as well as automotive factories and car parking. AS/RS is a complex system in which equipment and control systems combined together. This complex system offers automatically handles, storage, and retrieval of loads with ideal speed and high accuracy without labor assistance. Although the AS/RS technology initiated in the late 1970s and its development continued over the years [54], however, with the requirement of Flexible Manufacturing Systems (FMS) and the advent of smart manufacturing and Industry 4.0 era, the significance of adopting automation technology in warehousing becomes increasingly imminent. Techniques and tools are being developed to manage, control, design, and optimize the AS/RS. This paper aims to present a systematic review of research in the design and optimization of AS/RS in order to highlight and expand the awareness on the current best practices, progress and future research directions.

AS/RS Configurations

AS/RS play an essential role in warehouses due to transportation of loads, especially for some facilities such as hospital and libraries due to major advantages of AS/RS(s). Typical AS/RS Configurations are shown in Fig. 1 and work as follows: first of all, items to be stored are sequenced and allocated to the special bins, containers, or boxes. The containers with the items inside are taken to the weighting location for confirming the load weights are within limit requirements. In some cases, different parameters of loads, such as dimensions, danger level, and fragile status, should also be checked and tested in a specific station. Those who successfully passed all tests are transported to the Input/Output station. While transportation, testing, and evaluation processes are being processed, the status of loads is regularly and currently received by the central computer. The central computer assigns the decision of the next step of loads, and then the status of loads is saved in its memory. The loads are then moved to corresponding places with the help of the S/R machine. Upon receipt of a request

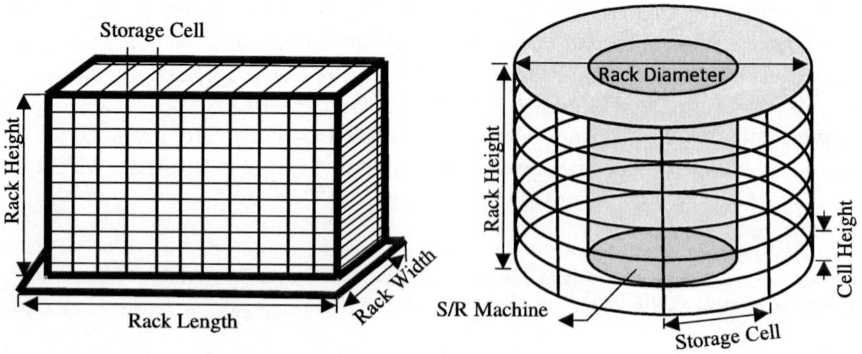

Fig. 1 Typical AS/RS configurations—rectangular and circular

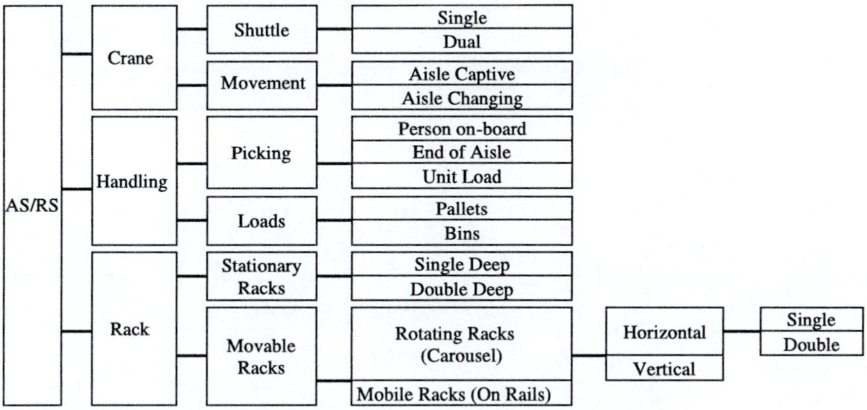

Fig. 2 Classification of AS/RS(s) adapted from Roodbergen and Vis [46]

for an item, the central computer gives a decision about loads, whether where to store or from which storage cell to retrieve and then sends a command to the crane to do the task. The loads are then taken from I/O station by the supporting transportation system to be transported to its final destination.

Classification of AS/RS(s)

Roodbergen et al. [45] categorized AS/RS(s) into three different essential classes according to the bin's arrangement, I/O capacity, and the number of S/R machines utilized in AS/RS. Vasili et al. [53] categorized AS/RS classes as unit load AS/RS, deep line AS/RS, multi-load AS/RS, mini-load AS/RS, the man on board AS/RS, vertical lift storage modules (VLSM), automated item-retrieval systems, multi aisles AS/RS, Carousel Systems, Mobil rack AS/RS (Fig. 2).

Research on AS/RS

A summary of research on AS/RS(s) is presented in this section. Based on SCOPUS data searched by "AS/RS" OR "Automated Storage and Retrieval System," year by year publication in increase trend and expected to have more than 1277 publications in the year of 2020 shown in Fig. 3b. Most of the publications are subjected to engineering (6129 publications) and mathematics (5820 publications), as shown in Fig. 3c. Figure 3d explains that based on the studied literature, most of the publications are from China (418 publications) and followed by Germany (1818 publications). Moreover, the highest publications related to AS/RS searched by affiliation

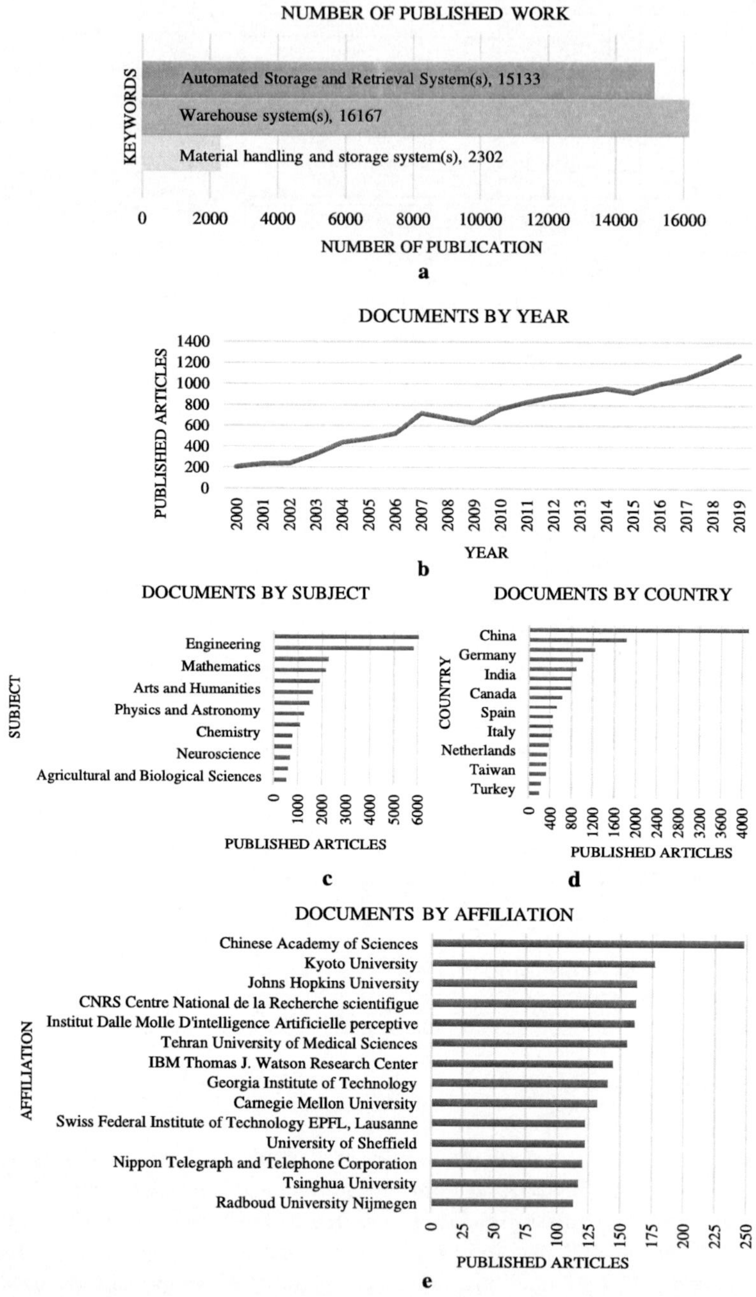

Fig. 3 Statistics from Scopus database (Date: 24.03.2020, keywords: "ASRS" OR "Automated Storage and Retrieval System"). **a** Published documents between 2000 and 2020, **b** Published documents per year, **c** Published documents by subject, **d** Published documents by country, **e** Published documents by affiliation

and presented in Fig. 3e. Also, comprehensive published work between "Automated Storage and Retrieval Systems," "Warehouse systems," and "Material Handling and Storage Systems" is presented in Fig. 3a.

Classification of Research

Systematic classification of research according to the design objectives, configurations, and optimization algorithms used in the design and optimization of AS/RS is presented in Tables 1 and 2.

Roodbergen et al. [45] studied that there are significant benefits such as improvement in efficiency and capacity of storage systems, improvement in quality and performance, better inventory security. Also, they studied the disadvantages of AS/RS, such as inflexibility of the system, high capital cost, fixed storage capacity, lack of visibility. Roodbergen and Vis [46] studied operating requirements of an AS/RS listed common subsystems: storage structure, S/R machine, storage modules, I/O station, control systems, and RFID systems. Storage structure is the main structure that accommodates the high weighted loads without significant deflection. The structure must have sufficient strength and rigidity. S/R machine provides transportation. Storage modules used to carry bins including stored products inside. Pallets, baskets, bins, containers, drawers are commonly used in AS/RS as storage modules. The station where the loads are taken in or out of the AS/RS termed as I/O stations. They are mostly accessed by external handling system that brings the loads into the AS/RS or takes out of AS/RS.

Vasili et al. [53] studied AS/RS in terms of specifically focused objectives and requirements, i.e., maximization of storage capacity, throughput, travel time, whereas minimization of the total cost, carbon footprint, maximization of storage density as well as improvement on inventory control and safety. Recently many researchers have focused on the design, analysis, and optimization of various AS/RS configurations [53]. However, there is no author found in the literature focusing on circular AS/RS.

Recently AS/RS are implemented to the automotive factories due to improved safety, inventory control, landscape utilization, cost and efficiency [12, 15, 20, 49].

A general overview of warehouse control and design haven been studied in the past years [3, 15, 48, 51]. The clarification of the current developments of the AS/RS design and design issues are presented by [46]. Sarker and Babu [49] studied the design aspects of an AS/RS and travel time model of the rectangular type AS/RS. In his research, Throughput capacity is explained as the inverse of the mean transaction time that is the expected travel time required for storage or retrieval process and P/D time. Therefore, the travel time of an AS/RS usually related to the S/R machine features as well as AS/RS rack configuration. Moreover, Sarker and Babu [49] made a list of top interesting design problems, as shown: (1) Assignment of the products to the storage locations in the storage structure. (2) Configurations of the storage structure (Ratio of length to height), (3) Operating policies for order storage, and

Table 1 Recent developments in the design and optimization of AS/RS (years between 1976 and 2012)

		Author	Hausman et al. (1976)	Graves et al. (1977)	Bozer and White (1984)	Ashayeri, J. et all. (1985)	Sarker, B. R. et al. (1995)	Van den Bergh, J. C. (2000)	Malmborg, C.J (2003)	Rene de Koster et all. (2006)	De Koster, R et all. (2007)	Roodbergen, K. J. et al. (2008)	Vasili, M. R. et all. (2008)	Roodbergen, K. J. et al. (2009)	Gagliardi, J.P. et al. (2010)	Felix T.S. et all.(2010)	Klaus Moellera et all. (2011)	Vasili, M. R. et all. (2012)	Khalid H. et all.(2012)	Haneyah, S. W. A. et al.(2012)	Dou, C. (2012)	Lerher, T. et all. (2012)
Objectives		Review Paper	•				•			•		•			•							
		Travel time					•	•		•	•	•	•	•		•	•		•			•
		Cost minimization				•			•	•	•					•		•			•	
		Comparison between models					•						•	•						•		•
		System Configuration				•		•	•			•	•	•		•		•			•	
		Requests sequencing		•						•	•	•		•					•			
		Storage Capacity / Assingment	•	•	•			•		•			•	•		•		•		•	•	•
		Performance optimization																				
		CO2							•													•
Modelling assumptions	Configurations	Flow-rack AS/RS																				
		Mobil racks																				
		Unit-Load AS/RS	•	•	•	•	•		•	•		•	•	•				•	•			
		Order picking system								•	•									•		
		All locations have same dimentions	•	•	•	•	•					•		•					•		•	
		Multi-load AS/RS				•	•	•				•		•							•	
		Single crane, Single Aisle	•	•	•		•		•	•			•					•	•			
		Symetrical Distances	•	•	•	•	•		•				•					•				
		Each I/O can perform S/R	•	•	•	•				•		•		•				•	•			
		AS/RS (AVS/RS)																				
		Square in time rack	•	•			•															
		Circular AS/RS (C-AS/RS)																				
		Rectangular in time rack			•	•			•	•	•	•	•	•				•			•	•
	Crane	Tchebychev time							•													
		Robotic load carrying carts																			•	
		Constant crane acceleration	•	•	•	•		•	•			•		•								•
		Constant pickup and deposit times	•	•	•	•		•	•			•		•					•			•
	Products	Constant item turnover	•	•	•			•		•		•	•	•				•	•		•	
		Various types					•			•	•										•	
		Items ordered EOQ model	•		•	•			•	•								•			•	
	Storage	Very narrow storage (VNA)																				
		Random storage assignment			•	•	•	•		•	•	•		•				•	•			
		Constant number of pallets	•	•	•	•	•	•		•								•			•	
		%100 Rack utilization	•	•	•	•			•			•		•					•			•
Optimization and Methodology	Analytical	Mathematical modelling	•	•	•	•			•										•			
		Statistical-Based																				
		Genetic algorithm															•			•		
		Dynamic sequencing						•														
		Eye ball technique																				
		Dwell-Point location										•		•								
		Bi-Level Optimization model																				
		Informed search algorithm																			•	
	Simulation	Pareto curve and UL mass dist.																				
		AMPL/CPLEX																	•			
		AutoMod																				
		AMCLOS																				
		ARENA														•						
Applica-tion		Baggage handling																		•		
		Automated parking																			•	
		Industrial warehousing system	•	•	•	•	•	•	•	•		•		•		•	•	•	•			

Table 2 Recent developments in the design and optimization of AS/RS (years between 2013 and 2020)

		Lerher, T. et all. (2013)	Guezzen, A. H. (2013)	Lisa M.Thomas et all. (2013)	Zollinger, H. (2014)	Lerher, T. et all. (2014)	Lerher, T. et all. (2015)	B.Y.Ekren rf.(2015)	Ghomri, L. et all. (2015)	Lerher, T (2016)	Bortolini, M. at all. (2016)	Cinar, D. et all. (2016)	Yue, L. et al. (2017)	Zrnić, N. et al. (2017)	Cinar, Z.M. (2017)	Xu, X. et al. (2018)	Oki, N. et al. (2019)	Xu, X. et al. (2019)	Tostani, H. H. et al. (2020)	Ekren, B. Y. (2020)
Objectives	Review Paper																			
	Travel time	•	•			•	•	•	•	•	•	•	•	•	•	•		•		
	Cost minimization	•				•					•			•	•				•	
	Comparison between models				•										•					
	System Configuration			•	•				•				•						•	
	Requests sequencing				•															
	Storage Capacity / Assingment	•	•	•		•	•			•	•	•			•	•	•			
	Performance optimization																			•
	CO2			•			•	•			•			•	•				•	
Modelling assumptions – Configurations	Flow-rack AS/RS								•											
	Mobil racks		•																	
	Unit-Load AS/RS	•			•			•		•	•							•		
	Order picking system	•			•			•			•		•							
	All locations have same dimentions	•			•	•	•	•	•	•	•	•								
	Multi-load AS/RS															•				
	Single crane, Single Aisle								•	•	•				•		•	•		
	Symetrical Distances				•						•				•					
	Each I/O can perform S/R					•	•	•	•			•								
	AS/RS (AVS/RS)																			•
	Square in time rack																			
	Circular AS/RS (C-AS/RS)														•					
	Rectangular in time rack	•	•		•	•	•	•	•	•	•	•	•	•		•	•	•		•
Modelling assumptions – Crane	Tchebychev time					•	•				•		•	•	•	•		•		
	Robotic load carrying carts				•					•					•					
	Constant crane acceleration	•			•						•									
	Constant pickup and deposit times	•				•	•	•	•		•	•								
Modelling assumptions – Products	Constant item turnover	•	•		•			•	•	•	•									
	Various types												•					•	•	
	Items ordered EOQ model	•							•		•									
Modelling assumptions – Storage	Very narrow storage (VNA)				•															
	Random storage assignment		•			•	•	•	•	•		•		•				•		
	Constant number of pallets		•					•		•		•								
	%100 Rack utilization		•			•	•	•	•								•			
Optimization and Methodology – Analytical	Mathematical modelling		•						•		•		•		•					
	Statistical-Based			•						•										•
	Genetic algorithm	•				•						•		•	•					
	Dynamic sequencing																			
	Eye ball technique							•												
	Dwell-Point location															•				
	Bi-Level Optimization model																		•	
	Informed search algorithm																			
Optimization and Methodology – Simulation	Pareto curve and UL mass dist.										•									
	AMPL/CPLEX																			
	AutoMod	•																		
	AMCLOS																			
	ARENA		•					•	•											
Application	Baggage handling																			
	Automated parking														•					
	Industrial warehousing system	•	•	•	•			•	•	•	•	•	•				•			•

retrieval. According to [9, 46] studied, the following five lists are the most recurrent assignment strategies for AS/RS warehouses.

1. Dedicated storage requires items to set a fixed storage location. It is also required to assure storage capacity for each product and for any time in the design phase by considering product features such as weight and shape [10].
2. Full-turnover based technique considers turnover frequencies to store the product. Products close to P&D point counted as quick-moving operation whereas, products far from P&D counted as slow-moving operation. Turnover frequency is evaluated through the cube per order index (COI) [26, 27].
3. Closest open location Technique requires product storage to empty locations, which are close to the P&D point. Therefore, storage can be distinguished by all locations close to R&D point, and empty locations are far from R&D points [47].
4. A class-based storage technique is utilized to overcome the disadvantages, and it splits the AS/RS locations into different classes. Based on the product's turnover frequencies this technique assigns products into classes [25].

Commonly literature focuses on average travel distance and time for S/R products [7, 18, 21, 23, 28, 31, 32, 53, 58]. In this context, AS/RS(s) differ from traditional handling tools such as a forklift. These devices/tools perform disjoint horizontal as well as vertical movements. However, AS/RS(s) perform concurrent movements in the two directions [2]. For standard storage locations, the required time for S/R to complete loading is maximum between the horizontal and vertical time intervals. Regarding the highlighted difference above the text, travel time is considered as the KPI metric for AS/RS [40]. Published articles that focus on minimization of travel time are presented as following. Bozer [8] developed for the first time a model to define the travel time of an AS/RS [29] proposed a strategy to minimize travel time for storage assignment. Both authors ignored crane acceleration/deceleration to simplify their models. Nevertheless, several articles presented travel time models, considering crane acceleration/deceleration [30]. Additionally, the dwell point is studied to minimize the travel time of AS/RS by Van den Berg [52]. Multiple goals are also focused on travel time. For instance, the operation cost of AS/RS and space requirement is selected as objective by [57]. Some researchers are studied minimization of energy consumption of AS/RS as well as minimization of travel time as multi-objective study such as [16, 30, 52].

Most of the published papers are about manufacturing environments, and a few papers are highlighted the AS/RS configuration designs as follows; Park and Webster [43] proposed an approach that synchronously picks the storage size and shape of storage of AS/RS. Summarily, almost all simulation models are addressed to physical design features, and only a few AS/RS(s) and their configurations are evaluated in combination with constant input values. Sarker and Babu [49] studied specific parameters of the physical design of an AS/RS and also defined that size of the storage bins, baskets or boxes is important to determine storage cell dimensions as well as expected travel time to a specific location. The shape factor is another parameter that deals with the AS/RS length and height. It is also used to determine the AS/RS structure as the square in time or rectangle in time. The shape factor is known as the time spent to reach an extreme location in the storage structure. The depth of the rack is another parameter for physical design and can be a single or double deep rack. The last parameter is the capacity and the no of S/R machines utilized in the system. As the number of S/R machine increases, faster product storage and retrieval process can be done. However, for the system performance, the no of S/R machines utilized in the system should be selected based on on-demand requirements. Lerher and Šraml [35] focused on the energy efficiency model for mini-load automated storage and retrieval systems. Crane velocities, accelerations, number of rows, and number of columns with the required number of cranes are set as design variables. In the paper, CO_2 consumption of the system is evaluated.

Bartlett and Strei [5] studied cost analysis of warehouse facility establishment at Ford. California. Lerher et al. [36] studied the total cost of an AS/RS and conducted Pareto optimization design to find optimal investment cost with respect to optimal travel time and reliability. Zrnić et al. [58] studied a multi-objective optimization model for minimizing cost, travel time, and energy consumption in an AS/RS.

Barbato et al. [4] identified significant factors affecting the performance of the shuttle-based storage and retrieval system (SBS/RS) by conducting an experimental study. Their research includes Tukey's test that is used to identify design factors, and then ARENA software is utilized for modeling. Consequently, the results of the proposed model are obtained by using Minitab software. Barbato et al. [4] research resulted that:

1. For travel time: SBS/RS design should have a high value of velocities for lifts and shuttles. Also, the number of bays should be less in the system.
2. For energy consumption: lifts and shuttle velocities should be high, whereas the number of tiers and acceleration/deceleration of lift should be low.

Bertolini et al. [6] proposed a meta-heuristic algorithm based on simulated annealing in which retrieval operation performance-optimized. In their proposed algorithm, a job list is created based upon customer demand, and retrieving performance according to missions per hour is evaluated to optimize the performance of retrieving operation. In order to validate this algorithm, the existing warehouse is utilized to test and then compared it with the current performance of the existing warehouse [6].

Ma and Wang [37] studied a single and dual command cycle time model for the Shuttle Based Storage and Retrieval System (SBS/RS) by considering the motion characteristics of the system. Different design configurations with four different velocity options are evaluated by the author. Summarily, it concluded that the dual command cycle time of SBS/RS decreased by up to 28.3%, and relatively throughput capacity is increased by up to 1.3% [37].

Rajković et al. [44] presented a new optimization model that contains three objectives; (1) minimization of investment expenses, (2) minimization of cycle time and (3) minimization of CO_2 consumption of the AS/RS(s). Also, author-defined constraints as; (1) capacity must be bigger than the required capacity, (2) throughput capacity must be higher than the minimum required throughput capacity, (3) warehouse capacity must not exceed the 20% of required capacity, and (4) number of storage and retrieval machines must be lower or equal to the number of hallways. Based on these objectives and constraints author applied NSGA II in order to optimize decision variables and model resulted that the best configuration has 25.68 acres of forest per year, cycle time is 26.32 s, and investment cost is 7.01 EUR/TUL.

Ekren [18] carried out a multi-objective optimization solution procedure for the design of Autonomous Vehicle-based Storage and Retrieval System (AVS/RS) by considering the minimization of average cycle time and average energy consumption. ARENA software is utilized by the author to perform simulation and simulation results presented as following;

1. Cycle time is affected by the footprint shape of the system. Also, the number of tiers has a direct impact on electricity consumption.
2. For the optimal solution, the number of tiers should be decreased, whereas the number of aisles increased.

Discussion and Conclusion

This paper presents a systematic literature review on the research in configuration design and optimization of AS/RS, by classifying the research according to the design objectives, configurations and optimization algorithms used, in order to highlight and expand the awareness on the current best practices, progress, and future research directions.

Based on the literature review, it is concluded that most of the researchers [1, 8, 11, 17, 18, 33, 34, 38, 42, 50, 53, 55–58] focused on rectangular AS/RS and some researchers such as [22, 25, 49] focused on square AS/RS. It is observed that over the years, many researchers have focused on the design, analysis, and optimization of various AS/RS configurations, as shown in Tables 1 and 2; however, less research is focused on the Circular AS/RS Configurations. Cinar [12] applied Genetic Algorithms (GAs) for the configuration design and optimization of Circular AS/RS with the objective to optimize the throughput, cost, and carbon footprint.

Common objectives from the studied articles found as; travel time optimization, total cost optimization, carbon footprint optimization. Also, critical parameters for the design of AS/RS are defined as crane, storage configuration, product types, and crane features [11].

Most authors [11, 12, 24, 33, 36, 58] used genetic algorithm to obtain optimal storage systems. Nevertheless, fewer considerations given by authors to the application of other optimization algorithms like Swarm Intelligence, Particle Swarm Optimization (PSO), and Ant Colony Optimization (ACO). Moreover, the application of machine learning and deep learning tools needs to be further explored. In addition, an uncertainty-based design approach can also be utilized to achieve more robust configurations.

Computer-based models and simulations are critical to the design, development, and optimization of smart manufacturing systems required for Industry 4.0. [17, 19,

21, 41] carried out simulation-based design in order to obtain optimal storage systems. Cinar et al. [13]. Virtual reality simulation platforms must be further explored for the design and optimization of AS/RS. In the context of the 4th industrial revolution, logistics and storage systems are required automatization and connectivity. However, there are challenges while implementing those technologies. Cinar et al. [13] defined the challenges of Industry 4.0 applications and also listed simulation techniques for smart factories, including logistics and storage systems.

Digital twins (DT) are the key enablers for virtual design and optimization of smart manufacturing systems required for transformation to Industry 4.0. Recently many researchers have contributed to the development of DTs for smart production processes and manufacturing systems. DTs are being applied to manage the performance, effectiveness, and quality of a manufacturer's fixed assets, such as manufacturing plants. DTs are also applied to shop-floor control, assembly production lines, and flexible manufacturing systems to optimize robustness, efficiency, autonomous failure detection, and production flow control [14]. Digital twin technology must be utilized for the design of AS/RS(s).

Rajković et al. [44] recommended that their presented model can be modified to optimize AS/RS with double deep SR machine by adding more variables and constraints. Ma and Wang [37] suggested having more than two I/O points and different assignment policies such as class-based storage policy and turnover based policy in order to understand the impact of I/O separation. Bertolini et al. [6] proposed an algorithm based on simulated annealing procedure to perform optimization for a more complex AS/RS where the storage compartments are double depth or two cranes share the same railway as a future study and also suggested that a new algorithm should be created and optimization can be performed for scheduling.

The scope of the study is not to include all the available articles about the theme, but to analyze a significant sample to give understandings about the recent best practices on design and optimization of AS/RS, assisting researchers and practitioners for further studies.

References

1. Ashayeri J, Gelders LF (1985) Warehouse design optimization. Eur J Oper Res 21(3):285–294
2. Atmaca E, Ozturk A (2013) Defining order picking policy: a storage assignment model and a simulated annealing solution in AS/RS systems. Appl Math Model 37(7):5069–5079
3. Baker P, Canessa M (2009) Warehouse design: a structured approach. Eur J Oper Res 193(2):425–436
4. Barbato M, Ceselli A, Premoli M (2019) Evaluating automated storage and retrieval system policies with simulation and optimization. In: Advances in optimization and decision science for society, services and enterprises. Springer, pp 127–137
5. Bartlett WW, Strei TJ (1990) A Cost Benefit—Analysis of Establishing A Warehouse Facility for the Services Division of the Directorate of Personnel And Community Activities at Fort Ord, California. (Master of Science), Naval Postgraduate School, Monterey, California
6. Bertolini M, Esposito G, Mezzogori D, Neroni MJPM (2019) Optimizing retrieving performance of an automated warehouse for unconventional stock keeping units 39:1681–1690

7. Bortolini M, Faccio M, Ferrari E, Gamberi M, Pilati F (2016) Time and energy optimal unit-load assignment for automatic S/R warehouses. Int J Prod Econ 190:133–145
8. Bozer YA, White JA (1984) Travel-time models for automated storage/retrieval systems. IIE Trans 16(4):329–338
9. Carlo HJ, Vis IF (2012) Sequencing dynamic storage systems with multiple lifts and shuttles. Int J Prod Econ 140(2):844–853
10. Choe K, Sharp G (2014) Small parts order picking: design and operation
11. Cinar D, Oliveira JA, Topcu YI, Pardalos PM (2016) Scheduling the truckload operations in automated warehouses with alternative aisles for pallets. Appl Soft Comput 52:566–574
12. Cinar ZM (2017) Configuration Design and Optimization of Circular Automated Storage and Retrieval System (C-AS/RS). Eastern Mediterranean University EMU-Doğu Akdeniz Üniversitesi (DAÜ)
13. Cinar ZM, Nuhu AA, Zeeshan Q, Korhan O (2019) Digital Twins for Industry 4.0: A Review. Paper presented at the Global Joint Conference on Industrial Engineering and Its Application Areas
14. Cinar ZM, Zeeshan Q, Solyali D, Korhan O (2019) Simulation of Factory 4.0: A Review. Paper presented at the Global Joint Conference on Industrial Engineering and Its Application Areas
15. De Koster R, Le-Duc T, Roodbergen KJ (2007) Design and control of warehouse order picking: a literature review. Eur J Oper Res 182(2):481–501
16. Dekker R, De Koster MBM, Roodbergen KJ, Van Kalleveen H (2004) Improving order-picking response time at Ankor's warehouse. Interfaces 34(4):303–313
17. Dou C (2012) Development of storage and retrieval algorithms for automated parking systems
18. Ekren BY (2020) A multi-objective optimisation study for the design of an AVS/RS warehouse. Int J Prod Res 1–20
19. Ekren YB, Sari Z, Rosi B (2015) Simulation analysis of shuttle based storage and retrieval systems
20. Gagliardi J-P, Renaud J, Ruiz A (2010) Simulation modeling approaches for automated storage and retrieval systems: a literature review: CIRRELT
21. Ghomri L, Sari Z (2015) Mathematical modeling of retrieval travel time for flow-rack automated storage and retrieval systems. IFAC-PapersOnLine 48(3):1906–1911
22. Graves SC, Hausman WH, Schwarz LB (1977) Storage-retrieval interleaving in automatic warehousing systems. Manage Sci 23(9):935–945
23. Guezzen AH, Sari Z, Castagna P, Cardin O (2013) Travel time modeling and simulation of a mobile racks automated storage/retrieval system. Int J Eng Technol 5(3):420
24. Haneyah SWA, Schutten JMJ, Schuur PC, Zijm WHM (2013) Generic planning and control of automated material handling systems: practical requirements versus existing theory. Comput Ind 64(3):177–190
25. Hausman WH, Schwarz LB, Graves SC (1976) Optimal storage assignment in automatic warehousing systems. Manage Sci 22(6):629–638
26. Heskett JL (1963) Cube-per-order index-a key to warehouse stock location. Transp Distribution Manage 3(1):27–31
27. Heskett JL (1964) Business logistics: management of physical supply and distribution. Ronald Press Co
28. Hu Y-H, HS, Chen C, Hsu W-J, Toh AC, Loh CK, Song T (2005). Travel time analysis of a new automated storage and retrieval system. Comput Ind Eng 32(6): 1515–1544
29. Hwang H, Ko CS (1988) A study on multi-aisle system served by a single storage/retrieval machine. Int J Prod Res 26(11): 1727–1737
30. Hwang H, Lee SB (1990) Travel-time models considering the operating characteristics of the storage and retrieval machine. Int J Prod Res 28(10):1779–1789
31. Kasemset C, Rinkham C (2011) Warehouse storage assignment: the case study of camera and lense manufacturer. Industrial Engineering and Engineering Management (IEEM), 2011 IEEE International Conference, pp 1108–1112
32. Lerher T (2016) Travel time model for double-deep shuttle-based storage and retrieval systems. Int J Prod Res 54(9):2519–2540

33. Lerher T, Edl M, Rosi B (2014) Energy efficiency model for the mini-load automated storage and retrieval systems. Int J Adv Manuf Technol 70(1–4):94–115
34. Lerher T, Ekren BY, Dukic G, Rosi B (2015) Travel time model for shuttle-based storage and retrieval systems. Int J Adv Manuf Technol 78(9–12):1705–1725
35. Lerher T, Šraml M (2012) Designing unit load automated storage and retrieval systems. Springer, London
36. Lerher T, Sraml M, Borovinsek M, Potrc I (2013) Multi-objective optimization of automated storage and retrieval systems. Ann Faculty Eng Hunedoara 11(1):187
37. Ma Y, Wang J (2019) Travel time analysis for shuttle-based storage and retrieval system with middle input/output location. Paper presented at the 2019 16th International Conference on Service Systems and Service Management (ICSSSM)
38. Malmborg CJ (2003) Design optimization models for storage and retrieval systems using rail guided vehicles. Appl Math Model 27(12):929–941
39. Mirhosseyni SHL, Webb P (2009) A hybrid fuzzy knowledge-based expert system and genetic algorithm for efficient selection and assignment of material handling equipment. Expert Syst Appl 36(9):11875–11887
40. Moon G, Kim G-P (2001) Effects of relocation to AS/RS storage location policy with production quantity variation. Comput Industrial Eng 40(1–2):1–13
41. Nagadi K (2016) A framework to generate a smart manufacturing system configurations using agents and optimization
42. Oki N, Ataka M, Sato C (2019). Bin-module based automated storage and retrieval system and method. In: Google Patents
43. Park YH, Webster DB (1989) Modelling of three-dimensional warehouse systems. Int J Prod Res 27(6):985
44. Rajković M, Zrnić N, Kosanić N, Borovinšek M, Lerher TJT (2019) A multi-objective optimization model for minimizing investment expenses, cycle times and CO_2 footprint of an automated storage and retrieval systems 34(2): 275–286
45. Roodbergen KJ, Sharp GP, Vis IF (2008) Designing the layout structure of manual order picking areas in warehouses. IIE Trans 40(11):1032–1045
46. Roodbergen KJ, Vis IF (2009) A survey of literature on automated storage and retrieval systems. Eur J Oper Res 194(2):343–362
47. Rosenblatt MJ, Roll Y, Vered Zyser D (1993) A combined optimization and simulation approach for designing automated storage/retrieval systems. IIE Trans 25(1):40–50
48. Rouwenhorst B, Reuter B, Stockrahm V, van Houtum GJ, Mantel RJ, Zijm WHM (2000) Warehouse design and control: framework and literature review. Eur J Oper Res 122(3):515–533
49. Sarker BR, Babu PS (1995) Travel time models in automated storage/retrieval systems: a critical review. Int J Prod Econ 40(2–3):173–184
50. Tostani HH, Haleh H, Molana SH, Sobhani F (2020) A Bi-Level Bi-Objective optimization model for the integrated storage classes and dual shuttle cranes scheduling in AS/RS with energy consumption, workload balance and time windows. J Cleaner Prod 120409
51. Van Den Berg JP, Gademann AJRM (2000) Simulation study of an automated storage/retrieval system. Int J Prod Res 38(6):1339–1356
52. Van den Berg R (2002) Teachers' meanings regarding educational practice. Rev Educ Res 72(4):577–625
53. Vasili MR, Tang SH, Vasili M (2012) Automated storage and retrieval systems: a review on travel time models and control policies. Springer, London
54. Wang J, Zhang N, He Q (2009) Application of automated warehouse logistics in manufacturing industry. Paper presented at the 2009 ISECS International colloquium on computing, communication, control, and management
55. Xu X, Zhao X, Zou B, Gong Y, Wang H (2019) Travel time models for a three-dimensional compact AS/RS considering different I/O point policies. Int J Prod Research 1–24
56. Yue L, Guan Z, He C, Luo D, Saif U (2017) Slotting optimization of automated storage and retrieval system (AS/RS) for efficient delivery of parts in an assembly shop using genetic algorithm: a case study. Paper presented at the IOP Conference Series: Materials Science and Engineering

57. Zollinger H (2014) AS/RS application, benefits and justification in comparison to other storage methods: A white paper
58. Zrnić N, Kosanić N, Borovinšek M, Lerher T (2017) A Multi-Objective Optimization model for minimizing cost, travel time and CO2 emission in an AS/RS. FME Transactions 45(4):621

An Experimental Design Study to Identify Significant Factors Affecting Tier-to-Tier SBS/RS Performance

M. Kucukyasar and B. Yetkin Ekren

Abstract The aim of this paper is to study a simulation-based experimental design to identify significant factors affecting the performance of the tier-to-tier shuttle based storage and retrieval system (SBS/RS). SBS/RS design is an automated warehousing technology mostly utilized for the storage of mini-loads. SBS/RS has emerged as a response to the recent order profile with low volume and wide variety and quick response. Although the traditional design of SBS/RS is comprised of multi-shuttles each dedicated in each tier of an aisle, there can be different designs in which shuttles can travel between tiers. We call this design as tier-to-tier SBS/RS where shuttles can travel between tiers by using a separate lifting mechanism located at an endpoint of each aisle. Note that in this new design, since shuttles are allowed to travel between tiers, there may be less number of shuttles running in the system. Hence, the shuttle's average utilization value might be larger compared to that traditional one where each tier has a dedicated shuttle. In this work, we study an experimental design to investigate which factors are statistically significant on a critical performance metric (i.e., average cycle time per transaction) in the studied tier-to-tier SBS/RS. First, we simulate the system by using the ARENA 16.0 commercial software and then implement a full factorial analysis. The results are analyzed in MINITAB 17.0 software.

Keywords Shuttle-based storage and retrieval system · Tier-to-tier shuttle · Automated warehousing · SBS/RS · DOE

M. Kucukyasar (✉) · B. Y. Ekren
Department of Industrial Engineering, Faculty of Engineering, Yasar University, Bornova, Izmir, Turkey

B. Y. Ekren
e-mail: banu.ekren@yasar.edu.tr

F. Calisir (ed.), *Industrial Engineering in the Internet-of-Things World*, Lecture Notes in Management and Industrial Engineering,
https://doi.org/10.1007/978-3-030-76724-2_15

Introduction

With the recent increase in e-commerce, the order profile has changed towards more product variety with decreased volume and short response time. This demand profile has also created more competitive business strategies. For instance, distribution centers tend to adopt recent warehousing technologies to cope with that recent order profile efficiently [4, 7, 8]. Automated storage and retrieval systems play a significant role in those technologies, one of which is shuttle-based storage and retrieval system (SBS/RS). An SBS/RS is typically composed of storage racks, storage, and retrieval (S/R) devices that are shuttles, and lifting mechanisms located at an endpoint of each aisle in the racking system. There is a dedicated shuttle in each tier of an aisle which cannot travel between different level of tiers and aisles. These shuttles are tier and aisle captive shuttles. In that design, due to having a dedicated shuttle in each tier, the throughput rate is high, while shuttle utilization is low. Lifts are usually bottlenecked in those designs. In an effort to reduce the total number of shuttles and increased utilization of them in the system, a new design of SBS/RS where shuttles can travel between tiers or aisles, etc. may worth studying to investigate how system performance is affected. Lerher et al. [11] show that SBS/RS works with a high throughput rate by reduced cycle time compared to its simulants. They focus on a traditional tier-captive SBS/RS design in which each tier of the racking system has a dedicated shuttle. Differently, in this paper, we study a tier-to-tier SBS/RS where shuttles can travel between tiers by using a lifting mechanism. The advantage of a tier-to-tier SBS/RS design is that there are fewer shuttles and hence higher utilization levels compared to a traditional tier-captive design.

Literature studies mostly focus on tier-captive SBS/RS design [4, 12]. One of the initial works is provided by Carlo and Vis [1]. They propose a heuristic approach for scheduling of non-passing lifting mechanisms in a tier-captive SBS/RS. Lehrer et al. [13] study analytical travel time models by considering characteristics of shuttles and lifts as acceleration, deceleration, and maximum velocity profiles. They study for both single and dual command scheduling rules in SBS/RS. Ekren [2] studied simulation modeling for the design of a traditional tier-captive SBS/RS. In order to evaluate the performance metrics based on pre-defined designs promptly, several graphs under various design concepts are drawn. Ekren et al. [3] introduced an analytical model-based tool that can estimate the mean and variance of travel time of shuttles and lifts as well as their energy consumption values by given parameters in tier-captive SBS/RS configuration. Recent studies are performed by Ekren [5] and Ekren [6]. Ekren [5] presents a simulation-based experimental design analysis to identify significant factors affecting the performance of the studied tier-captive SBS/RS. Ekren [6] proposes a multi-objective optimization procedure for the design of SBS/RS. In that study, as multi-objectives, simultaneously minimization of average cycle time and energy consumption per transaction is considered.

In literature, there are few studies considering tier-to-tier SBS/RS studies. One of the earliest studies is published by Ha and Chae [9]. They compare two operation approaches called free and non-free balancing. They aimed to prevent the collision of

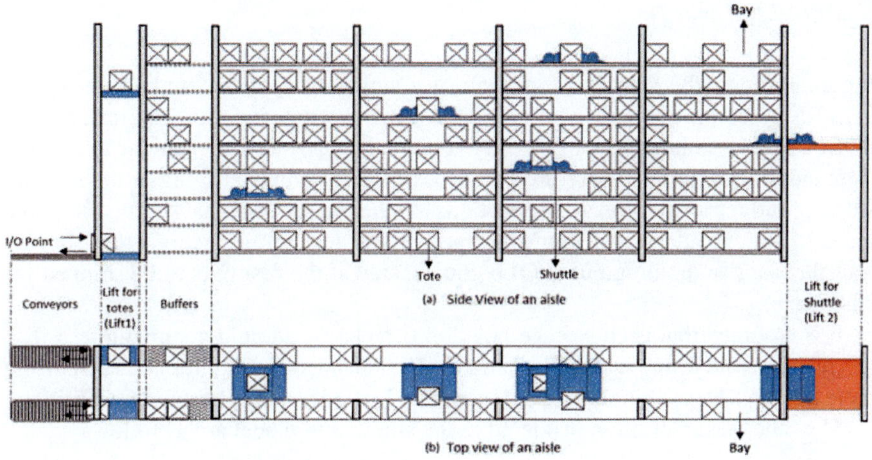

Fig. 1 The physical configuration of a tier-to-tier SBS/RS

shuttles by using these approaches. Later, Ha and Chae [10] studied a decision model to determine the number of shuttles dedicated to each aisle for tier-to-tier SBS/RS. Finally, Zhao et al. [14] introduced an integer programming scheduling model to minimize the idle time of lifts and shuttles. However, they only consider the retrieval process in this study. Differently, in this work, we conduct an experimental design to find out significant factors affecting the performance of the tier-to-tier SBS/RS.

Methodology

In an effort to decrease the number of shuttles and increase the utilization of shuttles in the system, we study a tier-to-tier SBS/RS design where shuttles can travel between tiers by using a separate lifting mechanism located at an endpoint of each aisle. Shuttles are considered to be aisle-captive, meaning that each is dedicated to a specific aisle. However, they can travel between tiers within that aisle. There are two lifting mechanisms serving for each aisle. One is for vertical travel of loads (i.e., totes), and its capacity is two; one is for vertical travel of shuttles, and its capacity is one. System assumptions, along with its definition, are given in the following section. Note that for the modeling purpose, the system is simulated by using the ARENA 16.0 simulation software.

System Definition

Figure 1 shows the side and top view of a single-aisle of the studied tier-to-tier SBS/RS. According to that, Lift 1 is dedicated to travel of totes, and Lift 2 is dedicated to the travel of shuttles. Note that there are two storage sides at each tier as left and right sides at each tier. There are two types of transactions arriving at the system, storage, and retrieval. Storage transactions are the transactions that shuttles store the loads to storage addresses. Retrieval transactions are the transactions that shuttles transfer loads to input/output (I/O) points located at the first tiers to be shipped for customers.

It is assumed that each storage position (i.e., bay) can hold a single tote. Lift 1 has two separate lifting tables that can travel independently. Shuttles and each lifting table of Lift 1 have a single tote capacity, and Lift 2 has a single shuttle capacity.

The other assumptions considered in the simulation model are as follows:

- Incoming transactions arrive at the Input/Output (I/O) points at the first tier of each aisle.
- For storage transactions, first, totes are dropped-off by Lift 1 at the buffers. Then, they are picked up by shuttles to be stored in their bay addresses.
- For retrieval transactions, totes are dropped-off by shuttles at the buffer locations. Then, they are picked up by Lift 1 to be released at the I/O points.
- If a transaction is at the first tier, Lift 1 is not utilized.
- The buffers at each tier can hold three totes at a time.
- Dwell point policy of lifts and shuttles is assumed to be the point where they complete their last process.
- When a shuttle is assigned for a transaction located at a tier different than the shuttle's current tier, the shuttle travels to the Lift 2 location immediately while it also enters the Lift 2 queue to seize it. The two devices travel simultaneously if Lift 2 is seized before the shuttle arrives at the Lift 2 location.
- To prevent the collisions of shuttles at the same tier, a single shuttle is allowed to travel within a tier.
- Storage and retrieval transactions arrive with equal mean and probability distributions. The mean arrival rate follows a Poisson distribution.
- Lift and shuttle acceleration and deceleration delays are considered to be 2 m/s^2. The maximum velocity that they can reach is also considered to be 2 m/s.
- A pure random storage policy is considered.
- Loading and unloading time delays for totes from/onto shuttles/lifts are ignored.
- Distance between two adjacent bays and distance between the buffer locations and the first bay is equal and 0.5 m in each tier. Besides, the distance between two adjacent tiers is equal and 0.35 m.
- The weight of a shuttle, Lift 1, Lift 2 and tote are considered to be 40 kg, 60 kg, 60 kg, and 20 kg, respectively.

Table 1 shows the notations that are used in the system model. The simulation flow charts of the system are given in Fig. 2.

Table 1 Notations used in system modeling

Notation	Description	Unit
C	Average cycle time	sec/transaction
T	Number of tiers of an aisle	
A	Number of aisles in the warehouse	
B	Number of bays in a tier	
N_s	Number of shuttles in each aisle	
λ	Mean arrival rate	Number of transactions/month
t_i	i-th transaction at the transaction queue	
s_j	j-th shuttle	

Note that there are three separate queues in the system. They are queues for Lift 1 tables, Lift 2, and shuttles. Each is processed based on the first-come, first-serve rule. In the simulation model, the transaction selects the most suitable shuttle for itself. In an effort to prevent the collisions of shuttles, first, the transaction checks whether or not there is a shuttle running at its targeting tier. If there is, that transaction enters the queue of that shuttle. If there is not a shuttle at that tier, then the transaction selects an available shuttle, randomly.

When the selected shuttle needs Lift 2 to change its tier, while it is traveling to the Lift 2 location, it also enters the Lift 2 queue simultaneously. Namely, while Lift 2 travels to the shuttle's tier, the shuttle also travels to the Lift 2 location at its current tier. To summarize, when a transaction selects a shuttle, Lift 1 and Lift 2 are triggered simultaneously, if necessary. The flow details can also be seen in Fig. 2.

The simulation model is run for forty days with ten days warm-up period. The warm-up period is determined by the eye-ball technique by checking the average cycle time per transaction performance metric. Verification is done by debugging the codes as well as animating the model. The validation is done by checking the simulation model results by an expert. The performance metric, average cycle time per transaction, is calculated for the averages of five independent replications. Those values are given in the following sections with their 95% confidence intervals.

Experimental Design

In this study, we conduct an experimental design to identify significant factors affecting the average cycle time per transaction. Design of experiment (DOE) is a statistical technique that analyzes relationships between independent (i.e., factor) and depending (i.e., response) variables in order to specify significant factors affecting

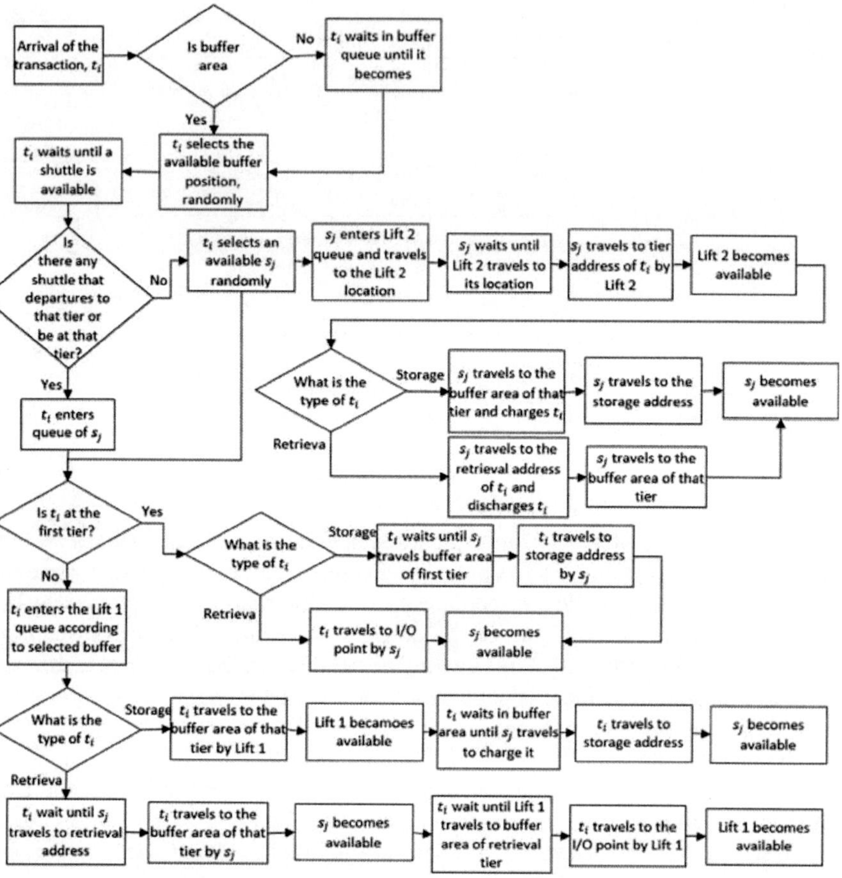

Fig. 2 Flow chat of tier-to-tier SBS/RS

the responses (i.e., average cycle time per transaction). Here, cycle time is defined as the time that a transaction spends in the system. Namely, it is the time between when it is created until disposed of. In DOE analysis, the factors that could affect the performance metric are defined to be:

- Number of tiers in each aisle, T
- Number of aisles in the warehouse, A
- Number of shuttles in each aisle, N_s

We consider that the total warehouse capacity is fixed with 12,000 storage positions. The number of bays is adjusted so that we fix the warehouse capacity to 12,000 bays as in (1). Because of the negative correlation between any two variables in Eq. (1) when the capacity is fixed, only T and A values are considered as design factors in the DOE.

Table 2 DOE for SBS/RS

Factor level	T	A	N_s
1	10	12	3
2	20	30	5

$$T \times A \times B = 12{,}000 \tag{1}$$

A full factorial design is considered in DOE, where simulation is run for all possible combinations of levels of factors. Table 2 shows the designs for DOE. Based on that, there will be $2^3 = 8$ combinations in DOE. The arrival rates for each combination are arranged such that we obtain a 95% average shuttle utilization. In other words, we check the throughput rate capacity of each design at a 95% utilization level of shuttles, which are the bottleneck in the system.

Remember that there are 8 combinations in the considered full factorial analysis of DOE. By five replications of each combination, we have 40 results to enter in Minitab. First, we complete an analysis of variance (ANOVA) to obtain significant factors affecting the performance metric, average cycle time per transaction in the system. Second, by using the information obtained by ANOVA, we develop a regression function based on significant factors. Analyses are given in the following section.

ANOVA and Regression Results

Table 3 shows the simulation averages of five replications for each combination from Table 2 designs. ANOVA and regression analyses are completed based on those results.

Table 3 DOE results for SBS/RS

Design no	T	A	N_s	C		
1	1	1	1	71.62	±	0.75
2	1	1	2	63.17	±	0.49
3	1	2	1	34.22	±	0.22
4	1	2	2	35.35	±	0.05
5	2	1	1	52.28	±	0.46
6	2	1	2	42.20	±	0.06
7	2	2	1	35.34	±	0.19
8	2	2	2	33.83	±	0.09

ANOVA Results for the C *Performance Metric*

Table 3 results are entered in the Minitab 17.0 software to complete the ANOVA. ANOVA test shows the statistically significant factors affecting the performance metric. However, ANOVA depends on some model adequacy assumptions. Model adequacy requires that ANOVA residuals should be normally distributed. They should have a mean of zero and have a constant variance. If one of them is not met, then the results are not confidential, and we may need a response transformation such as inverse, logarithm, square root, etc. In the current model, because there was a non-constant variance, the model adequacy was assumed not to be satisfied well. Hence, an inverse transformation on C is applied (e.g., $1/C$). The residual plots are given in Fig. 3 of that inversed transformed results. According to that, the ANOVA assumptions are assumed to be satisfied well. The ANOVA results in Table 4 are interpreted as follows:

- It is understood that all one-way, two-way, and three-way interactions have a significant effect on $1/C$. This is because, all the *p*-values are smaller than the desired α-value, 0.05.
- Since the F-value is the highest one in that factor, the most significant factor affecting $1/C$ is observed as A. The reason for this might be when A increases, the number of lifts in the system also increases.

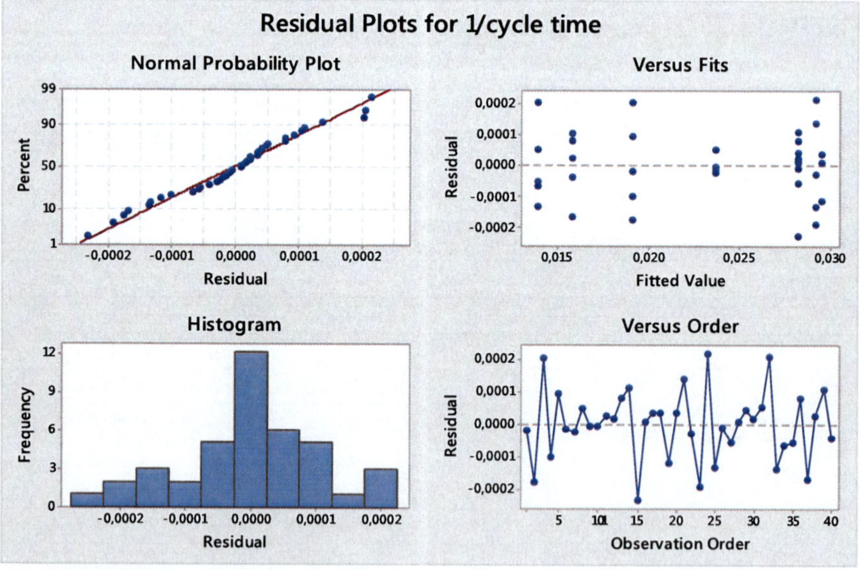

Fig. 3 Residuals distribution plots for $1/C$

Table 4 ANOVA results for $1/C$

Factor Information

Factor	Type	Levels	Values
Tier	Fixed	2	10; 20
Aisle	Fixed	2	12; 30
Shuttle	Fixed	2	3; 5

Analysis of Variance

Source	DF	Adj SS	Adj MS	F-Value	P-Value
Tier	1	0,000112	0,000112	8275,14	0,000
Aisle	1	0,001142	0,001142	84623,06	0,000
Shuttle	1	0,000029	0,000029	2112,03	0,000
Tier*Aisle	1	0,000101	0,000101	7458,87	0,000
Tier*Shuttle	1	0,000015	0,000015	1111,80	0,000
Aisle*Shuttle	1	0,000023	0,000023	1732,48	0,000
Tier*Aisle*Shuttle	1	0,000000	0,000000	11,96	0,002
Error	32	0,000000	0,000000		
Total	39	0,001422			

Regression Analysis for the C *Performance Metric*

After determining the significant design factors by ANOVA, we develop a regression function for $1/C$ to find out how $1/C$ changes based on the design factors, A, T, Ns. Regression analysis is a tool that builds a function to find the expected value of the dependent variable by using independent variables. First, we do ANOVA and find out significant factors that could be included in the regression. Then, we develop the regression function by using that information. By that function, one can estimate the performance metric for different levels of design parameters that are not already experimented. To check the adequacy of a regression model, the parameter, coefficient determination, R^2 is used. R^2 refers to the amount of variability of data. It is preferred that this value is high. Table 5 shows the Minitab regression analysis results. In that table, the $R^2 = 99.97\%$ and large enough to accept it as a good fit function. $R^2_{adj} = 99.96\%$ is also large and close enough to the R^2 value to accept the function as a good fit.

Table 5 Regression results for $1/C$

Model Summary

S	R-sq	R-sq(adj)	R-sq(pred)
0,0001162	99,97%	99,96%	99,95%

Coefficients

Term	Coef	SE Coef	T-Value	P-Value	VIF
Constant	-0,006472	0,000608	-10,64	0,000	
Tier	0,000466	0,000038	12,12	0,000	109,56
Aisle	0,001378	0,000027	51,77	0,000	170,00
Shuttle	0,000346	0,000147	2,35	0,025	64,44
Tier*Aisle	-0,000030	0,000002	-17,59	0,000	262,56
Tier*Shuttle	0,000152	0,000009	16,31	0,000	167,56
Aisle*Shuttle	-0,000064	0,000006	-9,88	0,000	224,44
Tier*Aisle*Shuttle	-0,000001	0,000000	-3,46	0,002	311,56

Regression Equation

1/cycle time = -0,006472 + 0,000466 Tier + 0,001378 Aisle + 0,000346 Shuttle
- 0,000030 Tier*Aisle + 0,000152 Tier*Shuttle - 0,000064 Aisle*Shuttle
- 0,000001 Tier*Aisle*Shuttle

Table 6 An example for prediction in regression model

```
Regression Equation

1/cycle time = -0,006472 + 0,000466 Tier + 0,001378 Aisle + 0,000346 Shuttle
               - 0,000030 Tier*Aisle + 0,000152 Tier*Shuttle - 0,000064 Aisle*Shuttle
               - 0,000001 Tier*Aisle*Shuttle


Variable  Setting
Tier           15
Aisle          20
Shuttle         4


      Fit    SE Fit            95% CI                    95% PI
0,0229041  0,0000185  (0,0228665; 0,0229418)  (0,0226645; 0,0231437)
```

By using the Table 5 equation, for instance, we can make a prediction for an unexperimented design that is: 15 tiers, 20 aisles, and 4 shuttles per aisle. The prediction result is summarized in Table 6 as an example. According to that, $1/C$ value is obtained to be 0.0229 s. whose 95% prediction interval is (0.0227, 0.0231) s. The function also suggests that a warehouse design with an increased number of aisle and tier works better with a high number of shuttles to reduce average cycle time per transaction.

Conclusion

In this paper, we study a simulation-based experimental design to identify statistically significant factors affecting the performance metric average cycle time per transaction in tier-to-tier SBS/RS. We consider three design factors with two levels: number of tiers per aisle, number of aisles, and number of shuttles per aisle. Warehouse configurations are designed based on a fixed warehouse capacity. We simulate each design under pre-defined factor levels under a 95% utilization of shuttles. First, we complete ANOVA. We observed that all one-way, two-way, and three-way interactions of design factors are statistically significant on the average cycle time per transaction performance metric. By using this information, we also build a regression function to conduct the relationship between design factors and the response. The function suggests that a warehouse with an increased number of aisle and shuttle works well to reduce average cycle time per transaction.

As future work, this study can be extended by considering more levels of factors as well as different performance metrics in the system.

Acknowledgements This work was supported by The Scientific and Technological Research Council of Turkey and Slovenian Research Agency: ARRS [grant number: 118M180].

References

1. Carlo HJ, Vis IF (2012) Sequencing dynamic storage systems with multiple lifts and shuttles. Int J Prod Econ 140(2):844–853
2. Ekren B (2017) Graph-based solution for performance evaluation of shuttle-based storage and retrieval system. Int J Prod Res 55:6516–6526
3. Ekren BY, Akpunar A, Sari Z, Lerher T (2018) A tool for time, variance and energy related performance estimations in a shuttle-based storage and retrieval system. Appl Math Model 63:109–127
4. Ekren BY, Heragu SS (2012) Performance comparison of two material handling systems: AVS/RS and CBAS/RS. Int J Prod Res 50(15):4061–4074
5. Ekren BY (2020a) A Simulation-based experimental design for SBS/RS warehouse design by considering energy related performance metrics. Simulation Modelling Pract Theor, in press, https://doi.org/10.1016/j.simpat.2019.101991
6. Ekren BY (2020b) A multi-objective optimisation study for the design of an AVS/RS warehouse. Int J Prod Res. in press, https://doi.org/10.1080/00207543.2020.1720927
7. Ekren BY, Heragu SS (2011) Simulation-based performance analysis of autonomous vehicle storage and retrieval system. Simulation Modeling Pract Theor 19:1640–1650
8. Ekren BY, Sari Z, Lerher T (2015) Warehouse design under class-based storage policy of shuttle-based storage and retrieval system. IFAC-PapersOnLine 48(3):1152–1154
9. Ha Y, Chae J (2018) Free balancing for a shuttle-based storage and retrieval system. Simul Model Pract Theor 82:12–31
10. Ha Y, Chae J (2018) A decision model to determine the number of shuttles in a tier-to-tier SBS/RS. Int J Prod Res 57(4):963–984
11. Lehrer T, Ekren YB, Sari Z, Rosi B (2015) Simulation analysis of shuttle based storage and retrieval systems. Int J Simulation Modelling 14(1):48–59
12. Lerher T, Ekren BY, Sari Z, Rosi B (2016) Method for evaluating the throughput performance of shuttle based storage and retrieval systems. Tehnicki Vjesnik – Technical Gazette 23: 715–723
13. Lerher T, Ekren BY, Dukic G, Rosi B (2015) Travel time model for shuttle-based storage and retrieval systems. Int J Adv Manuf Technol 78(9–12):1705–1725
14. Zhao X, Wang Y, Wang Y, Huang K (2019) Integer programming scheduling model for tier-to-tier shuttle-based storage and retrieval systems. Processes 7(4):223

Blockchain Technology for Supply Chain Management

Taner Dursun, Fatih Birinci, Busra Alptekin, Isa Sertkaya, Orkun Hasekioglu, Bahadir Tunaboylu, and Selim Zaim

Abstract Blockchain technology offers important opportunities for the supply chain management. This paper aims to overview the employment of blockchain technology in the field of the supply chain. Although the technology has been widely associated with cryptocurrencies, non-financial applications such as supply chain, power, and food industry are also promising. Blockchain can provide a permanent, shareable, auditable record of products through their supply chain, which improves product traceability, authenticity, and legality in a more cost-effective way. In this chapter, the potential improvement expectations via blockchain technology for the case of agribusiness were discussed. The proposed case for automotive manufacturing-micro factory with blockchain technology was also introduced.

Keywords Supply chain · Blockchain · Survey · Smart contract

T. Dursun (✉) · F. Birinci · I. Sertkaya
TÜBİTAK BİLGEM Blockchain Research Lab, Kocaeli, Turkey
e-mail: taner.dursun@tubitak.gov.tr

F. Birinci
e-mail: fatih.birinci@tubitak.gov.tr

I. Sertkaya
e-mail: isa.sertkaya@tubitak.gov.tr

B. Alptekin
Istanbul Şehir University, Istanbul, Turkey
e-mail: busraalptekin@std.sehir.edu.tr

O. Hasekioglu
TÜBİTAK, Kocaeli, Turkey
e-mail: orkun.hasekioglu@tubitak.gov.tr

B. Tunaboylu
Marmara University, Istanbul, Turkey
e-mail: bahadir.tunaboylu@marmara.edu.tr

S. Zaim
Istanbul Sabahattin Zaim University, Istanbul, Turkey
e-mail: selim.zaim@izu.edu.tr

F. Calisir (ed.), *Industrial Engineering in the Internet-of-Things World*,
Lecture Notes in Management and Industrial Engineering,
https://doi.org/10.1007/978-3-030-76724-2_16

Introduction

The definition of a supply chain is given as "the network of organizations that are involved, through upstream and downstream linkages, in the different processes and activities that produce value in the form of products and services in the hands of the ultimate customer" [1]. Products cross at least one border in global supply chains. The global supply chains are generally very large scale formations that may consist of complex patterns of production processes, transactions, and knowledge. World global merchandise exports totaled US$ 19.48 trillion in 2018 [2]. Based on World Bank data, more than $16 trillion in goods are shipped across international borders each year, and the yearly cost of global trade is estimated at $1.8 trillion annually (Maersk Strategy Group, May 2016). Global trade could increase by nearly 15%, boosting economies, and creating new job opportunities if barriers within the international supply chain are removed [3].

In a supply chain, ownership of products changes several times among participants until they are delivered to consumers. For low-added-value products such as agricultural commodities and certain types of mining commodities, supply chains function as an aggregation method by which goods are provided by many small-scale producers to larger-scale supply chain partners for further processing towards an end-product.

Existing supply chain models begin when two supply chain members, namely manufacturers and importers, send their products to the next stage of the supply chain. The next stage, also called the middle layer, includes the wholesaler, which processes the basic products received by the export, processor, and supply chain. In the last step, there is a retailer and foodservice that sells products. The main problem with this model is that the data is encapsulated in elements of the supply chain and shared less. For example, it is not possible for the consumer to verify the source of the food to be purchased [4].

Moreover, sustaining operations across a complex chain of resources, activities, and organizations can be hard for supply chain partners, especially when a large number of smallholders are involved. According to ISO standard 9000, the traceability concept is defined as "the ability to trace the history, application, use, and location of an item or its characteristics through recorded identification data". Lack of visibility and incentives may cause difficulties for sustainability.

Consumers cannot be sure about the reliability of data in current supply chain systems. The existing model becomes even more burdensome in the global supply chain. A reliable system is difficult, even impossible, on a global scale without building trust. With the advent of blockchain as a disruptive technology for most processes related to our daily lives, the transition to the use of blockchain technology [5] has begun to overcome all these challenges of supply chains. Participants and their roles in a typical blockchain integrated supply chain flow can be depicted as in Fig. 1. The tipping point for adoption is expected to be 10% of global gross domestic product (GDP) stored on blockchain ledgers by the expected date of 2027 [6].

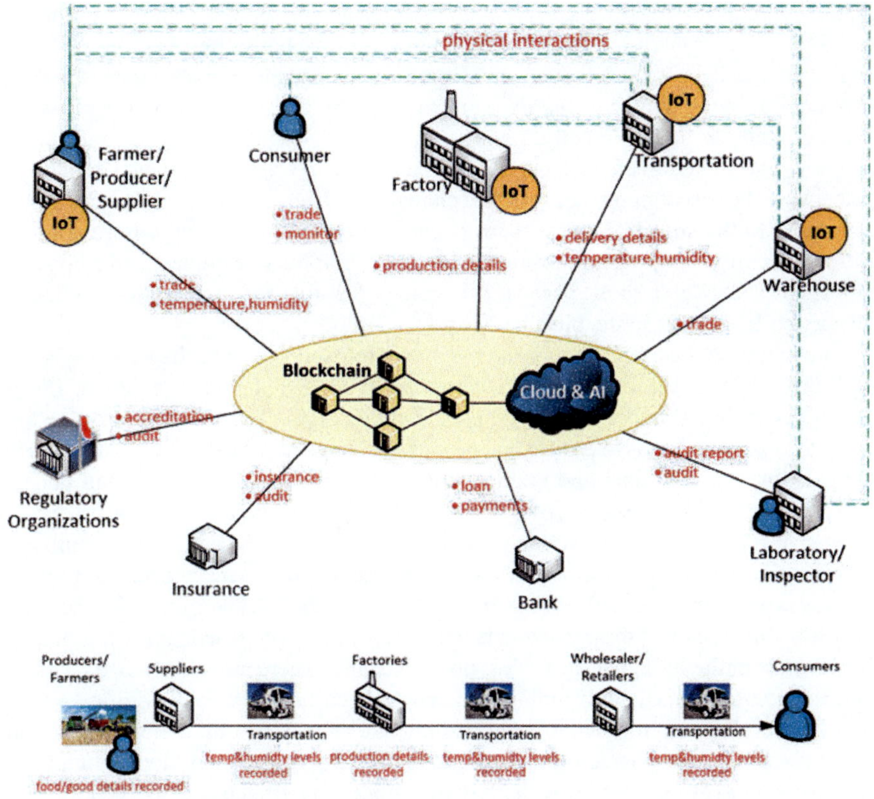

Fig. 1 Participants and their roles in a typical blockchain integrated supply chain flow

In this manuscript, opportunities, challenges, and possible research directions on blockchain-based supply chain solutions are surveyed. We believe this would be a starting point and guide for future studies.

Blockchain Background

The original idea and building blocks of blockchain technology are coming from crypto money and date back to the 1980s. Most recently, in 2008 December, the article by an author nicknamed Satoshi Nakamoto titled "Bitcoin: A Peer-to-peer Electronic Cash System" popularized the blockchain technology. The blockchain concept consists of a combination of mathematics, cryptology, computer, and monetary science.

Blockchain technology, in fact, is a type of parallel and distributed computing architecture [5]. It allows us to eliminate central servers or trusted authorities in the digital interactions of partners. Thus, it is classified as a disruptive technology that has the potential to transform most of the processes in our daily life radically. Simply, copies of the data, called ledger, are stored on thousands of computers working together, and all changes to the data are realized by consensus of partners. Every change made on data is recorded with a timestamp to ensure integrity and transparency. The stakeholders of the system do not have to trust each other. The factors that ensure the trust among them are that changes on the stored data can only be made according to the specified rules, these changes are kept in a ledger whose content is transparently open to audit, protected by cryptographic techniques, and a copy of this chain is available to all parties. It becomes possible for digital data to change ownership like assets in the physical world.

The main consensus protocols and mechanisms used to ensure the trust are Proof of Work (PoW), Byzantine Fault Tolerance (BFT), Proof of Stake (PoS), Proof of Authority (PoA) and Proof of Elapsed Time (PoET). The main purpose of the consensus mechanism is to ensure that proposed change requests are compatible with the existing status of data and defined rules. Blockchain computers, called nodes, perform these validations. Cryptography is mainly used to ensure the authenticity of change requests on data and the immutability of data in the ledger by organizing modification history as blocks that are cryptographically connected to each other. Privacy is another important issue in the blockchain. Thus, cryptography is also used to ensure the privacy of the participants. High availability of the ledger is provided by keeping the entire ledger at the nodes, not at a center. There are mainly two types of blockchain platforms namely, public and private according to the accesibility policy. In a public blockchain, anyone can send change requests to the network and can operate a node. In a private blockchain, also called permissioned blockchain, both sending requests to the network and having a node is restricted to a set of actors.

Problems of Supply Chain and Opportunities with Blockchain

The main objectives of the supply chain are listed as cost, quality, speed, dependability, risk reduction, sustainability, and flexibility [7]. Manufacturing has been globalized, leads well-defined supply chain management more crucial and valuable. In today's supply chain systems, it is difficult for customers to know exactly the value of a product due to a lack of transparency. In addition, investigating supply chains is mostly not feasible in case of suspicion of illegal or unethical activities. Heavy paperwork, process costs, and slow processes are other main challenges of the supply chain.

A literature survey on the research focuses on blockchain for the supply chain domain [2, 8–14] shows that the supply chain domain already benefits from blockchain technology because of its four main features (Fig. 2).

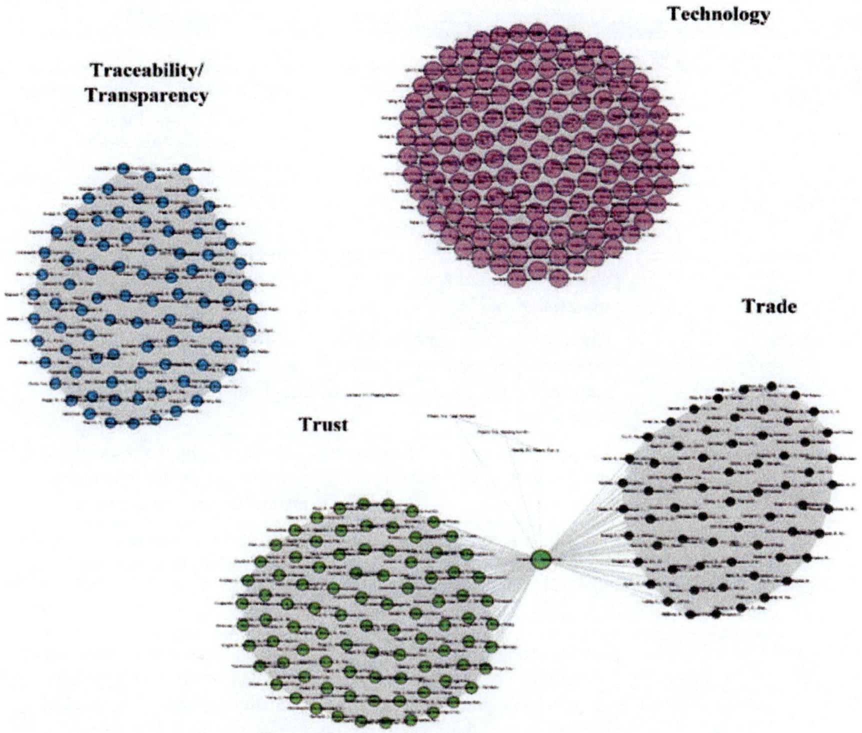

Fig. 2 Research focuses in literature for deployment of blockchain for supply chain [2]

In order to give an idea how blockchain might impact the needs of supply chain actors, we quoted Table 1 from literature. It presents Litke et al. [15] summary of how blockchain responds to the limitations that the actors of supply chains encounter today.

As a summary, Blockchain technology can benefit in many ways in the supply chain as it does in many other application areas. Employing blockchain in supply chain processes provides transparent, decentralized, secure, faster, and low-cost transactions. By eliminating unnecessary third parties and covering more daily life processes in digital systems minimizes paperwork. Blockchain establishes trust among trading partners. Making more detailed data available in blockchain improves supply chain monitoring ability and safety. This reduces insurance risks. Smart contracts and automated payments are game-changers. They add efficiency and remove bureaucracy, especially in insurance, and traceability. They also allow escrowed payment by keeping money until terms of the deal are met and agreed and then releasing automatically. Blockchain technology, in fact, provides missing infrastructure the cutting edge technologies need. Thus, increasing focus on providing integration and cooperation with technologies such as Artificial Intelligence, Big Data

Table 1 How blockchain can improve the existing limitations of supply chains [15]

Supply chain participants	Current limitations	Blockchain's impact
Producer	Lack of ability to prove the origin and quality metrics of products transparently	Benefits from increased trust of keep track of the production raw material and value chain from producer to consumer
Manufacturer	Limited ability to monitor the product to the final destination. Limited capabilities of checking quality measured from raw material	Added value from shared information system with raw material suppliers and distribution networks
Distributor	Custom tracking systems with poor collaboration capabilities. Limited certification ability and trust issues	Ability to have proof-of-location and conditions certifications registered in the ledger
Wholesaler	Lack of trust and certification of the products' path	Ability to check the origin of the goods and the transformation /transportation conditions
Retailer	Lack of trust and certification of the products' path, Tracking of products between consumers and wholesalers	Ability to handle effectively the return of malfunctioning products
Consumer	Lack of trust regarding the compliance of the product with respect to origin, quality and compliance of the product to the specified standards and origin	Full and transparent view on the product origin and its whole journey from raw material to final, purchased product

Analytics, Cloud Computing, and IoT will help to realize advanced supply chain systems.

IoT, for example, is already used in the supply chain domain to collect information about environmental conditions, to verify how long and under what conditions a cargo is loaded on a particular truck or port, and whether it is tampered with or affected by any incident that violates the shipping guidelines. IoT sensors also help companies to optimize and improve their supply chain systems. McKinsey Global Institutes research suggests that "Economic size of IoT applications globally will reach between 4 and 11 trillion by 2025.[1] However, using IoT and blockchain technology together will leverage systems thanks to the trust level achieved via blockchain technology. Matching data from the physical and digital world and validating their consistency will provide more trust between parties and prevent invoice disputes. Thanks to the robust digital infrastructure of IoT devices and sensors integrated with blockchain platforms, consumers can track the entire product life cycle throughout the supply chain.

[1] https://www.mckinsey.com/industries/semiconductors/our-insights/whats-new-with-the-internet-of-things.

As the recent situation by coronavirus pandemic showed the importance of supply chain infrastructures to communicate with multiple ecosystems. As the existing supply chain network has been highly affected by Covid-19, thus interoperability and compatibility seem to be crucial for the global supply chain after this pandemic. For the purpose, blockchain technology provides disconnected supply chain systems with low cost and maximum efficiency to interoperate securely [16].

Despite all its benefits, however, blockchain technology is not a "one size fits all" solution. New problems raise related to blockchain, and other assisting technologies should also be addressed to realize projects in the supply chain domain. Blockchain-based supply chain systems need various new legislative regulations. Current blockchain platforms cannot exactly fit the high level of transaction throughput requirements of supply chain systems. Supply chains combine diverse participants with varying interests. Thus, incentives need to be provided, such as efficiency gains, improved liquidity, and data security to motivate all participants. Security and privacy are other important issues. The data security concerns with IoT and lack of commonly accepted baseline protocol standards for IoT interaction. The current IoT ecosystem is built on a central model in which IoT devices are identified, connected, and validated. Thus, there is a need for transformation for blockchain adaptation.

Impact of Blockchain on the Logistics Industry

This subsection highlights the blockchain-based supply chain projects focused on rather Logistics subdomain.

Supply chain actors and startups focused on providing blockchain-based supply chain solutions to improve efficiencies and reduce operating costs. There has been a large number of ongoing projects. For example, Global Shipping Business Network (GSBN), created by nine ocean carriers and terminal operators, and global cross-border supply chains by IBM and Maersk collaboration, will be made available for the ocean shipping and logistics industry. United Parcel Service (UPS) employs blockchain to combine with the truckload pricing futures market, by enriching the real-time matching of loads and empty trucks with a data analytics and artificial intelligence. Similarly, Waltonchain is a blockchain/supply chain project to incorporate both digital and physical elements into one system. It employs IoT devices and RFID chips to ensure the security, traceability, and authenticity of the business.

For mining and jewelry industries, tracking assets from the mines to the consumers is important. Tracing an asset back to its provenance facilitates proving ownership in cases of thefts and attesting authenticity. Due to diamonds' high value, annually, $45 billion is lost to insurance fraud, and 65% of false claims pass undetected. EverLedger and De Beers, Jewelry Company, use blockchain to track assets throughout their lives with the aim of assuring the consumers for buying genuine articles. BHP Billiton, as the world's largest mining firm, plans to benefit from blockchain technology for better tracking and recording data throughout the mining process.

In addition to samples given, general-purpose, or logistic specific significant projects in the supply chain domain are summarized in Table 2.

Table 2 General-purpose or logistic specific significant projects in supply chain domain (extended version of Justin, 2019 [17])

Name/started in	Description	Partners
VeChain (VET) 2015	Smart contracts to track inventory. It tokenizes products and tracks each step through RFID labels as they navigate through the supply chain. The system allows us to view each historical detail of the product at any point in its lifecycle in the supply chain. Example of applications includes cold-chain logistics, automobiles, medical and healthcare, luxury and fashion	BMW, Haier, BIOS, BYD, DIG, DB Schenker
WaltonChain (WTC) 2016	It tracks objects via proprietary RFID technology. Detailed information is provided about locations a product passes, who handled it, and steps relating to the supply chain process	Fashionchain, MoneyNet, Huodull, Mitoq, Freyrchain
Ambrosus (AMB) 2017	Ethereum smart contract-based supply chain system, including proprietary IoT devices. It can be integrated easily into any industry or market that relies on supply chain or logistics. Ambrosus makes food and medicine safer by combining high-tech sensors and blockchain technology	BioFirm, Nestle, Cantone Group, Trek Therapeutics, Crypto Valley Association
OriginTrail (TRAC) 2013	A process sharing data throughout the supply chain. Businesses have an easy and effective way to exchange data, both in-house and across borders. A transparent way to increase accountability and efficiency	Ferdinand, Perutina, Natureta, Planika, H-Farm
Modum (MOD) 2016	A startup combining IoT devices and smart contracts to provide supply chain data to its clients as it becomes available in real-time. Modum users know if their product has been tampered with and whether or not they receive shipments on time	Swiss Commission for Technology, Institute for Supply Chain Management, and Communications Systems Group

(continued)

Table 2 (continued)

Name/started in	Description	Partners
Tael (WABI) 2017	A project which develops solutions for the authenticity of products. With the anti-counterfeit QR codes, a consumer can ensure that a product is valid and has not been tampered with prior to purchasing. It is popular in lower regulated countries like China	Blackmores, Nutrilon, Swisse, Nature's Fare
CargoCoin (CRGO) 2018	A project employing smart contracts to create a secure method of storage and transfer for tokenized goods in many different supply chain industries (shipping via land, sea, and air). It provides effective communication between cargo traders and transporters on a global scale. This allows providing a method for sending, receiving, rejecting, approving, or signing necessary documentation to all stakeholders of the supply chain process	Bancor, H&B, Bitrue, NoBar, CargoLine
Bext360 2016	Online platform to track goods on a public blockchain system throughout each step in the supply chain. This provides transparency. Assets are represented as tokens. and stored on Bext360 network to manage payments, smart contracts, and asset tracking throughout the entire lifecycle	Great Lakes Coffee, Coda Coffee, Moyee Coffee
ShipChain (SHIP) 2017	A project to provide end-to-end visibility for all products transported. The main goals are to lower costs, reduce theft and fraud, and reduce transaction times. Smart contracts are used to shorten the communication path between shippers and carriers. On an Ethereum blockchain, it only allows authorized parties to view details of the process	CaseStack, ScanLog, Sweetbridge, Direct Outbound

(continued)

Table 2 (continued)

Name/started in	Description	Partners
CargoX (CXO) 2015	A project using Ethereum blockchain technology to develop a cost-effective and smart bill of lading solutions that can process from anywhere in the world	TPG Logistics, MakerDAO, Europacific, Fracht AG, Mana, and DBA Group
ZetoChain 2018	It is the blockchain-enabled IoT solution to provide enhanced security, traceability, and scalability for food safety via IoT sensors	
Irene Energy 2018	Irene Energy operates a Stellar blockchain to create transparency in electrical supply chain management	
Devery 2018	Devery is a decentralized ecosystem providing developer tools to utilize Ethereum blockchain technology to secure supply chain & product verification	
EverLedger 2016	The platform enables buyers and sellers of high-value assets to trade with confidence. Its blockchain provides a secure record of an asset's origin and journey	Shairu & Atit Diamonds, Fred Meyer and Littman Jewelers

Impact of Blockchain on Agricultural and Food Industries

As summarized in Fig. 3, the Agriculture and Food Industries already take advantage of blockchain technology in order to get improvement in the various aspects of the supply chain such as traceability, insurance, finance, transaction, and optimization.

The projects and solutions employing blockchain are listed below as grouped by categories [18, 19, 13].

Agriculture trade and transactions.

- AgriChain—A blockchain company that focuses on end to end agricultural processing and reducing intermediaries.
- AgriDigital—A cloud, IoT, and blockchain-based commodity management solution for the global grains industry. It connects farmers, buyers, site operators, and financiers through a single platform. They are allowed to contract, deliver, and make payments securely and in real-time.
- AgriLedger—British social enterprise project that supports farmers' food origins, easier access to financing, and storage of transaction data.

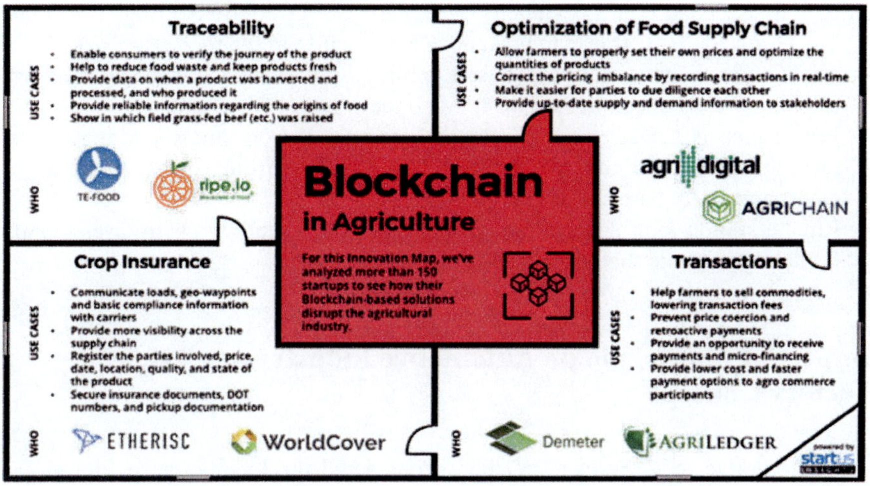

Fig. 3 The areas within the agricultural sector improved by blockchain (Startus Insights 2018 [13])

- AgUnity, FAO, Gates Foundation and International Finance Corporation (IFC) farming cooperatives run for and by local farmers. Collective bargaining, creating a network between farmers within which they can agree to share equipment and resources and helping farmers to select more profitable crops to plant.
- Agroxy- Ukrainian based agricultural product trading system.
- Demeter—A central hub for the leasing and cultivation of microdomains anywhere in the world without intermediaries, complexity, or the burden of a large organization.
  Agriculture finance and insurance.
- Worldcovr—Developing a product insurance system to protect against loss of yield by using satellites to monitor precipitation and trigger payments automatically.
- Trado—A consortium for innovative blockchain supply chain finance structure
- Etherisc—A blockchain company that offers farmers product insurance through decentralized insurance practices.
- FARMS, Financial and agricultural risk management platform integrated with satellite sensing for smallholders
  Food supply chains.
- One of the biggest retailers in the world, Walmart, is developing blockchain-based traceability for lettuce participation.
- Walmart is trying to keep track of fresh and leafy greens products back to the farm by using a blockchain platform developed by IBM. Walmart also uses blockchain technology to track its meat from China [20]. For this purpose, data such as cold chain processes and sales dates are stored in the blockchain.
- Global companies like Unilever, Nestle, and Dole also use blockchain for similar purposes.

- IBM Food Trust is a blockchain-based use case for cold chain logistics. Food Trust aims to create transparency and accountability in the food supply chain.
- TE-FOOD—With a Germany-based public permissioned blockchain platform (2016), customers can track their food all the way to their tables.
- Provenance is a UK startup (2018) to trace food. Consumers are able to get more information about where their products come from and if something is environmentally harmful or genuinely safe.
- Ripe, a startup aiming for a transparent digital food supply chain, uses quality food data to create the Food Blockchain that maps the food journey.

Impact of Blockchain on Automotive Industry and Micro Factory Concept

The automotive industry has long been considered the locomotive of the global supply chain. Blockchain-related supply chain use cases are also considerable for the automotive industry. Figure 4 summarizes the potential application areas for the automotive industry to use blockchain technology.

Since smart factories enable rapid retooling, small-scale manufacturing facilities may actually have a competitive advantage to produce customized products for specific local needs, at a local scale. Micro factory concept is a novel production facility to manufacture new generation electric vehicles.

Local Motors established automobile-building micro-factories across Phoenix, Las Vegas, and Knoxville and host an open library of vehicle designs to include and empower communities of global designers, engineers, manufacturers, and automotive

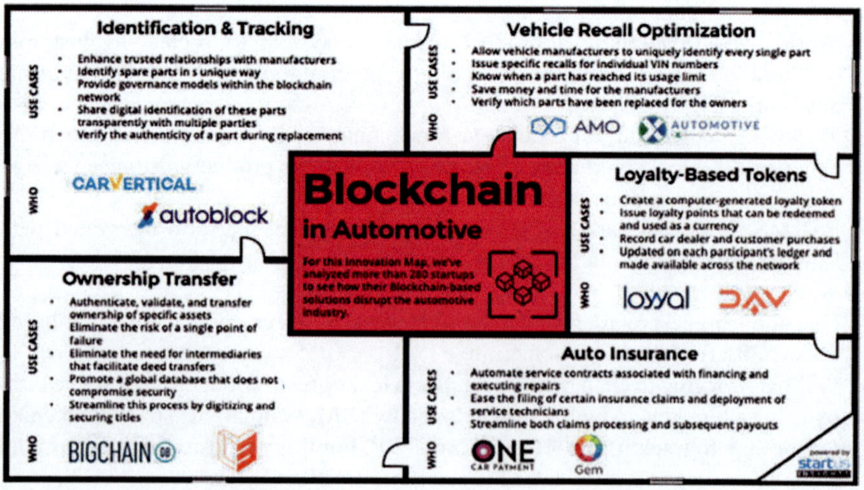

Fig. 4 The areas within the automotive sector improved by blockchain (Startus Insights 2019 [12])

enthusiasts. It can be claimed that, with the advanced technology, there is a potential for a decentralized production to democratize the manufacturing function, as the Internet makes the information accessible for everyone, designing and manufacturing would be more accessible to everybody within distributed manufacturing. However, weak supply chain infrastructure can lead to high distribution costs.

For this reason, supply chain security is critical for micro factory networks. The well-defined supply chain for components such as the motor/inverter, tubular SHSS, electronic board, lights, seats, battery cells, etc. is required. It can be improved by integrating blockchain technology into the supply chain for future studies.

Each micro factory/country has the capability to create its own brand with customization according to the demand of its local market. The proposed network model for micro-factories is the collaboration to make their local brand stronger and compatible in the already captured automotive industry. Therefore, a globally distributed network will be established with the replication of micro factories. To succeed in this highly distributed network, central authority is required. However, it is not defined in sharing economy concepts developed for micro factories. The distribution of data into the network securely without intermediaries is a challenge of this model. However, this can be overcome by implementing blockchain technology, which allows transacting peer-to-peer and trusts each other by using collaboration and cryptography without a third party [21]. It means all transactions will perform between related micro factories and engaged suppliers, a central authority to control and manage the system is not needed. The data of all micro factories distributed across a global ledger, using the highest level of cryptography. To hack a blockchain network, someone needs to reach all computers in that network, which is almost impossible. Thus, blockchain offers a more secure model, which can be adaptable to the micro factory network. Consequently, this micro factory network consists of a smart contract between parties that allow them programmatically to define the rules and steps that should be performed any time and a certain type of event is recorded in the blockchain.

Blockchain is a simple way of passing information in a fully automated and safe way without the need for third parties since 'central authority' does not exist. Besides, it provides transparency, speed, accessibility, and coherence so that the blockchain logic fits the business model of micro factory networks.

Discussion and Conclusion

This study suggests that blockchain technology has certainly an important role in enhancing and fundamentally transforming supply chains in many industries. It will be expected to create more sustainable solutions for supply chain bottlenecks experienced today in many industries such as logistics, agriculture, and automotive. By removing the intermediaries with blockchain-based transformations, the transactions will become faster and secure thanks to cryptography. Therefore, the infrastructures are evolving along with regulatory changes, technological advancements, new financial mechanisms that will facilitate blockchain-based supply-chain management systems.

References

1. Christopher M (1998) Logistics and supply chain management: strategies for reducing cost and improving service. Financial Times Professional Limited, London
2. World Trade Organization (2019) World trade statistical review 2019, 31 May 2019
3. World Economic Forum (2013) Enabling trade valuing growth opportunities
4. Vara RC, Prieto J, Prieta F, Corchado JM (2018) How blockchain improves supply chain: the case study alimentary supply chain. Procedia Computer Science 134:393–398
5. Yaga D, Mell P, Roby N, Scarfone K (2018) Blockchain technology overview NISTIR 8202, October
6. World Economic Forum (2015) Deep Shift Technology Tipping Points and Societal Impact, Survey Report, September
7. Kshetri N (2018) Blockchain's roles in meeting key supply chain management objectives. Int J Inform Manag 39:80–89
8. Hofmann E, Strewe M, Bosia N (2017) Supply chain finance and blockchain technology: the case of reverse securitisation. Springer
9. McKinsey (2017) Blockchain technology for supply chains—a must or a maybe? McKinsey.https://www.mckinsey.com/Business-Functions/Operations/Our-Insights/Blockchain-technology-for-supply-chainsA-must-or-a-maybe. Accessed 2018
10. Pawczuk L, Massey R, Schatsky D (2018) Breaking blockchain open: Deloitte's 2018 Global Blockchain Survey. https://www2.deloitte.com/content/dam/Deloitte/cz/Documents/financial-services/cz-2018-deloitte-global-blockchain-survey.pdf
11. Pournader M, Shi Y, Seuring S, Koh SL (2020) Blockchain applications in supply chains, transport and logistics: a systematic review of the literature. Int J Prod Res 58(7):2063–2081
12. Startus Insights (2019) 10 Blockchain startups disrupting the automotive industry. Technology Research, January 2019. https://www.startus-insights.com/innovators-guide/10-blockchain-startups-disrupting-the-automotive-industry/
13. Startus Insights (2018) Blockchain startups disrupting the agricultural industry, Technology Research, December 2018. https://www.startus-insights.com/innovators-guide/8-blockchain-startups-disrupting-the-agricultural-industry/
14. Vyas N, Beije A, Krishnamachari B (2019) Blockchain and the supply chain: concepts, strategies and practical applications. Kogan Page Publishers
15. Litke A, Anagnostopoulos D, Varvarigou T (2019) Blockchains for supply chain management: architectural elements and challenges towards a global scale deployment, MDPI Logistics
16. Liao R (2020) How interoperability establishes blockchain's utility and effectiveness for trade finance May 2020, WEF. https://www.weforum.org/agenda/2020/05/blockchain-interoperability-utility-effectiveness/
17. Caldwell J (2019) Top 10 supply chain blockchain projects, rated and reviewed. Bitcoin Mark J (2019). https://www.bitcoinmarketjournal.com/supply-chain-blockchain-projects/
18. CTA, GIZ (2019) Opportunities of Blockchain for agriculture, May 2019
19. UN FAO and ITU (2019) E-agriculture in action, blockchain for agriculture. http://www.fao.org/3/CA2906EN/ca2906en.pdf
20. Walmart (2018) Food traceability initiative fresh leafy greens, Charles Redfield, Russell Mounce, Martin Mundo, Frank Yiannas, 24 Sep 2018
21. Korpela K, Hallikas J, Dahlberg T (2017) Digital Supply Chain Transformation toward Blockchain Integration, Hawaii International Conference on System Sciences (HICSS), January
22. Brody P (2017) How blockchain is revolutionizing supply chain management. Ernst & Young. https://www.ey.com/Publication/vwLUAssets/ey-blockchain-and-the-supply-chain-three/$FILE/ey-blockchain-and-the-supply-chain-three.pdf. Accessed 2018

23. Gindner K, Jain S (2017) Supply chain meets blockchain: when two chains combine. Deloitte. https://www2.deloitte.com/ch/en/pages/technology/articles/supply-chain-meets-blockchain.html. Accessed 2018

A Strategy for Increasing the Employment Rate of Graduates Using a Compact Module

Hatice Camgoz Akdag and Cicek Ersoy

Abstract The aim is to increase the higher education employment rate in the first six months after graduation at Istanbul Technical University. The paper involves different systems and techniques used, such as; ITU Career and Skill Management System (KAYS), which is an internship- and job-seeking platform for Istanbul Technical University students, annual Career Summit performance analysis, workshops, interview simulations, company network, and mentoring modules. Initially, the process of determining the contents of the Career and Skill Management System platform has been addressed with a Management Engineering approach. In this process, the platform at hand was observed, and the details of the inner workings of the system were noted. Secondly, user groups of the system were interviewed, performance questionnaires of different activities from different user groups were analyzed, and the methodology has been determined. New strategies have been set according to the results, and the student employment rate has been recalculated after six months, which showed an acceptable increase in the new graduate employment rate.

Keywords Career · Skill · Job · University · Employment · Performance

Introduction

The dynamic change in technology and emerging new jobs with the new generation of humankind, it has been extensively important to corporate into career and skills operations as well as student recruitment policies.

Career counselors or career offices use computer software to help students in finding short-term or long-term internships and full-time or part-time jobs. Students

H. Camgoz Akdag (✉) · C. Ersoy
Management Engineering Department, Management Faculty, Istanbul Technical University, Istanbul, Turkey
e-mail: camgozakdag@itu.edu.tr

C. Ersoy
e-mail: cicek.ersoy@itu.edu.tr

F. Calisir (ed.), *Industrial Engineering in the Internet-of-Things World*,
Lecture Notes in Management and Industrial Engineering,
https://doi.org/10.1007/978-3-030-76724-2_17

now conduct their research on employers electronically, and employers have recognized the importance of attractive and informative websites. Even though companies have been using independent employee search websites, they have notices the importance of reaching the university directly and looking for specific students on campus. This actually decreases the time of the initial recruitment process, which will also be explained in the following sections.

The management of the entire campus recruiting process is handled electronically by the ITU Career and Skills Management System called KAYS, managed by the Career Center of Istanbul Technical University linked directly to the Rector.

Although technology makes employer-student communication easier, it also requires a significant investment in the career center in terms of equipment and staff with technical expertise.

This paper will explain a State University Career Center Model in which aims to increase student enrollment in six months after graduation.

Literature Review

Career counseling is a concept that is defined in six stages throughout its history, due to the change in business and working life over time. There are six stages in the development of career counseling in the United States [1]. In the first stage (1890–1919), placement services were offered for an increasingly urban and industrial society. In the second stage (1920–1939), educational guidance through the elementary and secondary schools became the focal point. The third stage (1940–1959) saw the focus shift to colleges and universities and the training of counselors. The fourth stage (1960–1979) was the boom for counseling, and the idea of work having meaning in a person's life came to the forefront; organizational career development began during this period. The fifth stage (1980–1989) saw the beginning of the transition from the industrial age to the information age and the growth of both the independent practice of career counseling and outplacement counseling. The 6th stage (1990—present), with its emphasis on technology and changing demographics, has seen increased sophistication in the uses of technology, the internationalization of career counseling, the beginnings of multicultural career counseling, and the focus on the school-to-job transition [1]. The National Career Development Association [2] defines career counseling as a process of helping individuals develop their lifelong careers by taking into account the definition of the worker role and the interaction of this role with other life roles. Generally, career counseling can be defined as the process of helping individuals to draw a route according to their wishes, interests, and skills in vocational education and work life. Even though the terms such as career counseling, career guidance, career development have different meanings, they are used in the same sense over time, especially because of their close relationship with the period in which they are used. On the other hand, the terms "professional development" or "professional development tasks" was introduced to the literature in Turkey by the end of the twentieth century, and the beginning of the twenty-first century [3].

Turkey has taken America primarily as a role model in the consulting career field after emerging efforts to join the European Union and has started its restructuring actions according to this model, the relevant standard [3].

Career counseling studies in America date back to the beginning of the twentieth century. Important names in this area are first mentioned by Frank Parsons, who is regarded as the pioneer of professional guidance from various sources [4, 5]. Parsons was entitled to receive this title, especially in the last years of his life.

Pope [1], the developments in the career counseling process in America in the following six stages:

(1) Job Placement Services (1890–1919): The important features of this period are the loss of jobs in the agricultural sector about the rapid industrialization and activities, the increasing demand for workers in the heavy industrial areas, and the increasing urbanization. At this point, the focus of various developing institutions is to help individuals who have lost their customers or who do not have a job at this point.

(2) Educational Guidance in Schools (1920–1939): The increase in the birth rate in the period following the end of the First World War led to an increase in school-age children. This situation, together with the increasing labor demand along with the industrial crisis, and the economic crisis in 1930, led to the transfer of vocational guidance activities to schools. The studies in this period are focused on both vocational education and support in the education process.

(3) Education of Universities and Consultants (1940–1959): This period has a more specific orientation than the previous two. One of the biggest reasons for this is that after returning from the Second World War, many soldiers have difficulty in keeping up with social life and work, and after the success of the USSR in space studies, the US decided to accelerate its technological developments. This led the American government to establish Advisory and Guidance Education Institutes, in particular in the fields of science and mathematics, in order to direct the individuals concerned to university education. The American Personnel Service and Guidance Organization (APGA), which currently serves as the American Psychological Counseling Association (ACA), was founded during this period. In this period, in order to bring a new dimension to vocational guidance, career selection and career issues have been directed towards a more individual and psychology oriented direction.

(4) Meaningful Work and Organizational Career Development (1960–1970): The political and social developments in this period, especially young people's interests, has shifted to a meaningful, work that would lead to a change in the world.

(5) Independent Consultancy of Career Counseling and Redeployment Counseling (1980–1989): The period of economic growth in the early 1960s was replaced by a decline in the late 1970s. This is the transition from the industrial age to the age of information and technology. Among the major problems created by this transition are the loss of jobs in the industrial field, the increase in the demand of employers with technological skills, and the loss of job security. Important

work during this period included the report Workforce 2000, published in 1987, which will affect career counseling policies in the next 12-year political process.

(6) 1990—Present: In the late 1980s and early 1990s, career counseling started to become a working area that shifted to a specific direction periodically and became an area that served various aspects of society. These aspects range from the reinstatement of employees in the company's top positions to helping homeless people prepare CVs and include social groups that did not have access to these services and/or did not need them until this time. Another feature of this period is the rapid developments in technology. Facilitating communication among individuals has helped to spread and implement career counseling services more easily. Career counselors in the United States have now been able to spread their services to different countries through these tools through telephone and internet access.

Career Counseling in Turkey

Yeşilyaprak [6], the main purpose of the study Psychological Counseling and Guidance in Turkey "to cultivate productive individuals to serve economic development assistance to choosing a suitable profession to his interests and abilities," have been announced. This view is in line with the rapid growth and economic development approach of the early 1950s, which was the period when the studies were started [7].

(1) 1953–1975: In this process, the first studies in the field of education were primarily aimed at helping vocational orientation. In this period, the interest, intelligence, and talent tests were generally taken from abroad were used by translating and adapting. Studies such as the opening of the Guidance and Research Centers and the studies such as the initiation of related undergraduate and graduate programs in our various universities are also important achievements of this period.

(2) 1976–1994: In this period of time, which Yeşilyaprak [6] called "unstable steps," studies in the field of central guidance and career counseling are aimed at reducing the agglomeration of universities by directing the students to the vocational and technical education institutions suitable for their interests and abilities or directly to business life. In line with this aim, students were introduced to the foreground, and the process of testing and inventory development or adaptation was accelerated. In these years, "Guidance in Education" has also been considered as a separate area of expertise, and undergraduate programs have been opened in various faculties of education in order to train experts in this field. When it comes to adult vocational guidance, studies are mostly carried out in cooperation with Germany. The Turkish Business Association (TEO) slowly started its operation in conjunction with, in 1992, in five provinces, depending on the employees' German Labor Agency of German

experts trained "Business and Professional Consulting Services" was established. The first "Vocational Counseling Center" within TEO (called IŞKUR in Turkish) was established in 1993 in Ankara.

(3) 1995–2010: As a result of efforts to reduce unemployment in the economic sphere and directly related to the efforts to enter the European Union in the political and social sphere, the process has begun to become more systematic, more result-oriented with the change observed in attitudes in this area. Before this point, vocational guidance studies, which were carried out with the aim of acquiring a profession, gained a new dimension and turned towards "holistic career development." This situation has led to various regulations in both primary and secondary education systems and in general. Apart from these, Vocational Guidance and Career Counseling activities in universities have been accelerated. In addition to the publication of the works in these areas, the publication of field-specific textbooks started in this period.

University Career Centers

The foundation of today's career centers is based on the professional guidance work initiated by Frank Parsons at the beginning of the twentieth century and the Central Guidance Bureau established by Parsons in 1908. Developments in the field of the career counseling from the process, and after that, both through the example of Turkey still America, are detailed in a previous section of this study.

In general, career centers can be defined as a part of the management unit of organizations such as schools and companies with a staff of staff providing various career services [8]. This section will focus on career centers focused on the planning and development of students' careers within the university.

The career development centers in Turkey are much newer. The first output of the current form career centers in Turkey, in 1980, "Bureau of Employment for Graduates" with the name of the Middle East Technical University, was established by the office. In 1988, Bilkent University became the first university to establish a career center [9]. This progress was followed by career centers established in different universities with various names and functions. In addition to these studies, a Career Center Meeting was hosted by the Higher Education Council in 2013. In this meeting, it was aimed to increase the awareness and awareness in universities about preparing students for work-life and effective internship practices. Some schools, including Istanbul Technical University, were given examples of the services provided in the career centers.

On the other hand, The Higher Education Counsel called "YÖK" in Turkish has announced on the official newspaper [10] and has sent it to all universities informing them that the National Employment Strategy (2014–2023) and Implication Plan (2017–2019) had come into force on the 7th of July 2017. In accordance with the

4.6 article of this plan, it was stated that the HEC (YÖK) is the responsible counsel to improve guidance on job search and discovery techniques at university career centers.

The good news was that Istanbul technical University Career Center was established within the Faculty of Management in 1997–1998 as Career Office. Since its establishment, it has undergone various processes of change, and in 2013 it was renamed as the ITU Career Center (ITU Kariyer Merkezi website). It still serves under this name today. In 2015, ITU Career Center established a platform called ITU Career and Talent Management System (ITU KAYS) to help ITU students find jobs and internships.

The career center units of the universities stated that their priority objectives were to improve the career planning and development competencies of the mentioned groups. This is followed by studies on the possibilities to be employed, studies on the problems that can be encountered in business life, scientific studies in career areas, contributing to regional development with the development of related institutions, and increasing the respectability and preferability of the university and its members.

As for the purpose of career centers, the services offered by the university also show changes related to the university. Although no specific study has been carried out on this subject, it is likely that this change is due to the resources of the career center, resources such as budget, the primary objectives of the career center, and the expectations of the management from these centers.

As with all individual-oriented services, the benefits and effectiveness of career centers in universities depend on themselves. As a result of the researches, no general study has been found on the interaction rates of students with career centers in the country, region, or university.

Methodology

As the aim of this paper is to increase the higher education employment rate in the first six months after graduation at Istanbul Technical University. Career and skill system electronic platform at hand was observed. The number of applications to job postings was determined. Secondly, user groups, which are companies registered to the university system, were interviewed. Performance questionnaires of different activities from different user groups were analyzed. Also, students one month prior to graduation were questioned to determine the employment rate before graduation. The students with no employment were invited to the career center. These students were counseled, and their CVs were be analyzed and rewritten if needed. Skill and job matches, previous applications, and characteristics of graduates and the culture of companies were analyzed before application, and necessary strategies were applied to increase the employment rate of graduates. Six months later, the same questionnaire was sent to the same sample group to figure out whether if there was an increase in the employment rate.

Sample and Data Collection

As there were different analysis done in this study, there was also a need for having different sample groups. The companies with job and internship postings on the Career and Skill system were selected. ITU KAYS leases the service modules it provides from Symplicity Corporation (https://www.symplicity.com). Student questionnaires were distributed to the students who were presenting their graduation design projects.

As the employment issue deals with complex data input, there were several different data sets collected.

1. Career and Skill System job posting data
2. Student application data to the above postings
3. Number of company participation to the summit, the number of the company, stands specifically in which faculty, number of workshops, number of student participation in each activity.
4. Employer questionnaire regarding the career summit
5. Student questionnaire regarding the career summit
6. Student employment questionnaire one month prior to graduation
7. Student employment questionnaire after six months of graduation.

Analysis and Results

Table 1 shows how many students applied to job postings in the year 2018. The grade division is also shown in the table. It is figured out that sophomore and juniors have the highest application number.

Table 2 shows the total number of applications made within these 1605 students. Year distribution is shown. It is interesting to see that new graduates are still looking for jobs almost at the same rate as junior students. Among these 12,546 applications to total 2017 job postings, Table 3 shows the departmental distribution and number of applications.

On the other hand, we have also analyzed the type of employer job postings, which is shown in Table 4.

The University Career Fair was held on 19–22 February 2018. The following tables show detailed information on participation information (Tables 5 and 6).

Table 1 Number and grade of students applied to job postings in the year 2018

Grade	Number of students applied to job postings
Prep	37
Freshman	108
Sophomore	269
Junior	346

Table 2 Number of applications of the above mentioned 1605 students

Grade	The grand total of applications to job postings
Senior	5263
Freshman	651
Sophomore	1677
Junior	2771
Graduate	2063
Prep	121
Total application to job postings	12,546

Table 3 Departmental distribution and number of applications

Department	Application	Department	Application
Industrial engineering	2587	Geology engineering	55
Electric engineering	1705	Telecommunication engineering	48
Computational science and eng.	1205	Molecular biology	36
Mechanical engineering	975	Ships and ocean engineering	36
Management engineering	908	Architecture	35
Chemical engineering	503	Defense technologies	35
Mathematical engineering	450	Information systems engineering	32
Metallurgical engineering	433	Industrial design engineering	28
Aeronautical engineering	361	Mining engineering	27
Control engineering	327	Textile engineering	26
Physics	228	Nano science	25
Environmental engineering	218	Polymer	24
Astronautica	189	Naval Arch. And engineering	19
Civil engineering	181	Information technologies appl	13
Economics	149	Food engineering	13
Mineral processing	132	Geographical engineering	10
Materials engineering	124	Urban and regional	8
Geomatics engineering	104	Bioengineering	7
Energy science and technology	90	Transport engineering	7
Business administration	80	Disaster and earthquake	6
Metaerogical	77	Coastal sciences engineering	3
Petroleum and natural gas	77	Department name unspecified	878
Mechatronics	72	Total	12,546

Table 4 Type of job postings

Job type	Job applications
Full time	2242
Internship	1536
Part-time	1282
IKZ'18 (ITU career summit)	8984
Total	12,546

Table 5 Number of employers attended career fair and number of touch the talent (TT) activities

Total employer attended to the career fair 2018	168
Number of touch the talent	175

Table 6 Number of employer stands opened in each building

Buildings included in the career fair	Number of employer stands opened
MED—general building on Ayazaga campus	28
Electric—electronic engineering faculty + computer and informatics engineering faculty building	73
Chemical and metallurgical engineering faculty building	24
Mining faculty building	8
Construction faculty building	5
Aeronautics and astronautics engineering faculty building	5
Gümüşsuyu campus (mechanical engineering, textile technology and design faculty building)	46
Maçka campus (management faculty building))	38
Taşkışla campus (architecture faculty building)	3

According to McGrath [12], career and job fairs are another way career services and employers can work together to their mutual benefit. ITU Career fair 2018 was a huge success, with 167 different companies with 230 stand in four different campuses and nine different buildings. The event was open to freshmen through seniors as well as graduates with no employment. This is a way for students to find out more about various employment opportunities. The purpose is for employers is not only to provide career information but to touch the talent on the campus.

The reason for having the highest employment rate is highly linked to student application rate as well as the number of stands of employers in the buildings of those students. This is proved with the employment rate questionnaire results done just before graduation. Table 7 shows the employment rate for each faculty just prior to graduation.

After the result of employment status prior to graduation, a new questionnaire was prepared and distributed among the employer network on the Career Center system.

Table 7 Employment rate for each faculty just prior to graduation

Faculties	Employment status before graduation (May 2018) (%)
Computer and informatics faculty	52.13
Management faculty	47.77
Maritime faculty	35.29
Aeronautics and astronautics faculty	34.48
Architecture and ocean engineering faculty	28.57
Science and letters faculty	27.85
Electric-electronic faculty	25.54
Textile technologies and design faculty	25.53
Mechanical engineering faculty	21.71
Chemical and metallurgical faculty	17.54
Architecture faculty	12.50
Construction faculty	11.55
Mining faculty	9.00

The results showed the strengths and weaknesses of ITU graduates very clearly. Table 8 shows the results of the employer questionnaire.

According to the results seen above, it was planned to start a talent touch program and touch the students who are unemployed but accepted as talented as they are graduates of one of the best three universities in Turkey.

Is it enough for the students to be good engineers when building their career ladder? For the previous generation, this question might be answered yes, but the conditions for being accepted to a job now vary considerably. The twenty-first century is a period of very different competencies. These competencies include factors such as the ability to demonstrate effective communication skills, the ability to cope effectively with stress, the ability to work in a team, and continuous development. We are in a period where we need to keep up with the changes in technology, and this requires us to be open to innovation and change, to be proactive, open to flexibility in job descriptions, and to be able to cope effectively with the inevitable stress situations brought about by the change in our roles and duties. In fact, we see that students do not want to be just "engineers." In the career counseling process, many students are talking about having a number of projects in mind and running their own business rather than working in a company. Therefore, besides engineering skills, skills such as organization and management, crisis management and leadership are becoming increasingly important.

Table 8 Results of the employer questionnaire (% of agreeing to the statement)

Question	Response (% of agree)
Capable of knowing the basic concepts in the professional context and of being able to evaluate the relationship between them	89.4
The development of ITU graduates (in terms of reputation, attractiveness and job opportunities in the market)	95.5
Ability to use necessary devices and computers for engineering applications/analysis/design	95.4
To be informed about contemporary issues	92.3
Ability to communicate in Turkish orally and in writing	90.9
Performance according to other university graduates	89.2
Being aware of professional and ethical responsibility	87.7
Defining the problems related to engineering/professional issues and taking into account economic factors	86.2
To be able to make evaluations regarding engineering/professional issues by considering the global and social impact	86.2
To have sufficient knowledge about quality and environmental issues	86.2
Ability to transfer theoretical knowledge to the application	84.9
Being willing to learn and open to innovation (having acquired the necessity of lifelong learning)	83.9
Ability to communicate in a foreign language (English) orally and in writing	81.8
Having the design skills to meet the system, product and process requirements	81.6
Teamwork/working with different people	80.3
I think students should improve their social skills	75
I think students should improve their foreign language skills	25.9

The Career Center decided to be more effective on soft skills as 75% of employers believed that the graduates are very successful in the technical background, but soft skills are missing. A new plan was set and immediately applied for improving the soft skills, which will also increase the employment rate of graduates. The plan included organizing seminars such as Career Awareness Psychoeducation, CV Preparation, interview performance simulations, personality tests, logical reasoning, self-motivation, coping up with Stress and Decision-Making Skills, time management.

Career counseling services were provided on the subjects such as recognizing the interests of the students, identifying the areas in which their skills are formed, taking steps for the career development steps, identifying which areas are more suitable for them and what skills they should acquire for these areas. Apart from this, necessary support was also provided in CV preparation and interview techniques.

After six months of all the above-mentioned seminars, workshops, face to face mentoring, and career coaching, the same questionnaire was sent to the same

Table 9 Employment rate for each faculty after graduation

Faculties	Employment status after 6 months of graduation (November 2018) (%)
Computer and informatics faculty	82.28
Management faculty	86.51
Textile technologies and design faculty	75.68
Aeronautics and astronautics faculty	70.83
Electric-electronic faculty	63.64
Architecture and ocean engineering faculty	61.11
Maritime faculty	53.85
Mechanical engineering faculty	49.11
Science and letters faculty	45.71
Chemical and metallurgical faculty	41.10
Mining faculty	37.65
Construction faculty	33.04
Architecture faculty	32.50

sample group. The results were satisfying. The table below shows the results of the employment rate after the six month one to one touch (Table 9).

The comparison and the rate of increase after six months of Career Center seminars workshops, talent touch, one to one career counseling, interview simulation, etc., the following table (Table 10) has been formed. The average rate of difference in employment rate after six months has shown almost a 30% increase, proving that Career Center seminars, workshops, talent touch activities, one to one career counseling, interview simulation, etc., have helped students be employed.

Discussion and Conclusion

The aim of the paper was to increase the higher education employment rate in the first six months after graduation at Istanbul Technical University. The importance of career centers at universities was highlighted. The stages of career counseling throughout history were mentioned. The introduction of career counseling in Turkey has been defined. Establishment of university career centers, its stages, and the level of ITU Career Center is touch upon. As the career center units of the universities stated that their priority objective is to improve the career plan and develop competencies of students and graduates, several studies, researches and questionnaires were done to figure out the rate of employment as well as the explanation of low and high employment rates before and after graduation. No general study has been found on the interaction rates of students with career centers and their employment rate.

Table 10 Rate of increase in employment after six months

Faculties	Employment status before graduation (June 2018) (%)	Employment status after 6 months of graduation (November 2018) (%)	Increase in employment after six months (%)
Computer and informatics faculty	52.13	82.28	30.15
Management faculty	47.77	86.51	38.74
Textile technologies and design faculty	25.53	75.68	50.15
Aeronautics and astronautics faculty	34.48	70.83	36.35
Electric-electronic faculty	25.54	63.64	38.10
Architecture and ocean engineering faculty	28.57	61.11	32.54
Maritime faculty	35.29	53.85	18.56
Mechanical engineering faculty	21.71	49.11	27.40
Science and letters faculty	27.85	45.71	17.86
Chemical and metallurgical faculty	17.54	41.10	23.56
Mining faculty	9.00	37.65	28.65
Construction faculty	11.55	33.04	21.49
Architecture faculty	12.50	32.50	20.00
		The average rate of increase	29,50

In this study, different researches were thrown, such as career and skill system job posting data, student application data to these job postings, number of companies attending to the career fair, number of the stands being opened in which building and which campus, number of student participation to each activity, employer questionnaire, student employment questionnaire just before graduation, student employment questionnaire after six months of graduation.

The findings show that soft skills are the missing parts of engineering graduates. In order to support the graduates and increase the employment rate, the ITU Career Center decided to be more effective on soft skills as 75% of employers believed that the graduates are very successful in the technical background, but soft skills are missing. Career Awareness Psychoeducation, CV Preparation, interview performance simulations, personality tests, logical reasoning, self-motivation, coping up with Stress and Decision-Making Skills, time management educations were held. After six months, the same questionnaire to the same sample of students was distributed once again. And the results were strongly positive. The average rate of increase in employment rate has been found as 30% overall after the ITU Career Center employment strategy.

For further implications, in order to increase the interaction of employers and students or newly graduates, the career fair may be organized twice a year instead of one.

References

1. Pope M (2000) A brief history of career counseling in the United States. Career Dev Q 48:194–211
2. NCDA (1997) Career counseling competencies (online). https://www.ncda.org/aws/NCDA/pt/sd/news_article/37798/_self/layout_ccmsearch/truey. Date of visit: 21 Nov 2018
3. Yeşilyaprak B (2012) Mesleki rehberlik ve kariyer danışmanlığında paradigma değişimi ve Türkiye açısından sonuçlar: geçmişten geleceğe yönelik bir değerlendirme. Kuram ve uygulamada eğitim bilimleri 12(1):97–118
4. Herr EL (2013) Trends in the history of vocational guidance. Career Dev Q 61:277–282
5. Williamson EG (1964) An historical perspective of the vocational guidance movement. Person Guidance J 42:854–859
6. Yeşilyaprak B (2010) Eğitimde rehberlik hizmetleri (19 bs.), Nobel, Ankara, 978-9755-915-579
7. Kanca OC (2012) 1950–1960 arası Türkiye'de uygulanan sosyo-ekonomik politikalar. Mustafa Kemal Üniversitesi Sosyal Bilimler Enstitüsü Dergisi 9(19):47–63
8. Garis J, Reardon RC, Lenz JG (2012) Current status and future development of career centers in the United States. Asian J Counselling 19(1 & 2):5–26
9. Özden MC (2015) Üniversite yönetmelikleri çerçevesinde kariyer merkezlerinin amaç ve faaliyetleri. Karadeniz Sosyal Bilimler Dergisi 7(12):146–170
10. Resmi Gazete (official newspaper) (2017). http://www.resmigazete.gov.tr/eskiler/2017/07/20170707M1-1.pdf. Date of visit: 21 Nov 2018
11. İTÜ Kariyer Merkez (online). http://www.ikm.itu.edu.tr/hakkimizda. Date of visit: 25 Nov 2018
12. McGrath GL (2002) The emergence of career services and their important role in working with employers. New Dir Stud Serv 100:69–84

Global E-commerce Market Segmentation by Using Fuzzy Clustering

Basar Oztaysi and Mert Kavi

Abstract Customer segmentation is essential for marketing, communication, and even operations management activities. E-commerce provides the data required for novel perspectives to customer segmentation. In this study, we focus on customer segmentation based on purchase variety. To this end, first, the data is preprocessed, and the optimal customer number is detected. Then the fuzzy c-means algorithm is applied, and the segments are formed.

Keywords Fuzzy segmentation · Fuzzy c-means · E-commerce

Introduction

The process of buying, selling, transferring, or exchanging products, services, and/or information via computer networks is defined as E-Commerce [1]. Starting from the invention of electronic fund transfer, and the development of internet e-commerce applications flourished, and various business models have emerged [2]. A standard classification of E-commerce is by the nature of the transactions or the relationship among the participants. The major types can be listed as business-to-business, business-to-consumer, consumers-to-business. As mobile technologies emerge and adapt to social media increases, new e-commerce models such as mobile commerce and social-commerce have become popular. E-commerce can take different forms in a company, depending on the technology.

Customer segmentation is the practice of separating customers into different, meaningful, and identical subsets according to various attributes. For marketing efforts, it is generally used for gathering insights about the customers and provide customized actions according to their characteristics. To achieve this, a company

B. Oztaysi (✉)
Industrial Engineering Department, Management Faculty, Istanbul Technical University, Istanbul, Turkey
e-mail: oztaysib@itu.edu.tr

M. Kavi
Modanisa.Com Head Office, Altunizade, Kuşbakışı Cd. No: 27/1, 34662 İstanbul, Turkey

F. Calisir (ed.), *Industrial Engineering in the Internet-of-Things World*,
Lecture Notes in Management and Industrial Engineering,
https://doi.org/10.1007/978-3-030-76724-2_18

needs to understand the customers' segments and the properties and desires of the segments [3]. Companies apply various forms of segmentation, such as value-based, lifetime period, or demographics based. Also, different sources of data can be included in the process, and segments can be formed based on demographics and characteristic properties, and their needs and loyalty levels [4].

Segmentation can be used in three decision levels, strategic, managerial, and operational level [5]. At the strategic level, segmentation is used to assist the corporate strategic plans and highlight product positions according to diversified ranges of buyers. At the managerial level, it is used to deal with resource assignments and targets setting based on customer segments in order to involve the customer groups into the marketing activities and compose organizational processes considering them.

This study aims to summarize potential segmentation perspectives and apply fuzzy clustering with a real-world case study. Fuzzy sets proposed by Zadeh [34] provide tools and operations for handling imprecision and vagueness in real-world problems. Fuzzy sets have been used for many other real-world problems such as, company selection Estralla et al. [29] technology selection [6], multi-criteria decision making Yatsalo et al. [33], public transportation [7], economic analysis [8], and risk management [9]. The case study is from Modanisa.com, which is one of the top e-commerce websites which focus on selling textile products globally.

The remaining of the paper is as follows, in the second section, a brief literature review is provided, then in the third section, fuzzy clustering is introduced. Section four presents the real-world application steps and the results of the study. Finally, the conclusion is given in the last section.

Literature Review

The literature provides various approaches and methods for customer segmentation, such as demographic-based segmentation, value-based segmentation, propensity based segmentation, life period segmentation. Some of the procedures use heuristics and expert opinions for segmentation, some of them use basic arithmetic operations. The most popular analytical tool for segmentation is cluster analysis. It is a data mining technique that assigns data elements to unknown groups with considering high similarity [10]. The literature also provides various clustering techniques that can be grouped as partitional or hierarchical algorithms. The partitional methods focus on grouping the data elements into a predefined number of clusters. On the other hand, hierarchical methods take advantage of enabling the optimization of a criterion considering the compatibilities of objects within clusters or incompatibilities between clusters.

In recent years various perspectives on market segmentation have been investigated in the literature. Ahani et al. [11] focus on integrating social media data to segmentation efforts to propose a market segmentation model and choice prediction model for SPA hotel market. Liu et al. [12] develop a market segmentation approach by integrating preference analysis and multi-criteria decision-making methods. In the proposed approach, additive value function and pairwise comparison matrices are used to gather preference data. Then this data is used with the hierarchical clustering method to form the segments. Lim et al. [13] develop a market segmentation approach by using a Bayesian spatial profile regression. By this approach, both spatial autocorrelations present in warehouse rents and multicollinearity among the known rent price determinants are handled for segmentation. Diaz Perez and Bethencourt-Cejas [14] focus on the segmentation of tourists that visit a location and propose using the Chi-square Automatic Interaction Detection method, which is a multivariate analysis technique. The results reveal that the proposed approach is superior to the traditional methods used in the tourism domain. Qin et al. [15] focus on market segmentation problem for demand-side platforms. In their study, the researchers model the optimal market segmentation granularity as an optimization problem and form a mathematical programming model to find the optimum granularity. Huerta-Munoz et al. [16] addresses a customer segmentation problem from a beverage distribution firm. Under a given requirement, the researchers try to form customer clusters where similar customers are involved in. The researchers propose a mathematical attribute formulation and use a greedy heuristic that iteratively destroys and reconstructs customer segmentations. Hong [17] proposes a novel segmentation method by employing the Taguchi method. The Taguchi method is used as a tool to select the initial seeds. The author compares the results with the results of a Self-Organizing Map and shows that the proposed method is superior. Oztaysi and Cevik Onar and Oztaysi [18] propose a user segmentation approach by using twitter data. The author uses the data from social media and uses a fuzzy c-means algorithm to segment the users. Oner [19] propose a two steps segmentation approach,in the first step, they use hesitant fuzzy sets to segment the retail locations, and then in the second step, they use customer segmentation by adding the results of the first step into the analysis. Murray et al. [20] deal with a real-world customer segmentation problem where the existing descriptive variables are not suitable for defining the similarities between customers. The authors employ data mining techniques and identify behavior patterns in historical data that involve noise. The authors claim that the proposed results are suitable for strategic decision-making. In a recent study [21] propose a multi-criteria decision-making model for determining the approach for customer segmentation. The authors use Neutrosophic sets to select the most propose approach. Oztaysi et al. [31] propose a segmentation approach based on customer location data. Dogan et al. [22] focus on customer segmentation based on indoor customer paths. Oner and Oztaysi [19] focus on retailer clustering based on hesitant fuzzy sets. Oztaysi [23] use fuzzy c-medoids clustering for gender prediction.

Methodology

In the domain of segmentation, the clustering procedure is used to define subgroups of customers who have common properties. Each customer is defined by a data point in a multi-dimensional space where each dimension represents different properties.

There are various techniques in the literature which convert input data into clusters. Chen et al. [24]. From one perspective, these techniques can be grouped as crisp and fuzzy. The main difference between the two groups is the definition of membership of an element to a cluster. Crisp clustering assigns a data element into a single specific cluster, while fuzzy clustering algorithms assign a data element to diversified clusters simultaneously with a membership degree [25]. In the literature, fuzzy clustering is used by using two points of view, either considering uncertain data or considering crisp data with uncertain clusters [26]. One of the most commonly adopted techniques for fuzzy clustering is the fuzzy *c*-means algorithm, in which data elements are assigned to a predefined number of clusters with different membership values [24].

The fuzzy *c*-means algorithm is based on similarity or dissimilarity measures, which are extracted from distance measurement such as Euclidean distance [27]. The main definition is the partition matrix, which represents the extracted clusters. A fuzzy partition matrix is defined from [28] with the conditions given in the following:

$$\mu_{ik} \in [0, 1], 1 \leq i \leq c, 1 \leq k \leq N,$$

$$\sum_{i=1}^{c} \mu_{ik} = 1, \quad \leq k \leq N,$$

$$0 < \sum_{k=1}^{N} \mu_{ik} < N, 1 \leq i \leq c$$

Equation (1b) defines the sum of each cluster should be equal to 1, and the membership degree should be represented with an interval [11].

Fuzzy *c*-means clustering utilizes an objective function and focuses on minimizing the objective function to find the appropriate clusters.

$$J(Z, U, V) = \sum_{i=1}^{c} \sum_{j=1}^{N} (\mu_{ij})^m \left\| z_j - v_i^2 \right\|$$

where *Z* is the set of data elements to be clustered, *U* shows the fuzzy partition matrix, *V* is the vector which indicates the cluster centers. As seen from the given formula, *N* represents the number of observations, μ denotes the related membership value, *c* is the number of appeared clusters, and *m* is the parameter called fuzzifier that identifies the fuzziness degree of the final clusters, and fuzzifier parameter can get values greater than 1. When the fuzzifier parameter equals to one, then the clusters are formed with crisp clustering. Besides that, $z_j - v_i$ denotes the distance between observation *j* and the center of cluster *i*.

The minimization of the given objective function comprised of a nonlinear optimization problem that can be calculated with a wide range of techniques such as iterative minimization and heuristic approaches such as simulated annealing or genetic algorithms. The steps of fuzzy c-means (FcM) clustering algorithm is defined as in the following [29]:

1. *Initialize* $U = [u_{ij}]$ *matrix*, $U^{(0)}$
2. *At k-step: calculate the centers vectors* $V^{(k)} = [v_i]$ *with* $U^{(k)}$

$$v_i = \frac{\sum_{i=1}^{N} \mu_{ij}^{m} . z_j}{\sum_{i=1}^{N} \mu_{ij}^{m}}$$

3. *Update* $U^{(k)}$, $U^{(k+1)}$

$$\mu_{ij} = \frac{1}{\sum_{k=1}^{c} \left(\frac{\|z_j - v_i\|}{\|z_j - v_k\|} \right)^{\frac{2}{m-1}}}$$

If $\left\| U^{(k+1)} - U^{(k)} \right\| < \delta$ *then STOP; otherwise, go to step 2.*

The number of clusters must be determined to reach meaningful clustering results. To this end, various clustering tests are accomplished with different cluster numbers. The results of the tests are compared by using the Xie-Beni index [30], and the clustering with the lowest value is selected for segmentation.

$$X(Z; U, V) = \frac{\sum_{i=1}^{c} \sum_{j=1}^{N} (\mu_{ij})^m \|z_j - v_i\|^2}{c \cdot \min\limits_{i \neq j} \|v_i - v_j\|^2}$$

Results

Modanisa.com is one of the leading e-commerce companies located in Turkey, which focuses on selling textile products to more than 150 countries worldwide. The company sells 70.000 products of 650 different brands. Segmenting customers from different countries is very important for marketing activities. While there can be various segmentation perspectives, segmentation based on purchase variety is selected for this study. Purchase variety refers to the purchase activities of the customers based on several categories the purchased, the number of brands they preferred, and the number of different products they purchased.

The first step of the application is data preparation. The source of the data needed for the segmentation analysis is purchase transactions. The first step of the application is the preprocessing step. The purchase transactions include Product Id, customer Id, quantity data, so Category Id and Brand Id values are added to the transaction table by using SQL commands. Then by raw data that will be used for segmentation are formed. A sample data set is given in Table 1.

The next step is outlier detection and normalization. The values in each column are checked for z values, and the values larger than 3 and lower than −3 are excluded from the data set. The z scores of the remaining data are used for fuzzy clustering.

Table 1 Sample dataset used for segmentation

Customer Id	#Categories	#Brands	#Products
1,000,000,001	23	24	342
1,000,000,002	5	34	167
1,000,000,003	13	45	87
1,000,000,004	8	6	59
1,000,000,005	2	3	26

Fuzzy c-means clustering is used for segmentation. In order to find the exact number of clusters, various different cluster number parameters are selected, and the results are compared by using Xei-Beni index (Table 2).

The lower values of the Xie-Beni index refer to better clustering results; thus, for this study, c value is selected as five, which means in the study, five clusters are formed. The results can be summarized by the centroid table given in Table 3.

The results present very clear and actionable results. As a summary, Cluster 1 is composed of customers with a very wide variety. They buy from various categories, brands, and products. Cluster 2 is loyal to some categories; they buy a variety of products from a very selected number of categories. Cluster 3 is composed of customers who are loyal to brands and products. Customers in Cluster 4 buy from different categories, but the variety of products does not change much. The final cluster, Cluster 5 is composed of customers with focused categories and brands.

Table 2 Clustering performances by using Xei-Beni index

Number of clusters	#Categories	Number of clusters	#Products
2	2.23×10^{-06}	9	4.41×10^{-05}
3	2.01×10^{-06}	10	8.13×10^{-05}
4	1.43×10^{-06}	11	6.27×10^{-05}
5	7.13×10^{-07}	12	3.52×10^{-05}
6	9.78×10^{-07}	13	4.95×10^{-05}
7	2.23×10^{-06}	14	6.73×10^{-05}
8	6.58×10^{-06}	15	9.33×10^{-05}

Table 3 Centroid table

Customer Id	#Categories	#Brands	#Products
1	1.76	1.93	2.02
2	−1,68	0.4	1.78
3	0.36	−2.01	−1.5
4	1.26	0.4	−0.91
5	−0.89	−1.45	0.21

Conclusion

In this study, a real-world case study is examined, and steps of segmentation by using the fuzzy c-means algorithm is given. After preprocessing, the data are normalized, and the outliers are eliminated. A set of c parameter is used for different runs of fuzzy c-means. Xie-Beni index is used to obtain the best parameter, and the result reveals that five clusters are best for segmentation.

For further studies, other perspectives of clustering, such as value-based clustering or intention based clustering, can be examined. Besides, other fuzzy and crisp clustering methods can be used with the same data, and the results can be compared with the outcome of this study.

References

1. Turban E, King D, Liang PL, Turban D (2012) Electronic commerce 2012: managerial and social networks perspectives. Prentice Hall, Upper Saddle River
2. Sharma V (2012) A Study about management and business ıssues of E-commerce. IOSR J Bus Manage 2(2):1–5. https://doi.org/10.9790/487X-0220105
3. Wind YJ, Bell DR (2007) Market segmentation. In Baker M, Hart S (eds) The marketing book. Butterworth Heinemann
4. Tsiptsis K, Chorianopoulos A (2009) Data mining techniques in CRM: inside customer segmentation. Wiley
5. Cravens WD, Piercy NF (2012) Strategic marketing. McGraw-Hill Higher Education
6. Dogan O, Oztaysi B (2018) In-store behavioral analytics technology selection using fuzzy decision making. J Enterp Inf Manag 31(4)
7. Kaya I, Oztaysi B, Kahraman C (2012) (2016) A two-phased fuzzy multi-criteria selection among public transportation investments for policy-making and risk governance. Int J Uncertain Fuzziness Knowl Based Syst 20(supp01):31–48
8. Kahraman C, Cevik Onar S, Oztaysi B (2016) A comparison of wind energy investment alternatives using interval-valued intuitionistic fuzzy benefit/cost analysis. Sustainability 8(2):118
9. Behret H, Öztayşi B, Kahraman C (2011) A fuzzy inference system for supply chain risk management. In: Wang Y, Li T (eds) Practical applications of intelligent systems. Advances in intelligent and soft computing, vol 124. Springer, Berlin
10. Han J, Kamber M, Pei J (2011) Data mining: concepts and techniques, 3rd edn. Morgan Kaufmann Publishers
11. Ahani A, Nilashi M, Ibrahim O, Sanzogni L, Weaven S (2019) Market segmentation and travel choice prediction in Spa hotels through TripAdvisor's online reviews 80:52–77
12. Liu J, Liao X, Huang W, Liao X (2019) Market segmentation: a multiple criteria approach combining preference analysis and segmentation decision. Omega 83:1–13
13. Lim H, Yoo HE, Park M (2018) Warehouse rental market segmentation using spatial profile regression. J Transp Geogr 73:64–74
14. Diaz-Perez FM, Bethencourt-Cejas M (2016) CHAID algorithm as an appropriate analytical method for tourism market segmentation. J Destin Mark Manage 5(3):275–282
15. Qin R, Yuan Y, Wang FY (2017) Exploring the optimal granularity for market segmentation in RTB advertising via computational experiment approach. Electron Commer Res Appl 24:68–83
16. Huerta-Munoz DL, Rios-Mercado RZ, Ruiz R (2017) An iterated greedy heuristic for a market segmentation problem with multiple attributes. Eur J Oper Res 261(1):75–87

17. Hong CW (2012) Using the Taguchi method for effective market segmentation. Expert Syst Appl 39(5):5451–5459
18. Oztaysi B, Cevik Onar S (2013) User segmentation based on twitter data using fuzzy clustering. Data Min Dyn Soc Netw Fuzzy Syst 316–333
19. Oner SC, Oztaysi B (2017) An interval valued hesitant fuzzy clustering approach for location clustering and customer segmentation. Adv Fuzzy Logic Technol 2017:56–70
20. Murray PW, Agard B, Barajas MA (2017) Market segmentation through data mining: a method to extract behaviors from a noisy data set. Comput Ind Eng 109:233–252
21. Kahraman C, Cevik Onar S, Oztaysi B (2019) Customer segmentation method determination using neutrosophic sets. In: International conference on intelligent and fuzzy systems, pp 517–526
22. Dogan O, Oztaysi B, Fernandez-Llatas C (2020) Segmentation of indoor customer paths using intuitionistic fuzzy clustering: process mining visualization. J Intell Fuzzy Syst 38(1):675–684
23. O Dogan, B Oztaysi (2019) Gender prediction from classified indoor customer paths by fuzzy C-medoids clustering. In: International conference on intelligent and fuzzy systems, pp 160–169
24. Chen S, Fern A, Todorovic S (2014) Multi-object tracking via constrained sequential label-ing. In: Paper presented at IEEE conference on computer vision and pattern recognition, Colum-bus, OH, 23–28 June
25. Oztaysi B, Isik M (2014) Supplier evaluation us-ing fuzzy clustering. In: Supply chain management under fuzziness, pp 61–79
26. Aliahmadipour L, Torra V, Eslami E (2017) On hesitant fuzzy clustering and clustering of hesitant fuzzy data. In: Torra V, Dahlbom A, Narukawa Y (eds) Fuzzy sets, rough sets, multisets and clustering. Studies in computational intelligence, vol 671. Springer, Cham
27. Babuska R (2009) Fuzzy and neural control disc course lecture notes. Retrieved 5 Oct 2012 from http://www.dcsc.tudelft.nl/~disc_fnc/transp/fncontrol.pdf
28. Ruspini EH (1970) Numerical methods for fuzzy clustering. Inf Sci 2(3):319–350. https://doi.org/10.1016/S0020-0255(70)80056-1
29. Estrella FJ, Cevik OS, Rodríguez RM, Oztaysi B, Martinez L, Kahraman C (2017) Selecting firms in University technoparks: a hesitant linguistic fuzzy TOPSIS model for heterogeneous contexts. J Intell Fuzzy Syst 33(2):1155–1172
30. Xie XL, Beni G (1991) A validity measure for fuzzy clustering. IEEE Trans Pattern Anal Mach Intell 13(8) (1991)
31. Öztayşi B, Gokdere U, Simsek EN, Oner CS (2017) A novel approach to segmentation using customer locations data and intelligent techniques. In: Handbook of research on intelligent techniques and modeling applications in marketing analytics
32. Oner SC (2018) B Oztaysi (2018) An interval type 2 hesitant fuzzy MCDM approach and a fuzzy c means clustering for retailer clustering. Soft Comput 22(15):4971–4987
33. Yatsalo B, Korobov A, Oztaysi B, Kahraman C, Martínez L (2020) A general approach to fuzzy TOPSIS based on the concept of fuzzy multi-criteria acceptability analysis. J Intell Fuzzy Syst 38(1):979–995
34. Zadeh LA (1980) Fuzzy sets and information granularity. Adv Fuzzy Set Theory App 11:3–18

A Review on Cell Balancing Techniques and Their Complexity Levels

Anas Faisal and Bahadir Tunaboylu

Abstract With the increasing adoption of battery-based energy storage systems, especially in areas such as e-mobility and on- and off-grid energy storage applications, techniques to manage these batteries are being developed to address various application-related challenges. Battery Management Systems (BMS) are used to provide reliable protection for the connected battery pack. One of the tasks of a BMS is Cell Balancing (CB), in which the BMS tries to ensure that each individual cell or cell module has the same voltage level during charging and discharging operations. This task can become critical in applications involving Li-ion batteries, due to their sensitivity to being overcharged or deeply discharged. The complexity of a CB technique, and thus its implementation cost, depends on the battery chemistry and the sensitivity of the application. Thus, different designs with different complexity levels are being developed to address the balancing issue. This paper explores the CB techniques found in the literature in the past 20 years and categorizes them based on their complexity level. Operational and feature comparisons were carried out between the different CB designs.

Keywords Battery management system (BMS) · Cell balancing · Batteries · Energy storage

Introduction

Observing the market, it is clear that the market rules are transforming to accommodate energy storage, even when the electricity markets are heavily regulated and influenced by politics. On the other hand, emerging markets, such as EVs, provide relatively economically-safer incentives for developing technologies to be used for

A. Faisal (✉) · B. Tunaboylu
Materials Engineering, Industrial Engineering, Graduate School of Natural and Applied Sciences, Marmara University, Istanbul, Turkey
e-mail: anas.bashar89@gmail.com

B. Tunaboylu
e-mail: bahadir.tunaboylu@marmara.edu.tr

F. Calisir (ed.), *Industrial Engineering in the Internet-of-Things World*,
Lecture Notes in Management and Industrial Engineering,
https://doi.org/10.1007/978-3-030-76724-2_19

energy storage applications in general. Even when the change in the current electricity markets is slow, energy storage systems (ESSs) will provide new profit revenues by addressing challenges produced by an increased adoption of distributed renewable energy sources [1]. The chemical deprecation of batteries affects their energy density and lifespan. This chemical degradation is mainly caused by both operating the battery under a wide range of thermal conditions and subjecting the battery to frequent deep charge/discharge cycles. The SOC of a battery pack can be estimated utilizing a combination of Coulomb counting, open-circuit voltage, and the Kalman filtering technique to improve the estimations [2].

Battery management systems (BMSs) are control and protection systems used in battery module designs. In general, a BMS monitors the performance and behavior of the battery module and then controls the charging, discharging, and possibly other operations related to the operation and protection of the battery module. The main goal behind a BMS is to protect the battery module by ensuring that the module operates in safe and acceptable conditions. This protection protects the main system by ensuring the safe operation of the utilized battery and can possibly prolong the lifespan of the battery by slowing down the chemical degradation in the battery composition. Different applications of BESs can require specific requirements from the BMS, the most sensitive of which are deep charge/discharge protection, accurate state of charge (SOC) determination, and state of health (SOH) estimations [3].

In order for the battery to provide the desired voltage output, multiple cells are stacked in a series string to accumulated voltage. The reason is that designing a single standalone battery for each voltage output level is technically and financially impractical. However, the main challenge of these cell strings is balancing, since no two cells are chemically perfectly identical. Over time and after multiple charge–discharge cycles, the cells in a string will develop differences in the amount of charge they hold at a certain time (SOC) and the individual cell's energy capacity (SOH). Therefore, cell balancing (CB) becomes the most important and sensitive role of a BMS. CB operations work on reducing the SOC differences between cells in a string. This will protect the cells from going through deep charging/discharging cycles, which protects the cells from chemical harms and prolongs and homogenize the SOH of cells [4].

Traditionally, BMSs were used in Lead-Acid battery systems to reduce the irreversible aging process on the batteries. For dynamic systems with rare stabilization periods, specific gravity and terminal voltage measurements were inaccurate; therefore, the most common method to determine the battery SOC was Ampere-hour (Ah) balancing. Attempts to calculate and the SOC of Batteries were made. An algorithm was developed to utilize the measured voltage, current, and temperature for the SOC calculations. However, the calculations were very sensitive to initial current measurement inaccuracies and thus required frequent periodic calibrations [5]. For multi-battery systems however, this was done by dividing the batteries into parallel battery strings, then monitoring and controlling each string individually so as not to lose the standalone operation capability of power supply systems [6].

Lithium Batteries are experiencing rapid adoption in the markets of both industry and consumer electronics due to their various chemistries, which can provide high power densities, energy densities, long life span, and additional safety if managed correctly [7, 8]. With the increasing adoption of Lithium and high-temperature batteries, which are more sensitive than traditional batteries, BMSs were long expected to adopt and provide more functionalities to monitor the power flow, control temperatures, and address safety concerns. These functionalities can include data acquisition, Communication, Battery state monitoring, safety management, alongside the typical electrical management [9]. EVs demand more from a BMS, due to nature rapidly varying charge and discharge conditions as the vehicle accelerates and brakes in addition to the real-time communication with various systems present on the vehicle. Furthermore, the thermal management of batteries in EVs is relatively more challenging due to the uncontrolled and possibly harsh environments in which vehicles can be operated [10].

Various designs for CB techniques have been developed. The variation between these designs is the result of not only the technological advancements made over time but also the fact that different applications require different design characteristics of the CB circuit. These design characteristics are mainly balancing efficiency, balancing speed, and manufacturing and maintenance costs. Balancing speed denotes how quick the CB response is for balancing a given amount of energy over a specific time period. Efficiency, on the other hand, measures the amount of energy wasted during the balancing operation. And Finally, the cost of a given CB circuit depends on the design complexity and the used components.

This paper provides a review of the CB techniques in the literature. Older designs are included for two reasons. Firstly, to provide a context for the balancing techniques, challenges, and tradeoffs. Secondly, recent advancements in the materials and electrochemical compositions of the Li-ion batteries reduced some of the disadvantages of the older, simpler designs [11]. In addition, the paper also includes published modifications and upgrades to these existing CB designs and discusses them when found in the literature. The methodology section categorizes the CB designs into three main groups based on the criteria discussed in the section. The discussion section includes a comparison of the advantages and disadvantages of the three CB categories. Finally, the paper is concluded with our thoughts on the current literature regarding CB techniques along with future plans.

Methodology

The CB techniques can be categorized depending on the method they utilize to achieve cell balancing within a cell string. Namely, we group the CB techniques into three categories: Dissipation, relocation, and prevention. Dissipation techniques identify the overcharged cells and drain the excess energy into passive electrical elements such as resisters or power transistors. Relocation techniques divert the excess charge from the overcharged cell and store it in other cells in the cell string; the charge route is either direct or through temporary storage elements such as capacitors or coils. The prevention techniques try to predict and prevent overcharging cells

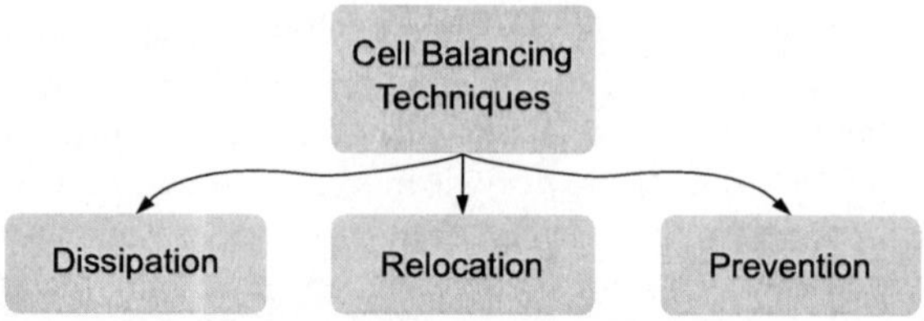

Fig. 1 Cell balancing techniques categories

before the cells actually get overcharged. Circuit illustrations are recreations from their respective cited literature. The three categories are shown in Fig. 1.

Dissipation Techniques

Shunt Resistors

This is the simplest form of cell balancing, and also the cheapest to implement. It involves using shunt ohmic resistors to dissipate the excess charge when detected on a cell in a string. This method is relatively simple to implement and cheap to manufacture. However, the excess charge is dissipated as heat through a shunt resistor. This means that the excess energy is lost rather than being relocated within the system. In addition, the generated heat will need to be removed from the system, especially in temperature-sensitive applications and/or batteries. Such designs can be found in some commercial battery systems [12]. There are two types of shunt resister methods; passive and active. The active method is achieved by adding a switch in series with the ohmic shunt resister to control the excess energy absorption rate [13]. Figure 2 illustrates an active shunt resister CB circuit.

The shunting Resistor can be replaced by transistors in the common-emitter mode. This eliminates the need for a centralized controller for the series switches by using comparators to control the base of the transistor and thus control the shunt collector current [14].

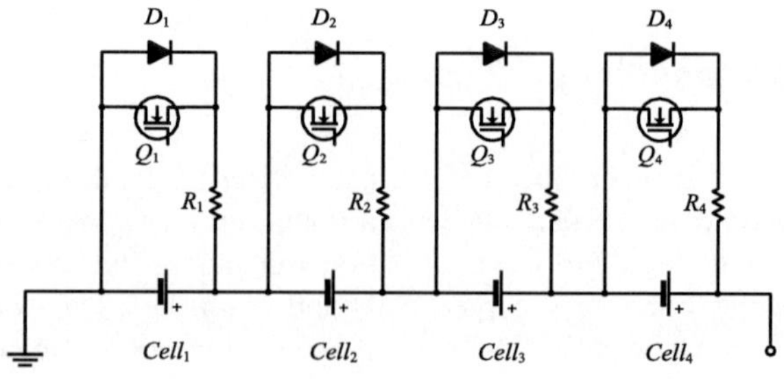

Fig. 2 An active shunt resister CB circuit

Controllable Current Sources:

The design utilizes MOSFETs connected in parallel to each cell in the string to achieve active energy dissipation cell equalization. The method is to operate gate-source MOSFETs in the saturation region where the MOSFETs behave as controllable current sources. A control algorithm is utilized to monitor the cells and identify the healthiest cell in the respective string. Healthier cells' online determination is based on aging estimation algorithms, which in turn is decided by evaluating the capacity fade and internal resistance of the cells. Once the healthy cell is identified, the controller uses this cell's voltage to determine the correct voltage level for the gate-sources of the MOSFETs that belong to the other cells in the string. Doing so, the voltage of each cell is changed by controlling the saturation current of its respective MOSFET, which in turn is controlled by changing the applied voltage on its gate-source.

The cell equalization is smooth and fast compared to the conventional passive dissipation methods. In addition, it is more cost-effective compared to the non-dissipative methods since it doesn't need additional components such as converters, capacitors, and inductors to temporarily store the excess energy [11]. Figure 3 illustrates a controllable current source CB circuit.

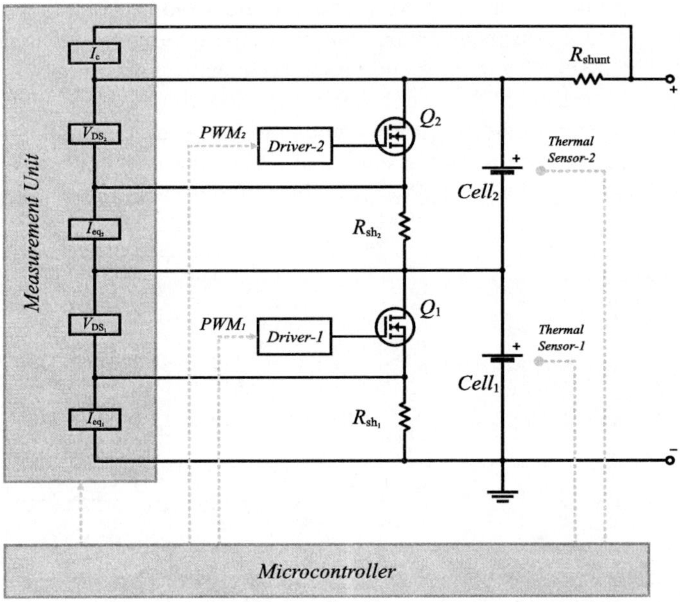

Fig. 3 A controllable current source CB circuit

Relocation Techniques

Capacitive Shuttling:

This design stores the excess charge in a capacitor and then redirects this charge to other less-charged cells in the string. In its simplest form, the design uses the switched capacitor topology, in which 2n switches and n − 1 capacitors are used to balance n cells in a string. The design is simple and cheap since it does not require a centralized intelligent controller. However, the switch count is high, and the cell equalization time is long [15]. Figure 4 illustrates a capacitive shuttling CB circuit.

The single switched capacitor topology was developed to speed up the equalization speed by using a single capacitor for the whole-cell string. Reducing the number of capacitors reduces the size and cost of the circuit to a certain extent. However, this design does require an intelligent centralized controller to control the switches and direct the balancing currents based on each cell's condition [16].

The double-tier capacitor topology combines the previous two designs for an even faster cell equalization. Naturally, the centralized intelligent controller is still needed. Compared to the switched capacitor topology, this design can reduce the equalization time to a quarter [17].

Modularized topologies for capacitive shuttling were proposed for long cell strings. This design increases the flexibility of the equalization operation by dividing the cell string into segments and then combining these segments using additional capacitors. The main advantage is shortening the equalization times and reducing the voltages and currents of the switches for long cell strings [18].

An automatic Switched-Capacitor novel topology was developed to address the main drawback in both the switched and double-tier capacitor topologies. Namely, to allow the balancing charge to flow from the cell with the highest voltage to the

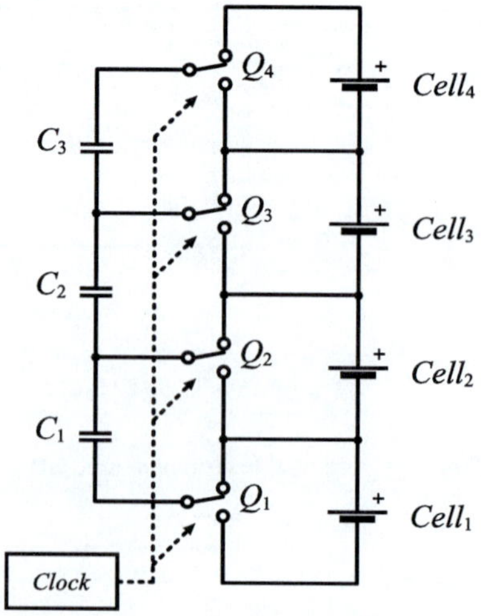

Fig. 4 A capacitive shuttling CB circuit

cell with the lowest voltage directly, rather than having to pass through adjacent cells first. The main advantage is to speed up the equalization process, especially for long cell strings. This design does not require an intelligent central controller, which helps keep the complexity and costs of the design low. A couple of pulse signal generators with fixed frequency and a duty ratio of 0.5 control the switches automatically. The main drawback of this design is the requirement for more switches since it requires double the number of switches found in the switched and double-tier capacitor topologies [19].

Energy Converters

Unidirectional flyback topology: The main goal of this design was to eliminate the individual cell equalizers (IECs) from individual cells and replace them with a converter for each cell string. A single flyback DC/DC converter with isolated topology is used to transfer the excess charge from a cell with a higher terminal voltage to the rest of the cell string. The path for each cell consists of MOSFETs and diodes. A digital controller in the BMS controls these MOSFET switches with PWM signals to control the current flow between the cells and the flyback converter. The design compares the cell voltage to the overall string voltage rather than the adjacent cells in the string. Thus, ICEs are not required for each cell. However, the design does require an accurate voltage sensing for each individual cell. This design is efficient for applications with a single service module; an additional higher-level controller is required for applications with multiple shared service modules. The design supports unidirectional current flow only, which forces the controller to perform the cell balancing during either charging or discharging operations [20]. Figure 5 illustrates a unidirectional flyback topology CB circuit.

Bidirectional flyback topology: This design modifies the selective flyback converter approach to accommodate bidirectional balancing current flow, which allows the balancing operation to run continuously. The used flyback converter has

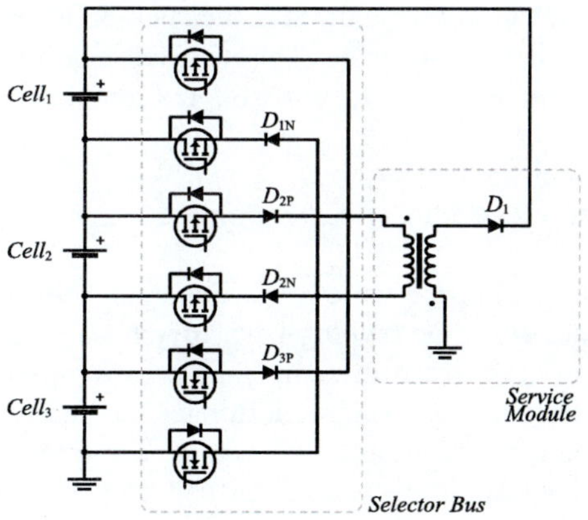

Fig. 5 A unidirectional flyback topology CB circuit

two windings on its secondary side to allow two modes of operation: pack-to-cell and cell-to-pack. For 22Ah high capacity Li-ion modules, this design was simulated to reduce the average active power loss due to balancing operations by more than %50 compared to the passive shunt resistor technique. The design provides high power transfer efficiency and quick balancing times. However, the use of a transformer which increases the size and weight of the circuit. The few diodes in use and the lack of inductors help decrease the production costs of the system [13]. Utilizing a soft-switching scheme for the switches for the same design can reduce the switching losses and increase the power transfer efficiency [21].

Ramp converters: Ramp converters are used to reduce the size and weight of the circuit. In addition, the small transformer design allows for soft-switching and high switching frequencies [22].

Full-bridge converters: This design is used in high-power applications that require flexible modularity of battery packs. The design is based on pulse-width modulation (PWM) full-bridge converter while using the SOC of cells as a control method for cell balancing. The circuit provides a high degree of control, especially with the AC-DC conversion capability of the utilized converter. However, the circuit requires an intelligent controller, which adds to the complexity of the control requirements and manufacturing costs [19].

Segmentation cell-to-cell converter: This method divides the cells string into smaller segments and provides two different paths for the balancing current to flow through. One path is a buck-boost converter and is used to transfer the energy between two cells in the same segment. The second path is a flyback operation and is utilized when the two cells are located in different segments. The overall goal of the design is to allow the excess energy to pass between any two cells in the string. However, the benefit of two modes of balancing within the same cell string is to shorten the balancing paths, which leads to faster balancing operations. Planar coupled inductors were included in the design for high power density industrial applications. To ensure the accurate driving of the switches, optocoupler isolated high-density driver circuits were utilized [23].

Direct cell-to-cell converter: An improvement to the segmentation topology was proposed to avoid the segmentation altogether by allowing any two adjacent cells to exchange charge through the buck-boost operation rather than limiting this path to predefined segments. This increases flexibility and, thus, the speed of the equalization operation without increasing the complexity of the controller [23].

Center-Cell Concentration

This design tries to achieve the balancing capabilities of the bidirectional flyback designs while keeping the main advantage of unidirectional designs, namely the use of one switch per cell rather than two. Comparing this design to a regular unidirectional design, this design concentrates the charge transfer in the middle cell on a string and then distributes the excess charge to the other cells in the string upwards and downwards from the central cell. The charge is transferred to the central cell and then

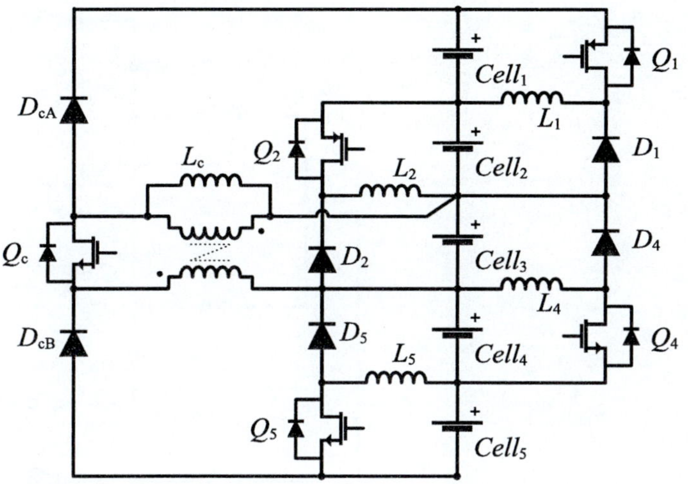

Fig. 6 A center-cell concentration CB circuit

from the central cell to the other cells in the string using three buck-boost converters. The design manages to achieve balancing speeds similar to a regular bidirectional flyback design while using a single switch per cell. Reduction of the number of switches by almost half results in a considerable decrease in the size and cost of a cell balancing circuit [24]. Figure 6 shows a center-cell concentration CB circuit.

LC Series Resonant Circuit

The design utilizes an LC resonant circuit to store the excess charge from strong cells and divert it to any other weaker cell in the string and not necessarily an adjacent cell. Balancing power flow through paths controlled by bilateral switches. The design utilizes zero-current switching to reduce the power switching losses and electromagnetic interference. No auxiliary converters are required since no power sources for the gate drive is needed. As a cell-to-cell balancing circuit, an advantage of this design over the bidirectional flyback circuit is the elimination of the transformer. This allows for small-size lightweight implementations of the circuit.

A string of 12 Li-ion cells was built for the experiment. The observed power transfer efficiencies were 93.2% and 78.9% for balancing 0.56 and 1.94 W of power, respectively. 85% and 61% SOCs were balanced within 92 min, with 1.94 W of balancing power [25]. Figure 7 shows an LC series resonant circuit CB design.

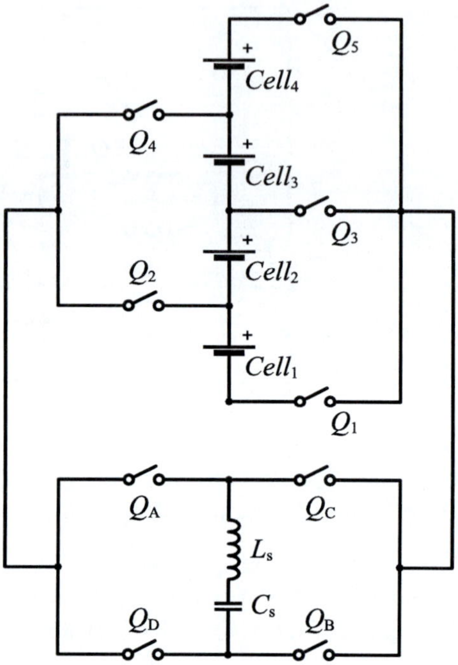

Fig. 7 An *LC* series resonant circuit CB design

Prevention Techniques

A Converter for Each Cell:

In this design, each cell is monitored and controlled by its own DC/DC converter. Current equalization is carried out by monitoring the current of each cell, which in turn is representative of the cell's own capacity during charging and discharging operations. Managing each cell's current ensures that all cells in the string have the same SOC, regardless of each cell's individual capacity. If one cell becomes incapacitated within a string, the string can continue operating in the limp-home mode, in which the string continues to produce the same voltage but with the reduced current. However, for the current equalization method to be effective, the capacities of all cells need to be estimated accurately.

Applications that utilize battery packs with short strings in addition to high-power applications can utilize this design for cell equalization since several small converters can sometimes be technically more viable than a single relatively bigger converter. Other less common applications in which cell strings consist of cells of various capacities, voltages, and chemistries can also utilize this design since each cell is monitored and controlled separately [26]. Figure 8 illustrates a converter for each cell CB design.

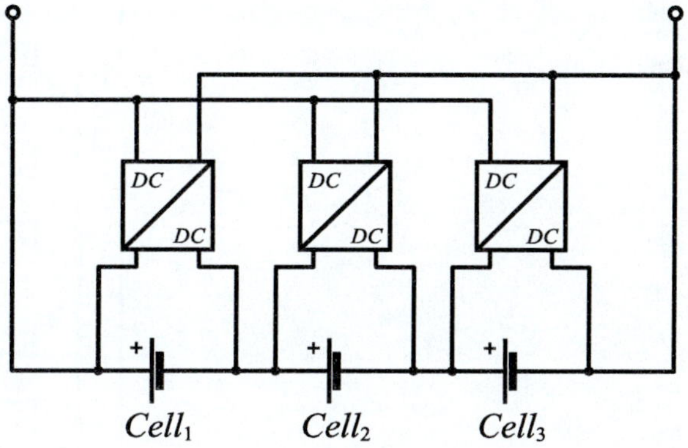

Fig. 8 A converter for each cell CB design

Estimation Algorithms:

This design uses a typical flyback equalizer in the pack-to-cell mode. However, the decision to start the balancing is taken before the voltage or SOC imbalances actually happen. The main goal is to predict which cell is on its way to be deeply charged/discharged. This is done by estimating the relative capacity of each cell while utilizing the power to energy ratio. Rather than comparing actual quantitative values, the algorithm monitors the rate of charging or discharging of each cell and compares it to the individual cell's power to capacity ratio. For the algorithm to work properly, the total energy capacity must be known, and all the series-connected cells having the same SOC versus open-circuit voltages curves are assumed. A drawback in this design is the lack of consideration of cells' initial SOC differences, which can have a negative impact on the balancing speeds. In addition, the power balancing circuit is capable of handling a limited balancing current [27]. Figure 9 shows the flyback equalizer used in the algorithm-driven CB design.

Decentralized Smart Cells:

Am more complex design for managing each cell individually is the concept of smart cell modules with internal standalone decentralized control. The main advantage here is the modularity of cells that can have different capacities and usage histories. Thus, the SOC of the cells is synchronized by loading the healthier cells more to make the degradation of the cells in the string more uniform. Each smart cell has a local controller, sensing equipment, and switching networks to monitor and control the cell. The design at hand uses a microcontroller that uses a series inductor to sense the voltage and then control the duty cycles of a half-bridge switching network [28]. Figure 10 shows a decentralized smart cell CB design.

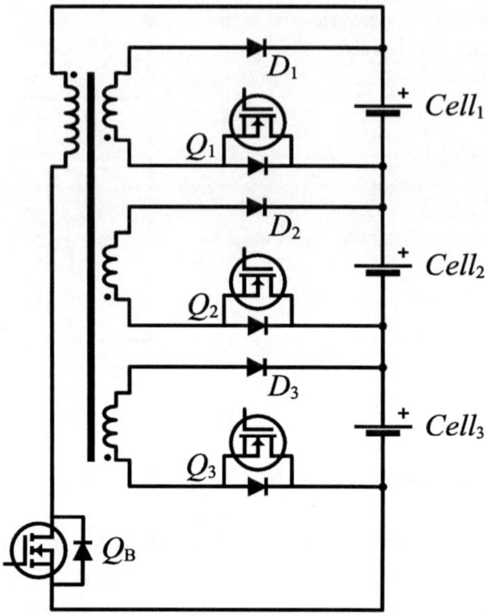

Fig. 9 The flyback equalizer used in the algorithm-driven CB design

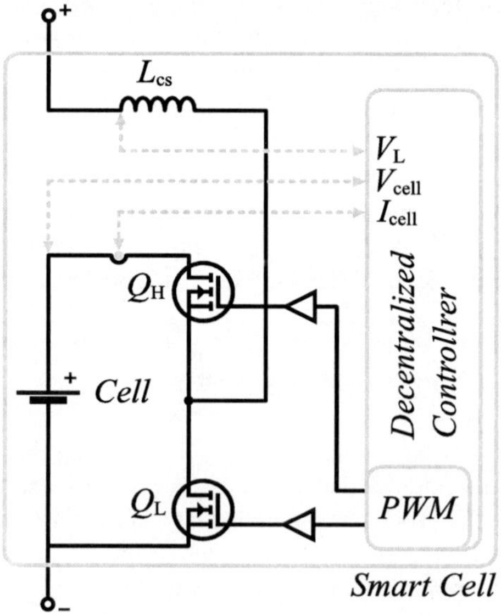

Fig. 10 A decentralized smart cell CB design

Discussion

Comparing the operational differences between the three categories, it becomes clear that dissipative techniques are simpler and cheaper, preventive ones are more complex and expensive, and relocation techniques sit somewhere in between.

Dissipative designs might look primitive, have limited functionality, and suffer some balancing energy losses. However, recent advancements in the materials and electrochemical compositions of the Li-ion batteries reduced some of the disadvantages of these older and simpler de-signs in terms of balancing efficiency and speed. Furthermore, these simple designs' lightweight nature provides an important advantage in applications where weight considerations are a priority.

The prevention methods are used in applications where the battery pack performance and lifetime longevity are relatively more important than the financial considerations. Cells will inevitably degrade and lose capacity over time. Therefore, in applications where replacing the degraded cells has relatively high logistical costs, investing in such designs reduces the operational costs of such systems in the long run. This becomes more viable when we take into consideration that such cell balancing techniques have the ability to manage and control each cell at its individual optimum, which in turn prolongs the effective lifespan of the cells and reduces the frequency of cell replacements.

Relocation designs try to strike a balance between the other two categories. Usually, the proposed designs focus on a certain aspect of the balancing operation and improve on it. The aspect of focus can be balancing speed, efficiency, or cost. Since different applications have different requirements and limitations, focusing on the priority aspect of the design is technically and financially more viable. The flexibility of these designs is their strength and advantage from a design point of view. This is the reason these designs are more diverse and common among the three categories. The main disadvantage of these designs is their reliance on additional components such as capacitors, coils, and converters, which in turn add their own weight and intrinsic power losses to the CB design.

Conclusion and Future Work

In this paper, we discussed the principles of cell balancing for cells in a series string. We reviewed the literature for CB designs for the past 20 years. The designs were categorized into three groups depending on their balancing method; Dissipation, Relocation, and Prevention. The advantages and disadvantages of each design were discussed for each design. Furthermore, updates and modifications to the reviewed designs were also included in their respective section for added clarity on the design progression over time. Each design proved its usefulness in certain applications by matching the application requirements and limitations. Some designs gained usefulness with recent advancements in the electrochemical compositions of the batteries, which was driven by the recent increase in the adoption of Li-ion batteries in various applications. The increasing pace of both BESSs adoption and battery chemistry development will require an equal pace for developments in the CB deigns. Developments in CB include both developing novel designs and exploring new applications for older designs.

For the future, we intend to perform simulation performance comparisons between these designs in light of both the addition of new CB designs and advancements in battery designs and performances.

References

1. Günter N, Marinopoulos A (2016) Energy storage for grid services and applications: Classification, market review, metrics, and methodology for evaluation of deployment cases. J Energy Storage 8:226–234. https://doi.org/10.1016/j.est.2016.08.011
2. Cheng KWE, Divakar BP, Wu H et al (2011) Battery-management system (BMS) and SOC development for electrical vehicles. IEEE Trans Veh Technol 60:76–88. https://doi.org/10.1109/TVT.2010.2089647
3. Rahimi-Eichi H, Ojha U, Baronti F, Chow MY (2013) Battery management system: An overview of its application in the smart grid and electric vehicles. IEEE Ind Electron Mag 7:4–16. https://doi.org/10.1109/MIE.2013.2250351
4. Choi SS, Lim HS (2002) Factors that affect cycle-life and possible degradation mechanisms of a Li-ion cell based on LiCoO2. J Power Sources 111:130–136. https://doi.org/10.1016/S0378-7753(02)00305-1
5. Duryea S, Islam S, Lawrance W (2001) A battery management system for standalone photovoltaic energy systems. IEEE Ind Appl Mag 7:67–72. https://doi.org/10.1109/2943.922452
6. Kaiser R (2007) Optimized battery-management system to improve storage lifetime in renewable energy systems. J Power Sources 168:58–65. https://doi.org/10.1016/j.jpowsour.2006.12.024
7. Hamid Vishkasougheh M, Tunaboylu B (2014) Charactrization of a Li-ion battery based standalone a-Si photovoltaic system. Appl Surf Sci 318:324–333. https://doi.org/10.1016/j.apsusc.2014.07.008
8. Lu L, Han X, Li J et al (2013) A review on the key issues for lithium-ion battery management in electric vehicles. J Power Sources 226:272–288. https://doi.org/10.1016/j.jpowsour.2012.10.060
9. Jossen A, Späth V, Döring H, Garche J (1999) Reliable battery operation - a challenge for the battery management system. J Power Sources 84:283–286. https://doi.org/10.1016/S0378-7753(99)00329-8
10. Pattipati B, Sankavaram C, Pattipati KR (2011) System identification and estimation framework for pivotal automotive battery management system characteristics. IEEE Trans Syst Man Cybern Part C Appl Rev 41:869–884. https://doi.org/10.1109/TSMCC.2010.2089979
11. Koseoglou M, Tsioumas E, Jabbour N, Mademlis C (2020) Highly Effective Cell Equalization in a Lithium-Ion Battery Management System. IEEE Trans Power Electron 35:2088–2099. https://doi.org/10.1109/TPEL.2019.2920728
12. Instruments T (2011) 3 to 6 Series Cell Lithium-Ion Battery Monitor and Secondary Protection IC. System
13. Lee WC, Drury D, Mellor P (2011) Comparison of passive cell balancing and active cell balancing for automotive batteries. 2011 IEEE Veh Power Propuls Conf VPPC 2011. https://doi.org/10.1109/VPPC.2011.6043108

14. Isaacson MJ, Hollandsworth RP, Giampaoli PJ, et al (2000) Advanced lithium ion battery charger. In: Proceedings of the Annual Battery Conference on Applications and Advances. pp 193–198
15. Pascual C, Krein PT (1997) Switched capacitor system for automatic series battery equalization. Conf Proc - IEEE Appl Power Electron Conf Expo - APEC 2:848–854. https://doi.org/10.1109/apec.1997.575744
16. Speltino C, Stefanopoulou A, Fiengo G (2010) Cell equalization in battery stacks through State Of Charge estimation polling. Proc 2010 Am Control Conf ACC 2010 5050–5055. https://doi.org/10.1109/acc.2010.5530710
17. Baughman AC, Ferdowsi M (2008) Double-tiered switched-capacitor battery charge equalization technique. IEEE Trans Ind Electron 55:2277–2285. https://doi.org/10.1109/TIE.2008.918401
18. Park HS, Kim CH, Park KB et al (2009) Design of a charge equalizer based on battery modularization. IEEE Trans Veh Technol 58:3216–3223. https://doi.org/10.1109/TVT.2009.2015331
19. Asakura J, Akagi H (2009) State-of-Charge (SOC)-Balancing Control of a Battery Energy Storage System Based on a Cascade PWM Converter. IEEE Trans Power Electron 24:1628–1636. https://doi.org/10.1109/TPEL.2009.2014868
20. Shin JW, Seo GS, Chun CY, Cho BH (2010) Selective Flyback balancing circuit with improved balancing speed for series connected Lithium-ion batteries. 2010 Int Power Electron Conf - ECCE Asia -. IPEC 2010:1180–1184. https://doi.org/10.1109/IPEC.2010.5543502
21. Yun J, Yeo T, Park J (2013) High efficiency active cell balancing circuit with soft-switching technique for series-connected battery string. In: IEEE applied power electronics conference and exposition (APEC), pp 3301–3304. https://doi.org/10.1109/APEC.2013.6520774
22. Gottwald T, Ye Z, Stuart T (1997) Equalization of EV and HEV batteries with a ramp converter. IEEE Trans Aerosp Electron Syst 33:307–312. https://doi.org/10.1109/7.570791
23. Han J, Yang S, Liu X, Yang W (2019) An active direct cell-to-cell balancing circuit in continuous current mode for series connected batteries. Energies 12. https://doi.org/10.3390/en12203978
24. Kim MY, Kim JH, Moon GW (2014) Center-cell concentration structure of a cell-to-cell balancing circuit with a reduced number of switches. IEEE Trans Power Electron 29:5285–5297. https://doi.org/10.1109/TPEL.2013.2292078
25. Lee KM, Chung YC, Sung CH, Kang B (2015) Active cell balancing of Li-Ion batteries using LC series resonant circuit. IEEE Trans Ind Electron 62:5491–5501. https://doi.org/10.1109/TIE.2015.2408573
26. Einhorn M, Guertlschmid W, Blochberger T et al (2011) A current equalization method for serially connected battery cells using a single power converter for each cell. IEEE Trans Veh Technol 60:4227–4237. https://doi.org/10.1109/TVT.2011.2168988
27. Min GH, Ha JI (2017) Active cell balancing algorithm for serially connected Li-Ion batteries based on power to energy ratio. 2017 IEEE Energy Convers Congr Expo ECCE 2017 2017-Janua:2748–2753. https://doi.org/10.1109/ECCE.2017.8096514
28. Frost DF, Howey DA (2018) Completely Decentralized Active Balancing Battery Management System. IEEE Trans Power Electron 33:729–738. https://doi.org/10.1109/TPEL.2017.2664922

A Distribution Network Design Model for Additive Manufacturing

Adnan Veysel Ertemel, Sinan Ertemel, Cemil Can Coktug, Ali Can Erk, and Ali Gorener

Abstract It is without a doubt that Additive Manufacturing (AM) is to radically change the way manufacturing processes take place. However, the vast majority of the extant literature deals with rather futuristic scenarios on the subject. This paper adopts a more realistic, short-term perspective by stating that new supply chain models have to be designed in a way adopting AM, but they should still incorporate Conventional Manufacturing (CM) as the de-facto means at least for a certain period. To this end, we formulate a new supply chain model where manufacturers are allowed to forward their orders to 3D shops. With some mild assumptions, our optimization model suggests that substantial cost savings could be attained when AM is adopted.

Keywords Additive manufacturing · 3D printing · Supply chain management · Distribution network design

---

A. V. Ertemel (✉) · S. Ertemel · C. C. Coktug · A. C. Erk · A. Gorener
Department of International Trade, Management Faculty, Istanbul Commerce University, Istanbul, Turkey
e-mail: avertemel@ticaret.edu.tr

S. Ertemel
e-mail: ertemels@itu.edu.tr

A. Gorener
e-mail: agorener@ticaret.edu.tr

S. Ertemel
Department of Economics, Management Faculty, Istanbul Technical University, Istanbul, Turkey

A. C. Erk
3Durak Technology Inc, Istanbul, Turkey

A. Gorener
Department of Business Administration, Management Faculty, Istanbul Commerce University, Istanbul, Turkey

F. Calisir (ed.), *Industrial Engineering in the Internet-of-Things World*,
Lecture Notes in Management and Industrial Engineering,
https://doi.org/10.1007/978-3-030-76724-2_20

Introduction

Looking from a sustainable development perspective, technology has often been regarded as both a panacea and root cause of society's problems (e.g. [1–4]). Besides finding solutions to previous problems, many inventions have also created newer problems. Today, products, services, and processes are renewing themselves through innovation, even replacing themselves with newer ones.

The recent advances in digital technology and communication networks have led to companies to be constantly compelled to renew their business models and innovate in order to survive in the competitive environment. Digital revolution has brought local competition to the global level and permanently changed the fundamental structure of various industries. This phenomenon resulted in an increased interest among nations in global economic competition through technology and innovation.

Businesses have renovated their business models to thrive in the highly digitalized world by adopting design thinking and lean startup concepts [5, 6]. New ventures that adopt this perspective could gain competitive advantage over their age-old, established counterparts. Through their innovative approach, such companies can disrupt other companies and even the entire market.

Various industries are to be disrupted by Additive Manufacturing (AM) technology. In many circumstances, new technologies are replacements of obsolete ones, and provide small to medium improvements [7]. However, Additive Manufacturing, also known as 3D printing technology, has the potential to cause the transition of the entire supply chain and enable the generation of newer production models.

In assessing the impact of AM on manufacturing processes, some scholars have concluded that AM processes will complement (as opposed to displace) CM processes [8–10]. Adopting this perspective, [11] have elaborated on new supply chain structures that incorporate AM. On the other hand, [12] stated that there will be a reduction in total supply chain costs due to the adoption of AM manufacturing processes. However, there is a shortage of studies on this subject with practical implications. Taking this fact as a starting point, this study approaches the subject as an optimization problem and adopts a practical view to make production decisions based on some well-reasoned assumptions.

Accordingly, the study discusses how AM may help develop new business models, along with the advantages and constraints of AM. Specifically, it analyzes how AM can change the supply chain models. The cost reduction effect of AM on existing production processes is taken into consideration to come up with a hybrid model that incorporates both conventional and additive manufacturing processes. In this regard, the time to manufacture 3D products is viewed as an additional cost variable. Hence, an optimization model is designed for balanced production with AM by taking its distinct characteristics into account.

Additive Manufacturing

Over the years, many studies on the effects of innovation, digitization and robots on manufacturing have addressed the disruptive nature of technology in relation to the way businesses operate (e.g. [13, 14]).

Additive manufacturing (AM), which is also known as 3D printing, has emerged as a new and disruptive manufacturing technology that has major implications for companies and industries at large [15]. In fact, the terms "3D printing" and "additive manufacturing" are used interchangeably for making solid structures layer-by-layer, depending on the architecture of the digital files [16].

The Additive Manufacturing technologies have such an enormous potential that the Additive Manufacturing market is projected to grow at a rate of 14.4% by 2026, to reach USD 23.33 billion in 2026 from USD 7.97 billion in 2018. These figures have serious implications for revising their business models, that is, their logic for creating and capturing value. As a hyper-flexible technology that can provide highly customized and personalized products and production, AM provides a specific set of opportunities and challenges for developing new business models.

Disruptive Nature of 3D Printers and Its Impacts on Industry

Additive manufacturing (AM) enables restructuring supply chains in terms of lead time, processes, distribution, investment capital, inventory management and manufacturing plant [17]. Additive manufacturing, which is one of the few areas where both the academic world and the industry has increased interest, is becoming the main subject of more research day by day. As AM becomes a common method for the manufacturing of end products, this computer-based technology needs further analysis and more research [18].

Several manufacturing technologies have been developed for AM during the past 30 years. Fused Filament Fabrication (FFF), Selective Laser Sintering (SLS), Stereolithography Apparatus (SLA, also: SL), Direct Light Processing (DLP), Polyjet Matrix (PJM) (also: Inkjet AM), Inkjet Zcorporation Technology (Zcorp) are among the mostly used AM technologies. Different plastic raw materials are being used in these technologies; and with the improvement of production technologies, metal alloyed end products such as carbon and onyx have begun to be produced in 3D printers.

The advancement of this technology at such a speed enabled the use of the technology even in niche markets like space industry and aviation. Today, 3D printers began to be used in the construction industry and houses printed by 3D printers have begun to be lived in. New technologies finding domains in such a speed have, consequently, posed a threat to many industries.

There are various studies in the literature on the use of AM in the manufacturing industry and the changes they will create. Silva and Rezende [19] conducted a study on the effect of additive manufacturing in logistics. Explaining applications for different AM techniques, researchers have focused on sectors such as healthcare, aeronautic and aerospace. In terms of logistics, they stated that storage and transportation will decrease and mass customization will increase. They claim that additive manufacturing will change the logistics industry, and competitive companies that see the opportunity will work toward this direction.

The Evolution of 3D Printers

Table 1 gives the historical development of 3D printers in the production process. It can be seen that with the final stage of adoption, Home Fabrication, involvement of consumers to production processes was introduced. But 3D printers have barely been integrated into production processes. Development of technology and printers continuously, advancement of production technologies and constraints about raw materials have limited their industrial adoption. However, in recent years, many companies have started to incorporate AM into their processes to utilize the advantages of this technology.

Mellor et al. [20] stated in their research that five basic factors were effective in the spread of AM applications. These are strategic factors related to personalization and business volume, technology considerations linked to machinery-materials and processes, organizational factors covering changes in the structure of the business, operational factors related to product design and production technique, supply chain factors that enable us to analyze the big picture. In the interviews they made in the company where they carried out the qualitative study, technology supplier, senior management, project engineer, maintenance manager, researchers who received feedback from the factory manager, they determined that contradictory issues such as cost, material use-variety, product limit, speed, design, flexibility should be clarified. The work of [21] was carried out under a more specialized group of materials. It has been investigated how feasible it is to produce ABS polymers-based products in various forms with low-cost three-dimensional printers. As a result of the study, it has been stated that the patient-specific brachytherapy mold design can be used by incorporating the repeatability features of the materials produced with a three-dimensional printer within certain physical measurements and usage conditions.

Table 1 Adoption stages of AM technologies and resulting involvement in production

Adoption stage	Started	Design	Tooling	Manufacturing	Distribution
Rapid prototyping	Early 1990s	+			
Rapid tooling	Late 1990s	+	+		
Direct manufacturing	Late 1990s	+	+	+	
Home fabrication	Early 2010s	+	+	+	+

Source Rayna et al. (2016)

Implementing 3D printers to manufacturing processes needs the reconstruction of business and manufacturing models. The opportunity to manufacture customized products notwithstanding to mold and standard products imprints manufacturing capacity and costs. Stocking of products digitally offers unlimited design chances to manufacturers as well as additional value and advantage to customers. It will change the nature of design, increase the interactivity between design and production, and radically localize manufacturing [16]. AM technology will also free designers from conventional manufacturing processes, and even many scholars argue that design for production will convert to production for design. Conventional manufacturing technology and industrial business models in a scale economy cannot compete with the advantages that customization provides. Offering customized products in a scale economy houses serious financial constructs.

To compete with new manufacturing technology, conventional business models need to address advantages and values to offer to customers and approach advantage saving or advance new advantages from AM disadvantages. Much as new manufacturing technology and one economy offers value with co-producing and on-site production models, a serious shift in cost and revenue structures is in question.

Chen and Lin [22], who carried out a study on the feasibility evaluation of AM-based smart manufacturing systems, discussed two main topics, which are "technical challenges" and "managerial concerns". Under the title of technical challenges, they evaluated the criteria of "time spent on product designs for additive manufacturing", "limited availability of materials available", "surface smoothness" and "efficiency". Under the heading of managerial concerns, they considered "database management", "linkage with lean manufacturing", "digital rights", "e-commerce", and "globalization and deglobalization" factors. The researchers who carried out the study, to make an AM based manufacturing system feasible, should have a productivity-based structure that is constantly attracting the attention of customers, connected with different production networks.

In [23] study, the concept of do-it-yourself (DIY) manufacturing is discussed. Based on a web platform that shares AM designs with an open source code, they analyzed how they can affect especially the Toy and Game Market. Using the open source distributed production model, the findings of this research suggest customers can generate higher value items for less cost. Eyers and Potter [24] considered Industrial AM systems in their studies. By "Industrial AM" expression, AM systems that have reached the level that can compete with traditional manufacturing systems and that can manufacture the parts and end products required by the sectors are expressed. Within the scope of the article, the cases with various product groups containing different process types are evaluated. Cases, considered as competitive elements of production, it has been evaluated for cost, dependability, quality, flexibility and speed. Ghiasian et al. [25], provide a system for practice that examines the feasibility of AM based manufacturing from three perspectives: geometric assessment, generation of orientation/support building, and resources needed. This study of manufacturing facilities is applied to 34 different parts of a wide variety of multipliers and valves, which are made using traditional factory techniques.

Ahtiluoto et al. [26] present a practical model of viability, analyzing the added value of using AM for output. In the feasibility model they have created, there are elements such as structural features of the part, manufacturing time and manufacturing cost. The researchers, who created a feasibility index using the specified evaluation factors, analyzed the validity of the model with sample applications on four different manufacturing products.

Shift in Cost and Revenue Structures and New Business Models with 3D Printers

The application of 3D printers in manufacturing systems will threaten many industries. Storing products not in physical forms but in digital form will allow designing manufacturing systems free from locations, while bringing about many threats. One of the major problems about this issue is converting linear, centralized supply chains to nonlinear decentralized supply chains, and eliminating stocking obligation. Therefore, this will trivialize place components from 4P's (product, price, promotion, place) of the marketing mix.

Manufacturing products free from locations in local places will accompany new limitations with new opportunities. Storing products digitally and producing in line with demand will accompany *time* limitation as is the case in all manufacturing processes. Henceforth, there will be a need for particular time to produce the products. This limitation constraints utilizing 3D printers in industry because for each production there will be the need for *unit production time,* and customers are obliged to wait prior customers *unit production times,* on demand increments. This situation will affect manufacturing depending on 3D printers and customer experience on high and unbalanced demand periods.

Awaiting customized products in one economy instead of standard products in stock in scale economy incorporates time dimension into the marketing mix in new economic model. There is *unit time cost* in producing with 3D printers. We think that for the calculating of this time cost and optimizing AM processes, there will be a need for mathematical models.

Demands can increase rapidly, in unbalanced, irregular and unexpected ways in AM processes. Stocking products digitally requires agile and flexible enhancing ability of production capacity up against rapid demand shocks. This should be the core fundamental that shapes the AM business models. To ensure flexible manufacturing capacity, manufacturers do not have so many alternatives.

Manufacturing companies;

(1) may own new equipment by agile investing,
(2) may rent different manufacturers AM equipment and use in manufacturing,
(3) may allow customers to join manufacturing process with their own 3D printers and make them *prosumers,*

(4) may be involved in joint venture with companies having 3D services by joining 3D printer hubs.

Each of the above-mentioned cases have unique opportunities and threats. Thus, companies should design their own model by choosing or mixing the above-mentioned conditions. The increasingly blurred line between the roles of users as consumer and producer (and, in particular, the growing importance of this latter role) has changed the view firms and academics have of users, who are now often branded as '*prosumers*' [27]. Consumers joining the production process will also affect distribution networks. The largest shift in distribution, however, may be the rise of printer hubs that directly support hobbyist and *prosumer* needs [16].

After all, by comparison to scale economy, fixed costs in one economy would be lower, by contrast unit costs in line with demand increases. Increasing demand rapidly and unpredictably may affect manufacturing and cost structures. Design, methods for manufacturing and supply chain are interlinked; and a successful development in one dimension will require scoping all three dimensions [28]. Thus, in the next section we will discuss time dimension cost on new manufacturing technology and try to design a model for network optimization.

Time Dimension in New Manufacturing Technologies

As mentioned above, storing products digitally in AM technology and on-site production has time constraints. The novelty of customization and personalization of this technology will require design process and additional time necessary depending on design. It will be illustrated in a basic model of three different customers' production process with the same raw material on one 3D printer with three different custom products.

Production process is discussed in two different time constraints as production process and design process. The variables of our model are given as follows:

C_i customer i for $i \in \{1, 2, 3\}$.
t_{di} design time for customer i.
t_{mi} manufacturing time for customer i.
t_{li} delivery time for customer i.

C_i customer i for $i \in \{1, 2, 3\}$.
t_{di} design time for customer i.
t_{mi} manufacturing time for customer i.
t_{li} delivery time for customer i.

As shown in the figure, manufacturing process in 3D printers consists of different time variables. Accordingly, for customer i delivery time is equal to the sum of design time and production time as well as delivery time for the customers preceding him.

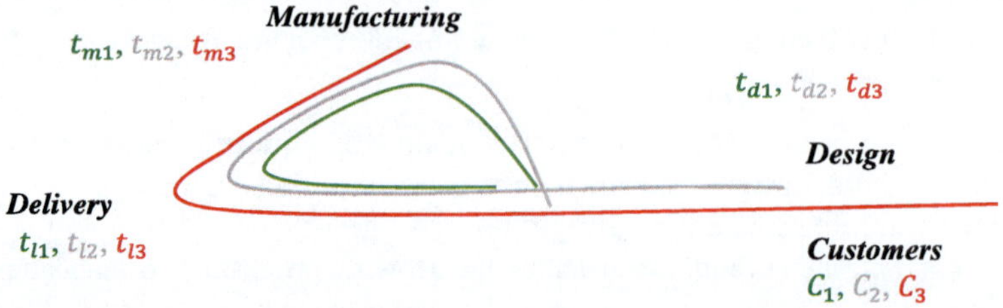

Fig. 1 Different time variables in AM production process for three different customers

$$t_{li} = \sum_{j=1}^{i-1} t_{lj} + (t_{di} + t_{mi})$$

For example, the delivery time of customer 3, t_{l3} depends on the production time as t_{m3} and design time as t_{d3} and total delivery time of the former customers prior orders as given by $t_{l1} + t_{l2}$. Therefore,

$$t_{l3} = t_{l1} + t_{l2} + (t_{d3} + t_{m3})$$

As shown in Fig. 1, if a customer does not want to get involved in the design process and order non-customized pre-designed product as $t_{d3} = 0$, delivery time as $t_{l3,}$ depends on the former orders delivery time free from order of customer 3.

As shown above, manufacturers utilizing 3D printers have time, resource and capacity constraints which impose costs on consumers. To deliver customized products rapidly to customers, agility and flexibility should be essential in the AM technology business model.

Even though many scholars offer innovative production liberation as *opportunities* such as producing with customers, renting 3D printers as a service, because of unbalanced demand, there will be an increase in decentralization of production to different locations. This could be in the form of renting 3D printers as a service customer getting involved in production processes are not *opportunities but imperativeness.*

Shifting production to unutilized printers at different locations may accelerate the production process; solely this will delay delivery time. This is because additional time will be needed for the transportation of the end product. In this context, opportunity of the new technology created would be diminished by itself. Thereby, this situation would create additional value and advantage to customers in delivering stocked products in scale economy in time and speed dimension.

A Distribution Network Model Design for the Supply Chain

Additive manufacturing model would enable the decentralization of production, lowering of production and transportation costs, improvement of product differentiation and more importantly disruption of the standard supply chain.

In the standard supply chain model, we have the following distribution network design:

Suppliers → Manufacturers → Retailers → Customers.

Additive Manufacturing model would change the supply chain as follows:

3D model creators → Manufacturers → 3D shops → Customers.
Eventually, the supply chain would evolve further where Customers become 3D shops as well.

As outlined in Chopra et al. [29], a distribution network design should account for two dimensions: (i) value provided to the customer, (ii) cost of meeting customer needs.

In this paper, we assume that manufacturers are using predesigned 3D models so our supply chain consists of manufacturers, 3D shops, and customers only. As the number of 3D shops increase, transportation costs to the customers become negligible. Accordingly, we focus on production costs as the essential cost parameter. Each customer submits a demand to a manufacturer for each order type. Each manufacturer decides whether to produce each order type in-house or send it to 3D shops. If a manufacturer produces any quantity, then it incurs a fixed cost. On the other hand, 3D shops only entail variable costs. We introduce our optimization problem below:

Indices

I number of manufacturers ($i = 1, 2, \ldots, I$).
J number of 3D shops ($j = 1, 2, \ldots, J$).
K number of customers ($k = 1, 2, \ldots, K$).
L number of orders ($l = 1, 2, \ldots, L$).

I number of manufacturers ($i = 1, 2, \ldots, I$).
J number of 3D shops ($j = 1, 2, \ldots, J$).
K number of customers ($k = 1, 2, \ldots, K$).
L number of orders ($l = 1, 2, \ldots, L$).

Parameters

D_{kl} total demand of customer k for order l.
t_{ikl} cycle time of manufacturer i for customer k for order l.
t_{ijkl} cycle time of 3D shop j from the dispatch of manufacturer i for customer k for order l
T_i time capacity of manufacturer i

T_j time capacity of 3D shop j
c_{ikl} unit cost of manufacturer i for customer k for order l
c_{jkl} unit cost of 3D shop j for customer k for order l
F_i fixed cost of manufacturer i.

D_{kl} total demand of customer k for order l.
t_{ikl} cycle time of manufacturer i for customer k for order l.
t_{ijkl} cycle time of 3D shop j from the dispatch of manufacturer i for customer k for order l
T_i time capacity of manufacturer i
T_j time capacity of 3D shop j
c_{ikl} unit cost of manufacturer i for customer k for order l
c_{jkl} unit cost of 3D shop j for customer k for order l
F_i fixed cost of manufacturer i.

Variables

x_{ikl} production of manufacturer i for customer k for order l
y_{ijkl} dispatch of manufacturer i to 3D shop j for customer k for order l
$z_i = 1$ if manufacturer i is open, 0 otherwise

x_{ikl} production of manufacturer i for customer k for order l
y_{ijkl} dispatch of manufacturer i to 3D shop j for customer k for order l
$z_i = 1$ if manufacturer i is open, 0 otherwise

Our model can be defined by a mixed integer program as follows:

$$\text{Min} \sum_{i=1}^{I}(F_i z_i) + \sum_{i=1}^{I}\sum_{k=1}^{K}\sum_{l=1}^{L}(c_{ikl}x_{ikl}) + \sum_{i=1}^{I}\sum_{j=1}^{J}\sum_{k=1}^{K}\sum_{l=1}^{L}\left(c_{jkl}y_{ijkl}\right)$$

subject to

$$\sum_{i=1}^{I}\sum_{j=1}^{J} x_{ijkl} + \sum_{i=1}^{I} y_{ikl} = D_{kl} \quad \text{for all } k \text{ and } l \tag{1}$$

$$\sum_{k=1}^{K}\sum_{l=1}^{L} x_{ikl} \le \left(\sum_{k=1}^{K}\sum_{l=1}^{L} D_{kl}\right) z_i \quad \text{for all } i \tag{2}$$

$$\sum_{k=1}^{K}\sum_{l=1}^{L} (t_{ikl}x_{ikl}) \le T_i \quad \text{for all } i \tag{3}$$

$$\sum_{i=1}^{I}\sum_{k=1}^{K}\sum_{l=1}^{L}\left(t_{ijkl}y_{ijkl}\right) \le T_j \quad \text{for all } j \tag{4}$$

$$x_{ikl}, y_{ijkl} \ge 0 \quad \text{for all } i, j, k, \quad \text{and } l \tag{5}$$

$$z_i = \{0, 1\} \quad \text{for all } i \tag{6}$$

The objective function is to minimize the total production costs, which constitute fixed costs of the manufacturers and variable costs of the manufacturers and 3D shops.

Constraint Eq. (1) ensures total demand is met by the manufacturers and 3D shops. Constraint Eq. (2) says that total production of a manufacturer cannot exceed total demand provided that it is open. Constraints Eqs. (3) and (4) present total time constraints of the manufacturers and 3D shops, respectively. Finally, Constraints Eqs. (5) and (6) represent the nonnegativity of the production and binary variable whether the manufacturer is open or not, respectively.

Discussion and Conclusion

Additive Manufacturing is in the midst of creating a revolution in design, manufacturing, logistics and supply chain to end-users [30]. As AM is becoming more common in industrial environments, there is an increased need to accurately identify appropriate production strategies [24]. While AM has made major strides recently, there are still constraints [17]. The main areas of interest are the variety of materials that can be used in manufacturing, production speed, capacity, surface quality of the end product, product design needs for AM and part cost.

AM technology is poised to radically change the way production and distribution take place. However, many papers on 3D printer revolution fail to adopt a realistic approach that considers the pros and cons of conventional versus additive manufacturing processes, besides the fact that production decision is an optimization problem for many cases. As such, this paper addresses this issue by developing a model for a typical challenge that incorporates tradeoffs in the light of different parameters like variable versus fixed costs, waiting time, quantity, etc. A model is developed to accommodate different scenarios.

Evidently, Additive Manufacturing is disrupting the standard supply chain in many industries. In our model, we focus on the minimization of the total production costs. And we conjecture that some substantial cost savings can be obtained for many industries. Our preliminary findings suggest that manufacturers with higher fixed costs are better off forwarding their orders to the 3D shops. On the other hand, a manufacturer with a lower cost might prefer to keep some of the production in-house. Our model assumes away transportation costs, which would become even more transparent as customers themselves become 3D printers. To this end, as a future work, we are planning to update the supply chain model accordingly and investigate further reduction in cost savings. Furthermore, our model can also be extended to accommodate transportation costs along with the production costs.

The proposed model could be run against real-world data in a specific production process. Furthermore, the model can be elaborated further to incorporate the distance between supply chain entities as an additional cost variable.

In conclusion, 3D printer revolution is definitely to change production and supply chain processes. However, this revolution will not completely replace conventional manufacturing systems in many instances. As such, models like the one proposed in this paper are undoubtedly to be developed in order to build such hybrid systems.

References

1. Davidson A (2001) Technology and the contested meanings of sustainability. SUNY Press, New York
2. Dunn PD (1979) Appropriate technology: technology with a human face, Palgrave
3. Goldsmith EAR (1972) A blueprint for survival,. The Ecol 2(1)
4. Lau A (2010) Sustainable design: a new paradigm for engineering education. Int J Eng Edu 26(2):252–259. Available at: www.ijee.ie.ezproxy.lib.purdue.edu/articles/HMUDD/Vol26-2/05_Ijee2255.pdf
5. Blank S (2013) Why the lean start-up changes everything. Harv Bus Rev 91(5):63–72
6. Ries E (2011) The lean startup: how today's entrepreneurs use continuous innovation to create radically successful businesses. Crown Business
7. Anderson P, Tushman ML (1990) Technological discontinuities and dominant designs: a cyclical model of technological change. Adm Sci Q 35(4):604–633. Available at: https://doi.org/10.2307/2393511
8. Durach CF, Kurpjuweit S, Wagner SM (2017) The impact of additive manufacturing on supply chains. Int J Phys Distrib Logistics Manage
9. Holweg M (2015) The limits of 3D printing, Harvard Business Review, Digital Article, Available at: https://hbr.org/2015/06/the-limits-of-3d-printing. Accessed 21 Nov 2016
10. Rogers H, Baricz N, Pawar KS (2016) 3D printing services: classification, supply chain implications and research agenda. Int J Phys Distrib Logist Manag 46(10):886–907
11. Holmström J, Partanen J (2014) Digital manufacturing-driven transformations of service supply chains for complex products. Supply Chain Manage Int J 19(4):421–430
12. Barz A, Buer T, Haasis H-D (2016) A study on the effects of additive manufacturing on the structure of supply networks. IFAC-PapersOnLine 49(2):72–77
13. Abernathy W, Clark K (1993) Innovation: mapping the winds of creative destruction. Res Policy 22(2):102. https://doi.org/10.1016/0048-7333(93)90040-O
14. Brettel M, Friederichsen N (2014) How virtualization, decentralization and network building change the manufacturing landscape: an industry 4.0 perspective. Int J Mech Aerosp Indus Mech Eng 8(1):37–44. Available at: www.waset.org/publications/9997144
15. Phaal R, O'Sullivan E, Routley M, Ford S, Probert D (2011) A framework for mapping industrial emergence. Technol Forecast Soc Chang 78(2):217–230
16. Petrick IJ, Simpson TW (2013) 3D printing disrupts manufacturing: how economies of one create new rules of competition. Res Technol Manag 56(6):12–16https://doi.org/10.5437/08956308X5606193
17. Achillas C, Aidonis D, Iakovou E, Thymianidis M, Tzetzis D (2015) A methodological framework for the inclusion of modern additive manufacturing into the production portfolio of a focused factory. J Manuf Syst 37:328–339

18. Khajavi SH, Partasnen J, Holmström J (2014) Additive manufacturing in the spare parts supply chain. Comput Ind 65(1):50–63
19. Silva JV, Rezende RA (2013) Additive manufacturing and its future impact in logistics. IFAC Proc Vol 46(24):277–282
20. Mellor S, Hao L, Zhang D (2014) Additive manufacturing: a framework for implementation. Int J Prod Econ 149:194–201
21. Harris BD, Nilsson S, Poole CM (2015) A feasibility study for using ABS plastic and a low-cost 3D printer for patient-specific brachytherapy mould design. Australas Phys Eng Sci Med 38(3):399–412
22. Chen T, Lin YC (2017) Feasibility evaluation and optimization of a smart manufacturing system based on 3D printing: a review. Int J Intell Syst 32(4):394–413
23. Petersen EE, Kidd RW, Pearce JM (2017) Impact of DIY home manufacturing with 3D printing on the toy and game market. Technologies 5(3):45
24. Eyers DR, Potter AT (2017) Industrial additive manufacturing: a manufacturing systems perspective. Comput Ind 92:208–218
25. Ghiasian SE, Jaiswal P, Rai R, Lewis K (2018) From conventional to additive manufacturing: Determining component fabrication feasibility. In: ASME 2018 international design engineering technical conferences and computers and information in engineering conference. American Society of Mechanical Engineers Digital Collection.Quebec, Canada, pp 1–16
26. Ahtiluoto M, Ellman AU, Coatanea E (2019) Model for evaluating additive manufacturing feasibility in end-use production. In: Proceedings of the design society: international conference on engineering design, vol 1, no 1. Cambridge University Press, pp 799–808
27. Rayna T, Striukova L (2015) Open innovation 2.0: is co-creation the ultimate challenge? Int J Technol Manage 69(1):38–53
28. Fixson SK (2005) Product architecture assessment: a tool to link product, process, and supply chain design decisions. J Oper Manage 23(3–4):345–369. Available at: https://doi.org/10.1016/j.jom.2004.08.006
29. Chopra S, Meindl P, Kalra DV (2013) Supply chain management: strategy, planning, and operation, vol 232. Pearson, Boston
30. Gao W, Zhang Y, Ramanujan D, Ramani K, Chen Y, Williams CB, Wang CC, Shin YC, Zhang S, Zavattieri PD (2015) The status, challenges, and future of additive manufacturing in engineering. Comput Aided Des 69:65–89

Brain Drain: A Multi-criteria Decision Model

Ayse Basak Incekas and Cigdem Kadaifci

Abstract In this study, a multi-criteria decision model is proposed to examine the criteria affecting the decision on brain drain, which is the immigration of highly educated and competent individuals. Due to the interconnections among the criteria, the Analytic Network Process (ANP) is selected to model the decision problem and to obtain priorities. Based on the assessments of twenty-six participants who immigrated temporarily or permanently to different countries, economic situations, career opportunities and, political situations are the most important criteria pushing educated people (e.g., students and academicians) to immigrate. Brain drain has a significant effect on economic, political, and socio-cultural drivers, especially in the sending countries, and causes a considerable change in the population. Thus, the proposed model will be a guide for policymakers to develop effective policies regarding the mitigation and management of brain drain.

Keywords Migration · Brain drain · Multi-criteria decision making · Analytic network process

Introduction

Migration is a process of people traveling to a new place or country in order to find a job and live there either temporarily or permanently [5]. Migration can be classified into two groups: internal migration is moving from one place to another in the same country, while international migration represents moving from one country to another [44]. Since the primary causes of these types may differ, researchers have been investigating the criteria affecting internal or international migration decisions separately.

A. B. Incekas (✉) · C. Kadaifci
Industrial Engineering Department, Management Faculty, Istanbul Technical University, Istanbul, Turkey
e-mail: incekas15@itu.edu.tr

C. Kadaifci
e-mail: kadaifci@itu.edu.tr

F. Calisir (ed.), *Industrial Engineering in the Internet-of-Things World*,
Lecture Notes in Management and Industrial Engineering,
https://doi.org/10.1007/978-3-030-76724-2_21

Characteristics of sending and receiving places, gravity criteria (i.e., population size and density), employment-related criteria such as starting a new career or labor market conditions, economic criteria, environmental criteria including public safety, social services, environmental quality, and personal issues including marriage, divorce, birth, retirement, graduation are among the decision criteria for internal migration ([7, 15, 19, 18, 28, 39], among others).

In the literature, different classifications have been made for international migration. The most common classes cover labor migration, force migration (i.e., refugees), and free migration (for more details refer to [46]). Even though the most important groups are economic, socio-cultural, political, and demographic criteria [26], the determinants of the migration decision change depending on the migration type and the discipline that the researcher belongs to [8]. Economists claim that there are push factors forcing people to leave their home countries, pull factors attracting them to settle, and network factors linking the home country and destination [9]. Better working conditions, public services, employment, higher wages, religious tolerance can be considered as pull factors, and unemployment, underemployment, poor economic conditions, poverty, political corruption, disagreement with politics are the push factors [2, 9, 10, 23, 30, 44]. Sociologists focus on the chain migration, which starts with an immigrant promoting the receiving country and causing more people to immigrate to the same area [9, 46]. Personal criteria are examined by psychologists, and changes in the standard of living and culture are the concern of anthropologists [9].

Besides fertility and mortality, migration seems like an important factor causing significant changes in the population. Hence, the concept is attracting researchers since the 1960s, when considerable movements among countries started [8, 11]. According to the United Nations recent report, international migration increases from 153 million to 271.6 million (almost 3.5% of the world population) between 1990 and 2019 [44]. In the same report, after defining migration corridors, it is indicated that two-thirds of international migrants are hosted by 20 countries [44]. In other words, while emigration rates are similar, the density of immigration is relatively high in particular countries, including North America, Western and Eastern Europe, and the oil-exporting Gulf Cooperation Council (GCC) countries [47]. Note that, among top receiving countries, the rank of Turkey is increasing due to the refugees [44, 47]. In 2010 the total migrant stock was 1.3 million and rose to 5.9 million in 2019 [44].

People generally migrate to be a part of the labor force as a production factor [24]. However, there is another type of migration, "brain drain," which is a human capital loss [6, 47]. Brain drain can be defined as a region's loss of highly skilled individuals due to permanent emigration, which might be reversible [6]. The concept refers to the migration of highly educated and competent people from developing to developed countries [11]. The possible consequences of brain drain vary and are widely discussed in the literature [3, 16]. For instance, the movement of people results in the loss of talented workforce for the sending countries, and at the same time this situation creates a "brain gain" through an increase in innovation, economic growth, competitive advantage and the consumption of services for the receiving ones [6], and

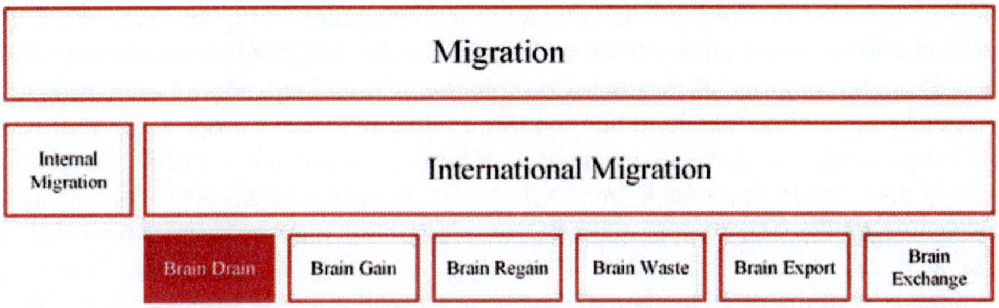

Fig. 1 The focus of this study

might result in a "brain waste" if the migrants turn out to be unemployed or underemployed [31]. "Brain export" occurs if a country willingly sends its citizens in exchange for remittances, and "brain exchange" can be applied for scholars, researchers, and students in order to gain mutual benefits [22, 48]. It is always possible to "regain" emigrated people by providing the appropriate conditions they desire [6]. The brain regains investigated in the literature as a return intention ([13, 20, 42, 49] among others).

In this study, a multi-criteria decision model is proposed to examine the criteria affecting brain drain decision. The set of decision criteria is determined based on a detailed literature review, and as will be explained later, the Analytic Network Process (ANP) is selected as the appropriate multi-criteria decision making (MCDM) method due to the existence of the dependence and feedback structure among the determining criteria. The graphical representation is given in Fig. 1 to emphasize the type of migration the model focuses on. Considering the role of the concept in changing the population of countries and in affecting the economic, political, socio-cultural, and even environmental drivers, brain drain deserves to be examined.

MCDM methods are not common in brain drain literature. The internal migration potential of Serbian municipalities is determined using ELECTRE Tri-C, together with the Multiple Criteria Hierarchy Process [1]. In another study, a Geographical Information System based multi-criteria model is proposed to model internal migration, and Analytic Hierarchy Process (AHP) is used to determine migration potential [32]. To the best of our knowledge, this is the first study examining the important criteria which may cause brain drain using an MCDM method.

The remainder of the paper is organized as follows. The theoretical background of the method is given in the following section. Then, the steps of the application are provided. Finally, the conclusion and further suggestions are given in the last section.

Methodology

Analytic Network Process, developed by Thomas Saaty, is the extended form of the Analytic Hierarchy Process [36]. In a hierarchical structure to be examined using

AHP, a strict assumption is made about the independence of higher-level elements from the lower level elements as well as about the independence within a level [37, 35]. The majority of the real-life problems involve dependence and feedback that makes AHP impossible to be used. So, to deal with these properties, ANP was proposed. ANP allows the representation of both the dependence within a level and among different levels, which are the inner dependence and the outer dependence, respectively [29]. The comparison between a hierarchy and a network is provided in Fig. 2.

The steps of the ANP method are explained as follows [32–35]:

- *Structuring the problem and constructing the model*: In this step, the decision problem is clarified and, the criteria, sub-criteria and, alternatives are determined. The criteria are grouped into clusters. The structure of the problem is built, and the dependencies and feedbacks are stated in it. In other words, a network model is constructed.
- *Pairwise comparisons*: Decision-makers are asked to compare two elements at a time considering their contribution to a particular element. A sample question for pairwise comparisons is: "for a given criterion, which of the following criteria is more dominant and how much more?". Saaty's nine-point scale is used to elicit the assessments. The elicited individual assessments are aggregated into a group assessment using an appropriate method (e.g., geometric mean). At this step, the inconsistency of individual (or group) assessments are checked whether being higher than the threshold value [34] suggested.

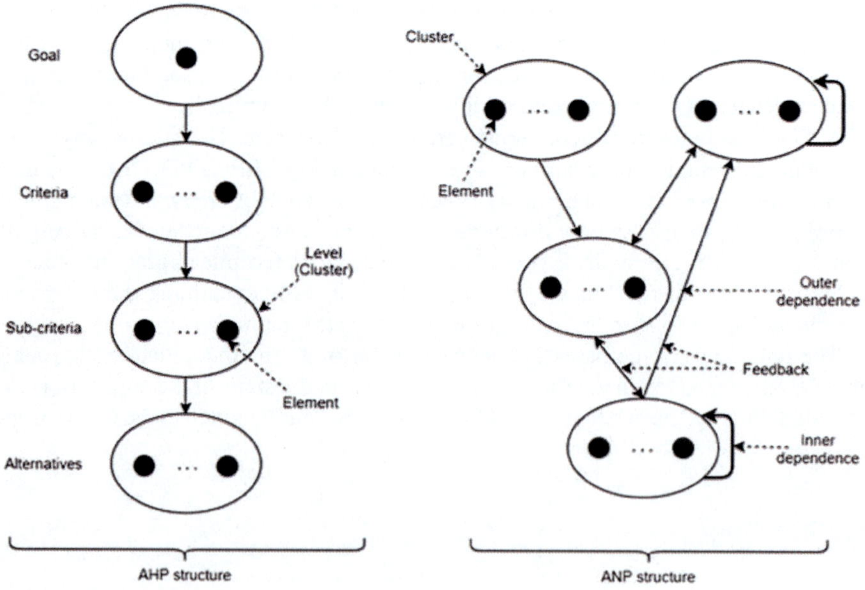

Fig. 2 Comparison of AHP and ANP structures. (adapted from Saaty [34])

- *Obtaining supermatrix*: Supermatrix is a two-dimensional matrix where each element is represented at one column and one respective column. The entries are the eigenvectors which are obtained in the previous step using paired comparisons. The sum of each column equals the number of comparison sets. In order to obtain limiting priorities, first, the supermatrix is reduced to a column stochastic matrix and turns into a weighted supermatrix, then the weighted supermatrix is raised to its arbitrarily large power.

Application

Brain drain is a type of international migration and symbolizes the movement of highly skilled people. When the causes pushing (or forcing) people to leave their home country and immigrate to another place are analyzed, it's observed that there are several conflicting criteria to be examined. In the literature, brain drain has been studied by using different theoretical and empirical methods. However, after a detailed literature review, a requirement for a comprehensive multi-criteria decision-making model for brain drain is identified. To address this issue, the purpose of this study is to offer a tool to examine and prioritize criteria affecting people's decisions on living abroad using ANP. In other words, a decision-making framework is proposed to assess the reasons behind the brain drain.

The steps of the application will be explained using the main stages of a decision-making process: (*i*) structuring the problem, (*ii*) constructing the decision model, and (*iii*) analyzing the model.

Structuring the Problem

The set of criteria affecting brain drain decision was constructed based on a detailed literature review. As a result, a list of 24 criteria was identified as relevant. After a final examination of the interconnections among those, 18 criteria that are grouped into four clusters remained (see Table 1). While the career-related set of criteria includes career opportunities, working environment, wages policy, and employee benefits, education cluster covers the quality of education, desire to have an international education, scholarship opportunities, and improving language skills. Government-related criteria set includes the economic and political situation, restriction of the fundamental rights and freedom, migrant integration policy, and strategies of foreign policy. Finally, personal criteria are religious freedom, personal interest in living abroad, standard of living, social/cultural wealth, and networking opportunities.

Table 1 The set of criteria

Clusters	Criteria	References
Career (C)	C1. Career opportunities	[2, 3, 6, 17]
	C2. Working environment	[42]
	C3. Wages policy	[6, 12, 24, 42]
	C4. Employee benefits	[24]
Education (E)	E1. Quality of education	[11, 24, 41, 42]
	E2. Desire to have an international education	[2, 41]
	E3. Scholarship opportunities	[14, 47]
	E4. Improving language skills	[9, 21, 25, 47]
Government (G)	G1. Economic situation	[2, 3, 6, 21, 42]
	G2. Political situation	[2, 3, 11, 14, 24, 42]
	G3. Restriction of the fundamental rights and freedom	[6]
	G4. Migrant integration policy	[24, 39, 47]
	G5. Strategies of foreign policy	
Personal (P)	P1. Religious freedom	[14]
	P2. Personal interest in living abroad	[14, 24]
	P3. Standard of living	[9, 24]
	P4. Social/Cultural wealth	[2]
	P5. Networking opportunities	[6, 24]

Constructing the Decision Model

Since the existence of interconnections among the criteria, ANP is selected as the appropriate tool for the analysis. In general, the network model may include alternatives, but in this case, the decision model represents only the criteria to be prioritized. As given in Fig. 3, both internal and external dependencies exist.

To elicit the decision-makers' judgments, an online pairwise comparison questionnaire is prepared using an open-source software which is developed at the Centre for Social Informatics [45]. Participants are asked to perform pairwise comparisons using the nine-point scale (Table 2) proposed by Saaty [29] to assess the relative importance of criteria pairs on the affected criterion.

The questionnaire includes 104 pairwise comparisons and 10 demographic questions. The demographic information of twenty-six participants, who are immigrated temporarily or permanently to different countries, is summarized in Table 3. According to Table 3, 57.7% of the participants are female. The percentage of participants who are students is 61.5%. More than half of the participants have a bachelor's degree. Half of the participants have less than two years of work experience. In the questionnaire, participants are asked which country they immigrated to. Germany, Italy, Sweden, and France are the most frequent answers, which means Europe is the continent where participants immigrated to the most, with a rate of 61.6%. Other

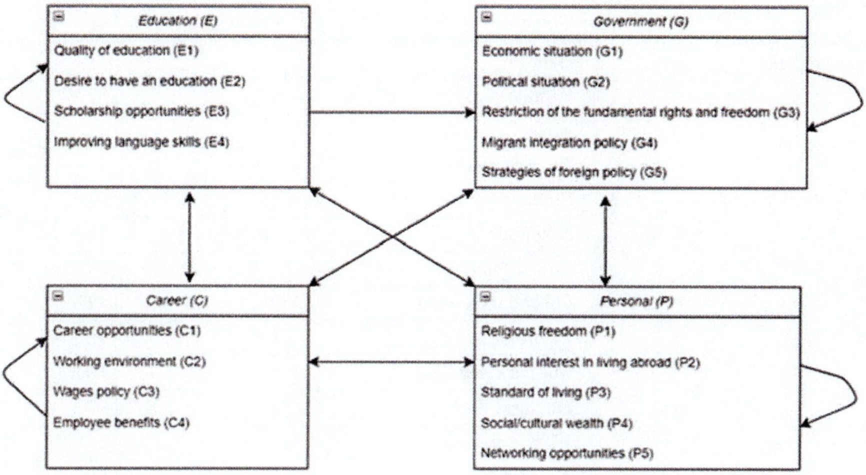

Fig. 3 Decision model

Table 2 Saaty's nine-point scale [29]

Intensity of importance	Definition
1	Equal importance
3	Moderate importance
5	Strong importance
7	Very strong importance
9	Extreme importance
2, 4, 6, 8	Intermediate values

countries are South Korea, Russia, and Iraq, with 19.2%. It has been more than one year since 65.4% of the participants emigrated from their home country. Lastly, the return intention is asked, and forty-six percent of the participants indicated that they are likely to return.

Analyzing the Model

Super Decisions v3.2 software (www.superdecisions.com) is used for the mathematical computations. The geometric means of 26 different answers of 104 comparisons are calculated to obtain the aggregated assessments. After entering the aggregated values into comparison matrices, inconsistencies are checked. Note that, even though Saaty indicated (2008) the allowable limit for inconsistencies as 10%, in this study, the assessments having lower than 15% inconsistency are accepted as consistent. If the inconsistency is higher than the present limit, the individual assessments that violate

Table 3 Demographic information of the participants

Category	Frequency	%
Gender		
Female	15	57.7
Male	11	42.3
Current status		
Student	16	61.5
Not student	10	38.5
Education level		
Secondary school	2	7.7
Bachelor's degree	14	53.8
Master's degree	8	30.8
Doctoral degree	2	7.7
Work experience		
<2 years	13	50.0
2–5 years	7	26.9
6–10 years	2	7.7
>10 years	4	15.4
Continent		
Europe	16	61.6
USA	5	19.2
Other	5	19.2
Duration of migration		
Less than a year	9	34.6
More than a year	17	65.4
Return intention		
Probably will return	12	46.2
Probably will not return	14	53.8

the rule are removed. Then, geometric means are recalculated for the remaining assessments. The limiting priorities are given in Table 4.

According to the findings, the most important clusters affecting brain drain decisions are career and government. "Career opportunities" sub-criterion has the highest priority in the career cluster, while "Quality of education" has the highest priority in the education cluster. "Economic situation" is the most important criterion in government cluster, and "religious freedom" is the most important one among personal related criteria.

The limiting priorities indicate that "economic situation," "career opportunities," and "political situation" are the most important criteria with weights of 0.122, 0.121 and, 0.120, respectively. On the other hand, "scholarship opportunities" is the least important criterion with the lowest priority.

Table 4 Comparison of the limiting priorities (darker cells represent higher priorities)

	Overall	Return+	Return-	Student+	Student-	Europe	USA	Other	<1 year	>1 year
G1	0.122	0.119	0.148	0.139	0.131	0.123	0.107	0.138	0.124	0.117
C1	0.121	0.114	0.136	0.126	0.127	0.119	0.137	0.102	0.128	0.112
G2	0.120	0.124	0.113	0.116	0.117	0.127	0.111	0.108	0.109	0.118
C2	0.111	0.116	0.097	0.111	0.101	0.121	0.090	0.108	0.121	0.124
C3	0.089	0.089	0.075	0.076	0.089	0.081	0.052	0.086	0.116	0.069
G3	0.070	0.076	0.054	0.063	0.060	0.069	0.089	0.080	0.055	0.092
P4	0.064	0.067	0.063	0.067	0.054	0.061	0.100	0.076	0.053	0.093
E1	0.048	0.051	0.046	0.053	0.047	0.050	0.023	0.052	0.052	0.040
P3	0.042	0.043	0.037	0.034	0.042	0.037	0.046	0.042	0.038	0.040
P5	0.038	0.040	0.035	0.032	0.042	0.036	0.041	0.032	0.026	0.036
C4	0.036	0.046	0.030	0.027	0.037	0.032	0.047	0.069	0.036	0.046
E4	0.031	0.027	0.038	0.022	0.044	0.026	0.071	0.014	0.027	0.034
G5	0.031	0.022	0.048	0.061	0.021	0.035	0.012	0.040	0.038	0.021
P1	0.028	0.030	0.025	0.023	0.025	0.031	0.029	0.025	0.025	0.027
P2	0.016	0.010	0.023	0.017	0.035	0.019	0.009	0.002	0.019	0.005
G4	0.014	0.012	0.012	0.012	0.011	0.013	0.024	0.013	0.017	0.013
E2	0.014	0.012	0.016	0.014	0.014	0.015	0.010	0.010	0.011	0.012
E3	0.004	0.003	0.005	0.007	0.002	0.005	0.001	0.005	0.005	0.003

After completing the overall analysis, the participants are classified based on their demographic information, and for each class, the priorities are calculated separately. Table 4 represents the comparison of the priorities. The first classification is made considering the return intention: probably will return and probably will not return. Similar rank orders for priorities are obtained for both classes. Different from the overall assessments, "political situation" turns out to be the most important criterion for people who are not likely to return when the two classes are examined separately. Classification of participants based on studentship yields quite similar rank orders to the overall results. It is interesting that "scholarship opportunities" are considered as the least important factor for both classes.

According to the participants immigrating to Europe, "political situation" is the most important criterion, while people immigrating to the USA considers the "career opportunities." The decision on migration to somewhere else is mostly affected by the "economic situation." It appears that people who want to improve their language skills prefer the USA as well as the ones looking for a social/cultural wealth.

"Working environment" is the most important criterion for the people who have been staying abroad for over a year. Whereas, it has the third rank for the ones who have been staying for less than a year. Instead, "career opportunities" and "economic situation" have the first and second ranks for newly immigrated people.

Conclusion

This study proposes a multi-criteria decision model to examine the criteria affecting brain drain decision. Due to the network structure of the decision problem, ANP is used to obtain the priorities. Among the determined clusters, career and government were the most important clusters having a significant role in this decision. Political and economic situations and the restriction of fundamental rights and freedom are the most important criteria in the government cluster. Career opportunities, working environment, and wages policy have the highest priorities in the career cluster. According to the findings, employee benefits are not important relative to other career-related criteria. Although not being as important as others, social/cultural wealth and the quality of education are the prominent criteria in the remaining clusters.

Brain drain causes a human capital loss and will eventually affect the socio-economic situation, especially in the sending countries. To circumvent the problem of losing talented and competent individuals, local authorities, and the government need to develop effective policies regarding the mitigation and/or management of brain drain [6]. The proposed decision model will guide the policymakers to identify the needs of skilled individuals planning to immigrate. Even though the required policies may change depending on the characteristics of the countries, strengthening the relevant systems in accordance with the findings the proposed model provides will help to deal with the consequences of brain drain and, at least to reduce the negative effects. For instance, in the short term, handling the challenges of the political and economic climate and dealing with employment issues should be the primary concern. Having the sixteenth rank, migration integration policy is not the primary concern of the participants. Nevertheless, this criteria deserves to be examined in detail, because it is directly related to the labor market, education, health, politics, anti-discrimination, and permanent residency policies [47].

Inevitably, there are some limitations of the proposed approach. The response rate was low, therefore increasing the number of participants and reanalyzing the assessments may provide more representative results. Also, the online survey was distributed to a group of people from different nationalities with different characteristics, but the majority of the participants are Turkish as expected. Since the characteristics of different countries vary, applying this study to different countries and comparing the findings will provide a better perspective. Finally, the list of criteria can be expanded.

References

1. Arandarenko M, Corrente S, Jandrić M, Stamenković M (2020) Multiple criteria decision aiding as a prediction tool for migration potential of regions. Eur J Operat Res
2. Bilecen B (2012) Trends in student mobility from turkey to Germany. Percept: J Int Affairs 17:61–84
3. Boeri T, Brücker H, Docquier F (2012) Brain drain and brain gain: The global competition to attract high-skilled migrants. Oxford University Press
4. Boncea I (2014) Brain drain in Romania: factors influencing physicians' emigration. J Commun Positive Pract 14:64–74
5. Cambridge Dictionary: https://dictionary.cambridge.org
6. Cavallini S, Soldi R, Di Matteo L, Utma MA, Errico B (2018) Addressing brain drain: the local and regional dimension. European Committee of the Regions
7. Cebula RJ (2005) Internal migration determinants: recent evidence. Int Adv Econ Res 11:267–274
8. Chi G, Voss P (2005) Migration decision-making: a hierarchical regression approach. J Reg Anal Pol 35:11–22
9. Ciarniene R, Kumpikaite V (2011) International labour migration: students viewpoint. Eng Econom 22:527–533
10. Dibeh G, Fakih A, Marrouch W (2018) Decision to emigrate amongst the youth in Lebanon. Int Migr 56:5–22
11. Docquier F (2014) The brain drain from developing countries: the brain drain produces many more losers than winners in developing countries. IZA World of Labor 31:1–10
12. Dodani S, La Porte RE (2005) Brain drain from developing countries: how can brain drain be converted into wisdom gain? J R Soc Med 98:487–491
13. Dustmann C, Fadlon I, Weiss Y (2011) Return migration, human capital accumulation and the brain drain. J Dev Econ 95:58–67
14. Erguvan D (2015) Transnational education in Turkey. J Educ and Soc Res 5:227–240
15. Etzo I (2008) Internal migration: a review of the literature. MPRA
16. Gibson J, McKenzie D (2012) The economic consequences of 'brain drain'of the best and brightest: microeconomic evidence from five countries. Econ J 122:339–375
17. Gouda P, Kitt K, Evans DS, Goggin D, McGrath D, Last J, Hennessy M, Arnett R, O'Flynn S, Dunne F, O'Donovan D (2015) Ireland's medical brain drain: migration intentions of Irish medical students. Hum Resour Health 13:11
18. Greenwood MJ (1997) Internal migration in developed countries. Handb Popul Fam Econ 1:647–720
19. Gökhan A (2008) The determinants of internal migration in Turkey (Doctoral dissertation)
20. Güngör ND, Tansel A (2014) Brain drain from Turkey: return intentions of skilled migrants. Int Migr 52:208–226
21. Hoppe A, Fujishiro K (2015) Anticipated job benefits, career aspiration, and generalized self-efficacy as predictors for migration decision-making. Int J Intercult Relat 47:13–27
22. Iravani MR (2011) Brain drain problem: a review. Int J Bus Soc Sci 2:284–289
23. Kanayo O, Anjofui P, Stiegler N (2019) Push and pull factors of international migration: evidence from migrants in South Africa. J African Union Stud 8:219–250
24. Karaduman HA, Çoban E (2019) Brain drain in Turkey: an investigation on leading motives of skilled migration. Avrasya Sosyal ve Ekonomi Araştırmaları Dergisi 6:322–339
25. Kaya M (2002) Beyin göçü / erozyonu (Brain drain/ erosion). Osmangazi University: Eskisehir: Technology Research Centre Report November 5, 2002
26. Kumar N, Sidhu AS (2005) Pull and push factors in labour migration: a study of brick-Kiln Workers in Punjab. Ind J Ind Rel 221–232
27. Molloy MP, Saaty TL (2004) An analytic network process model for financial-crisis forecasting. Int J Forecast 20:573–587
28. Molloy R, Smith CL, Wozniak A (2011) Internal migration in the United States. J Econom Perspect 25:173–196

29. Niemira MP, Saaty TL (2004) An analytic network process model for financial-crisis forecasting. Int J Forecast 20:573–587
30. Parkins NC (2010) Push and pull factors of migration. Am Rev Polit Economy 8:6
31. Pires AJG (2015) Brain drain and brain waste. J Econom Dev 40:1–34
32. Rashid MFA (2019) Capabilities of a GIS-based multi-criteria decision analysis approach in modelling migration. GeoJournal 84:483–496
33. Saaty TL (2001) Decision making with dependence and feedback: the analytic network process, 2nd edn. RWS Publications, Pittsburgh
34. Saaty TL (2008) The analytic hierarchy and analytic network measurement process: applications to decisions under risk. Eur J Pure Appl Math 1:122–196
35. Saaty TL, Vargas LG (2013) Decision making with the analytic network process: economic, political, social and technological applications with benefits, opportunities, costs and risks, 2nd edn. Springer, US
36. Saaty TL (1996) decision making with dependence and feedback: the analytic network process. RWS Publications, 4922 Ellsworth Avenue, Pittsburgh, PA. 15213
37. Saaty TL (1999) Fundamentals of the analytic network process. ISAHP 12–14
38. Sell RR, DeJong GF (1978) Toward a motivational theory of migration decision making. J Popul 1:313–335
39. Sert D (2012) Integration and/or transnationalism? The case of Turkish-German transnational space. Perceptions 17:85
40. Super Decisions (2020) http://www.superdecisions.com
41. Süoğlu BB (2012) Trends in student mobility from Turkey to Germany. Perceptions 17:61–84
42. Tansel A, Güngör ND (2003) Brain drain from Turkey: survey evidence of student non-return. Career Development International
43. Thet KK (2014) Pull and push factors of migration: a case study in the urban area of Monywa Township, Myanmar. News from the World of Stat 1:1–14
44. UN International Migration (2019) https://www.un.org/en/development/desa/population/migration/publications/migrationreport/docs/InternationalMigration2019_Report.pdf
45. University of Ljubljana (2020) http://www.1ka.si
46. Wickramasinghe AAIN, Wimalaratana W (2016) International migration and migration theories. Social Affairs 1:13–32
47. World Bank (2019) Migration and brain drain. International Bank for Reconstruction and Development, The World Bank, Washington, D.C., USA
48. Yılmaz EA (2019) Uluslararası Beyin Göçü Hareketleri Bağlamında Türkiye'deki Beyin Göçünün Durumu. LAÜ Sosyal Bilimler Dergisi 10:220–232
49. Zweig D (1997) To return or not to return? Politics versus economics in China's brain drain. Stud Comp Int Dev 32:92–125

The Effect of Social Media Usage on Loneliness

Murat Durucu, Ozgenur Tuncer, and Cahit Ali Bayraktar

Abstract This study has been done in order to determine the effect of social media usage on loneliness perception. At the same time, it has been examined whether there was a significant difference in terms of loneliness for gender, age, and frequency of use. One hundred eighteen women and 83 men took part in the study. Social Media Use Scale [3] and the UCLA Loneliness Scale [11] have been used together with the Personal Information Form developed by the researcher. Correlation Analysis and Multiple Linear Regression Analysis have been used to determine the relationships between variables (social media use and loneliness). The relationship of loneliness according to gender, age, and frequency of use has been determined used one-way ANOVA. As a result of the analysis, social integration and emotional connection factor of social media use scale positively affect the sociality variance of loneliness scale. Furthermore, the loneliness scale of both factors of social media use scale has an effect on high human communication. High human communication is influenced positively by social integration and emotional engagement and is influenced negatively by social routines. At the same time, it was determined that the perception of loneliness did not differ according to gender and age, but differed significantly according to the frequency of internet use. The results of the study have shown that those who use social media for 5–6 h a day differ significantly on asocial variable compared to those who use less than 1 h or 1–2 h a day.

Keywords Social media · Loneliness · Regression · ANOVA · Frequency of use

M. Durucu · O. Tuncer · C. A. Bayraktar (✉)
Industrial Engineering Department, Management Faculty, Istanbul Technical University, Istanbul, Turkey
e-mail: cahitali@itu.edu.tr

M. Durucu
e-mail: durucumur@itu.edu.tr

F. Calisir (ed.), *Industrial Engineering in the Internet-of-Things World*,
Lecture Notes in Management and Industrial Engineering,
https://doi.org/10.1007/978-3-030-76724-2_22

Introduction

Nowadays, with the development of technology, the social platforms are increasing in number; The number of individuals who are members of these platforms and spend their time in these areas is rapidly increasing thanks to its features such as being easily updated, multi-use, and open to sharing. With the decline in the age of using mobile phones, tablets, and computers, the use of social networks such as Facebook, Youtube, Instagram, and the time they spend in these environments is increasing. Although the use of social media is seen as beneficial for some individuals and suitable for socializing, it is increasingly replacing face-to-face communication among people. Studies reveal the idea that this increase in the use of social media leads people to loneliness.

According to the 2019 Global Digital Report, the use of social media in the world increased by 9% when compared to the previous year [23], while in Turkey, it is explained that an increase of 2% [23]. Professor Dr. Turker Bas emphasized that this increase sees social media as a tool to get rid of loneliness today, but this preference has the opposite effect and argued that social media use increasingly increases individuality, selfishness, and loneliness [20].

While the studies conducted are aimed at proving the accuracy of this statement in young people between the ages of 16–25, in this study, in order to reveal the relationship between social media and loneliness in general and to reveal this situation for people of all ages, the concepts of social media and loneliness an attempt was made to reveal the relationship.

Literature Review

Social Media

When the concept of social media is examined in the literature, there are many definitions. Boyd [7] defines web-based environments as social media, which increase communication between individuals, facilitate the sharing of individuals with common interests, create a list of friends with everyone's own personal profile, and communicate with friends [35].

On the other hand, Hazar [18] describes social media as sites that provide deeper social interaction, community formation, and collaboration projects, while Belin and Yildiz [6] refer to online social networks that support the maintenance of existing social ties and the establishment of new connections [35].

According to Balci and Kocak [5], social media is the medium where the number of users is increasing, where individuals are messaging with each other, sharing their videos and photos, and making their own content. These channels are social networking platforms where users communicate with other users, develop and maintain their relationships, give them the opportunity to meet people they already know or meet new people, bring together those whose ideas are similar, and allow emotional relationships to form [26].

Social media has different features than traditional media, although it is media in its name. The most important difference that creates its originality is that any person can create the content of social media, comment, and contribute [32]. People can express their instantaneous opinions on social media, and they can discuss new ideas and come up with new ideas. It can even share various photos, videos, and find jobs. In short, it enables individuals to live in the real world in a virtual environment [19].

Mayfield (2010) lists the features of social media as follows:

- Participants: Social media encourages participants and gets feedback from each interested user.
- Openness: Most social media services are open to feedback and participants. These services instill courage on issues such as voting, commenting, and sharing information. These rarely set barriers to transportation.
- Speaking: While traditional media is about broadcasting (content transfer or information to the audience), social media is better in terms of enabling two-way speech.
- Community: Social media allows communities for a quick and effective formation. Communities thus share what they are interested in, such as their favorite photos, political values, favorite TV shows.
- Connectivity: Most types of social media do connected things; other sites allow researches and links to any subject related to people [32].

The foundations of the historical development of social media were laid in 1991 with the opening of the world wide web (WWW) to the public. With the interest-oriented forums that started to occur in the late 1990s, the individual has now started to actively share his views with other individuals. However, it can be stated that the history of modern social media, where the user can be more effective, started in 1997 with the SixDegrees site, which offers the opportunity to create profiles, list friends, and then review their friends' lists. The LiveJournal site, launched in 1999, is built on blogs that can be renewed continuously. Thanks to this network, users started to be effective in following each other, creating a group and interacting with others. Wikipedia was launched in 2001, and after 2003, as a result of the growing interest in social media, a significant number of applications such as Myspace, Linkedln, Photobucket have been included in the network. Flickr and Facebook in 2004, Yahoo! 360, and Youtube in 2005 and Twitter in 2006 were among the noteworthy applications in social media [19].

With the increase in internet usage, the number of individuals entering social media has increased in the world and in our country. Morgan Stanley, who conducted research on social media entry rate in 2004, found that reaching 50 million users was 38 years for radio, 13 years for television, and 5 years for internet [19].

Examples of social media that affect people and spend most of their time are Facebook, Twitter, Instagram, and Youtube.

Facebook was founded in 2004 for Harvard University students. Facebook, which also included schools around Boston, covered all Ivy League schools within two months and all schools in the United States in the first year. The site provides users with opportunities to make new friends, share their photos, thoughts, and personal information. The site, which receives income from advertisements and sponsors, is free for users [32].

Launched in 2006, Twitter is a software that allows the possibility of a 140-letter speech act, in which these speech acts are defined as "tweets." Reasons for using Twitter; chat and dialogue feature can be explained as enabling solidarity and exchange with certain users, enabling self-expression and self-communication, updating and controlling status, sharing information and news, marketing, and advertising [32].

Instagram is a photo-sharing application that was founded in October 2010 by two entrepreneurs named Kevin Systrom and Mike Krieger. The Instagram application, which has achieved significant success by exceeding the number of 100 million subscribers in a short time, is very popular and was purchased by Facebook for an enormous fee of $1 billion in 2012. It is possible to say that the number of users exceeded 500 million in 2016 [15].

Youtube was first established in 2005, and it was a social media platform that allows video uploading, sharing, and user interaction purchased by Google in 2006. Youtube, which is a combination of the words "you" and "tube," means "your TV" or "your channel." Thanks to this platform, each user has become both a content producer and a consumer [29].

Loneliness

Loneliness in the literature: it is the psychological aspect of a person who lives alone or despite living with the community in his/her social environment for any reason.

According to Ponzetti [28], loneliness is an unpleasant subjective psychological condition that occurs as a result of the difference between the individual's existing social relationship and the desired social relationship [27]. According to Cacioppo et al. [9], loneliness is defined as a complex series of emotional experiences, consisting of negative emotions and cognitions experienced by the person's lack of social and personal relationships [14].

Loneliness is a very serious problem that increases the complexity of life within time and negatively affects people's life. Loneliness is usually physically alone in society and can be seen at certain times in individuals, regardless of age, gender, and social status. Therefore, the main reason for loneliness is the lack of social relationships and the low level of satisfaction with these relationships [36].

Seyyar and Genc [30] examined loneliness in three ways. The first is objective loneliness, in which the person is isolated from the social relationships he/she needs. The other is the subjective loneliness that the individual experiences, despite not being alone, because he/she is unable to catch the intimacy and close behavior he/she expects from his/her environment. The last one is conscious loneliness, where even one chooses loneliness with his/her own consent [16].

Loneliness has gained a distinct quality with the modernization of life. Facilitating the communication of people with rapidly changing and renewed technology causes the concept of togetherness to decrease gradually. Thus, individuals can communicate virtually with technological tools. Relationships established in this way are now considered routine for everyone. Although people are social entities, as they use virtual uses, they become more and more isolated in daily life. This situation is called dialectical loneliness. In this way, people are doomed to be alone. In this study, loneliness is considered as dialectical loneliness [34].

Relationship Between Social Media and Loneliness

Social media platforms such as Facebook, Instagram, Twitter, which have entered our lives rapidly with the spread of the internet today, can cause negative effects besides positive effects on individuals. Even though individuals, especially young people, see these social networking channels as ideal channels to socialize for reasons such as sharing information, expressing themselves and having a good time, it has been a matter of debate how individuals can socialize because it causes them to get away with themselves, their family and the society they live in. In fact, internet games, chat in the virtual environment, long phone calls quickly drag individuals towards individualization, loneliness, and unmanned communication.

Sucu [33] argues that internet games, which have become a social activity, both reduce social distance and destroy social contact. While the use of social media and games played on social media change the perception of reality of the real world, it also conveys the communication forms of individuals to virtual environments [27].

Aktas and Cayci [2] believe that people can communicate with people whom they are afraid of even talking in their lives, in social networks where communication is not mandatory, they can improve their relationships, and they will not be alone. For this reason, they believe that people spend a lot of time in social networking media, and this leads to the development of the concept of "sofalising" (being so lonely while socializing) derived from the word sofa and socialising [1]. Sofalising causes the realization of people's real-life relationships, the existence of a new communication process that depends on symbols, icons, and signs. Along with these, the socialization process is also changing. Going to traditional culture-based visits, going to parks, cafes, and associations is replaced by social media [26].

Cayci and Karagulle [10] argue that individuals who spend most of their time in social networks in order to socialize are also alienated from their cultural values and identities, especially their families and their environment. Individuals who share online on these networks become unfamiliar with this new environment they interact with, and become alien to itself and its environment over time, allowing them to replace their own cultures. Digital environments are far from the reality due to their structure and create dependency on technology and screen in users. Due to social networking channels, while the individual becomes alone in the crowds, alienation problems for themself and their environment also arise. Kara and Ozgen [22]

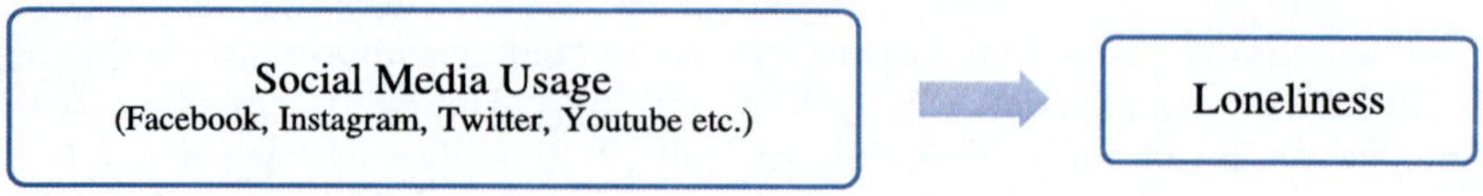

Fig. 1 Conceptual model for the study

state that the individuals involved in friendship relationships created through social networks are different individuals from their real characters, and that they make friends based on artificial and misleading characters created in line with their own wishes. Friendships with this unrealistic character can also affect individuals' real-life social relationships and cause situations such as making their relationships difficult [26].

Diker and Tasdelen [13], although social networking channels are considered as the media used to eliminate the problems such as loneliness, stress, and depression, in some cases, it makes users addicted when they are used too much, and as a result, it can cause behavioral and psychological disturbances and also self-anxiety in young people. Also, they say that it can create changes in the closure, dissatisfaction, and emotional states. Ocal [25] says that individuals socialize on social media and move away from their real lives [26].

Conceptual models and hypotheses in Fig. 1 have been introduced by taking into consideration the literature studies conducted to determine the relationship between social media use and loneliness.

H1: The use of social media has an impact on loneliness.

Methodology

The study tried to find an answer to the question of whether social media channels used in society have an effect on the perception of loneliness. In this context, it has been acted with the assumption that there is a significant relationship between the use of social media and the perception of loneliness and that it positively affects.

Sampling

The study population consists of individuals with membership to any social media platform in Turkey. The sample of the study consists of 201 people. Attention has been paid to ensure that there is diversity in people who applied the survey and not to focus on a certain audience.

Data Collection

The study was carried out with the application of a questionnaire prepared to determine the role of the social media use of individuals on their perception of loneliness. Since it offers the opportunity to reach a large audience in a short time, an internet questionnaire was preferred in the study. In addition to the Social Media Usage Scale, the UCLA Loneliness Scale, and the Personal Information Form, the questionnaire includes three questions that measure whether people meet with people they meet on social media in their external environment.

UCLA Loneliness Scale is a Likert type self-evaluation scale that helps to determine the general degree of loneliness of the individual. The original of the scale has a grading between "I NEVER experience this situation" (1) and "I experience this situation FREQUENTLY" (4) with a total of 20 items, 10 items of which are positive, meaning that they do not contain semantic loneliness, 10 items are negative, meaning semantically only individuals. The highest score that can be obtained from the scale is 80, and the lowest score is 20. High scores indicate that individuals experience more loneliness. Questions 1, 4, 5, 6, 8, 10, 15, 16, 19, 20 in the scale were asked as inverse questions. The validity and reliability studies of the scale in our country were carried out by Demir [11]. It is observed that the UCLA Loneliness Scale was used in many studies conducted in the world and in our country about loneliness.

Social Media Usage Scale was developed by Jenkins-Guarnieri et al. [21] and adapted to Turkish by Akin et al. [3]. The scale ranks between "I totally disagree" (1) and "I totally agree" (6). High scores indicate a high level of social media use. 1–6 questions are explained as integration and emotional connection. 7–10 questions are explained as integration with social routines. The 8th question is the inverse item.

The personal information form consists of 8 questions about the participants' gender, age range, occupation, educational level, marital status, employment status, which social network they use the most, and how often they use social media.

The data collected within the scope of the research were analyzed using SPSS 20.0 Package Program.

Findings

In Table 1, according to the statistical findings regarding the participants' gender, age, current working status, occupation, educational level, marital status, the most preferred social media platform, and time spent in social media; 58.7% of the participants are women, and 41.3% are men. It was determined that 31.3% were between the ages of 18–24, 55.2% were working, 22.4% were students, 45.3% were university graduates, and 68.7% were single. In addition, the majority of 201 respondents prefer to use Instagram with 60.2, and 38.3% spend 1–2 h a day on social media platforms.

Table 1 Distribution of participants' demographic information

Variables		Frequency	%
Gender	Women	118	58.7
	Men	83	41.3
Age	Younger than 17	9	4.5
	18–24	63	31.3
	25–29	47	23.4
	30–39	32	15.9
	40–49	26	12.9
	50–59	21	10.4
	Over 60	3	1.5
Current occupation	Employed	111	55.2
	Unemployed	90	48.8
Profession	Housewife	13	6.5
	Student	45	22.4
	Engineer	32	15.9
	Retired	15	7.5
	Other	96	47.8
Education level	Primary school graduate	6	3
	High school student	10	5
	High school graduate	28	13.9
	University student	46	22.9
	University graduate	91	45.3
	Graduate student	10	5
	Graduate level	10	5
Marital status	Merried	63	31.3
	Single	138	68.7
Most preferred social media platform	Facebook	27	13.4
	Instagram	121	60.2
	Twitter	24	11.9
	Youtube	22	10.9
	Diğer	7	3.5
Time on social media	Less than 1 h/day	33	16.4
	1–2 h/day	77	38.3
	3–4 h/day	49	24.4
	5–6 h/day	26	12.9
	More than 6 h/day	14	7
	1–2 h/week	2	1

Table 2 Cronbach's alpha values of scales

Scales	Item	Cronbach's alpha
Social media usage scale	10	0.84
UCLA loneliness scale	20	0.873
Meeting outside with people you meet on social media	3	0.852

Normal distribution was examined for both scales used in the study. Indeed, skewness and kurtosis values indicate whether the data show normal distribution. The fact that the z statistics found by dividing the skewness and kurtosis values by the standard error is less than 1.96 for $\alpha = 0.05$ and 2.58 for $\alpha = 0.01$ is evidence of the normal distribution of data [12]. According to this, all the data in the study are less than 2.58 for $\alpha = 0.01$, that is, it is possible to say that it fits the normal distribution.

Reliability analysis is the analysis of the consistency of the answers given. Cronbach's alpha coefficient, which is a measure of the internal consistency of the items, is used to explain the homogeneous structure of the items. As Cronbach's alpha coefficient approaches 1, it is interpreted that the items in the scale are consistent and measure the same feature (Yildiz and Uzunsakal [12]. According to this,In the study, the results of the reliability analysis of using social media, loneliness, and meeting with people you meet in social networks in real life are given in Table 2.

When the reliability analysis results are examined in Table 2, it is seen that the reliability coefficient of all scales is $\alpha = 0.70$ and above in general. Accordingly, it is understood that the scales are at a reliable level. Since the sub-dimensions in the scales are not defined in the literature, Cronbach's Alpha values were interpreted over the scales in general.

Factor analysis is a method of obtaining a small number of conceptually meaningful variables by bringing together many related variables [8]. Factor analysis was applied because there were many questions measuring the same subject in the study, and its sub-variables were not fully expressed in the literature.

Firstly, the KMO criterion expressed in Table 3 was examined in the scales. The result is good for values between 0.7 and 0.8 (Karaalioglu 2015), that is, the sample size in the research is sufficient for factor analysis.

According to Table 3, it is seen that Barlett test significance level values are 0.00. The Barlett Sphericity test was found to be significant because this value was less than a 5% margin of error. The data set is suitable for factor analysis.

Table 3 Kaiser–Meyer–Olkin values and barlett's test significance of scales

Scales	KMO	Barlett's test reliability
Social media usage scale	0.834	0.00
UCLA loneliness scale	0.896	0.00
Meeting outside with people you meet on social media	0.732	0.00

Table 4 Variance explanation percentages of total variance and factors

Total variance explained									
Component	Initial eigenvalues			Extraction sums of squared loadings			Rotation sums of squared loadings		
	Total	% of Variance	Cumulative %	Total	% of Variance	Cumulative %	Total	% of Variance	Cumulative %
1	4.209	42.087	42.087	4.209	42.087	42.087	3.366	33.665	33.665
2	1.300	13.003	55.090	1.300	13.003	55.090	2.143	21.425	55.090
3	0.798	7.983	63.073						
4	0.702	7.017	70.090						
5	0.694	6.936	77.026						
6	0.631	6.312	83.338						
7	0.570	5.695	89.033						
8	0.513	5.134	94.166						
9	0.389	3.889	98.055						
10	0.195	1.945	100.000						

Extraction Method: Principal Component Analysis

Table 5 Distribution of social media usage scale to factors

Rotated component matrix		
	Component	
	1	2
SMED3- I am disappointed when I have never been able to use social media	0.824	
SMED4- I get nervous when I can't connect to social media	0.779	
SMED1- I feel disconnected from my friends when I can't connect to social media	0.719	
SMED5- I prefer to communicate with other people mostly through social media	0.668	
SMED2- I wish everyone used social media to communicate	0.656	
SMED6- Social media plays an important role in my social relationships	0.643	
SMED7- I like to check my social media account		0.734
rSMED8- I don't like using social media		0.732
SMED10- I reply to others' social media posts		0.656
SMED9- Using social media is a routine part of my daily life		0.564

As a result of the factor analysis, according to Table 4, 2 factors with an eigenvalue greater than 1 ($\lambda > 1$) were selected for the Social Media Usage Scale. Thus, 55.090% of the total variance was explained with two factors instead of the initial ten variables.

Factor rotation was performed to interpret the factors. While performing factor rotation, the Varimax method was preferred. As a result, the alternating factor loads matrix obtained from 10 items, and two factors are shown in Table 5.

According to the distribution of the questions seen in Table 5, the first factor was named as social integration and emotional connection, and the second factor was integration with social routines.

Four factors with an eigenvalue greater than 1 ($\lambda > 1$) were selected for the UCLA Loneliness Scale. Thus, 60.898% of the total variance was explained with four factors instead of the original 20 variables. The factor loads matrix obtained by the Varimax rotation is shown in Table 6.

According to Table 6, the factors included in the questions were examined, and the first factor was named as sociality, the second factor was high human communication, the third factor was popularity, and the fourth factor belonged to the society.

Three questions that question whether or not they meet with the people they meet on social media in real life in external environments were merged in 1 factor as a result of factor analysis and named as the interview factor. Accordingly, factors were obtained for all scales with 'Main Component Analysis.'

As a result of the correlation analysis, a significant relationship was found between the social integration and emotional connection independent variable and the sociality dependent variable according to 1% significance level, and between the

Table 6 Distribution of UCLA Loneliness Scale by Factors

Rotated Component Matrix				
	Component			
	1	2	3	4
YLNZ18- There are people around me, but they are not with me	0.781			
YLNZ17- I am unhappy with this degree of being closed in me	0.740			
YLNZ14- I feel isolated from other people	0.728			
YLNZ12- I think my social relationships are weak	0.679			
YLNZ11- I feel pushed out of the group	0.672			
YLNZ7- I'm not sincere with anyone anymore	0.668			
YLNZ13- Nobody really knows me well	0.645			
YLNZ3- There is no one I can apply to	0.562			
YLNZ2- I do not have any friends	0.557			
rYLNZ20- There are people I can talk about		0.808		
rYLNZ19- There are people I can talk to		0.791		
rYLNZ16- There are people who really understand me		0.788		
rYLNZ10- There are people I feel close to		0.699		
rYLNZ6- I have something in common with the people around me		0.535		0.498
rYLNZ5- I feel like part of a group of friends		0.504		
rYLNZ15- I can make friends whenever I want			0.755	
YLNZ9- I'm an extrovert person			−0.648	
rYLNZ8- My interests and ideas are shared by those around me			0.620	
rYLNZ4- I don't feel like I'm alone				0.742
rYLNZ1- I feel in harmony with the people around me				0.479

social routines independent variable and the high human communication dependent variable.

Multivariate linear regression analysis was performed to estimate the effect of social media use on loneliness. The stepwise method has been applied to explain the model with the least number of variables. First of all, regression analysis between social integration and emotional connection and independent variables of social routines and sociality dependent variables was examined. As a result of this analysis, the Durbin Watson coefficient is between 2.220 and 1.5–2.5, so there is no problem. VIF value is below 5.3, and tolerance is above 0.10; There is no collinearity. Regression equation was found to be significant with $R^2 = 0.094$, although its explanatory level was low ($F = 20.749$; $p < 0.01$). Social integration and emotional connection argument make sense. According to the results of the analysis, the objectivity is

'0.307 sbutduybag'. When social integration and emotional connectivity increases by one unit, nobility will increase by 0.307 units. In other words, only the social integration and emotional connection independent variable have a positive, meaningful effect on the sociality dependent variable.

According to the result in the regression analysis between social integration and emotional connection and independent variables of social routines and high human communication, Durbin-Watson is in the desired range with 1.916. There is no collinearity in terms of VIF and tolerance values. Regression equation was found to be significant with $R^2 = 0.053$, although its explanatory level was low ($F = 5.531$; $p < 0.01$). Both independent variables are meaningful for the model. According to the result of the analysis, high human communication is 0 −0.170 srutinbut + 0.154 sbutduybag. When the integration with social routines increases by one-unit, high human communication will decrease by 0.170 units, when social integration and emotional connection increase one-unit, high human communication increase by 0.154 units. In this case, integration with social routines negatively affects high human communication, and social integration and emotional connection positively affect high human communication.

As a result of regression analysis for popularity and community-dependent variables, no significant effect was detected.

The frequency and asociality of the participants using social media were compared with ANOVA analysis, and a statistically significant difference was found ($p = 0.000$). Tamhane's T2 test, which is one of the multiple comparison tests, was used to compare that the average frequency of using social media is different from each other. According to the results of the analysis, the average of people who spend 5–6 h a day on social media and those who use social media less than 1 h a day and 1–2 h a day are statistically different from each other. Accordingly, the lowest average of nobility is those who use less than 1 h of social media per day, then 1–2 h of social media per day. The highest average of nobility belongs to the participants who use social media for 5–6 h a day.

It was determined as a result of ANOVA analysis that the education level and age of the participants did not make a significant difference in the sociality.

Independent sample t-test between gender, marital status, and employment status variables, and nobility and high human communication could not be applied since the data did not conform to the normal distribution. As a result of the Mann–Whitney U test, a significant difference could not be reached. So, the H0 hypothesis was not rejected.

As a result of all these analyzes, the use of social media has an effect on loneliness, albeit low. The social integration and emotional connection factor of the social media use scale positively affect the sociality variable of the loneliness scale. Also, the loneliness scale of both factors of the social media use scale has an impact on high human communication. Social integration and emotional connection are positive. Integration with social routines negatively affects high communication. There was no effect on the variables of popularity and community ownership. The hypothesis given in the study is evaluated as partially accepted.

Results and Discussion

Social media platforms provide positive developments such as the loss of importance of time and space unity, enabling communication with one end of the world, easier access to information, globalization and the expansion of the established social relationship network, but according to another segment, it reveals problems such as the feeling of loneliness of the individual [31].

This study was carried out by applying an internet questionnaire to 201 participants to determine the role of people's social media usage on their perception of loneliness. Survey application; It includes a total of 41 questions, including the Social Media Usage Scale, the UCLA Loneliness Scale, the Personal Information Form, and three questions about the participants' questioning in which environment they met with the people they met through social media.

In the literature, it is seen that the number of studies on this subject is quite low. It is possible to say that the reason for this is the evaluation of the positive reflections of social media and the idea that it may be negative occurred in new people. For this reason, it was aimed to investigate the claim that the use of social media isolates people and to create a source for the literature.

In the research, the data were transferred to IBM SPSS.20.0 (Statistical Package for Social Sciences) program, and analyzes were carried out. Correlation and regression analysis were used to evaluate the hypothesis in the study. In addition, the suitability of the data for normal distribution, reliability of the scales and factor analysis, independent sample t-test, and Mann–Whitney U test and ANOVA analysis are other applied analyzes.

As a result of the analysis, the data fit the normal distribution. The scales used are reliable, according to Cronbach's alpha values. As a result of factor analysis, a total of 33 scale questions were reduced to 7 factors. Social media use scale as the first factor, social integration and emotional connection, and the second factor as integration with social routines; The four factors of the UCLA loneliness scale were named as sociality, high human communication, popularity and community ownership. Finally, the three questions that were asked whether people met in social life with the people they met through social media were combined under the interview factors.

With the increasing use of tablets, computers, and smartphones and internet access opportunities, the number of members using social media is also increasing. 100% of the participants use at least one of the social media platforms. As stated by Solmaz et al. [32], social media has become an important part of our lives. The fact that all of the participants in the survey use social media is proof of this.

The majority of the participants prefer to use Instagram with 60.2%. High school and university students who participated in the survey also stated that they prefer Instagram the most. In the study of Ozsari and Karaduman [27], it was stated that 60% of the education faculty students prefer Facebook compared to other sites.

38.3% of the participants spend 1–2 h a day on social media platforms. In addition, 99% say that they visit social media platforms during the day, even if it is less than

1 h a day. Similar to the study, Solmaz et al. [32] stated that the participants spend an average of 1–3 h a day on social networks.

As a result of ANOVA analysis, no statistically significant difference was found between age and factors of the UCLA loneliness scale. Yenturk and Baslevent [37] and Yalcin [36] achieved similar results in their studies.

In the ANOVA analysis, the frequency of the use and sociality of the participants was compared with the Tamhane's T2 test and the average of those who spent 5–6 h a day on social media and those who used social media less than 1 h a day and 1–2 h a day were found statistically different from each other. The highest average of nobility belongs to the participants who use social media for 5–6 h a day. The study supports the meaningful relationship that Bozkurt (2018) finds between lonely social media users, especially 5 h a day,In other words, as the duration of the participants' use of social media increases, their loneliness increases. Ayazseven and Cenkseven Onder [4] conducted a study that supports this statement in the literature. Considering that the excessive time spent by the participants on social media during the day indicates internet addiction, they stated that there is a relationship between internet addiction and loneliness.

According to the study of Ayazseven and Cenkseven Onder [4], there is also no significant difference between the frequent use of social media and gender. According to the results of this study, no significant difference was found between the frequent use of social media and gender.

Bozkurt (2018) states that women use social media longer than men; however, he stated in his study that he could not achieve a significant relationship between gender and loneliness. This result of Bozkurt (2018) is similar to the result of this study.

As a result of the correlation and regression analyzes applied to test the main purpose of the study, a low and statistically significant relationship was found between the sense of loneliness of social media use. It is possible to say that the hypothesis is partially accepted. Dogan and Karakus [14] found a low and statistically insignificant relationship between the use of social sites and loneliness. They have reached a conclusion that supports the work of Guven and Kovanlikaya [17]. But it differs from the result of this study as statistical significance. Similar to this study, Yalcin [36] found that there was a significant relationship between loneliness levels with the presence of Facebook, Instagram, and Twitter accounts of students who participated in the study.

As a result, social media platforms appeal to the emotions that people actually want to be. Even though people have the feeling that they are lonely thanks to these networks, they will be faced with the fact that they are alone again when they turn off the phone and computer. This will make the feeling of loneliness even stronger.

References

1. Akinci Yuksel N (2017) Yeni Medyanin ve Sosyal Ağlarin Yaşamimiza Soktuğu Yeni Bir Yaşam Pratiği: Sofalising. Anadolu Üniversite İletişim Bilimleri Fakültesi Uluslararası Hakemli Dergisi 25(1)
2. Aktas C, Cayci B (2013) Yeni enformasyon ve iletişim teknolojilerinin sosyal hayattaki rolü. Uluslararası Medya Çalışmaları Sempozyumu
3. Akın A, Ozbay A, Baykut I (2015) Sosyal Medya Kullanım Ölçeği'nin Türkçe Formu'nun Geçerliği ve Güvenirliği. Uluslararası Sosyal Araştırmalar Dergisi 8(38):647–651
4. Ayazseven O, Cenkseven Onder F (2018) Ergenlerde İnternet Bagimliligi: Saldirganlik ve Duygusal Başa Çikmanin Rolü. Uluslararası Eğitim Bilimleri Dergisi
5. Balci S, Kocak MC (2017) Sosyal Medya Kullanımı ile Yaşam Doyumu Arasındaki İlişki: Üniversite Öğrencileri Üzerinde Bir Araştırma. 1. Uluslararası İletişimde Yeni Yönelimler Konferansı, Ankara: Nobel Akademik, 34–45
6. Belin MN, Yildiz H (2011) Lise öğrencilerinin Facebook adlı sosyal paylaşım sitesini kullanma amaçları üzerine sosyolojik bir inceleme: Eskişehir Örneği. Türkiye Bilimsel ve Teknolojik Araştırma Kurumu. http://mebk12.meb.gov.tr
7. Boyd J (2003) The rhetorical construction of trust online. Commun Theory 13:392–410. https://doi.org/10.1111/j.1468-2885.2003.tb00298.x
8. Buyukozturk S (2002) Faktör Analizi Temel Kavramlar ve Ölçek Geliştirmede Kullanımı (32):470–480
9. Cacioppo JT, Ernst JM, Burleson MH, McClintock MK, Malarkey WB, Hawkley LC, Spiegel D et al (2000) Lonely traits and concomitant physiological processes: The MacArthur social neuroscience studies. Int J Psychophysiol 35(2):143–154. https://doi.org/10.1016/S0167-8760(99)00049-5
10. Cayci B, Karagulle AE (2014) X kuşağından Z kuşağına değişen mahremiyet algısı. In: International Trends and Issues in Communication & Media Conference, pp. 190–196
11. Demir A (1989) UCLA Yalnızlık Ölçeği'nin geçerlik ve güvenirliği. Psikoloji dergisi 7(23):14–18
12. Demir E, Saatcioglu O, Imrol F (2016) Uluslararası Dergilerde Yayımlanan Eğitim Araştırmalarının Normallik Varsayımları Açısından İncelenmesi s. 134
13. Diker E, Tasdelen B (2017) Sosyal medya olmasaydı ne olurdu? Sosyal medya bağımlısı gençlerin görüşlerine ilişkin nitel bir araştırma. İleti-şim ve Edebiyat Araştırmaları Dergisi 17:189–206. https://doi.org/10.17361/UHIVE.2017.4.17
14. Dogan U, Karakus Y (2016) Lise Öğrencilerinin Soyal Ağ Siteleri Kullanımının Yordayıcısı Olarak Çok Boyutlu Yalnızlık. Sakarya Üniversitesi Eğitim Dergisi, pp 57–71
15. Esen A (2018, Mart) Sosyal Medyada Mahremiyetinifşasi: Namik Kemal Üniversitesi Akademisyenleri Üzerine Yapilan Bir Araştirma. Yüksek Lisans Tezi, Kayseri
16. Genc Y, Durgun A, Kara HZ, Cakir R (2018) İnternet Kullaniminin Üniversite Öğrencilerinin Yalnizlik Algilarina Etkileri. Akademik İncelemeler Dergisi pp 301–336
17. Guven SK, Kovanlikaya Ç (2008) Postmodern rear window: Facebook. In: 6th International Symposium Communication in The Millenium-Proceedings Book, vol 1, pp 419–431
18. Hazar M (2011) Sosyal Medya Bağımlılığı-Bir Alan Çalışması. Gazi Üniversitesi İletişim Fakültesi, 151
19. Ili K (2013) Sosyal Medya Ortam Ve Araçlarinin Öğrenci Davranişlarina Etkisi (Gaziçiftliği Anadolu Lisesi Örneği). Yüksek Lisans Tezi. Ankara
20. Independent Turkish (2019, August 29) Independent Turkish Web site: https://www.independentturkish.com/node/65616/haber/g%C3%BC%C3%A7l%C3%BC-ili%C5%9Fkiler-mutluluk-sosyal-medya-yaln%C4%B1zl%C4%B1k-getiriyor
21. Jenkins-Guarnieri MA, Wright SL, Johnson B (2013) Development and validation of a social media use integration scale. Psychol Popular Media Cult 2(1):38–50. https://doi.org/10.1037/a0030277
22. Kara T, Ozgen E (2015) Sahte Düşler: Benliğin Araçsallaşması ve Sosyalleşme Yanılgısı. Ağdaki Şüphe. Beta, İstanbul, pp 59–80

23. Kemp S (2019, January 31) Digital 2019: Turkey. DATAREPORTAL. Datareportal Web Site: https://datareportal.com/reports/digital-2019-turkey?rq=turkey%20digital%202019
24. Kemp S (2019, January 31) Digital 2019. "We are social". We are social Web Site: https://wearesocial.com/global-digital-report-2019
25. Ocal D (2018) Instagram Örneğinde Sosyal Medyadan Yansıyanlar. INIF İnönü Üniversitesi İletişim Fakültesi Elektronik Dergisi, vol 3, no 1, pp 23–36
26. Ozkan N (2018) Sosyal Medyada Sosyalleşmeme Üzerine Bir Değerlendirme. İzmir. http://community.ege.edu.tr/comm5/wp-content/uploads/2019/02/Oturum02.pdf
27. Ozsari I, Batdal Karaduman G (2016, December) Eğitim Fakültesi Öğrencilerinin Sosyal Medya Kullanimi ve Yalnizliklarinin İncelenmesi. Eğitim ve Öğretim Araştırmaları Dergisi 5
28. Ponzetti J (1990) Loneliness among college students. Fam Relat 39(3):336–340. https://doi.org/10.2307/584881
29. Sabuncu OB (2019, May) Sosyal Medya Pazarlamasinda Ünlü Ve Fenomen Kullanimi Tercihi: Hizli Moda Sektörü Örneği. Yüksek Lisans Tezi. İstanbul
30. Seyyar A, Genc Y (2010) Sosyal hizmet terimleri. Sakarya Yayıncılık (6), 143–200
31. Sezerer Albayrak E (2019, April) Sosyal Ağlar ve Modern İnsanın Yalnızlaşması: Coca-Cola Reklam Örneği. Elektronik Sosyal Bilimler Dergisi 18(70):929–942
32. Solmaz B, Tekin G, Herzem Z, Demir M (2013, April 4) İnternet ve Sosyal Medya Kullanımı Üzerine Bir Uygulama. Selçuk İletişim
33. Sucu I (2012) Sosyal Medya Oyunlarinda Gerçeklik Olgusunun Yön Değiştirmesi: Smeet Oyunu Örneği. Gümüşhane Üniversitesi İletişim Fakültesi Elektronik Dergisi 1(3). https://doi.org/10.19145/gumuscomm.98392
34. Turkel Z (2019) Ergenlerde Sosyal Medya Kullanimi, Yalnizlik ve Yaşam Doyumu Arasindaki Yordayici İlişkiler. Necmettin Erbakan Üniversitesi Ereğli Eğitim Fakültesi Dergisi 1(1):57–70
35. Tutgun Unal A (2015) Sosyal Medya Bağimliliği: Üniversite Öğrencileri Üzerine Bir Araştirma. Doktora Tezi. İstanbul
36. Yalcin G (2015) Sosyal Medyanin Yoğun Kullaniminin İleri Ergenlite Yalnizlik ve Depesyona Etkkisi. Yüksek Lisans Tezi. İstanbul
37. Yenturk N, Baslevent C (2007) Türkiye'de Genç İşsizliği Raporu. İstanbul Bilgi Üniversitesi Gençlik Çalışmaları Birimi (2), 10
38. Yildiz D, Uzunsakal E (2018) Alan Araştırmalarında Güvenilirlik Testlerinin Karşılaştırılması ve Tarımsal Veriler Üzerine Bir Uygulama. Uygulamalı Sosyal Bilimler Dergisi 1:19

Engineering and Technology Management

A Children Retail Company's e-Commerce Warehouse Optimization

Burcu Altuntas, Meryem Bulut, Miray Turhan, Bahadir Tunaboylu, and Selim Zaim

Abstract The efficient and well-designed warehouses can provide critical advantages for retail stores. They also play a crucial role in the achievement of supply chain management. Nowadays, warehouses have become centers where fast access to the customer is provided apart from the storage and protection of the goods. The objective of this study is to find the more efficient and useful warehouse design of A Children Retail Company, which is used for e-Commerce. The best place for the products will be determined by using a combination of such tools like the Pareto Principle, ABC Analysis, Technical Drawing, Simulation, and 5S Model. The optimum layout determined from the analyses enables more efficient work for warehouse employees, and orders to the logistics department are faster and more accurate for the facility. Moreover, an automation system is proposed with the barcode scanning and packaging part for the orders to simplify the existing complexity of the process.

Keywords Supply chain · Warehouse · Optimization · e-Commerce · Layout · Automation

---

B. Altuntas (✉) · M. Bulut · M. Turhan · B. Tunaboylu
Marmara University, Istanbul, Turkey
e-mail: burcu199779@gmail.com

M. Bulut
e-mail: meryembulut97@gmail.com

M. Turhan
e-mail: mirayturhan.mt@gmail.com

B. Tunaboylu
e-mail: bahadir.tunaboylu@marmara.edu.tr

S. Zaim
Istanbul Sabahattin Zaim University, Istanbul, Turkey
e-mail: selim.zaim@izu.edu.tr

F. Calisir (ed.), *Industrial Engineering in the Internet-of-Things World*,
Lecture Notes in Management and Industrial Engineering,
https://doi.org/10.1007/978-3-030-76724-2_23

Introduction

As the internet use increases dramatically, and virtually all individuals are capable of shopping with quick clicks online with many benefits. Companies need to satisfy this rising demand by proper reallocation of time and resources in this critical process. An optimum designed warehouse is a key element in a successful supply chain. In the past, the warehouses were used with the sole purpose of protection and storage of the goods, but now it has more nuanced purposes. Delivering the products to the customer has to be fast in order to achieve customers' satisfaction, save time and money. To deliver the products faster, a correctly-designed warehouse is necessary. In this study, a company is selected, which is a retailer company targeting a customer segment of 0–14-year-old children. When the company was established, it only offered products from a physical store. Then it started to sell the products online to catch the trend and has created an e-commerce warehouse. In the meantime, online sales were increased very fast before the company could carry out any optimization in the warehouse. Therefore, the aim of this study is to perform the Company's warehouse optimization.

There are many problems and complaints coming from the customers and employees, which mainly stems from the warehouse. While the main complaints from employees are related to having to carry heavy loads regularly, unknown locations of inventory, and inefficient ergonomic conditions, customers' complaints are about missing or late product delivery. Figure 1 shows the ratio of the collection of orders.

As seen from Fig. 1, The Company can collect just 33% of the orders in a day, which means that The Company is unable to satisfy 67% of the orders. There are many more problems, which The Company faces. After receiving the information about the problems from The Company, the warehouse was further to verify these issues. The critical problems are determined and analyzed using Fishbone Diagram (Ishikawa Diagram, Fig. 2), which is provided below.

Fig. 1 The ratio of collected and cannot be collected orders

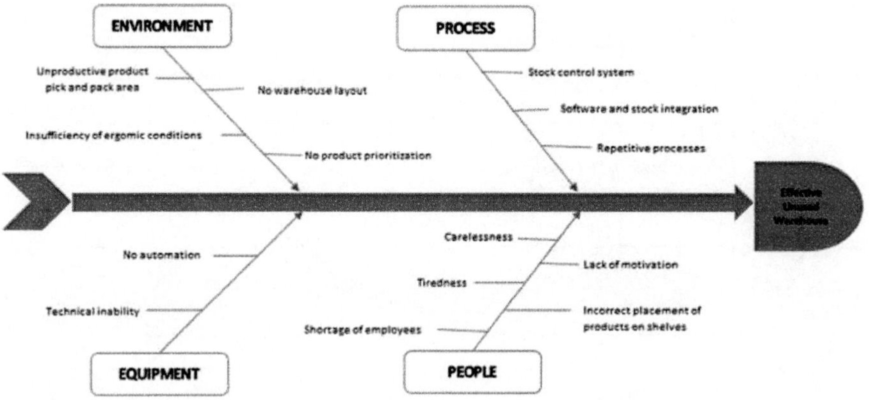

Fig. 2 Fishbone diagram

From the diagram, it is seen very clearly that The Company uses its warehouse very inefficiently. Therefore, the study's goal is to make an efficient and optimum layout for e-Commerce Warehouse by considering the ergonomic necessities. At the end of this study, a new layout for warehouse planning was offered to The Company.

Methodology

After the literature review conducted by examining the past studies, the methods that would be convenient for the study were determined, and all were implemented. Figure 3 briefly shows the steps and methods. There are many more methods that can be used for warehouse optimization, but these methods are not appropriate to be implemented for this particular warehouse. For example, Linear Programming Formulation is widely used for the warehouse design. Hung and Fisk [4] reveal that this formulation gives the optimal size warehouse when the demand fluctuates and space in the warehouse available changes frequently. Although this method provides the optimum warehouse design, it is not achievable to implement in The Company's e-Commerce warehouse because the space and warehouse area are not sufficient to change over time and employees in the warehouse are not qualified professionals so that they may not be able to manage this method every month or season. Another calculation that is important for the warehouse design is Inventory Turnover Rate. However, since The Company's system is not well-functioning, this calculation cannot be implemented for the warehouse.

Fishbone Diagram method was used to determine the problems of The Company's e-Commerce warehouse, as stated in the introduction. Then, all the necessary information was gathered, like the sales amount, etc. When The Company's warehouse was visited, ergonomic conditions of the work environment are observed. According to literature, there should be a controlled range of temperature and humidity levels

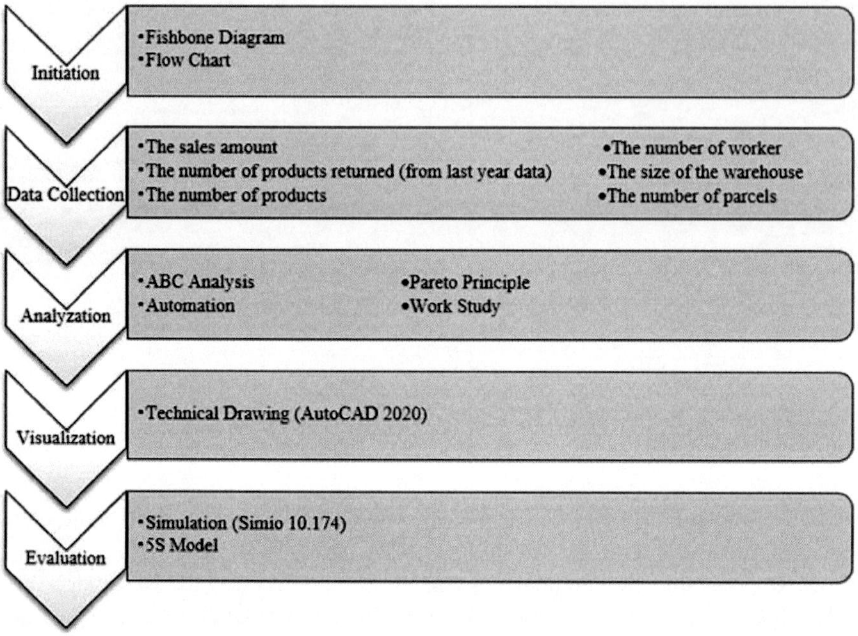

Fig. 3 Methodology design

to have good ergonomic conditions for the employees. OSHA (Occupational Safety and Health Act) suggests a temperature level in the range of 20–24 °C and humidity control between 20 and 60% [1, p. 1]. However, in The Company's warehouse, there was no temperature and humidity control, which means the temperature and humidity levels are unknown. Therefore, in this study, the suggestions were made on adding these devices to the new warehouse design. On the other hand, there were no problems with the lighting system and security-warning against a fire disaster. The warehouse has adequate firefighting equipment and fire exit. In the study, precise measurements of the halls of the existing warehouse were performed because there was very limited data, just the data of the square meter of the warehouse, no visual drawing, etc. Therefore, the warehouse drawing in AutoCAD 2020 was generated, which is seen in Fig. 4.

Another issue from observations in the warehouse is that the orders are given for packaging after being taken from the cabinet and the chaos in the packaging process.

As seen from Fig. 5, although 84% of the orders' cash transactions are made, 16% of the orders are waiting in the packaging area for cash transactions even though these orders are collected. For analyzing that process, a flow chart was drawn to understand the packaging process in more detail. The processes here were as follows; the collected products were left behind the packaging personnel. The packaging staff did not have their own space, so there were boxes that looked quite complicated around. Therefore, the boxes of packaging staff could be mixed up. In addition,

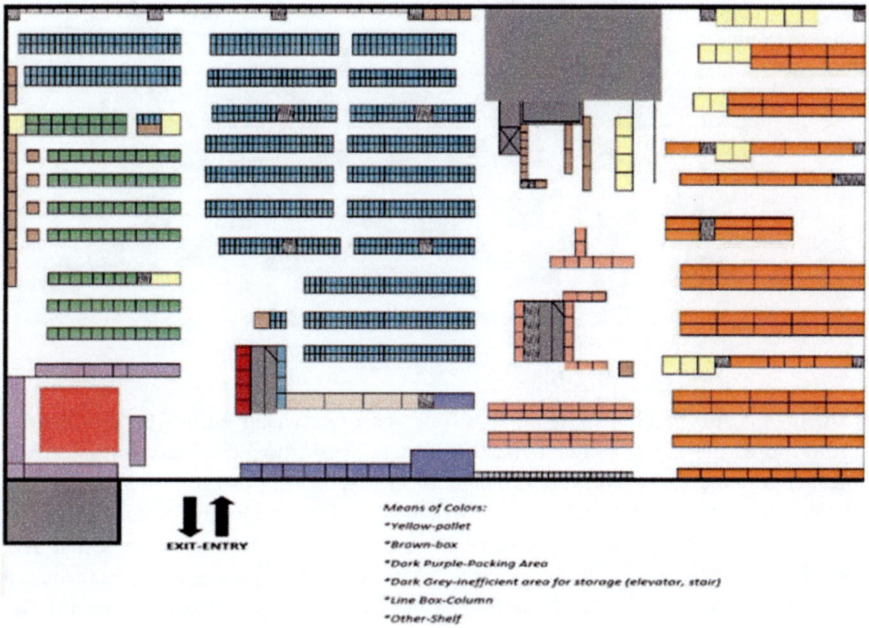

Fig. 4 The company's e-Commerce warehouse AutoCAD drawing

Fig. 5 The ratio of orders collected and cash transaction

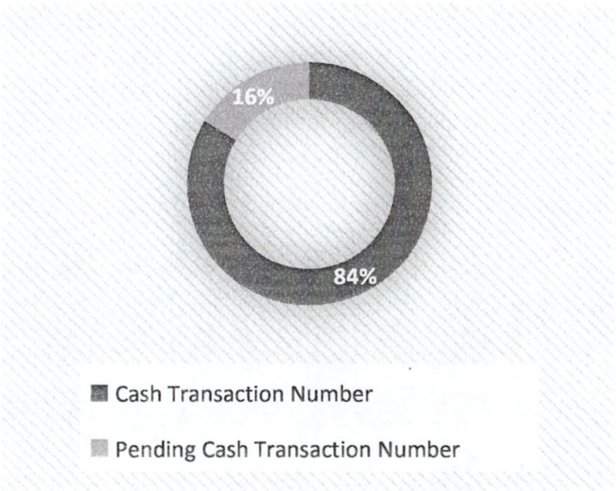

although some orders consisted of bulk parts (like 2–3 parcels), some parts could be lost due to confusion in the packaging area. In addition, the employees leaving the orders they collected in the packaging area closest to the collection area caused an unequal workload among the packaging personnel, as seen in Fig. 6. All these

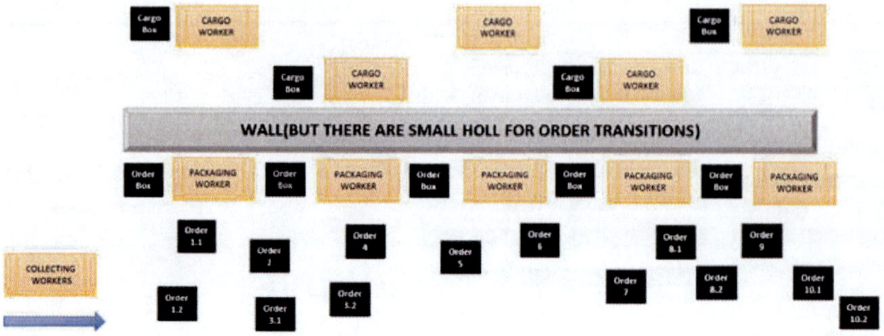

Fig. 6 Current packaging and cargo area

resulted in a mixed environment, lost time, and decreased employee motivation. Furthermore, the study noticed that the orders were moved unnecessarily a few times. Heavy manual labor, repetitive handling, awkward postures, and previous or existing injuries or conditions are all risk factors for developing Musculoskeletal disorders [3, p. 2]. For all these problems, a Conveyor System was suggested as a solution, as shown in Fig. 7. The suggested system in the packaging area provided some gains that are listed as; reduction of the number of processes, decreasing the walking distance, the collection time, and the number of sending wrong products to the customer. Also, the ergonomic conditions of the employees are improved.

The last issue of the study is the storage of food products. Food is an important product because it can cause health problems to the people because of outdated, wrong storage conditions such as temperature, etc. The study realized that the food was not stored in suitable places, for example, a place susceptible to dust under the stairs. In addition, products are not stored according to the expiry date. Some complaints were discovered, such were about the delivery of outdated food products. For controlling an outdated problem, many companies use First in First Flow. Manohar and Appaiah

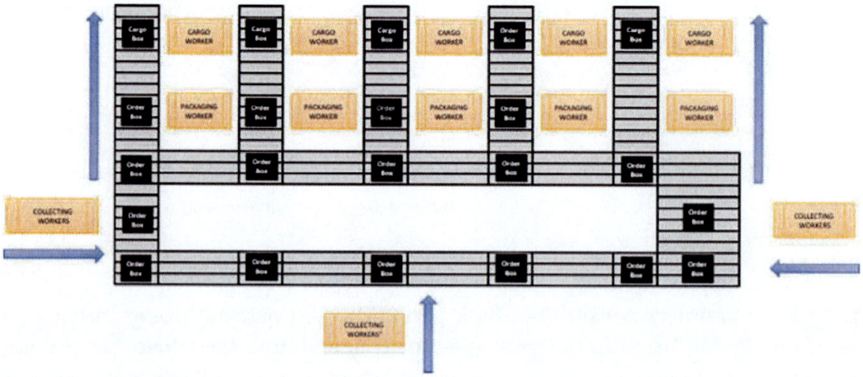

Fig. 7 Suggested packaging and cargo area

[5] summaries FIFO as "The component which comes first has to dispatch or issue first." (p. 5632). Therefore, the study suggested new cabinets, which are using FIFO, with these new cabinets system, first in products will stand at the front of the shelf so the expiry date' early product will go to the customer.

For placing the products in an efficient way, ABC Analysis and Pareto Principle are some of the widespread methods so that the study used these analyses. According to the sales amount of category-based products, the methods are shown in Table 1.

As seen from the above table, the most sold three products in a year are clothing products, followed by the diaper, toy, etc. According to these rates, available products are decided and then will be placed in the possible nearest place to the product

Table 1 ABC analysis and Pareto principle

Category	Sales amount	Rate (%)	Cumulative rate (%)	ABC analysis	Pareto principle
Top knitting	Hidden	29	29	A	
Set knitting	Hidden	13	42	A	
Bottom knitting	Hidden	12	53	A	
Diaper	Hidden	7	61	A	
Toy	Hidden	7	68	A	%20
Cosmetic	Hidden	7	75	A	(A)
Foot textile	Hidden	4	79	A	
Top weaving	Hidden	3	82	B	
Baby's bootle and pacifier	Hidden	2	84	B	
Breast-feeding	Hidden	2	86	B	
Underwear	Hidden	2	88	B	
Bottom weaving	Hidden	2	90	B	
Set weaving	Hidden	2	92	B	
Baby food and nutrition	Hidden	2	94	B	%80
Bathroom and toilet	Hidden	2	95	C	(B and C)
Tool accessory	Hidden	1	96	C	
Shoe	Hidden	1	97	C	
Baby carrier	Hidden	1	98	C	
Hat-scarf-glove	Hidden	1	99	C	
Sleep group and home textile	Hidden	1	99	C	
Accessories and room decoration	Hidden	0.3	100	C	
Maternity clothes	Hidden	0.1	100	C	
Books-stationary-school supplies	Hidden	0.1	100	C	
Total		100			

packaging area. To make an efficient warehouse layout, one of the predominant aims is decreasing distinction in the warehouse's areas.

Heragu concludes that experts should make an evaluation and determine the products, which have higher mobility. Then, since those products remain for a short period, they should be positioned near to inbound and outbound processes. Heragu called this area as "Forward Area" and "Reserved Area" for the storage of less frequent products [2, p. 317]. The Pareto Principle was first looked at in order to select the forward area. This principle shows that 20% of the products covered 80% of the products sold in the last year. In addition, the ABC analysis, it is seen that the products belonging to the 20% ratio are those belonging to the group A. When the group A product in the ABC analysis is examined in more detail, it is seen that the top three categories are textile products. However, in ABC analysis, the selected diapers for the forward area because the diapers have a larger volume than textile products. Textile products that are the top three categories were placed in the nearest location to the packaging area according to the design of the warehouse. Thus, the employees will have used less distance when collecting the products, and the transportation of the diapers with a large volume will be easier for the employees. It will be close to the packaging area. After making all observations and calculations, a new design is suggested shown in Fig. 8.

By looking at ABC analysis, the most selling products are placed near to each other and packaging area to make the order-picking process more productive, the categories' priorities are seen in Fig. 8 with A, B and C letters. As observed, in the current warehouse, there are not enough spaces between the cabinets, which make hard the transition between the products. Therefore, the length of the cabinets has

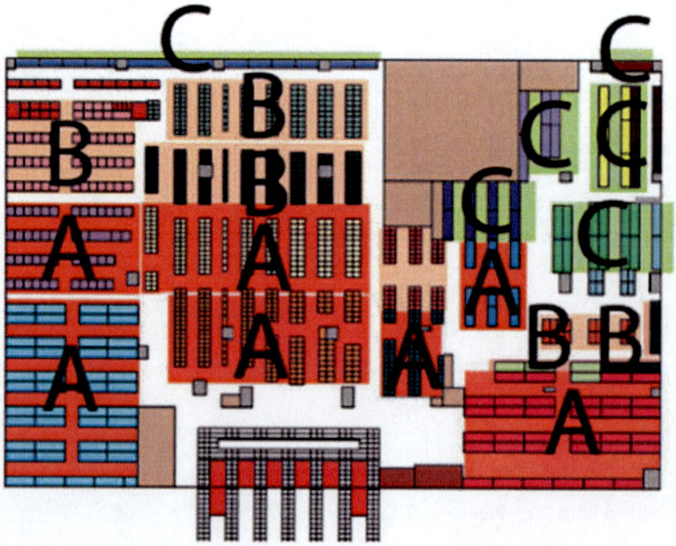

Fig. 8 Category placement according to ABC analysis in suggested warehouse layout

been shortened such that the workers who collect the orders are less likely to come across. Other changes, suggestions, and the pros and cons of the new layout will be examined more in upcoming titles, Current and Suggested Warehouse Layout Comparison Part and Simulation Part.

Results

The advantages and limitations of the proposed redesigned-warehouse layout are analyzed as compared to existing warehouse design in different aspects.

The first comparison between the proposed and the existing layout is category-based products square meter calculation, which is provided in Table 2.

According to the above comparison, the new m^2 is larger than the old one. By placing the products in a productive way, the capacity increase is achieved, which is almost 14% more than the current layout.

Another comparison is the distance traveled from one end of the warehouse to the other for the left side of the warehouse. In the current warehouse layout, the cabinets are placed in a row with no space, while in the suggested layout, there is a space between the cabinets, which facilitates the transition between the products. As seen from below Fig. 9, in the current warehouse layout, the distance traveled from one end of the warehouse to the other for the left side of the warehouse, which is the

Table 2 Category-based products square meter comparison

Categories	Current m^2	Suggested m^2	Number of cabinets
Diaper	82.9	87.99	45
Textile	233.89	239.62	1015
Maternity clothes	13.68	16.68	15
Toy	64.19	67.32	34
Baby's bootle	5.2	7.92	4
Breast feeding	5.12	7.56	4
Baby's food	5.1	9.12	16
Cosmetics	9.95	12.06	9
Bathroom and toilet	23.14	23.76	12
Tool accessory	16.42	24.06	11
Sleep group and home textile	7.32	8.52	10
Baby carrier	10.65	11.88	6
Book	3.12	3.43	1
Shoe	10.59	13.92	8
Total	257.38	294.22	1190

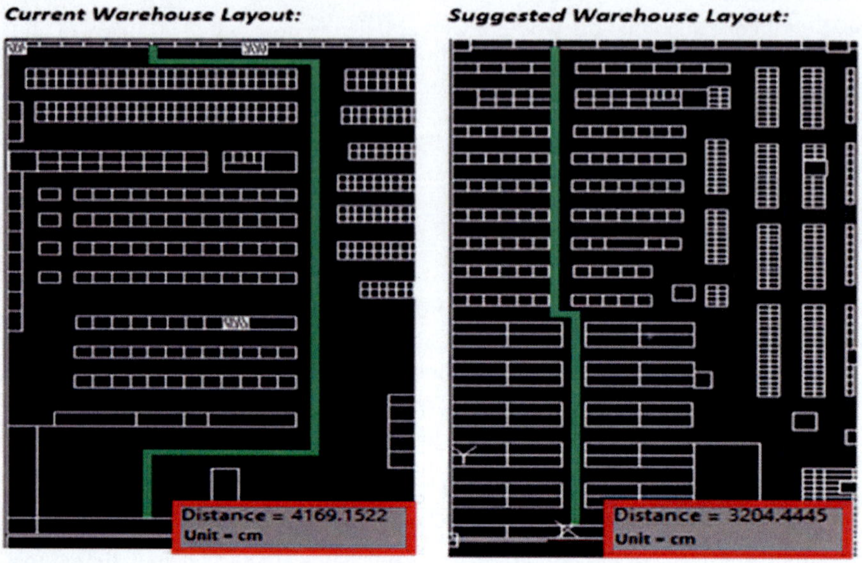

Fig. 9 Distance traveled from one end of the warehouse to the other (for the left side of the warehouse)

green line in Fig. 10, is 4.169 cm and in suggested warehouse layout, it is 3.204 cm, which is very less than the current one. With the addition of the spaces, the distance, and in the meantime, the collection time of the orders will be able to be reduced.

On the right side of the warehouse, while in the current warehouse layout, the distance traveled from one end of the warehouse to the other is 4.374 cm, in suggested warehouse layout, it is 3.091 cm, which is again very less than the current one. In the current layout's right side of the warehouse, the same 'no space' problem is available like the left side of the warehouse. In addition to that, the cabinets are all placed horizontally. In the suggested layout, with the addition of the spaces and placing some of the cabinets horizontally and some vertically, the distance is again decreased.

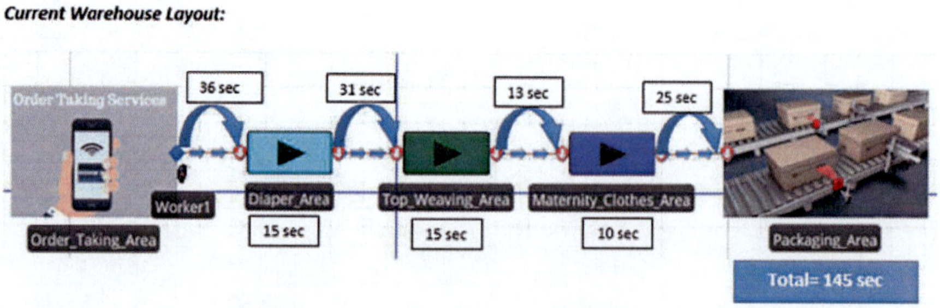

Fig. 10 The process of the order-picking model for current warehouse layout

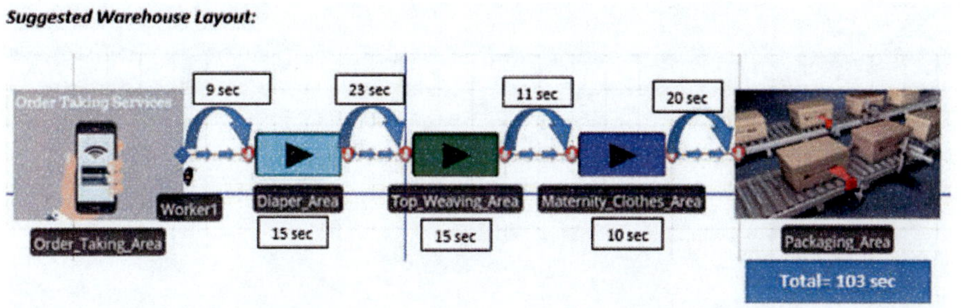

Fig. 11 The process of the order-picking model for suggested warehouse layout

In Simulation Part, Simio 10.174 program is used for visualization and to illustrate the total time of collecting an order. The order samples were taken from The Company. For example, a worker takes an order, which includes diapers, top weaving, and maternity clothes shown in Fig. 10. The worker starts walking from the order-taking area to the locations of the products one by one, collects the products, and finally goes to the packaging area.

The distances between the order-taking area, products, and packaging area were entered by looking at the distances from the AutoCAD program for both currents shown in Fig. 10 and suggested warehouse layout shown in Fig. 11. Then, which one has less product collection time will be determined. The worker's speed is assumed to be 1.4 m/s, which is the average person walking speed. In this model, the service time of product categories 1, 2, and 3 is the time of finding the product and putting in the product collection vehicle. Since there is a terrible event, Coronavirus condition, the study could not visit the warehouse again to calculate the exact service times, but as observed, finding textile products, diapers and toys is harder to find so that the study made assumptions accordingly. As it is revealed in Comparison Part, finding products is easier with the suggested layout, which means the service times are less in the suggested layout. However, since the study's aim of the simulation is to find if the product categories are placed in the right places or not, the same service times were given for both current and suggested layout.

As seen from the results, the total order collection time is 145 s in the current layout and 103 s for the suggested layout. Since the products are replaced according to their categories, the order collection time is decreased. Another point that plays a role in reducing that time is about the packaging area, where the orders are collected. In the current layout, the orders are collected in the middle side of the packaging area like a red arrow in Fig. 12. With a new conveyor system, orders can become together in the nearest place possible to the product area like green arrows Fig. 12. It can save 777 cm within walking distance.

By looking at ABC analysis, the most selling products are placed near to each other and packaging area in order to fasten order-picking time. Undoubtedly, it is not possible to obtain better results from all of the orders made up of a combination of different products. In the current warehouse, there is no such systematic placement. Therefore, in most cases, the suggested layout is more advantageous than the current

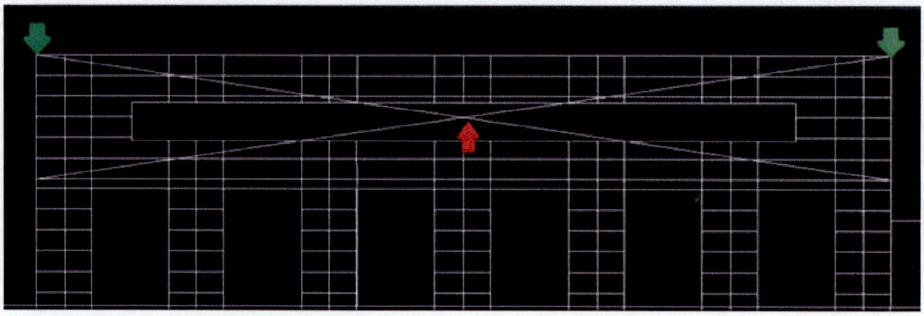

Fig. 12 Collection points of orders

one, especially the orders that contain A and B group products and the orders that include just one group product.

Discussion and Conclusion

On average, e-Commerce warehouse employees are able to collect 2113 orders per day. Below orders (1, 2… 8) shown in Table 3 are the examples of these orders, which include different combinations of the products. By looking at ABC analysis, the possibility of the orders and according to these possibilities, order quantity per day is found as follows:

Collecting 2113 orders for these eight orders takes 44.8 h in the current warehouse layout and 40 h in suggested warehouse layout, which means that everyday new warehouse design can save 4.8 h of work to The Company if one worker tries to collect all orders. The e-Commerce warehouse's employees work five days a week.

Table 3 Total saving with suggested layout

Orders	Possibility (%)	Order quantity per day	Current layout's time (h)	Suggested layout's time (h)	Saving (h)
Order 1	11	227	6.6	4.0	2.6
Order 2	17	359	5.4	9.9	−4.5
Order 3	22	471	11.5	8.5	3.0
Order 4	12	247	4.2	2.7	1.6
Order 5	11	224	5.9	5.5	0.4
Order 6	9	182	3.2	3.7	−0.6
Order 7	16	336	6.5	4.4	2.1
Order 8	3	67	1.4	1.3	0.1
Total	100	2113	44.8	40.0	4.8

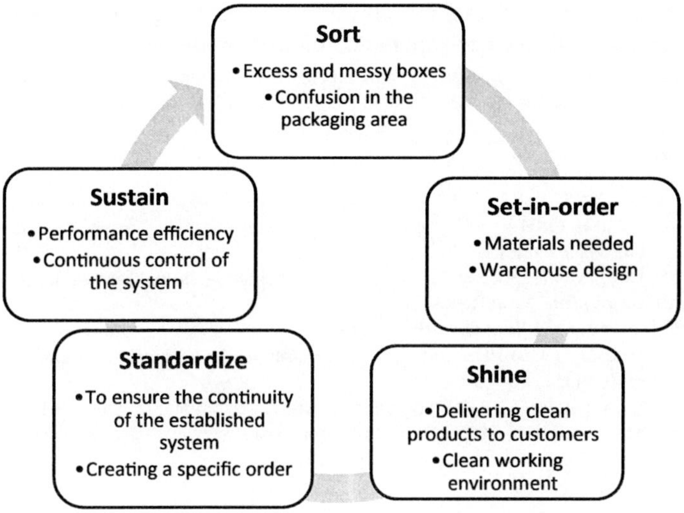

Fig. 13. 5S Model

Then, the saving is 115 h for a month and 58 days for a year, which will be worth to change the layout.

During the optimization process, 5S Model regulations (Sort, Set-in-Order, and Shine) are implemented shown in Fig. 13. After all improvements, it is necessary to ensure the continuity of these processes. Therefore, one employee from each department will be selected to follow the new system. Within this process, the system controls by the manager at certain periods and monitoring the performance of the employees will significantly contribute to this process.

The objective of this study is to offer a new design for The Company's e-Commerce warehouse, which is more profitable than the current design by decreasing time for an order to giving Logistic Department and providing an environment for employees to operate more ergonomically and happily. In the beginning, the warehouse was visited to detect the problems by making a Fishbone Diagram and collect necessary data from The Company to demonstrate the problems numerically. After that, in this study, measurements of the hall warehouse from the length of the cabinets to the length of the spaces were made and illustrated in the warehouse drawing in AutoCAD. Then, the most critical process, with Flow Chart Diagram, was determined, which is happening in Order Collection and Packaging Area, and the study has suggested a conveyor system to that process to solve the conflict.

ABC Analysis and Pareto Principle were implemented to offer a more profitable layout. By that, the most selling products are aimed to be placed near to each other to fasten the order-picking process. Forward Area's product was determined by looking at ABC Analysis. After all these analyses, the new warehouse layout's drawing was made with the AutoCAD program, and with the Simio program, the saving was revealed. The final step was implementing the 5S Model to standardize all these improvements. To sum up, in this study, The Company's e-Commerce warehouse problems were solved, and new and more efficient design was proposed.

References

1. Fairfax RE (2003) OSHA policy on indoor air quality: office temperature/humidity and environmental tobacco smoke
2. Heragu SS, Du L, Mantel RJ, Schuur PC (2005) Mathematical model for warehouse design and product allocation. Int J Prod Rep 317–321
3. HSE (2020) Manual handling at work a brief guide.
4. Hung MS, Fisk CJ (1984) Economic sizing of warehouses: a linear programming approach. Comput Oper Res 11(1):13–18
5. Manohar HM, Appaiah S (2017) Stabilization of FIFO system and Inventory Management. Int Res J Eng Technol (IRJET) 04(06) Retrieved from https://www.irjet.net/archives/V4/i6/IRJET-V4I6512.pdf

Churn Prediction in the Payment Services Industry: An Application at Token Financial Technologies for IoT Devices

Dicle Aslan and Umut Asan

Abstract This paper presents a study on churn prediction in the payment services industry. Churn prediction is crucial in preventing customers from switching to competition and also in keeping and developing relationships with the customers. Previous studies mainly focus on the problem of churn prediction for customers. Using machine learning algorithms, this study, for the first time, predicts churn rates for IoT devices and, thereby, generalizes the concept to the usage of devices. A dataset on POS devices provided by Token Financial Technologies—a company that aims to develop a churn prediction system—was used for the analyses. The methods Naïve Bayes, Generalized Linear Model, Logistic Regression, Deep Learning, Decision Tree, Random Forest, and Gradient Boosted Trees were applied. The experimental results show that the best predictions are obtained by the algorithms Random Forest and Fast Large Margin. According to experimental results, battery life has a significant effect on the device churn as well as lifetime value of device has a major impact. The predictions helped Token Financial Technologies to save more than 60% of the usage of EFT-POS devices from potential churns by changing batteries and EFT-POS devices in the last quarter of 2019.

Keywords Churn prediction · Fintech · Machine learning · Payment services · IoT devices

Introduction

In recent years, dynamic market conditions, change in customer expectations, entry of new competitors with better offerings made it harder for companies to survive. To

D. Aslan (✉)
Token Financial Technologies, Istanbul, Turkey
e-mail: dicle.aslan@tokeninc.com

U. Asan
Industrial Engineering Department, Management Faculty, Istanbul Technical University, Istanbul, Turkey
e-mail: asanu@itu.edu.tr

F. Calisir (ed.), *Industrial Engineering in the Internet-of-Things World*,
Lecture Notes in Management and Industrial Engineering,
https://doi.org/10.1007/978-3-030-76724-2_24

retain and increase their customers, companies need to define new business strategies, objectives, and tactics. Gartner Group statistics show that almost 80% of the future revenue of companies will come from 20% of the existing customers like the Pareto principle [14]. Bain and co. also reports that an improvement of 5% in customer retention, might increase the profitability by 75% [14].

Customer Relationship Management (CRM) is a strategic approach to manage, maintain, and improve customer relationships and create loyal customers for the products or services. CRM as a tool provides more information about customers' needs and preferences to create a strong relation [1].

It is essential to note that customer acquisition is much harder and costly than customer retention. More effort, money, and time are required for customer acquisition. Attracting and bringing new customers to your business from the competitors is not an easy task for the companies. Therefore, customer retention is crucial to surviving in a competitive market. Studies in the literature show that customer acquisition is almost 20 times more expensive than customer retention [17].

A considerable number of studies have been conducted both in theory and practice about customer churn analysis, in terms of retention and prediction. Customer retention shows the company's skill to sustain its customers. After acquiring new customers, companies should implement customer relationship management systems to satisfy and retain these customers. Customer retention also involves actions and strategies to grow the existing customer base. Retention activities, such as improved customer services and relationships, well-defined, and successful onboarding processes, also facilitates the creation of loyal customers.

Customer churn prediction is a major instrument in customer retention. Churn refers to the loss of customers who switch to competition or cancel their subscription [11]. Customer churn prediction helps to detect customers who are likely to leave the system. Customer churn is a major problem for companies under economic constraints. For this reason, churn analysis and prediction, widely studied in the literature, have become a critical issue for industries in recent years.

Customer churn (attrition) might be voluntary or involuntary [12]. In voluntary churn, customers prefer terminating the relationship with the brand and company. The voluntary churn might be the results of customer dissatisfaction, new competitive offers, regulations, and so on. Bad customer experience, as a critical issue, may influence only a one-time purchase decision or a long-term relationship with the customer. On the other hand, involuntary churn, also known as passive churn, happens without any action taken by the customers to cancel their subscription. Customers experiencing involuntary churn are likely to continue preferring the same brand or company.

Customer churn analysis requires companies to analyze historical data to be able to predict the probability of customer churn. It enables companies to cluster their existing customers according to their characteristics (demographic, psychographic and geographic), preferences, and risk levels (high risky customers, risky customers, moderate customers, low risky customers, very low risky customers for churn) and, thereby, to predict whether they will leave or not the company. The analysis helps ranking customers from the most to the least likely to churn before customers switch to competition or cancel their subscription. Especially attractive customers who are likely to churn may receive a direct incentive or customized service plan to stay. Therefore, it is crucial to implement CRM programs that allow us to predict potential churns and develop customer retention strategies to improve the total lifetime.

In theory and practice, customer churn analysis and prediction are often performed in subscription markets such as telecommunications, insurance, banking, and retail market, since subscribers might switch to another provider easily. Moreover, limited research has been conducted on churn analysis in the business to business context.

Token Financial Technology is a subsidiary of Arçelik, which is a leading brand of Turkey's durable consumer goods market. Having more than three decades of experience, Token provides value-added retail and payment services/solutions for the Fintech ecosystem. Token, which is the leader in the payment industry in Turkey, operates more than 550,000 active payment terminals and 600,000 applications, including banking, meal card, and other retail applications. Therefore, churn analysis for IoT devices is a crucial task for Token.

To the best of our knowledge, in the literature, there is no study reporting a churn analysis in the payment services industry. Furthermore, churn analyses have mainly focused on consumers. This study predicts for the first-time churn rates of POS devices and generalizes the concept to the usage of IoT devices by merchants in a business to business context. Since merchants might have more than one EFT-POS (Electronic Funds Transfer at Point of Sale) device, the analyses have been performed for each device. The machine learning methods Naïve Bayes, Generalized Linear Model, Logistic Regression, Deep Learning, Decision Tree, Random Forest, and Gradient Boosted Trees have been applied in an experimental framework.

The remainder of the paper is structured as follows: In section two, past studies on churn prediction are summarized. In the next section, the methodology is presented. In section three, device data history and the application of the methods for churn prediction are examined. In the next section, a discussion of the experimental results and the performance of each method is presented. Concluding remarks and further research directions are provided in the last section of the paper.

Literature Review

A considerable amount of research has been devoted to customer relationship management in general and churn prediction and retention in particular. Previous studies have examined customer churn in two different ways. While one group of studies tends to improve the performances of churn prediction models, the other group focuses on the reasons for customer churn and the ways to satisfy customers with improving retention strategies [5].

Especially, telecom and banking industries have been widely analyzed to predict customer churn. In the literature and business, machine learning techniques have been mostly applied in order to predict churners. In these studies, prediction models have been commonly evaluated using the area under the curve (AUC) and top decile

Table 1 Review of studies in the literature

Author	Year	Technique(s)	Dataset
Au et al	2003	DMEL	100,000 records and 251 variables
Hung et al	2006	Decision tree, neural network	160,000 subscribers with 14,000 Churns
Dasgupta et al	2008	Social network analysis (SNA), J48 decision tree	60 GB data, which contains voice calls, SMS, etc
Lu et al	2014	Boosting algorithms	7000 customers with 70 variables
Farquad et al	2014	Support vector machine (SVM)	Business Intelligence Cup data (2014)
Vafeiadis et al	2015	ANN, SVM, DT, NB, LR, Boosting algorithm	5000 records
Coussement et al	2017	Data preparation treatment logistic regression model	30,104 customers with 1361 churns
Amin et al	2017	Just-in-time customer churn prediction (JIT-CCP)	5784 samples and 250 variables
Caigny et al	2018	Logit Leaf model	14 different data sets
Zhang et al	2018	PVTREE (Profit model based on decision-tree)	30,748 customers

lift (TDL). Especially large datasets are preferred to yield more accurate predictions. Previous studies reviewed in this article are summarized in Table 1.

Most of the studies in the literature model customer churn as a classification problem. For example, [2] apply the evolutionary learning (DMEL) technique to predict the likelihood of subscribers' churn accurately under different churn rates. Hung et al. [13] try to find the best data mining technique and show the effectiveness of the techniques in making accurate churn predictions in the wireless telecommunications market. In another study in the telecommunications industry, Coussement et al. [6] show that data preparation techniques significantly improve the prediction performance of logistic regression by 14.5% for AUC and 34% for TDL. These findings support the arguments of [21] that 50–80% of data mining effort should be devoted to data preparation, which helps to increase the data quality and thereby the model performance directly. Data cleaning, removing bias from the data set, imputation of missing values, and transformation operations are carried out in this phase [10]. Feature selection and finding relevant features as part of data preparation are also critical issues in churn prediction. The common features that have delivered accurate churn prediction in the mentioned studies are customer demographics, billing information, contract/service status, call detail records, and service changelog (see also Azeem et al. [3]).

In recent years, hybrid and ensemble models, which combine methods in order to achieve higher accuracy in churn prediction, are studied. In a recent one, [17] compare the state of the art machine learning techniques for customer churn prediction in the telecommunications industry. According to the results of the study, boosting improves the techniques' performances and boosted Support Vector Machine (SVM) obtained the highest accuracy (approximately 97%) and F-measure (over 84%). Lu et al. [15] also use boosting algorithms to enhance prediction and show that the suggested algorithms enable better separation of churn data. Caigny et al. [5] develop a new hybrid algorithm named as Logit Leaf Model (LLM) that is based on decision trees and logistic regression. They showed that the performance is significantly better than the plain versions of the techniques. Farquad et al. [9] also proposed a hybrid approach to improve the performance of SVM in churn prediction. In this study, first, the feature dataset is reduced and, then, used in the SVM model, and then rules are extracted using Naïve Bayes Tree. Amin et al. [1] use a cross-company dataset that enables historical data by developing an ensemble-based JIT-CCP model to increase the accuracy level of churn prediction.

Dasgupta et al. [7] show that the effect of the social relation to be a potential churner and Social Network Analysis (SNA) method's results have powerful to predict the churn. Zhang et al. [20] propose a novel classification method, PVTREE based on profit-maximizing decision tree model and shows that increase profit whereas common performance indicators being the same.

The current study presents a first attempt to analyze churn data for POS devices in the payment services industry. Therefore, an important and challenging issue is to determine the most important features influencing the churn rates of POS devices. To address this issue, both widely used and advanced machine learning techniques have been applied. The choice of the methods and measures used to compare the performance of these methods was based on the review of the literature.

Methodology

The methodology used in this study is depicted in Fig. 1. As seen from the diagram, the initial step of the methodology consists of data collection and data preprocessing. This step is often the most time consuming and challenging one since an accurate prediction of churn behavior requires high-quality data and relevant features. In the next step, widely used and advanced machine learning algorithms have been applied using RapidMiner. Naive Bayes, Generalized Linear Model, Logistic Regression, Fast Large Margin, Deep Learning, Decision Tree, Random Forest, Gradient Boosted Trees algorithms are suggested to predict the churn rates of devices. Finally, common measures are used to evaluate and compare the performances of the suggested methods.

Data Collection, Preprocessing and Modeling

A dataset on POS devices provided by Token Financial Technologies was used for the analyses. The company offers three different models, and each device has a unique

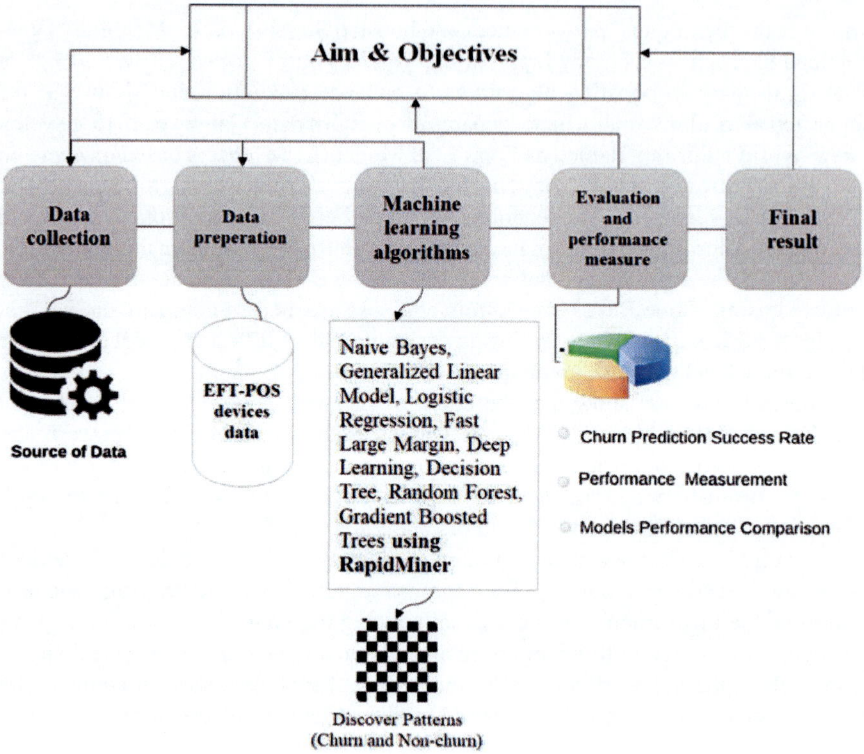

Fig. 1 Flowchart of the methodology

identification number. The dataset contains historical records of 8557 devices with a churn incidence of almost 50%. In the data preprocessing stage, incomplete and irrelevant parts of the data were identified. Expectation–Maximization was preferred to impute missing values, and boxplots were used to detect outliers.

The dataset includes the following five independent variables (features): Device Lifetime Value (LTV), Number of Active Applications in Device, Number of Passive Applications in Device, Number of Total Applications, Battery Life. The dependent variable in this study is Churn Decision (see Table 2). While all five independent variables are metric scaled, the target variable Churn Decision is nonmetric scaled and has two categories: active and churn.

To avoid problems of overfitting, the original dataset was split into a training set (60% or 5134 customers) and test set (40% or 3423 customers). The distributions of the features are given in Figs. 2, 3, 4, 5, 6 and 7.

Figure 2 shows the number of active and churn devices in the dataset as 4298 and 4259, respectively. Devices described as churn are no longer used, and banking applications are deleted from these devices.

Table 2 Features in the dataset

No	Feature	Scale
1	Device lifetime value	Metric
2	Active app number	Metric
3	Passive app number	Metric
4	Total app number	Metric
5	Battery life	Metric
6	Decision (active/churn)	Nonmetric (Binary)

Fig. 2 Active/churn devices in the dataset

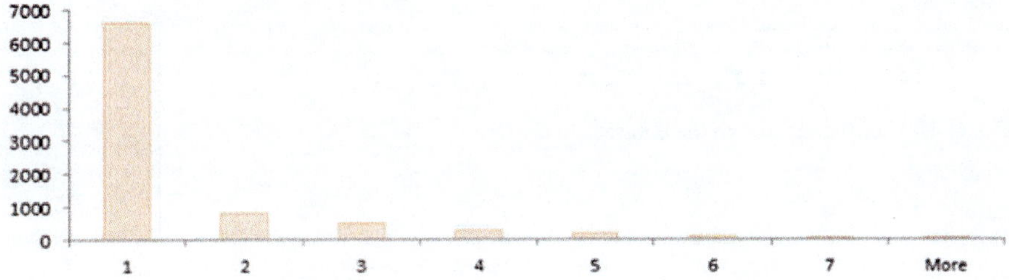

Fig. 3 Number of active applications in the dataset

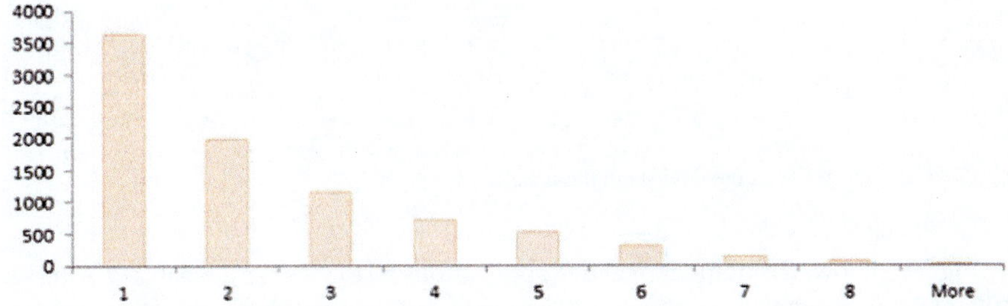

Fig. 4 Number of passive applications in the dataset

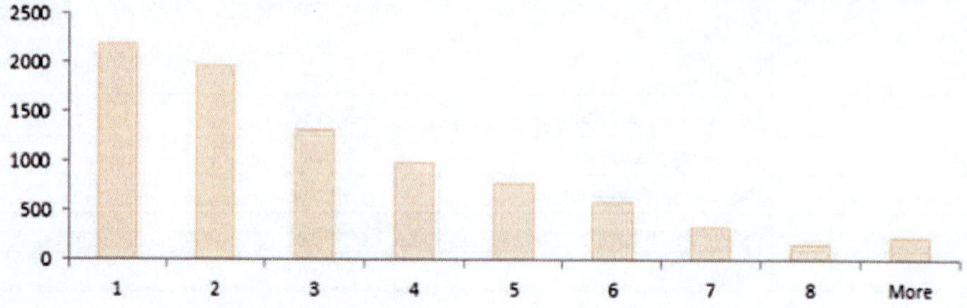

Fig. 5 Number of total applications in the dataset

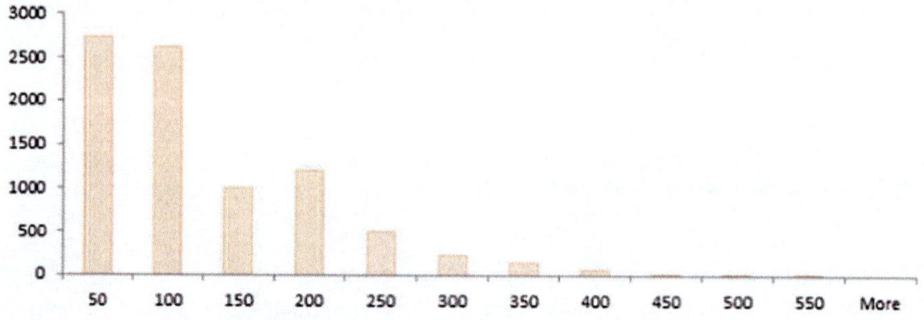

Fig. 6 LTV of the devices in the dataset

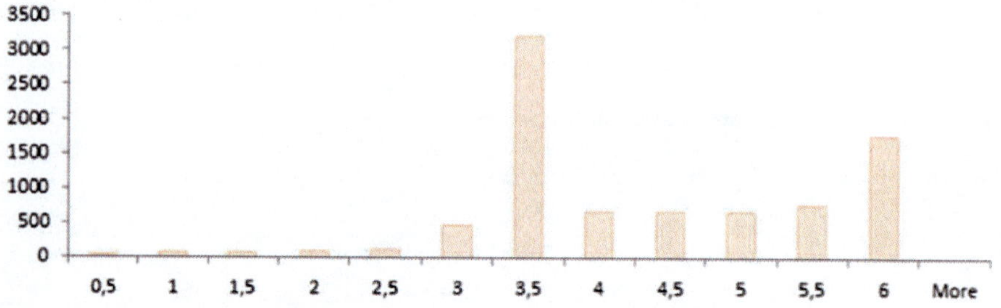

Fig. 7 The battery life of the devices in the dataset

Figure 3 shows the number of active applications on devices in the dataset. Almost 80% of devices have one application. The remaining devices in the dataset have more than two applications (at most ten).

Figure 4 presents the number of passive applications on devices in the dataset.

Figure 5 shows the total number of applications on devices in the dataset.

Figure 6 presents the lifetime value (LTV) of devices in the dataset. LTV presents the total amount of time of the applications installed on the devices. LTV differs from the number of times a device application is used.

Figure 7 shows the battery life of the devices in the dataset. Most devices have more than 3.5 years old battery life. In fact, the batteries of the devices in the market owned by Token Financial Technologies are more than three years old. Battery life was calculated by considering the device installation date.

In order to build and analyze the classification models and measure their performance, the software RapidMiner was used. The software is one of the world-wide leading open-source platforms for data science and machine learning [16].

Machine Learning Algorithms

Naive Bayes, Generalized Linear Model, Logistic Regression, Fast Large Margin, Deep Learning, Decision Tree, Random Forest, Gradient Boosted Trees algorithms were suggested to predict the churn rates of devices. The parameter settings used for each algorithm are shown in the tables given below (Tables 3, 4, 5, 6 and 7). Parameters were set to default values provided by the Software.

Table 3 Decision tree parameters

Parameters	Value
criterion	Gini
splitter	Best
max_depth	None
min_samples_split	2
min_samples_leaf max_leaf_nodes min_impurity_split	1 None None

Table 4 Random forest parameters

Parameters	Value
n_estimators	100
criterion	Gini
max_depth	None
min_samples_split	2
min_samples_leaf max_leaf_nodes min_impurity_split	1 None None

Table 5 Logistic regression parameters

Parameters	Value
penalty	‘12’
dual	False
tol	0.0001
C	1.0
intercept_scaling	1
max_iter	100
solver	‘lbfgs’

Table 6 Gradient boosted trees parameters

Parameters	Value
loss	‘Deviance’
learning_rate	0.1
n_estimators	100
subsample	1.0
criterion	‘friedman_mse’
min_samples_split	2
min_samples_lift	1
max_depth	3
tol	0.0001

Table 7 Deep learning parameters

Parameters	Value
activatin	maxout
hidden_layer_size	3
hidden_dropout_ratios	0.5
epochs	1e3
rho	0.9
epsilon	1e−4
distribution function	auto
momentum_start	0.5
momentum_stable	0.99

Results

The Mean Decrease of Impurity, also known as the impurity-based feature importance measure, is used to find out the impact of each attribute on the accuracy of the model. The results of the analysis show that battery life has a significant influence on the churn decision. The devices which have lower LTV might tend to be potential

Table 8 Weights of the attributes in the dataset

Attribute	Weight
Battery life	0.35
LTV	0.28
Number of active applications	0.22
Number of passive applications	0.12
Number of total applications	0.03

churners. Past records on the number of active and passive applications on devices also demonstrate an influential impact on the likelihood of churning.

Table 8 shows the weights of each attribute having an effect on the churn decision.

In the evaluation phase, the performances of the suggested algorithms are evaluated and compared to each other.

Figure 8 shows the resulting Receiver Operating Characteristic (ROC) curves of the algorithms.

The ROC curve shows the proportion of true positives to false positives, and the area under this curve (i.e., AUC) represents the performance of a model in distinguishing between classes [4]. The algorithm with a larger area under the curve works better. In this study, Fast Large Margin and Random Forest algorithms produced the best results.

Figure 9 shows the results of the Random Forest model. According to the model, if LTV is greater than 132.6 and the number of active application numbers less than or equal to 0.5 and the total application number is less than or equal to 11.5 and battery life is greater than 5.7, this device will be churn.

Figures 10, 11, 12, 13 and 14 shows the results of each algorithm with respect to the following performance measures: accuracy, precision, recall, F-measure, and specificity, respectively. The winning approach is highlighted in green.

Accuracy is the proportion of the total number of correct predictions and shows the correct classification rate. The accuracy value of each algorithm is given in Fig. 10. Random Forest approach has the highest accuracy level as 87.9%.

Precision, which describes how good a model is at predicting the positive (i.e., churn) class, is the proportion of correctly predicted positive cases to the total number of cases predicted as positive [8]. The precision of each algorithm is given in Fig. 11. Fast Large Margin approach has the highest precision level as 90.1%.

Recall (Sensitivity) describes the proportion of the number of correctly predicted churns to the total number of churns that occurred; i.e., it gives the rate of correctly predicted churns (see Fig. 12). Random Forest approach has the highest recall level as 89.3%.

F-measure is the harmonic mean of precision and recall (see Fig. 13). If an algorithm is optimized to increase only precision or recall and disfavor the other, the harmonic mean tends to decrease. But it takes the greatest value when both precision and recall are equal to 1 [18]. Random Forest approach has the highest F-measure level as 88.1%.

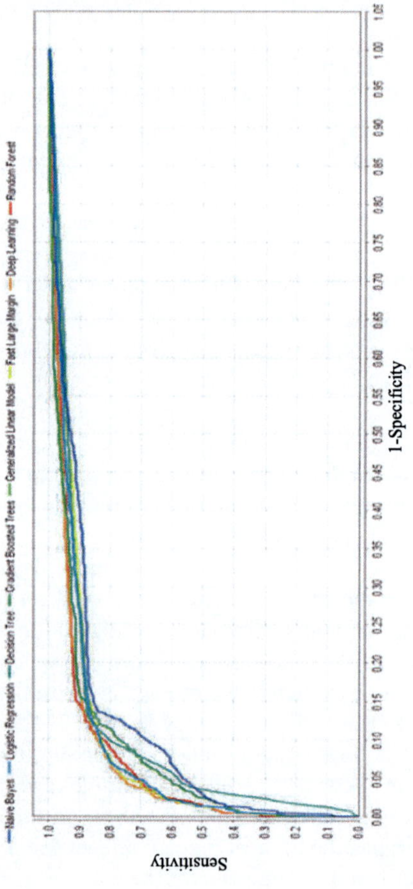

Fig. 8 ROC curve comparison

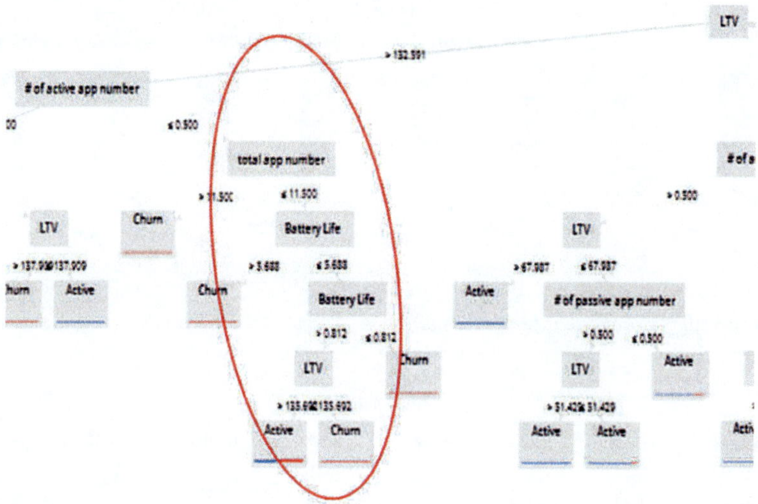

Fig. 9 Random forest model result

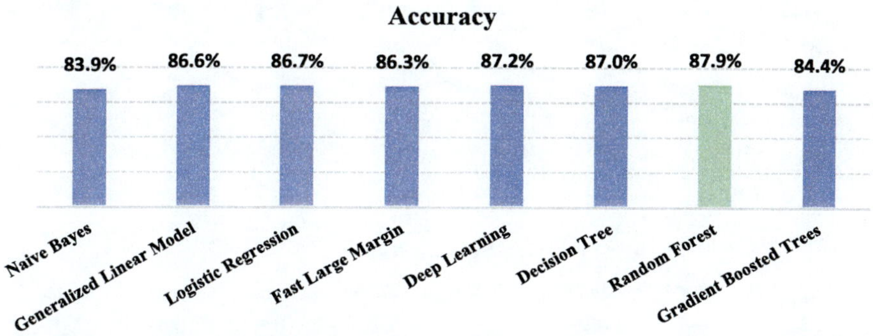

Fig. 10 Accuracy performance measure results

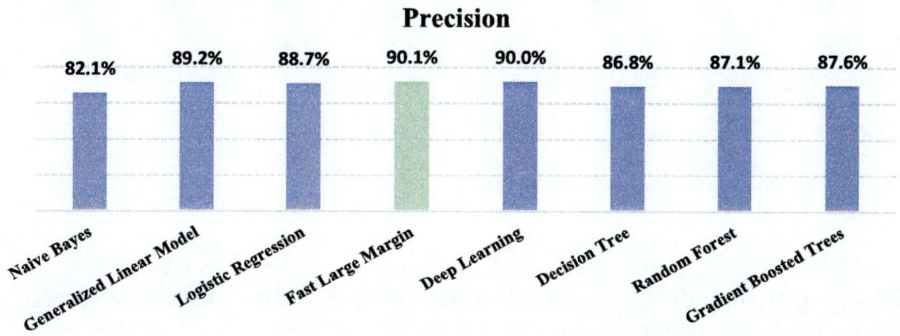

Fig. 11 Precision performance measure results

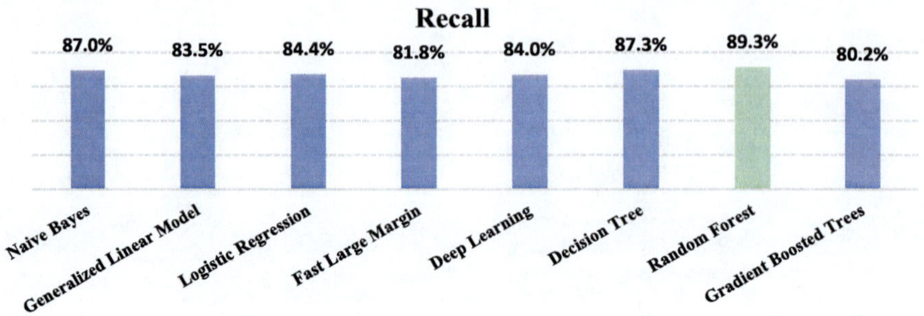

Fig. 12 Recall performance measure results

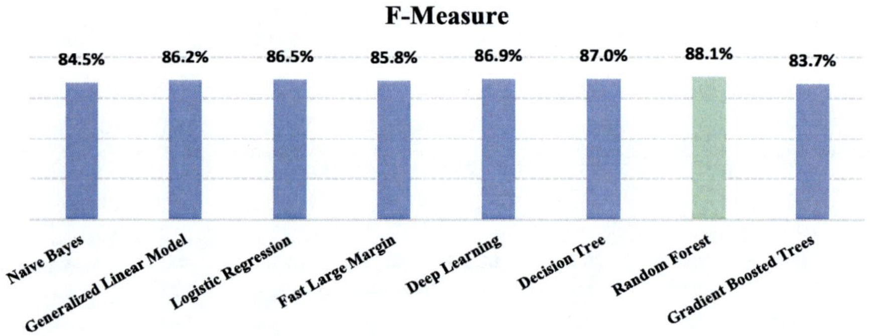

Fig. 13 F-measure performance measure results

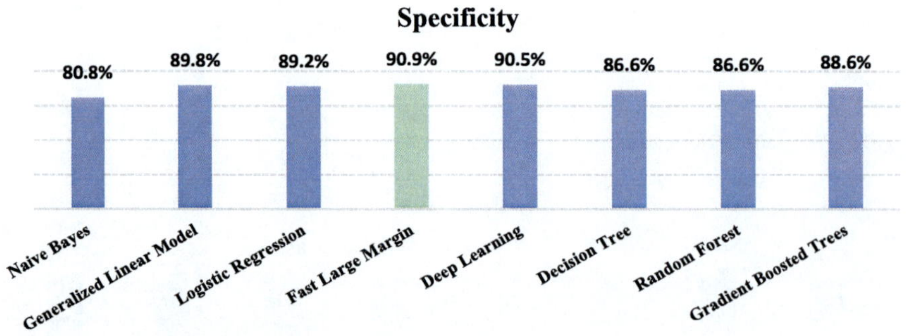

Fig. 14 Specificity performance measure results

Finally, specificity measures the proportion of real negatives (i.e., active POS devices) that are correctly predicted as negative (i.e., active). Fast Large Margin approach has the highest accuracy level as 90.9%.

To summarize the performance measure results, Table 9 shows the best algorithm with respect to each performance measure.

Table 9 Performance results of the algorithms

Performance Measure	Algorithm	Values (%)
Accuracy	Random forest	87.9
Precision	Fast large margin	90.1
Recall	Random forest	89.3
F-measure	Random forest	88.1
Specificity	Fast large margin	90.9

Random Forest performs better on Accuracy, Recall, and F-measure, while Fast Large Margin on Precision and Specificity, instead. Random Forest approach is the winner since optimizing the Recall measure is more important in our setting than optimizing Precision. In other words, discovering a new potential churn is more important than precisely recognizing a true churn instead.

Conclusions

Since customer acquisition is very costly in terms of time, money and effort, churn prediction, and customer retention strategies have become a crucial issue for the companies in recent years. Especially telecommunications, banking services, and insurance companies tend to predict the potential churns in advance to offer new competitive strategies and tactics that will help to keep their profitable customers.

In this study, after preparation of the dataset of POS devices, we applied common (e.g., Decision Trees, Logistic Regression) and advanced (e.g., Random Forest, Gradient Boosted Trees) machine learning algorithms to predict the churns. The Random Forest algorithm achieved around 88% of accuracy and F-measure. According to the results, the battery life among the other attributes has a significant effect on the reason for the churn as well as LTV has a major impact.

One contribution of this study to the literature is that it is the first attempt to predict churn behavior in the payment services industry. Another contribution is that analyzing churn data for POS devices adds value to the literature in predicting the likelihood of merchant churn. Thanks to the study, in the last quarter of 2019, Token Financial Technology saved more than 60% of the usage of EFT-POS devices from potential churns by replacing the battery and device. Replacement with high technology batteries and devices guarantee that merchants keep the banking application(s) on the devices.

For further research, expanding the feature set to include transaction values, number of service & part requests, the total amount of failure requests might increase the performance of the suggested algorithms. Also, to further improve the performance, it is planned to use advanced ensemble methods. Finally, for Token Financial Technologies, retention strategies will be integrated into the churn prediction system to achieve less churn rates for the current devices. For our payment services sector, the complaints and relations of merchants could be crucial, and because of this, we plan to add this effect in the next step with Social Network Analysis method.

Acknowledgements This work was supported by the Token Financial Technologies company.

References

1. Amin A et al (2017) Just-in-time customer churn prediction in the telecommunication sector. J Supercomput 1–25.
2. Au W-H, Chan KCC, Yao X (2003) A novel evolutionary data mining algorithm with applications to churn prediction. IEEE Trans Evol Comput 7:532–545
3. Azeem M, Usman M, Fong ACM (2017) A churn prediction model for prepaid customers in telecom using fuzzy classifiers. Telecommun Syst 66:603–614
4. Bradley AP (1997) The use of the area under the ROC curve in the evaluation of machine learning algorithms. Pattern Recogn 30:1145–1159
5. Caigny A, Coussement K, Bock KW (2018) A new hybrid classification algorithm for customer churn prediction based on logistic regression and decision trees. Eur J Oper Res 269:760–772
6. Coussement K, Lessman S, Verstraeten G (2017) A comparative analysis of data preparation algorithms for customer churn prediction: a case study in the telecommunication industry. Decis Support Syst 95:27–36
7. Dasgupta K, Singh R, Viswanathan B, Chakraborty D, Mukherjea S, Nanavati AA, Joshi A (2008) Social ties and their relevance to churn in mobile telecom networks. In: Proceedings of the 11th international conference on Extending database technology advances in database technology—EDBT pp 668–677
8. Davis J, Goadrich M (2006) The relationship between precision-recall and ROC curves. In: Proceedings of the 23rd international conference on Machine learning, pp 233–240
9. Farquad H, Vadlamani R, Bapi RS (2014) Churn prediction using comprehensible support vector machine: an analytical CRM application. Appl Soft Comput 19:31–40
10. Federico C (2014) Data preparation in the big data era. Igarss 1:1–5
11. Genachowski J, Copps MJ, McDowell RM, Clyburn M (2011) A comprehensive compilation of decisions, reports, public notices and other documents of federal communications commission (FCC) 26:9173–10152
12. Haenlein M (2013) Social interactions in customer churn decisions: The impact of relationship directionality. Int J Res Mark 30:236–248
13. Hung S-Y, Yen D-C, Yen DC, Wang H-Y (2006) Applying data mining to telecom churn management. Expert Syst Appl 31:515–524
14. Lawrance A (2012) Five customer retention tips for entrepreneurs. Forbes
15. Lu N, Lin H, Lu J, Zhang G (2014) A customer churn prediction model in telecom industry using boosting. IEEE Trans Industr Inf 10:1659–1665
16. RapidMiner Studio (9.1). Retrieved Dec 15, 2019, from https://rapidminer.com/
17. Vafeiadis T, Diamantaras KI, Sarigiannidis G, Chatzisavvas KCh (2015) A comparison of machine learning techniques for customer churn prediction. Simul Model Pract Theory 55:1–9
18. Vicente JJB, Klett F (2011) A generic evaluation framework for knowledge-based infrastructures: design and applications. Int J Comput Inf Syst Ind Manage Appl (IJCISIM) 3:290–297
19. Wang G, Liu L, Peng Y, Nie G, Kou G, Shi Y (2010) Predicting credit card holder churn in banks of China using data mining and MCDM. In: IEEE/WIC/ACM international conference on web intelligence and intelligent agent technology (WI-IAT), vol 3, pp 215–218

20. Zhang X, Zhang Z, Liang D, Jin H (2018) A novel decision tree based on profit variance maximization criterion for customer churn problem. In: International conference on intelligent human-machine systems and cybernetics, pp 20–23
21. Zhang S, Zhang C, Yang Q (2010) Data preparation for data mining. Appl Artif Intell 17:375–381

Virtual Reality: A Possibility for Training Operator 4.0

Mohamad Fallaha, Orhan Korhan, and Qasim Zeeshan

Abstract As technological advancements have led to the onset of industry 4.0, there must be a shift in the organizational frameworks, wherein manufacturing propensities and the traditional workplaces must transform to adopt a system wherein humans and machines can work together towards a rise in productivity and flexibility. These recent developments call for an up-gradation of the traditional operator to Operator 4.0 or Smart Operator as there has been a rise in a typical, unique work relationships which will require interactions between the operators and machines. These are challenging circumstances for engineers and operators as they must enhance their skills and abilities to keep up with the changing trend. Smart operators will also need to update their cognitive skills parallel with the innovations in the field of virtual reality and wearable technology equipment, as the equipment's augment the abilities of the Operator 4.0. This article aims to (1) Highlight the cognitive ergonomics skills required by operator 4.0 for Virtual Reality VR applications (2) Demonstrate how VR can be used to train future operator 4.0 to do his future tasks (3) Suggest how VR application in manufacturing can be improved for operator 4.0. The methodology in this research involves the usage of Virtual Reality to simulate a factory setup emulating Industry 4.0 factory. Based on the feedback of the participants, the efficiency of VR in training workers in the manufacturing industry can be determined. Also, this exercise can aid designers in avoiding Graphical User Interface issues and further develop the system to maximize efficiency and to provide a positive user experience.

Keywords Industry 4.0 · Operator 4.0 · Cognitive ergonomics · Virtual reality (VR) · Virtual operator · Artificial intelligence (AI) · Smart factory

---

M. Fallaha (✉) · O. Korhan · Q. Zeeshan
Industrial Engineering Departement, Eastern Mediterranean University (EMU), via Mersin 10, Famagusta, Northern Cyprus, Turkey

O. Korhan
e-mail: orhan.korhan@emu.edu.tr

Q. Zeeshan
e-mail: qasim.zeeshan@emu.edu.tr

F. Calisir (ed.), *Industrial Engineering in the Internet-of-Things World*,
Lecture Notes in Management and Industrial Engineering,
https://doi.org/10.1007/978-3-030-76724-2_25

Introduction and Background

The aim of the I4.0 concept is to integrate lots of consumers' daily life aspects into the elements of the value chain process, where this integration leads to one sole system [7]. I4.0 permits new ways of cooperation amongst software systems, machines, and humans. This phenomenon is called the Industrial Internet of Things, Services, and People (IoTSP) [5]. The main elements of industry 4.0 are Cyber-Physical Systems (CPS), Internet of Things (IoT), Big Data, Artificial Intelligence (AI), Intelligent Factories.

Operator 4.0 comes as the human factor within the Industry 4.0 paradigm. Researchers agree that there are eight different types of Operator 4.0, as demonstrated in Fig. 1. This article focuses more on the Virtual Operator type.

All types of Operator 4.0 must have the skills to cooperate with robots and also work aided with machines whenever required, through human cyber-physical systems (H-CPS), advanced human–machine interaction technologies, and adaptive automation. According to [13].

The concept of Operator 4.0 has arisen to achieve a balanced interaction between machines and humans, and apart from the physical and organizational ergonomics skills required by Operator 4.0, cognitive ergonomics skill comes as the most important factor. Since it is related to human exchange with machine constituents. Cognitive ergonomics typically affect mental activities, e.g., memory, reasoning, perception, and response, which occurs as a result of the exchange between humans, and diverse basics of the system [6].

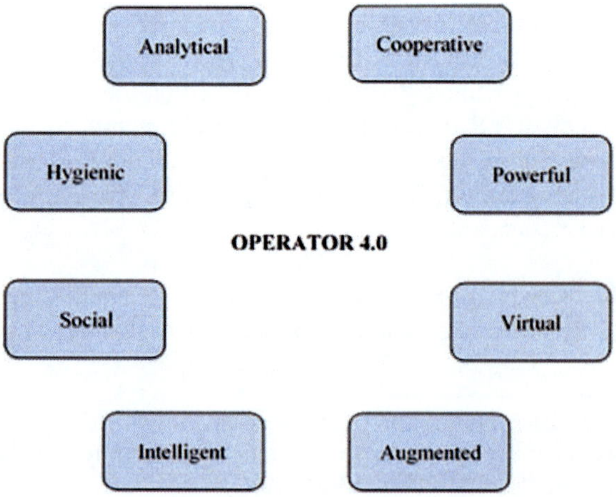

Fig. 1 Operator 4.0 types. Adapted from Ruppert et al. [15]

Table 1 VR applications in manufacturing systems

VR application	Details
Virtual prototyping	VR prototyping is a method used to test different product designs before manufacturing. This way is much cheaper than making a physical prototype [2]
Virtual Assembly (VA)	Virtual assembly facilitates design faults allocation process for designers as they are able to take a closer look at product parts during the assembly process virtually [11]
Virtual Manufacturing (VM)	It is virtually manufacturing a product within the simulation of a virtual factory. This Virtual Manufacturing technique utilizes processes for manufacturing a desktop VR system for Computer-Aided Design (CAD)
Virtual training	Virtual training is used instead of traditional training with simulation technology so each type of operator can focus on their own field comfortably [12]
Virtual reality approach in immersive design	The combination of VR tools and CAD software facilitates the work for designers to make immersive designs
Virtual reality approach for hierarchical path planning	For this to commence, a tool capable of swiftly testing the product throughout the PLM using VR is needed [3]

Virtual Reality in Manufacturing

Based on a study that is conducted by Aniruddha Talekar [1], virtual reality has many applications related to manufacturing, as demonstrated in Table 1.

Virtual Reality and Operator 4.0

Virtual Operator can apply all applications mentioned in Table 1 since VR allows operators to fully involve themselves, experience, and interact with cyberspace that provides insights within virtual and augmented reality or directly communicates with a personal digital assistant on topics such as man and machine interactions which fulfill safety standards. [8].

One of the main important aspects of VR in manufacturing is that it provides operators training with the necessary supportive information that usually cannot be found in the work environment. Such information can be described as expected maintenance activities, warnings on unexpected dangers, machine productivity, unforeseen risks that might happen, suggestions for increasing productivity, etc.

VR is a magnificent tool; however, it does necessitate more cognitive effort when communicating as opposed to talking face to face [14].

Cognitive Ergonomics Skills Required by Operator 4.0 for VR

According to [4], the modification of cognitive functions occurs by age; this necessitates a vital need for rehabilitation programs in this aging society.

VR can ameliorate the trustworthiness of neuropsychological analysis by placing intricate test stimuli under constant manipulation and accurate measure of participants' reactions. The legitimacy of the measurement of different cognitive areas could increase due to the capability of the VR technology to quantify responses more discreetly. These cognitive domains comprise visuospatial abilities, executive processes, and an enhanced level for problem-solving, attention, and memory. A second approach to this is to directly analyze performance in a simulated human environment. Thus, VR provides the possibility for cognitive analysis within a simulation of real-world practical testing spaces whose sole goal is to enhance the ecological legitimacy of neuropsychological analysis [9].

The required cognitive skills for operators for using VR technology are shown in Table 2.

Based on the important role of VR in manufacturing in processes such as design, assembly, and prototyping. VR can be used to train Operator 4.0 through 3D simulations. To further understand how training can be done with VR, a survey is conducted with university students. The survey involves a virtual simulation of a factory where students have full control over machines.

Table 2 Cognitive skills required by Operator 4.0 within VR

Skill	Explanation
Memory	The individual is requested to gradually remember responses occurring at a specific time, settings on gauges, locations of tools, and behavioral patterns to accomplish several duties
Sensory processing: visual and auditory	Touch, taste, and smell are extremely crucial senses to humans, and difficulties pertaining to audition and vision are relevant in day to day operations. A VR space would be beneficial when addressing these difficulties
Higher cognitive functions	The ability to solve problems, conceptual and organization thinking, executive functioning and critical thinking, and more

Virtual Simulation Design

A built model was used from Game4automation Company, which is specialized in making 3D Simulation shareware frameworks of automated systems. The simulation engine for this project is Unity game designing software. Game4automation provides all required elements which will power up a digital twin with Unity software—software that is needed in the creation and animation/programming of 2D, 3D, VR, and AR environments. Although the 3D simulation is available, it still required some programming to work with the VR kit. Figure 2 shows the user interface of the Unity Engine in VR modeling.

As demonstrated in Fig. 3, the experiment's 3D environment includes five main sections, control box, conveyor belt, sorting machine, product container, and a 5-axis robot. The control box is mainly a power supply which is used to power up the production line. The Conveyor belt is mainly used to move the product, which is the beverage-can to a specific position. The sorting machine then moves and picks the product from the conveyor belt to the blue container. Lastly, once the blue container is full, the 5-axis robot holds the container and empty's it.

As the machines are communicating with each other as it is in the I4.0 Smart factory, the Operator 4.0 job is to only observe and control the machines once necessary. Basic on/off controllers are available for all the machines, so it facilitates the worker's ability to understand the relation between the machines. As shown in Fig. 4, a tablet is added to the 3D simulation. The tablet allows operators to control the machines, get notifications, and see warning messages.

As shown in Fig. 4, the green-colored button controls the conveyer, yellow-colored button controls the gantry, and the red-colored button controls the five axis robot.

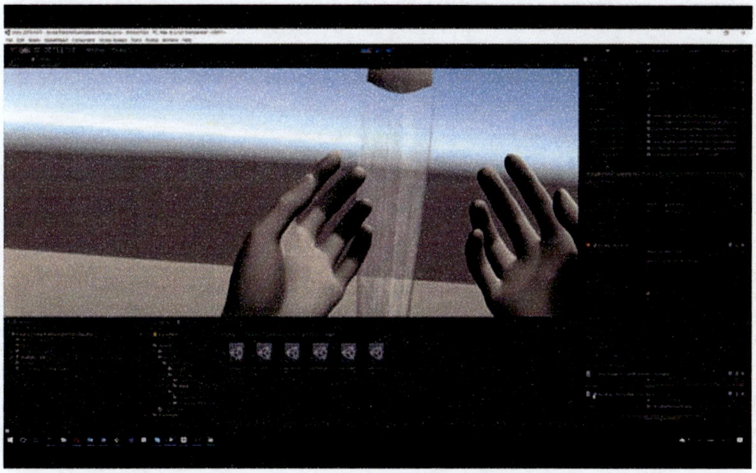

Fig. 2 VR modeling inside the Unity engine. Adapted from Medium.com (2019)

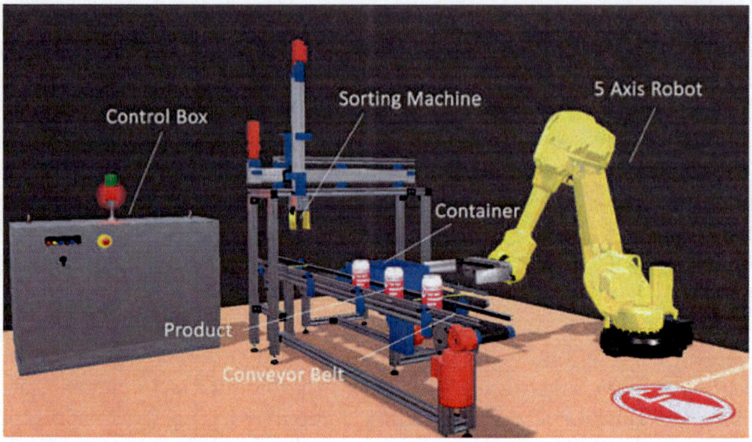

Fig. 3 Game4automation 3D simulation

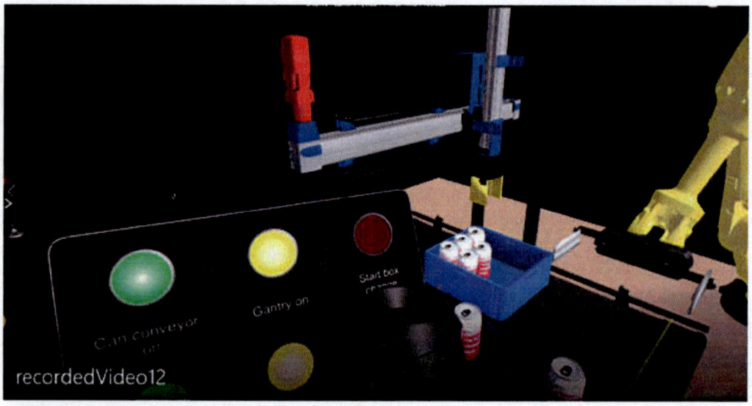

Fig. 4 Virtual reality simulation

Experiment and Task

The experiment can be demonstrated as a series of steps that the operator/participant has to take in order to complete the task. Each participant is requested to fill a survey questionnaire related to the task right after finishing the steps in Table 3.

All selected volunteers were above 18 years old. Participants were divided into three groups, bachelor, masters, and PhD (see Table 4). 10 participants of each group participated, which made 30 participants in total. The age range was between 20 and 40 years (average of 30 years). To ensure the validity of the survey, prior to doing the experiment, participants were asked if they had any medical injury that might affect their ability to participate, such as physical or mental injuries. Half of the participants

Table 3 Steps to complete the experiment tasks

Steps	Details	Importance
Step 1	Wear the VR apparatus and confirm that the vision of 3D simulation is clear. Then it is proven that the participant is ready for the next step	• Ensure the comforts of the participant and avoid unnecessary distractions
Step 2	The participant is requested to move inside the factory's 3D simulation (e.g., forward, backward, rotation… etc.)	• To ensure the participant's ability to observe the surroundings • Allows participants to sense the reality of the situation
Step 3	Find the tablet, on top of the control box, hold it, and check for any notifications	• Assists the operator to further understand the status of the machines
Step 4	A warning appears on the tablet (unplanned change) for one of the machines about pressure loss, and the operator is required to turn off all machines and visually observe if anything is wrong	• Understand the operator's situational awareness level
Step 5	After concluding that machines are in good working conditions, the operator is required to turn on all machines	• Assists in observing the memory level of each participant. Which button controls which machine
Step 6	Put the tablet controller back to its place and end the experiment	

Table 4 Participants' classification table

Degree level	Age range	VR Experience
Bachelor	20–26	60% gaming experience in VR 40% no previous experience in VR
Masters	22–31	30% gaming experience in VR 70% no previous experience in VR
PhD	28–40	20% gaming experience in VR 80% no previous experience in VR

were below or equal to 25 years old, and the other half were above 25 years old. Any participant who has mental disabilities, eyesight weaknesses, or any other physical disability that might affect the results of the research were excluded.

A total of 30 volunteers did the experiment. Statistical Package for the Social Sciences (SPSS) version 23 was used to analyze collected quantitative data. Participants answered all the questions. Excluding participants names from the data analysis part, the data analysis included mainly numerical values except for the last question which was related to any suggestions for future developments.

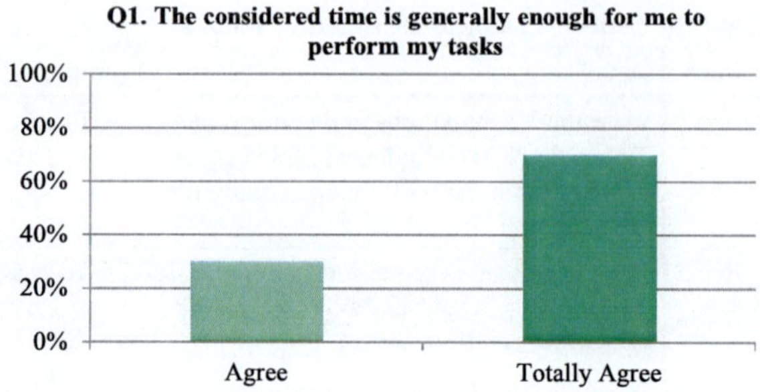

Fig. 5 Results of survey question 1

Survey Questionnaire

The survey was planned based on the CompleXity Index method (CXI). This method aims to assist production companies to identify and reduce working complexities that could affect operators' performance [10]. The questionnaire includes ten questions that are presented using the Likert scale. The scale range is from 1 (Do not agree at all) to 5 (Totally agree).

The questionnaire was divided into two main sections. The first section has questions related to the task and operation. The first question was related to the time limit given to participants to do the task, three questions related to the warning message (unplanned change), and a question related to whether participants felt any stress during the task.

The second section of the questionnaire was mainly related to the software design of the tablet controller, starting from the icon shapes and functionality to the general design of the software.

Results

All participants agree that the given time given for them to perform their tasks during the experiment was enough (see Fig. 5). The rest of the results of the survey questionnaire are all shown in Fig. 6.

Discussion

According to the survey results, all of the participants agreed that the considered time (5 min) for the given task was enough for them to learn and complete the task. However, some participants exceeded the time limit given to them. In Fig. 7, an analysis of the effect of unplanned changes to participants' stress and time duration to complete the task was made. The word stress here refers to the participant not

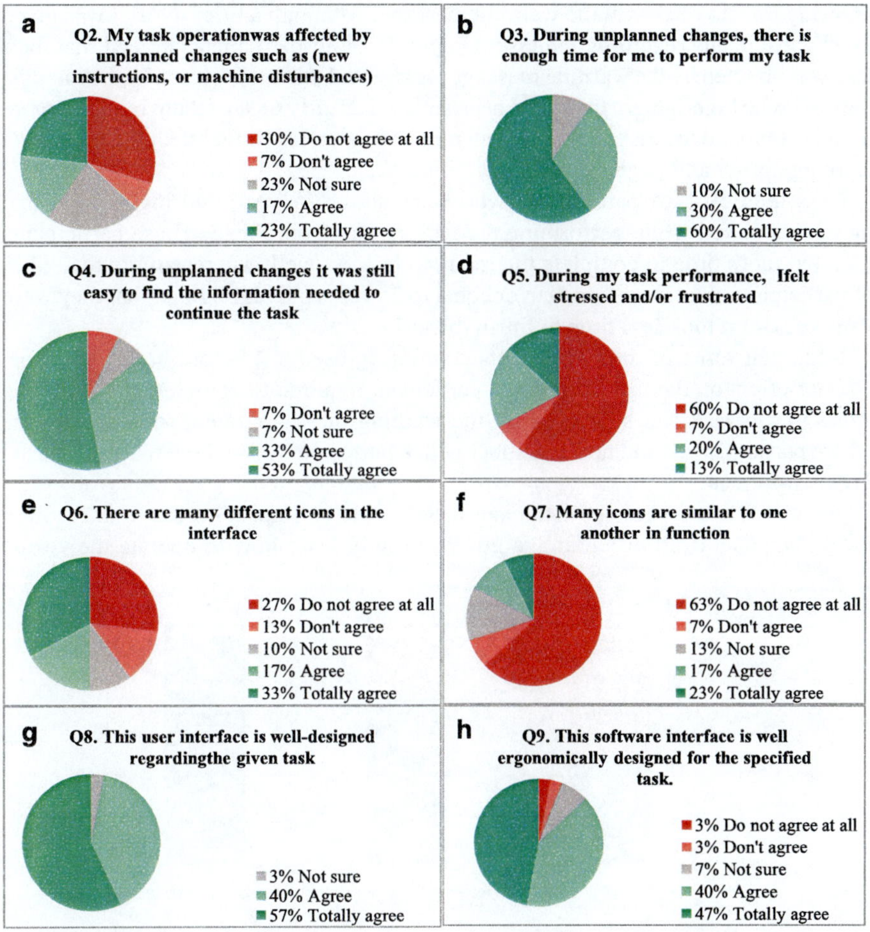

Fig. 6 Data elaboration for **a** Survey question 2, **b** Survey question 3, **c** Survey question 4, **d)** Survey question 5, **e** Survey question 6, **f** Survey question 7, **g** Survey question 8, **h** Survey question 9

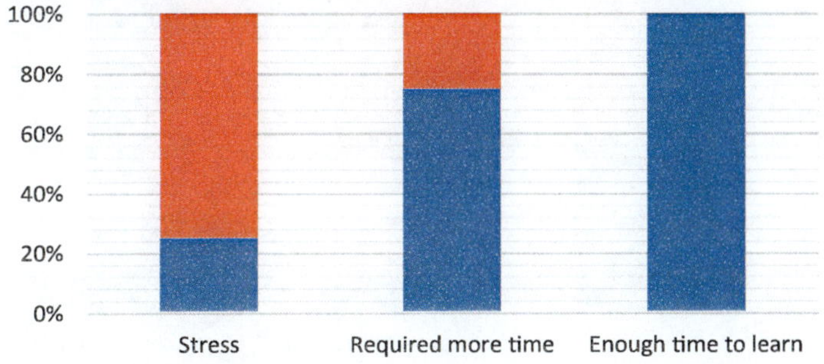

Fig. 7 Unplanned changes and stress analysis

knowing how to deal with the warning message. Although a brief tutorial was given to all participants regarding the task. Only 25% of all participants agreed that their task was affected by the warning message highlighted that they felt stressed. Amongst the ones who highlighted, they have been stressed. Nearly 65% of them required more time to complete the task. This gives an idea of how stress could affect the operator's learning ability and cognition at the working place.

Less than 50% of participants who were above 25 years old highlighted that they felt stressed while performing the task, as shown in Fig. 8. These participants required more time to complete the required task, as well, whereas more than 50% of participants who were less than or equal to 25 years old highlighted that they were stressed. But it took less time to finish the task.

Although some of the participants could have used VR before, it did not show a major effect on the time duration a participant required to complete the task. The stress in this task can be caused by the warning message, tablet software design, where participants could not remember which button controls which machine, or the 3D environment.

Figures 8 and 9 show how the age factor made an impact on the time required for participants from different age groups to fully learn how to operate the virtual

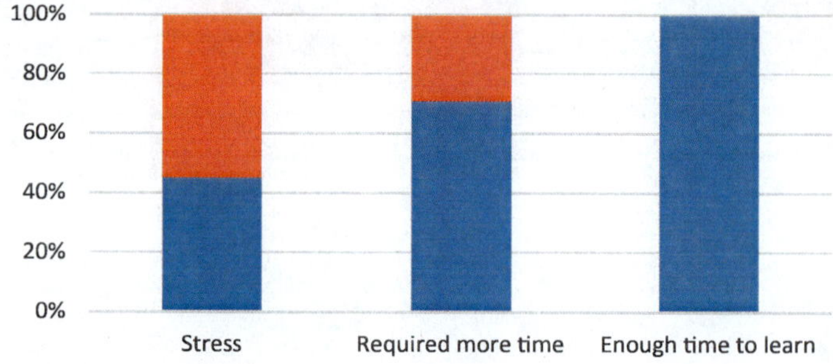

Fig. 8 Age analysis for participants above 25 years old

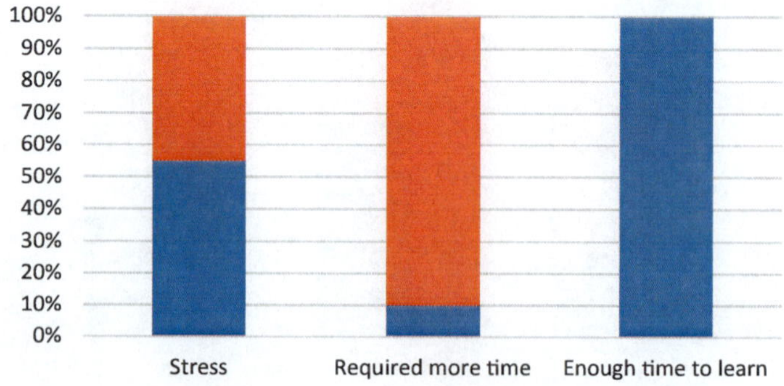

Fig. 9 Age analysis for participants below or equal to 25 years old

factory. That can be described as a result of age over cognition. However, although it took older participants more time to learn the task, they were still able to complete it.

In general, all participants agreed that such an application is a useful method to teach future workers on how to perform their tasks in a smart factory. As for the survey question that asked about any comments and future proposal on how to improve this application, the list below sums up important comments by participants:

- Add visual assistance, such as comments on the screen, for better instructions.
- Audio assistance for better understanding.
- Extend it to a complete production line.
- Add a seat for better physical ergonomics.
- Positive comments by participants:
- Workers can have their training before starting their jobs.
- It can be implemented in the education system.

Conclusion

The concept of smart factory systems can be easily presented through innovative techniques that are manifested in simulation, learning in action, and the VR. Moreover, these innovative techniques can further promote future development in the field.

Through research, the required cognitive ergonomics skills for Industry 4.0 and VR in manufacturing were identified in the literature. Furthermore, with the designed VR experiment, it was possible to show how VR can be used to train future workers. Such application could help Operator 4.0 to adapt to new industrial paradigms such as Industry 4.0. 30 students from the faculty of engineering participated in the survey. Participants' feedback on the survey questions showed important findings. One of those findings is that age had its effect on the learning duration of participants indeed. However, in the end, all of the participants were successfully able to complete the given task. This shows that older workers can have equal chances to be hired in the future if they had sufficient training.

Moreover, participants' feedback regarding the experiment helped in finding out the required improvements to the application model regarding the VR environment, user interface, and task instructions. Hence, the application used was useful to analyze how VR in manufacturing could be progressed to make similar applications more convenient, particularly that the experiment needed workers' physical and mental interaction with the machines and tools required to control the machines.

According to participants' responses, the experiment reinforced thesis goals in the following:

- Design a manufacturing environment which assists Operator 4.0 to learn and have a visual-based training on a 3D based production line.
- Search on different methods Operator 4.0 would interact with a VR training program and study the ergonomics of the software user interface.

- Observe the type of communication that can be achieved between Operator 4.0 and machines through technologies like IoT and CPS.

This paper showed the impact of implementing new technologies such as VR in the education and training of future operators. Adapting such techniques in educational curriculums like engineering departments can be the first step to prepare current engineering students on the evolving Industry 4.0. Furthermore, Future factories 'human resources sectors must be involved with the improvement of the human asset through teaching, learning, and qualifying worker, rather than just being responsible for recruiting workers.

References

1. Aniruddha Talekar SP, Thakre P, Rajkumar E (2017) Virtual reality and its applications in manufacturing industries. J Chem Pharmaceut Sci (JCHPS), 10(1):151
2. Arrighi P-A, Mougenot C (2019) Towards user empowerment in product design: a mixed reality tool for interactive virtual prototyping. J Intell Manuf 30(2):743–754
3. Cailhol S, Fillatreau P, Fourquet J-Y, Zhao Y (2015) A hierarchic approach for path planning in virtual reality. Int J Interact Des Manuf (IJIDeM) 9(4):291–302
4. Grewe P, Kohsik A, Flentge D, Dyck E, Botsch M, Winter Y, Markowitsch HJ, Bien CG Piefke M (2013) Learning real-life cognitive abilities in a novel 360°-virtual reality supermarket: a neuropsychological study of healthy participants and patients with epilepsy. J NeuroEng Rehab 10(1):42. https://doi.org/10.1186/1743-0003-10-42
5. Kaasinen E, Schmalfuß F, Özturk C, Aromaa, S, Boubekeur M, Heilal, J, Heikkilä P, Kuula T, Liinasuo M, Mach S, Mehta R,Walter T (2020) Empowering and engaging industrial workers with Operator 4.0 solutions. Comput Ind Eng 139.https://doi.org/10.1016/j.cie.2019.01.052
6. Kim IJ (2016) Cognitive ergonomics and its role for industry safety enhancements. J Ergonom 6(4). https://doi.org/10.4172/2165-7556.1000e158
7. Kinzel H (2017) Industry 4.0–Where does this leave the Human Factor? J Urban Cult Res 15:70–83
8. Longo F, Nicoletti L, Padovano A (2017) Smart operators in industry 4.0: a human-centered approach to enhance operators' capabilities and competencies within the new smart factory context. Comput Ind Eng 113:144–159. https://doi.org/10.1016/j.cie.2017.09.016
9. Martens MA, Antley A, Freeman D, Slater M, Harrison PJ, Tunbridge EM (2019) It feels real: physiological responses to a stressful virtual reality environment and its impact on working memory. J Psychopharmacol 33(10):1264–1273
10. Mattsson S, Gullander P, Harlin U, Bäckstrand G, Fasth Å, Davidsson A (2012) Testing complexity index–a method for measuring perceived production complexity. Procedia CIRP 3:394–399
11. O'Neill G, Ball M, Liu Y, Noghabaei M, Han K (2020) Toward automated virtual assembly for prefabricated construction: construction sequencing through simulated BIM. arXiv preprint arXiv:2003.06695
12. Ordaz N, Romero D, Gorecky D, Siller HR (2015) Serious games and virtual simulator for automotive manufacturing education and training. Procedia Comput Sci 75:267–274
13. Rabelo RJ, Romero D, Zambiasi SP (2018) Softbots supporting the operator 4.0 at smart factory environments 536:456–464. https://doi.org/10.1007/978-3-319-99707-0_57

14. Stadnicka D, Litwin P, Antonelli D (2019) Human factor in intelligent manufacturing systems - knowledge acquisition and motivation. Procedia CIRP 79:718–723. https://doi.org/10.1016/j.procir.2019.02.023
15. Ruppert T, Jaskó S, Holczinger T, Abonyi J (2018) Enabling technologies for operator 4.0: A survey. Appl. Sci 8(9):1650

Transition from AC to DC Powered Homes

Bahadir Tunaboylu, Tolga Erkmen, Selim Zaim, and Fatma Serra Ciftci

Abstract Increasing electronic appliances in residential buildings have caused researchers to study direct current (DC) powered systems in homes. In this paper, we state a scenario that a home with traditional alternating current (AC) powered appliances are replaced with DC devices. The questions which devices can be replaced, how much savings in terms of energy and electric bill is provided, what are economical, technical, social and environmental problems in this AC-DC transition and which system between DC and AC for residential buildings are much preferable based on the AHP method are answered in the presented study. Results show that 74.126-percentage cost saving in the electric bill is provided. However, the problems that nonexistence in some residential DC devices in the market, immaturity of DC system in terms of technical issue, added system costs, less information about DC by experts and related people, and a public bias that DC is unsafe **to** exist. In this study, based on these defined advantages and disadvantages of DC and AC system, criteria and sub-criteria have been determined for the analytic hierarchy process (AHP) method. By way of AHP, the result that DC system is much preferable than the AC system for residential buildings has been reached.

Keywords Direct current · Alternating current · Analytic hierarchy process

---

B. Tunaboylu · S. Zaim · F. S. Ciftci (✉)
Department of Materials Engineering, Marmara University, Istanbul, Turkey
e-mail: fatmaserra.ciftci@uskudar.edu.tr; ciftcifatmaserra@gmail.com

B. Tunaboylu
e-mail: bahadir.tunaboylu@marmara.edu.tr

S. Zaim
e-mail: selimzaim@izu.edu.tr

T. Erkmen
Ecoland Institute, Cary, USA
e-mail: tolga@ecolandinstitute.org

F. S. Ciftci
Uskudar University, İstanbul, Turkey

F. Calisir (ed.), *Industrial Engineering in the Internet-of-Things World*,
Lecture Notes in Management and Industrial Engineering,
https://doi.org/10.1007/978-3-030-76724-2_26

Introduction

From past to present, people pay attention to the fact that it is benefitting from energy resources in the most efficient way. Especially in terms of electricity production, it is a considerable fact that using renewable sources such as solar energy is eco-friendly. When solar energy is used as an electrical source, electricity is generated as a direct current (DC). However, electrical appliances in our homes are AC (alternating current)-powered. So, when the direct current comes from the solar panels to the loads, it has to be converted to AC. This conversion means power losses. Power losses mean too much-consumed energy. It causes our electric bills to increase. In order to change this situation, researchers have done many studies about using direct current directly without conversions. They have found that DC is efficient for data centers, commercial buildings, appliances such that LEDs, laptops, smartphones. In terms of residential buildings, there have been studies in some countries such as the USA and Japan.

In our presented study, a case is evaluated for a residential building in İstanbul in Turkey. In this case, all electrical appliances are discussed in terms of replacement to DC-powered ones. Some advantages are energy savings and cost savings in the electric bill, and some disadvantages that are financial, technical, and social problems are defined. Based on these defined benefits and problems, criteria, and sub-criteria for the projected AHP method have been identified.

The presented study's difference from other ones is asserting the preferable system between DC and AC system in residential buildings by way of benefitting from AHP.

Literature Review

If we think about designing sustainable houses, villages, or cities, the fact that benefits from DC rather than AC comes to mind for the electric system. In the future, when smart grids become widespread, DC systems may be implemented. Because direct current goes in a fixed direction, while alternating current shifts direction, the number of power converters of DC systems is fewer than the number of power converters of AC systems. It can be given as the main advantage of DC systems. It provides cost and energy savings, which becomes more of an issue in terms of sustainability [1]. In reference to Wunder, DC distribution eliminates fifty percent of power conversion losses and seventy percent of volume and weight in internal switch-mode power supplies. According to Denkenberger, when market growth of DC products happens in the advancing years, then their cost will decrease. Therefore, direct current distribution will be cost-effective for zero net energy buildings [2]. Also, DC distribution does not bring the problem of harmonics and reactive power. In addition, according to the research, AC distribution has more negative impacts on health when compared to DC distribution. If we proceed to mention the advantages of DC systems, the efficiency of DC-DC converters for home electrical appliances (HEAs) can be given. Correspondingly, direct current electrical power is used in many fields. For instance, many systems of renewable energy, such as wind turbines and photovoltaic solar systems, generate DC electrical power. Electrical devices in homes such as computers, laptops, and TVs use DC electrical power. Also, sensitive electrical appliances in space and naval operations benefit from DC distribution. In addition, DC is a secondary power supply for energy storage systems [3]. When the economic feasibility study was done for using DC circuits in LEDs and fluorescents in commercial buildings, it was seen that it could reduce the total unsubsidized capital costs of the system. However, it is important that this system should be grid-connected and solar photovoltaic [1].

To continue to mention about fields where DC electrical power is used, DC is beneficial in buildings with large battery capacity. In this process, batteries are used as sources of DC. Also, motors with changeable frequency drives operate internally in DC. Moreover, systems of DC distribution have been asserted and put into practice in data centers. In this way, 7–28% of electricity has been saved for the 380 V DC distribution system by comparison with the 208 V AC distribution system. In addition, according to a comparison between DC distribution systems at varying voltage levels and AC distribution system at 230 V AC in terms of the economic and technical situation, the distribution system at 326 V DC was the optimum from among the systems at 48, 120, 230, 326 V DC [4].

When we look at the side of residential buildings in terms of direct current, a study in 2016 uses 120 homes that have AC distribution in Texas in order to understand DC circuits technically and economically. Totally, there are ten scenarios. In the first case, DC distribution is given to all loads. In the second one, it is applied to a lighting circuit. Thirdly, DC distribution is given to a condensing unit. In the fourth and fifth cases, it is applied to PEV charger and refrigerator, respectively. All 5 cases are applied with and without storage. So, the total number of scenarios is ten. Also, all of the ten scenarios include solar PV. According to the results of the study, when there is battery storage, then energy savings is between 14–25%. When there is not, then it is between 9–20%. To observe the economic results, converting all equipment to DC equipment makes the levelized annual costs of electricity double. However, only one scenario is cost-effective. It is transforming air conditioning condensing units to direct current. Thus, costs are decreased, and the energy savings of 7–16% is produced in that case. Averagely, for each home, a PV array makes the levelized annual cost go up from $1200 to $2300. Also, the added costs of the solar array, bidirectional inverter, and DC devices are $770, $380 and $900 respectively and averagely. The financial advantage arising from PV generation and conversion to DC makes electric bill decline as approximately $950. However, it is too few in comparison to the added costs. When DC distribution is given to all loads, then the median of annual saved electrical energy per home is approximately 1400 kWh /yr without storage. When storage happens, then it is 1900 kWh/yr. Most of these energy savings is because of air conditioning condensing units. They produce savings of between 1100–1200 kWh/yr by oneself [5].

When we think about DC equipment installations for new constructions and renovations in residential buildings, according to King and Brodrick, new constructions' cost is half of the renovations' cost. So, their claim shows that new constructions are more cost-efficient than renovations [2].

With reference to a techno-economic analysis of DC power distribution in commercial buildings, if an onsite solar system has a large capacity of battery storage, then this DC system can be financially advantageous. In this analysis, two different climate regions which are Los Angeles and San Francisco, three types of buildings which are office, restaurant and retail, two different market conditions that are current and future, six types of configurations (50% PV and no battery, 50% PV and 50% battery, 50% PV and 100% battery, 100% PV and no battery, 100% PV and 50% battery, 100% PV and 100% battery) are used. Therefore, this techno-economic analysis of DC power distribution has 72 (2*3*2*6) different scenarios. In the future market scenarios, the electricity tariff, which is currently applied in Hawaii, is benefited from. Since the fact that there will be increasing penetrations of solar generation in the future, Hawaii, where takes higher insolation, is observed. Again, in future market scenarios, the question of how electricity will be priced in the future is answered according to electricity price trends to the annual electricity bills in U.S Energy Information Administration's Annual Energy Outlook. Also, in this study, the PV capacity of the San Francisco buildings is higher than the buildings' PV capacity in Los Angeles because San Francisco has less sun exposure than Los Angeles in a year. Thus, balancing annual PV generation and load is provided. In all 72 scenarios, the voltage level for HVAC, refrigeration, and lighting is 380 V in the modeled DC building, while the voltage level for electronics and other plug loads is 48 V. In conclusion, for the current market scenario, offices and restaurants having DC systems and battery storage have positive life cycle cost savings and payback periods of four years. When they have maximum battery storage, then the buildings with DC systems have a lower first cost than the same buildings with the AC system. Then, it creates instant payback periods. On the contrary, in most scenarios, when buildings with DC systems do not include battery storage, then they have negative life cycle cost savings and long payback periods. For the future market scenario, there will be improvements in converter efficiencies for AC systems as much as DC systems, and the DC systems have lower efficiency savings. To look at another conclusion of this analysis, if PV generation and load match with each other, then it will make efficiency savings and economic advantages increase [2].

Methodology

In this presented study, 4-person residential home is considered with conventional AC distributed appliances. Its location is in Istanbul in Turkey. According to the wattage values gained from the labels on the devices and user manuals, total weekly energy consumption is found as 373,293.85 Wh, as can be seen in Table 1.

So, daily energy consumption:

$$= 373293.85/7$$

Table 1 Energy consumption with the conventional AC devices in the presented home

Device as brand and model	Operating time as (h)	Device wattage (W)	Number in home	Operating days per week	Weekly energy consumption (Wh)
Air conditioner samsung Ar24nsjxbwk/sk	4	2319.375	1	7	64,942.5
Television samsung ue40f6340ss	5	122	1	7	4270
Washing machine siemens wm12t48str	3	2100	1	2	12,600
Dishwasher siemens sn 258I06jt	2	1300	1	7	18,200
Stove siemens hr 746545t	3	910	1	7	19,110
Refrigerator siemens kg57np00ne	24	1230	1	7	206,640
Combi boiler Arçelik dgk y 24 db	14	180	1	7	17,640
Coffee maker blue house bh 295 tc	1/6	800	1	7	933.3
Kettle fakir armilla	1/2	2200	1	7	7700
Toaster siemens tg23331v	1/2	2000	1	2	2000
LED Bulb Era light ea10	6	10	6	7	2520
Iron Philips gc 9240	3	2400	1	1	7200
Laptop 1 Hp pavilion dv6	2	120	1	7	1680
Laptop 2 Hp 250 g6	2	45	1	7	630
Wireless modem netmaster infinity 401	24	9.5	1	7	1596

(continued)

Table 1 (continued)

Device as brand and model	Operating time as (h)	Device wattage (W)	Number in home	Operating days per week	Weekly energy consumption (Wh)
Hairdryer philips hp 4876	1/6	2000	1	7	2333.3
Vacuum cleaner dc29 db origin	1	1400	1	2	2800
Iphone 7 Plus	3	5	1	7	105
Samsung note 5	3/2	12.5	3	7	393.75
Total					373,293.85

$$= 53327.69\text{Wh/day}$$
$$= 53.32769\text{kWh/day}\ (53327.69/1000)$$

Thus, monthly energy consumption:

$$= 53.32769 * 30$$
$$= 1599.8307\text{kWh/month}$$

Since the sale price of one kWh electricity is 0.14336486 TL in Istanbul, the monthly electric bill of the presented home is approximately 229.36 TL (1599.8307*0.14336486). Additionally, operating time and operating days of the devices are defined based on the household. The projected scenario is that electricity of the home is produced from photovoltaic cells and produced DC electricity is not converted to AC electricity thanks to DC-powered appliances.

From Existing AC Devices to Prospective DC Devices

When the market is examined for DC-powered appliances, the table of devices that can be substituted for existing AC-powered devices is given (Table 2).

So, decrease in weekly energy consumption is:

$$309419.1 - 32712 = 276707.1$$

So, decrease in daily energy consumption is:

$$276707.1/7 = 39529.6\ \text{Wh/day}$$
$$= 39.5296\ \text{kWh/day}$$

Table 2 Weekly energy consumption with substituted DC-powered devices [6–17]

	Power consumption (W)	Power consumption of corresponding AC-powered device (W)	Weekly energy consumption (Wh)	Weekly energy consumption of corresponding AC-powered device (Wh)
DC air conditioner	545.75	2319.375	15,281	64,942.5
DC television	49	122	1715	4270
DC washing machine	320	2100	1920	12,600
DC refrigerator	72	1230	12,096	206,640
DC coffee maker	150	800	175	933.3
DC kettle	150	2200	525	7700
DC iron	150	2200	450	7200
DC hairdryer	300	2000	350	2333.3
DC vacuum cleaner	100	1400	200	2800
Total			32,712	309,419.1

Thus, decrease in monthly energy consumption is:

$$39.5296^{*}30 = 1185.888\,\text{kWh/month}$$

Since sale price of one kWh electricity is 0.14336486 TL in Istanbul, decrease in monthly electric bill of the presented home is:

$$1185.888^{*}0.14336486 = 170.015\,\text{TL}$$

Therefore cost savings percentage is:

$$\text{Saved cost/Original cost}^{*}100$$
$$= 170.015/229.36^{*}100 = 74.126\%$$

The applied AHP model is presented in Fig. 1 (Theodorou 2010) (Figs. 2, 3, 4).

Results

Based on the proposed AHP method, three decision-makers have taken their votes for determining their preferences about two alternatives, which are DC and AC

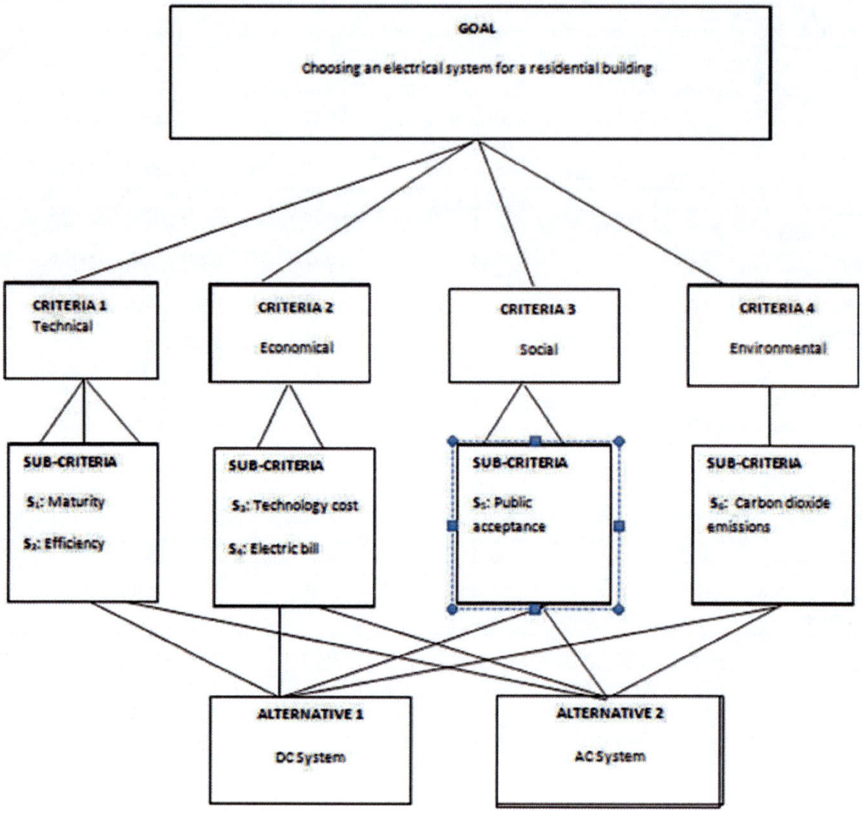

Fig. 1 The proposed AHP method

	S_1	S_2	S_3	S_4	S_5	S_6
A_1	0.1667	0.8333	0.1667	0.8333	0.5	0.5
A_2	0.8333	0.1667	0.8333	0.1667	0.5	0.5

Sub-criteria ranking

$$* \begin{pmatrix} 0.1073 \\ 0.2618 \\ 0.0345 \\ 0.0989 \\ 0.3301 \\ 0.1674 \end{pmatrix} = \begin{matrix} A_1 \\ A_2 \end{matrix} \begin{pmatrix} 0.57295937 \\ 0.42704063 \end{pmatrix}$$

Fig. 2 The first decision maker's voting

	S_1	S_2	S_3	S_4	S_5	S_6
A_1	0.25	0.875	0.875	0.8333	0.875	0.8333
A_2	0.75	0.125	0.125	0.1667	0.125	0.1667

Sub-criteria ranking

$$\cdot \begin{pmatrix} 0.0357 \\ 0.2328 \\ 0.1711 \\ 0.4262 \\ 0.0697 \\ 0.0645 \end{pmatrix} = \begin{array}{|c|} \hline A_1 \\ \hline A_2 \\ \hline \end{array} \begin{pmatrix} 0.83222531 \\ 0.16777469 \end{pmatrix}$$

Fig. 3 The second decision maker's voting

	S_1	S_2	S_3	S_4	S_5	S_6
A_1	0.125	0.1667	0.875	0.8333	0.875	0.875
A_2	0.875	0.8333	0.125	0.1667	0.125	0.125

Sub-criteria ranking

$$\cdot \begin{pmatrix} 0.2772 \\ 0.2144 \\ 0.0966 \\ 0.3214 \\ 0.0234 \\ 0.0670 \end{pmatrix} = \begin{array}{|c|} \hline A_1 \\ \hline A_2 \\ \hline \end{array} \begin{pmatrix} 0.5018381 \\ 0.4981619 \end{pmatrix}$$

Fig. 4 The third decision maker's voting

systems. At the end of this technique, the same result has been reached for the three decision-makers separately. The first result for the first decision-maker is:

$$\begin{pmatrix} 0.57295937 \\ 0.42704063 \end{pmatrix} = \begin{array}{|c|} \hline \text{DC} \\ \hline \text{AC} \\ \hline \end{array}$$

It means that the DC system, which has a percentage of 57.3, is much preferable than the AC system, which has a percentage of 42.7.

The second result for the second decision maker is:

$$\begin{pmatrix} 0.83222531 \\ 0.16777469 \end{pmatrix} = \begin{array}{|c|} \hline \text{DC} \\ \hline \text{AC} \\ \hline \end{array}$$

It means that the DC system, which has a percentage of 83.22, is much preferable than the AC system, which has a percentage of 16.78.

The third result for the third decision maker is:

$$\begin{pmatrix} 0.5018381 \\ 0.4981619 \end{pmatrix} = \begin{array}{|c|} \hline DC \\ \\ \hline AC \\ \\ \hline \end{array}$$

It means that the DC system, which has a percentage of 50.2, is much preferable than the AC system, which has a percentage of 49.8.

Totally, these three separate results create a consensus in terms of the experts' decision. As a result of the AHP method, it can be said that the DC system is much preferable than the AC system.

Discussion and Conclusion

In the presented study, based on a scenario that a residential building with conventional AC devices turns into a residential building generating its electricity as a direct current from solar PV panels and using it directly by DC electronic devices without converting it to AC, potential advantages such as energy-saving and low electric bill and potential problems such as maturity problem, high technology cost, and low public acceptance have been defined. Thanks to literature and these defined advantages and problems, criteria and sub-criteria are determined for the proposed AHP method. With this method, three decision-makers have reached the same result that the DC system is much preferable than the AC system.

Direct current can be produced from eco-friendly renewable energy sources such as solar energy, wind energy, and hydropower. Also, it causes a reduction in electricity bills thanks to decreasing power conversions. However, for a long time, alternating current has been used so that electricity can be transmitted to long distances. It has a broad market in terms of its own devices. It has been tested and applied for a long time. Nevertheless, in the future, when detached and nature-friendly houses become widespread, when houses produce their own electricity and when technology cost of DC system gets low by way of a growing market of DC devices and when experts have enough technical information about DC system, the most preferred current system maybe DC, as our decision-makers in the presented study have indicated.

This study may be an inspiration for people who want to study on the topic of direct current in residential buildings. They may benefit from the proposed AHP method with more criteria than the criteria of this study. They may diversify these used criteria in the study, and also, they may benefit from more decision makers than decision-makers of this study.

References

1. Thomas, Azevedo, Morgan (2012) Edison revisited: Should we use DC circuits for lighting in commercial buildings? Energy Policy 45:399-411
2. Bennani, Brown, Gerber, Marray, Vossos (2018) Techno-economic analysis of DC power distribution in commercial buildings. Applied Energy 230:663–678
3. Keles K, Akcin K, Alagöz G (2015) A smart building power management concept: smart socket applications with DC distribution. Electr Power Energy Syst 64:679–688
4. Gerber V, Feng M, Nordman B (2018) A simulation-based efficiency comparison of AC and DC power distribution networks in commercial buildings. Appl Energy 210:1167–1187
5. Azevedo, Glasgo, Hendrickson (2016). How much electricity can we save by using direct current circuits in homes? Understanding the potential for electricity savings and assessing feasibility of a transition towards DC powered buildings. Applied Energy 180:66–75
6. Cordless Car Vacuum Cleaner (2019). https://www.amazon.com/Kensun-Rechargeable-Cordless-Portable-Cleaning/dp/B07FJWRX2Y
7. Cordless Portable Hair Dryer (2020). https://www.alibaba.com/product-detail/Cordless-Portable-Hair-Dryer-Wireless-Compact_62509896276.html?spm=a2700.7724857.normalList.7.45f46c29zEH3DE&s=p&fullFirstScreen=true
8. Free Signal TV Transit 32" 12 Volt DC Powered LED Flat Screen HDTV for RV Camper and Mobile Use (2018). https://www.amazon.com/Free-Signal-TV-Transit-Powered/dp/B07CBSGR5F
9. Glasgo, Azevedo, Hendrickson (2018) Expert assessments on the future of direct current in buildings. Environmental Research Letters 13
10. New design 12V DC solar power washing machine (2020). https://www.alibaba.com/product-detail/New-design-12v-DC-solar-power_60544795634.html?spm=a2700.7724857.normalList.11.10753762bicdHU
11. RAGGIE DC 12V Upright Solar Refrigerator 125L Solar Fridge (2020). https://www.alibaba.com/product-detail/RAGGIE-DC-12V-Upright-Solar-Refrigerator_60839679095.html?spm=a2700.7724857.normalList.54.14d03c48vsfOhR
12. Sıemens (2020). https://www.siemens-home.bsh-group.com/tr/urun-listesi/arsiv/KG57NP00NE#/Tabs=section-technicalspecs/Togglebox=-1816749319/Togglebox=469219655/Togglebox=469222032/Togglebox=892725271/Togglebox=204286259/
13. Solar/DC Air Conditioner (2016). https://www.hotspotenergy.com/DC-air-conditioner/Specs-DC48.pdf
14. Theodorou, Florides, Tassou (2010) The use of multiple criteria decision-making methodologies for the promotion of RES through funding schemes in Cyprus a review. Energy Policy 38:7783-7792
15. Vatan COMPUTER (2019). https://www.vatanbilgisayar.com/samsung-ue40f6340-40102-cm-3d-smart-led-tv200-hzhd-uydu-alicisi.html
16. Volt coffee makers (2018). https://www.amazon.com/1000ML-Travel-Kettle-Electric-Heater/dp/B07L3T1B48/ref=sr_1_13?keywords=12+volt+coffee+makers&qid=1585760690&sr=8-13.
17. 12V DC electric IRON (2020). https://www.alibaba.com/product-detail/12V-DC-electric-IRON-SPRAY-IRON150W_60798686471.html?spm=a2700.7724857.normalList.17.197368d2pgvpQW

Analysis of the Relationship Between Organizational Justice and Job Satisfaction in the Airline Industry

Elif Okan and Cahit Ali Bayraktar

Abstract The purpose of this study is to determine the relationship between organizational justice and job satisfaction. For this purpose, three main components of organizational justice, which are distributive, procedural, and interactional justice and three main components of job satisfaction, which are internal, external, and overall satisfaction, have been analyzed. Organizational justice means employees' perception of job applications in workplaces, and recently it has been seen as a critical term that can produce valuable results for employees and organizations. Though there have been many studies in international literature about the relationship between organizational justice and job satisfaction, there is only a limited number of studies in the national literature. To the best of our knowledge, this is the first study that searches the relationship between organizational justice and job satisfaction in the airline industry. For this purpose, we conducted a survey involving 101 pilots. As a result of statistical analysis and data interpretation, a positive relationship was found between organizational justice and job satisfaction. The highest level of relationship was found between procedural justice and external job satisfaction.

Keywords Organizational justice · Distributive justice · Procedural justice · Interactional justice · Job satisfaction · People management · Human resources management

E. Okan
Graduate School of Science Engineering and Technology, Istanbul Technical University, Istanbul, Turkey
e-mail: okane@itu.edu.tr

C. A. Bayraktar (✉)
Industrial Engineering Department, Management Faculty, Istanbul Technical University, Istanbul, Turkey
e-mail: cahitali@itu.edu.tr

F. Calisir (ed.), *Industrial Engineering in the Internet-of-Things World*,
Lecture Notes in Management and Industrial Engineering,
https://doi.org/10.1007/978-3-030-76724-2_27

Introduction

Employees need to have job satisfaction in order to contribute to the profitability and efficiency goals of the firms that they work for. The factors that affect job satisfaction have been categorized into three different sub-topics as internal, external, and overall satisfaction [1]. Moreover, it has been stated that the characteristics of the job, salary, improvement, and development opportunities' and working conditions also have a direct effect on job satisfaction [2]. Organization justice refers to the employee's perception of the means used in the work environment. Organization justice has been studied in the areas of human resources and organizational behavior, and it has a major role in performance assessment, career management, and performance management [3]. Organizational justice is vital for the quality of work-life and needed for building a healthy work environment.

Organizational Justice

According to Al-Zu'bi [4], organizational justice includes the employee's perception of fairness related to social and economic interactions in the organization that they work for and employee's relationship with their managers and colleagues. Organizational justice is classified into three different sub-categories, which are distributive, procedural, and interactional justice.

Distributive Justice

According to Çakmak [5], distributive justice refers to the perception of fairness towards the distribution of organizational outputs such as rewards, punishment, and promotion. Outputs can be distributed based on equality, need, or contribution, and employees can perceive distributional fairness by comparing their shares with others. According to Özdevecioglu [6], essential of distributive justice is the employee's belief that they get their shares fairly. However, the importance of the procedure used in the distribution has changed the focus from distributive justice to procedural justice.

Procedural Justice

According to Özler [7], procedural justice is the level of perception about the procedure and policies used in the determination of organizational elements such as salary, promotion, working conditions, and performance assessment. According to Keklik and Coskun [8], while distributive justice is about the results of distribution, procedural justice is more related to the means of distribution. Therefore, transparency of the distribution process, the behavior of the participants, and the reliability of the authority who makes the decision become extremely important in improving the perception of procedural justice. According to Al-Zu'bi [4], there is a positive relationship between the perception of fairness of organizational process and procedures and the feeling of job satisfaction. The employees who have the perception of distributive justice feel more motivated to accept the solutions required by procedures and processes, and their attitudes are likely to be more positive.

Interactional Justice

Interactional justice refers to the fairness level of attitudes and behaviors of the managers when they apply the procedures and make them accepted among employees. Interactional justice is more about the relationships among individuals during operations [9], s.40). Greenberg gave a new dimension to the interactional justice with his study in [10] and broke it down into two sub-components as interpersonal justice and informational justice. Interpersonal justice is the reflection of courtesy, respect, and dignity, which is shown by managers to their subordinates, and it is related to the distribution of outcomes and information and explanations given to the employees during the distribution process.

Job Satisfaction

According to Locke, job satisfaction is described as the positive emotions which occur as a result of work-related activities. In other terms, job satisfaction or dissatisfaction means the general attitudes and thoughts of the employees directed to their jobs, colleagues, and work environments ([11], p.186).

The meaning of the job which maintains people's life by meeting the social, economic, and psychological needs determines satisfaction that they get from their job. Additionally, physical (noise level, air conditioning, lightning, crowdedness), individual (amount of responsibility, workload, preciseness on roles descriptions, etc.), inter-personal (the relationship between superiors and colleagues) and corporate factors (ambiguousness, administrative policies, too little or too much hierarchy) in a work environment also affect the rises and falls on job satisfaction [12].

According to the researches in the literature, job satisfaction is classified into three dimensions as internal, external, and overall satisfaction [13]: 129).

Internal Satisfaction

Internal satisfaction is about the negative or positive attitudes of an individual towards his/her job. Meeting the quality objectives or applying new techniques successfully in the job is expected to increase the internal satisfaction of an employee [14]. Employees who have internal job satisfaction think that they have the power to affect the happenings that affect their lives, and they take the responsibility of their lives [15], p. 531). The factors that provide internal job satisfaction can be list as follows: the job itself, scope, degree of freedom it gives, the importance of the job, relationship with the management, taking responsibility, opportunities for creativity, consistency of job with the talent and competencies that employee has, opportunities for growth and improvement (Bektas 2012, p. 630).

External Satisfaction

External satisfaction arises according to the outer factors which affect job satisfaction. Determination, desire, individual talents, and skills are not self enough to form external satisfaction. Physical work conditions, salary, reward and punishment systems, job security, being part of a team, relations with superior, manager and colleagues are some of the factors which affect the external satisfaction of an employee (Bektas 2012, p. 631).

One of the differences between employees who have internal job satisfaction other than external job satisfaction is the tendency to search for information about their work environment. Employees who have internal job satisfaction are likely to feel the need more to search for information about their environment and to claim their rights in social events compared to the employees who have external job satisfaction [15], p. 531).

Overall Satisfaction

Overall satisfaction includes both internal satisfaction and external satisfaction, which affect employees' lives. It is expected that an employee who has a high degree of internal and external satisfaction should have a high degree of overall satisfaction [8], p.145).

The Relationship between Organizational Justice and Job Satisfaction

According to several studies in the literature, a positive relationship has been determined between organizational justice and job satisfaction. According to Schappe [16], Colquitt [17], and Bakhshi [18], distributive justice has an important effect on job satisfaction. Keskin et al. [19] also studied the relationship between distributive justice and job satisfaction and found that when employees believe that outputs are distributed fairly, their job satisfaction increases, and vice versa, when they perceive unfairness in the distribution process, their job satisfaction and commitment decrease. Congruently, Hodgetts [20], p. 88) states that one of the most important factors that determine job satisfaction are the material and non-material benefits that one gets in return for his/her work. He also states that employees who think they are paid less than their effort will be dissatisfied with their job. According to Bingöl ([21], p. 267–271), who studied the factors that affect job satisfaction, satisfaction is proportional to the material benefits that one gets from his/her job. An employee becomes satisfied in the level of salary or other monetary payments that s/he gets in return for his/her work effort.

According to another research, Yaghoubi et al. [22] determined a positive relationship between procedural justice and job satisfaction. Keskin et al. [19] also stated that when the organizational process and procedures are perceived fairly, employees' satisfaction increases, and they tend to accept the solutions more willingly and to act in a more positive manner about the organization. In other research, it has been determined that job satisfaction increases when employees become a part of the decision-making process [23].

Cropanzano et al. [24] p. 38 studied interactional justice as a phenomenon that determines how individuals behave each other, and he stated that if one share his/her information properly and interacts with others gently, s/he has interactional justice attitude. According to Bingöl who has studied the factors which effect job satisfaction, one of the factors is about the attitudes of the superiors ([21], 267–271). For example, positive attitudes of a superior towards his/her subordinates and the positive relationship between them have a direct effect on job satisfaction of the employee.

Considering the relationships between managers and their subordinates as an external factor that affects job satisfaction, it can be stated that there is a positive relationship between interactional justice and external satisfaction [20], p. 89.

Research Model and Hypotheses

The aim of this study is to determine the relationship between organizational justice and job satisfaction. There are 3 dependent and 3 independent variables in this research. While distributive justice, procedural justice, and interaction justice are the independent variables of the research, internal job satisfaction, external and overall job satisfaction are the dependent variables.

The relationship between those dependent and independent variables are shown in the following model (Fig. 1). The relationship between three dimensions of organizational justice and job satisfaction and the hypotheses which have formed regards to this frame explained in order.

Distributive justice is one of the essential components of organizational justice. Keskin et al. [19], who study the relationship between distributional justice and job satisfaction, stated that employees' job satisfaction increases when they believe that organizational outputs are distributed fairly. According to Al-Zu'bi [4], and Yaghoubi et al. [22], there is a positive relationship between distributional justice and job satisfaction.

Therefore, in the light of theoretical information and application results above, the following hypotheses have been formed:

H1: There is a positive relationship between distributive justice and job satisfaction.

H2: There is a positive relationship between distributive justice and internal job satisfaction.

H3: There is a positive relationship between distributive justice and external job satisfaction.

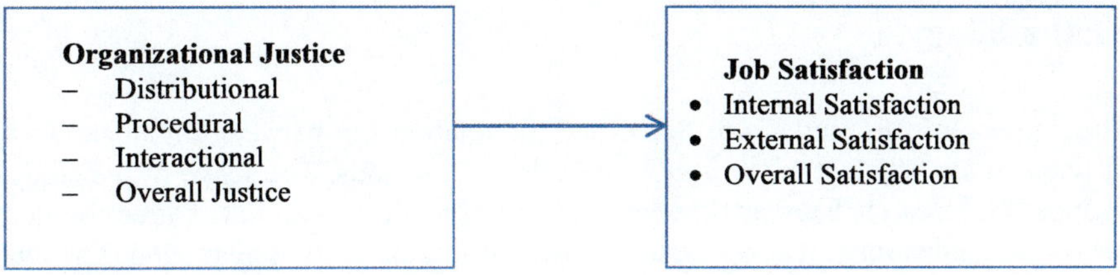

Fig. 1 Research model

Procedural justice is assessed as another essential component of organizational justice and analyzed as an important factor that affects the job satisfaction of employees.

According to Judeh ([25], p. 216), procedural justice is a component of organizational justice, which refers to the fairness of the processes and procedures that are used in the creation of outputs. Özgen and Yalçin state that contradictory and ambiguous organization and management rules have a negative effect on job satisfaction. Therefore, it can be stated that efficient policies and rules, as well as the fair application of them, can increase job satisfaction.

In the frame of the information and explanations above, the following hypotheses can be formed about the relationship between procedural justice and job satisfaction:

H4: There is a positive relationship between procedural justice and job satisfaction.
H5: There is a positive relationship between procedural justice and internal job satisfaction.
H6: There is a positive relationship between procedural justice and external job satisfaction.

Interactional justice is another essential component of organizational justice, and it has effects on job satisfaction of employees.

Özgen and Yalcin stated that having colleagues in the workplace who are supportive of their colleagues to reach business goals and objectives increase job satisfaction [26], p. 357). Reaching goals, achieving success, and having relations with management are other factors that increase internal job satisfaction.

In the light of theoretical information and findings above, the following hypotheses have been formed:

H7: There is a positive relationship between interactional justice and job satisfaction.
H8: There is a positive relationship between interactional justice and internal job satisfaction.
H9: There is a positive relationship between interactional justice and external job satisfaction.

Methodology

The purpose of this study is to determine the relationship between organizational justice and job satisfaction by applying statistical methods. The scope of the study includes the research that seeks knowledge on whether the perception of three components of organizational justice provides any contribution to job satisfaction. The job satisfaction term has been discussed in three dimensions as internal, external, and overall satisfaction.

The population of the study consists of an organization that operates in the airline industry in Istanbul and the pilots who work for this organization. In the scope of this research, 160 pilots have been surveyed, and 101 valid data has been collected.

The sample consisted of 16.4% female and 83.6% male pilots. Among the participants, 20% are between ages 20–29, 51% are aged 30–39, 22% are aged 40–49, 7% are aged 50 or older. Also, 37% of the participants' fleet type is Boeing, and 63% are Airbus. 2% of the participants are training pilots, 25% are chief commander, 73% is the first pilot. The work experience of the participants is as follows: 30% lower than 2 years, 28% between 3-5 years, 27% between 6–10 years, 15% above 10 years. The flight hours of the participants are as follows: 21% lower than 1000 h, 42% between 1000–4000 h, 29% between 4000–10000 h, 8% above 10,000 h.

In the scope of the study, the questionnaire method has been applied in physical forms. Personal information form, organizational justice scale, and job satisfaction scale have been used.

The first part of the questionnaire includes personal questions in order to gather demographic characteristics such as gender, marital status, years of experience, and flight hours.

The perception of organizational justice of the flight crew has been assessed in 3 different justice dimensions, which have been included in Robert Moorman's 20 propositional organizational justice scale.

Minnesota Job Satisfaction Survey (long form), which was developed by Weiss and others, is used in order to measure job satisfaction. The scale measures not only internal job factors related to psychological needs such as recognition, taking responsibility, achievement, and development but also external factors related to the workplace such as salary, control, promotion, and working conditions. Minnesota Job Satisfaction Survey-long form consists of 100 questions that cover 20 criteria related to organizational justice, and each of them is represented with 5 questions. By assessing the whole of the scale, the most efficient 11 factors in the airline industry are determined, and 3 questions for each factor have been included in the study. According to Lee et al., the most related factors which affect the flight crew's job satisfaction are the social dimensions of the job, such as success, development, organizational procedures and applications, salary, responsibility, and role and giving service (2012, p.164).

Analysis

Regarding the used data set, the asymmetry and kurtosis values for the entire survey must be in the range of $\pm$1,96 for 0.01 significance [27], and the values from the set are found to be in this range.

Papacharissi and Rubin [28] indicate that Cronbach's Alpha must be between 0.6–1. When the organizational justice and organizational commitment dimensions were tested for confidence between themselves, all dimensions were found to have Cronbach's Alpha values above 0.6.

Hair et al. [27] state that to be able to perform factor analysis on a sample, the sample size must be at least $n = 100$. This study is suitable for factor analysis with

Table 1 Reliability test related to sub-dimensions of organizational justice

Sub-dimensions of organizational justice	Cronbach's alpha	KMO	% Variance
Distributional justice	0.862	0.785	61.2
Interactional justice	0.907	0.801	50.41
Informational justice	0.892	0.806	75.85
Workload justice	0.939	0.5	94.2
Overall justice	0.873	0.738	79.77

*p < 0.001

Table 2 Reliability test related to sub-dimensions of job satisfaction

Sub-dimensions of satisfaction	Cronbach's alpha
External satisfaction	0.883
Internal satisfaction	0.809
Overall satisfaction	0.796

*p < 0.001

its sample consisting of 101 subjects. In the factor analysis, it was confirmed that the correlation matrices were above 0.30, and the anti-image matrices were above 0.50.

Reliability analysis of the scale has been done according to the subgroups of organizational justice. Cronbach Alpha reliability coefficients and factor analysis results are summarized in Table 1.

Reliability analysis of the scale has been done according to the subgroups of satisfaction. Cronbach Alpha reliability coefficients and factor analysis results are summarized in Table 2.

In the light of the information above comes from the literature and as a result of principal component analysis, four subgroups have been formed: organizational rules and policies and interpersonal relations from hygiene factors [29], job security and responsibility from motivators factors and power motivator from Mc Clelland's theory of needs.

Fifteen questions which belong to external satisfaction will be represented with 4 sub-factors. When motivational theories are assessed, it is determined that those 4 sub-factors are related to both Herzberg's theory and Mc Clelland's theory of needs (Table 3).

Table 3 Sub dimensions of external satisfaction

Sub-dimensions of external satisfaction	KMO	% Variance
Interpersonal relations	0.667	64.336
Organizational rules and policies	0.701	69.840
Job security	0.678	58.006
Responsibility	0.700	58.089

*p < 0.001

Table 4 Sub dimensions of internal satisfaction

Sub-dimensions of internal satisfaction	% Variance	Cumulative %
Salary and security	39.117	39.117
Growth and development	14.127	53.244

*p < 0.001, KMO 0.778

Table 5 Sub dimensions of interactive justice

Sub-dimensions of interactional justice	% Variance	Cumulative %
Informational justice	50.410	50.410
Workload justice	31.939	82.350

*p < 0.001, KMO 0.801

Twelve questions which belong to internal satisfaction will be represented with 2 sub-factors (Table 4). Those factors are related to growth and development, which are in the motivators group in Herzberg's theory, and salary and security, which are in the hygiene factors group in Herzberg's theory.

Two questions that belong to overall satisfaction are reduced into one factor. Six questions belong to distributive justice are reduced into one factor. Seven questions belong to interactional justice are reduced to 2 sub-factors (Table 5). 4 questions belong to procedural justice, and 3 questions belong to overall justice have been reduced to a single factor.

Results

The data collected via the survey method have been analyzed statistically with 't-test,' 'one-way variance analysis (ANOVA),' 'Pearson correlation analysis and regression analysis using SPSS 23.0.

According to the result of ANOVA analysis, for 'Organizational rules and policies, 'Interactions between people,' 'Job security,' 'Monetary benefits and status' dimensions of job satisfaction there is a meaningful difference between age and pilot's procedural justice perception.

Pilots' level of satisfaction is lower among pilots with 20–29 years and higher among pilots with other years (30–39 years, 40–49 years, and 50–59 years) for 'Organizational rules and policies' and 'Monetary benefits and status' factors. For 'Interactions between people' and 'Job security' factors, pilots' level of satisfaction is higher among the pilots with 30–39 years compared to the pilots among 20–29 years.

According to the Post hoc test results, there is not a meaningful difference between age and dimensions' of organizational justice ($p>0.05$). According to the Tamhane test, there is a significant difference ($p<0.05$) between procedural justice and age.

Pilots' level of procedural justice perception is lower among the pilots with 20–29 years and higher among pilots with 30–39 and 40–49 years.

There is a significant relationship ($p<0.05$) between organizational justice and overall satisfaction with the position. The overall satisfaction level of pilots' who have a captain or first officer position is higher than pilots with the Instructor position. However, there is not a meaningful relationship ($p>0.05$) between captain pilots and first officers for the position variable.

The ANOVA test did not show any meaningful difference between groups with different work experiences, fleet types, and gender in terms of organizational justice and job satisfaction ($p > 0.05$).

According to the Tamhane test result, there is a significant relationship between procedural justice and flight hours. The pilots who fly 1000–4000 h have a higher degree of procedural justice perception than the pilots who fly less than 1000 h. The effect of correlation between dependent and independent variables is shown in Table 6.

The results prove that there is a significant positive correlation between all dependent and independent variables. Distributional Justice, Informational Justice, Workload justice, and Procedural Justice are positively correlated with all the factors of job satisfaction. When pilots' organizational justice perceptions increase, their job satisfaction level rises as well, and also, when their job satisfaction level increases, their organizational justice perceptions rise too.

The value of the correlation coefficient between procedural justice and interpersonal relationships is 0.726, which is the highest correlation among all relationships. Similarly, the value of the correlation coefficient between and interpersonal relationships is 0.624, which indicates a strong correlation.

Table 6 Correlation analysis results

	Distributional justice	Informational justice	Workload justice	Procedural justice
Responsibilty	0.615**	0.516**	0.389**	0.539**
Organizational rules and polices	0.599**	0.442**	0.434**	0.705**
Interpersonal relations	0.553**	0.624**	0.225*	0.726**
Job security	0.643**	0.609**	0.383**	0.624**
Growth and development	0.388**	0.416**	0.298**	0.352**
Salary and security	0.600**	0.458**	0.396**	0.468**
Overall satisfaction	0.636**	0.556**	0.251*	0.669**

**Correlation is significant at the 0.01 level (2-tailed)
*Correlation is significant at the 0.05 level (2-tailed)

Table 7 Regression analysis results

Variable regressed	Beta coefficients	t value	Sig.	Tolerance	VIF
(Constant)	0.041	0.476	0.636		
Distributive justice	0.491	5.767	0.000	1.000	1.000
R	0.522				
R square	0.272				
Adjusted R Square	0.264				
Durbin-watson	1.929				
F	33.26				
Sig.	0.000				

a Predictors: (Constant), Distributive Justice
Excluded Variables: Informational Justice, Workload Justice, Procedural Justice
b Dependent Variable: Overall Satisfaction

In order to further explain these correlational findings, a regression analysis was conducted for seven factors to examine the relationship between sub-dimensions of organizational justice and job satisfaction of pilots. The Durbin Watson coefficient calculated for each regression analysis has a value of between 1.5–2.5, which indicates that there is no autocorrelation. ANOVA tests have a significant result ($p = 0.000$), which means the regression models are significant for each factor. VIF and Tolerance value are used to detect multicollinearity. Variance inflation factor (VIF) values less than 10, and tolerance value higher than 0.10 indicates the absence of multicollinearity. VIF values less than 10, and the tolerance values greater than 0.1, which means the independent variables are not highly correlated, and therefore the data set do not have a multicollinearity problem. Only one of the regression analysis result is shown in Table 7 as an example. The correlation coefficient between procedural justice, distributional justice, informational justice, workload justice, and overall satisfaction of 0.522 indicates the strength of the relationship. The R squared of 0.272 or 27.2% is the proportion of variation in the dependent variable overall job satisfaction, as explained by the regression model. The adjusted R squared is the coefficient of determination, which tells the variation in the dependent variable due to changes in the independent variable. There is a 26.4% variation in pilots' overall job satisfaction due to changes in procedural justice, distributive justice, information justice, and workload justice. However, significance values revealed that only distributive justice perception has a meaningful significance ($p>0.05$) that adds support to the model.

For the rest of the factors of job satisfaction, regression analysis results are explained below.

The correlation coefficient between procedural justice, distributional justice, informational justice, workload justice, and overall justice and responsibility factor of job satisfaction of 0.660 indicates the strength of the relationship. The R squared of 0.436 or 43.6% is the proportion of variation in the dependent variable responsibility, as explained by the regression model. The adjusted R square is 40.2%; hence there is a 40.2% variation in responsibility factor due to changes in procedural justice, distributive justice, information justice, and workload justice.

The correlation coefficient between procedural justice, distributional justice, informational justice, workload justice, and overall justice and organizational rules and policies factor of job satisfaction of 0.742 indicates the strength of the relationship. The R squared of 0.550 or 55.5% is the proportion of variation in the dependent variable responsibility, as explained by the regression model. The adjusted R square is 52.3%; hence there is a 52.3% variation in responsibility factor due to changes in procedural justice, distributive justice, information justice, and workload justice.

The correlation coefficient between procedural justice, distributional justice, informational justice, workload justice, and overall justice and interpersonal relations factor of job satisfaction of 0.746 indicates the strength of the relationship. The R squared of 0.556 or 55.6% is the proportion of variation in the dependent variable responsibility, as explained by the regression model. The adjusted R square is 52.9%; hence there is a 52.9% variation in interpersonal relations factor due to changes in procedural justice, distributive justice, information justice, and workload justice.

The correlation coefficient between procedural justice, distributional justice, informational justice, workload justice, and overall justice and job security factor of job satisfaction of 0.712 indicates the strength of the relationship. The R squared of 0.507 or 50.7% is the proportion of variation in the dependent variable responsibility, as explained by the regression model. The adjusted R square is 47.8%; hence there is a 47.8% variation in job security factor due to changes in procedural justice, distributive justice, information justice, and workload justice.

The correlation coefficient between procedural justice, distributional justice, informational justice, workload justice, and overall justice and growth and development factor of job satisfaction of 0.479 indicates the strength of the relationship. The R squared of 0.229 or 22.9% is the proportion of variation in the dependent variable responsibility, as explained by the regression model. The adjusted R square is 18.3%; hence there is an 18.3% variation in growth and development factor due to changes in procedural justice, distributive justice, information justice, and workload justice.

The correlation coefficient between procedural justice, distributional justice, informational justice, workload justice, and overall justice and monetary opportunities and status factor of job satisfaction of 0.695 indicates the strength of the relationship. The R squared of 0.483 or 48.3% is the proportion of variation in the dependent variable responsibility, as explained by the regression model. The adjusted R square is 45.3%; hence there is a 45.3% variation in monetary opportunities and status factor due to changes in procedural justice, distributive justice, information justice, and workload justice.

Discussion and Conclusion

The results obtained from this study showed that there is not a meaningful difference between groups with different work experiences, fleet types, gender, and position in terms of organizational justice perception. This finding is in line with the findings from earlier research for different sectors, which states there is not a significant difference between groups with work experiences, gender, and position in terms of organizational justice perception. Since there is no research found in the aviation industry to the best of our knowledge, it can not be stated any discussion or conclusion about the fleet type.

The relationship between age and 'Organizational rules and policies,' 'Monetary opportunities and status' factors, and procedural organizational justice perception of pilots in this study appear to be significant. It might be explained with the generation differences. The pilots between 20–29 years are also called Y generations and have started their first job; hence they have high expectations but the low commitment and low satisfaction. These findings are in line with the findings from previous research, which suggest that employees' satisfaction level rises as they become older. However, the finding of this research about procedural organizational justice perception is different from prior research, which suggests there is not a significant relationship between components of organizational justice perception and age. The results of this study about distributional and interactional justice are in accord with the earlier research.

There is a significant relationship between procedural justice and flight hours. The pilots who fly 1000–4000 h have a higher degree of procedural justice perception than the pilots who fly less than 1000 h. This finding is in accord with the findings of procedural justice perception, which is higher in pilots who are 30–40 years of age than those who are 20–29 years of age. The results of age and flight hours variables show consistency.

The results of the correlation analysis revealed that the correlation has differed between 23–73% between all factors of job satisfaction and sub-components of organizational justice. The value of the correlation coefficient between procedural justice and interpersonal relationships is 0.726, which is the highest correlation among all relationships. Similarly, the value of the correlation coefficient between and interpersonal relationships is 0.624, which indicates a strong correlation. Informational justice is related to the distribution of outputs and the information and explanations given to the employees about these distributions. Informational justice has a direct relationship with procedural justice. This finding is in line with the findings from earlier research. In other words, when the pilots' perception of informational and procedural justice increases, their job satisfaction coming from interpersonal relations also increases.

The preparation of crew flight programs fairly and determination of resting duration according to the limits stated by international authorities and regulations is vital in the airline industry in terms of justice perception of the crew. Hence, workload justice is crucial for the physical and mental health of the crew, and it has a positive and significant relationship with organizational rules and policies with a value of 43.4%.

The results of the regression analysis revealed that there is a positive relationship between 4 sub-dimensions of organizational justice and each factor of job satisfaction. The results of this research are in accordance with the past research done in other industries. The reason for the lower variation (18.3%) in growth and development factor due to changes in procedural justice, distributive justice, information justice, workload justice, and overall justice might be explained by monotonous characteristics of flight operations. Pilots who fly the same type of aircraft on similar routes into similar destinations might feel less satisfied in terms of growth and development due to the monotonous tasks.

In conclusion, most of the results of the current study were consistent with previous research. Moreover, it added value to the literature on organizational justice and job satisfaction since there is a lack of study in the airline industry in the existing literature. On the other hand, although this study includes many vital insights about the perceptions of pilots about organizational justice, the number of sample space can be stated as a limitation of the study and therefore a recommendation for the future researchers is to improve the application of this study by replicating these results using larger samples.

References

1. Özkalp (1996) Örgütsel Davranış, Ankara: Anadolu Üniversitesi Yayını
2. Balcı A (1985) Eğitim Yöneticisinin İş Doyumu. Yayımlanmamış Doktora Tezi, Ankara Üniversitesi
3. Gelens J, Dries N, Hofmans J, Pepermans R (2013) The role of perceived organizational justice in shaping the outcomes of talent management: a research agenda. Human Resource Manage Rev 1–13
4. AL-ZU'BI HA (2010) A study of relationship between organizational justice and job satisfaction. Int J Business Manage 5(12)
5. Çakmak KÖ (2005) Performans Değerlendirme Sistemlerinde Örgütsel Adalet Algısı ve Bir Örnek Olay Çalışması. İstanbul Üniversitesi, Sosyal Bilimler Enstitüsü, Yüksek Lisans Tezi, İstanbul
6. Özdevecioğlu M (2003) Algılanan Örgütsel Adaletin Bireylerarası Saldırgan Davranışlar Üzerindeki Etkilerinin Belirlenmesine Yönelik Bir Araştırma. Erciyes Üniversitesi İİBF Dergisi (21):ss 77–96
7. Özler ED (2012) Örgütsel Vatandaşlık Davranışı. (Edt: Derya Ergun Özler) Örgütsel Davranışta Güncel Konular, Bursa: Ekin Basım Yayın Dağıtım
8. Keklik B, Coşkun Us N (2013) Örgütsel Adalet Algılamalarının İş Tatminine Etkisi: Hastane Çalışanları Üzerinde Bir Araştırma. Süleyman Demirel Üniversitesi İktisadi ve İdari Bilimler Fakültesi Dergisi 18(2):143–161
9. Durmuş R (2015) Örgütsel Adalet Ve Tükenmişlik Arasındaki Etkileşimin Değerlendirilmesine Yönelik Bir Araştırma- Eskişehir'de Bir Kamu Hastanesi Örneği (Yayınlanmamış Yüksek Lisans Tezi). Toros Üniversitesi, Sosyal Bilimler Enstitüsü, Mersin
10. Greenberg J (1990) Organizational justice: yesterday, today and tomorrow. J Manage 16(2):s.400
11. Solmuş T (2004) İş Yaşamında Duygular ve Kişilerarası İlişkiler, İstanbul, Beta

12. Yıldırım F (2007) İş Doyumu ile Örgütsel Adalet İlişkisi, Ankara Üniversitesi SBF Dergisi. 62.1:s.253–278
13. Özkalp E, ve Kirel Ç (2001) Örgütsel Davranış, Anadolu Üniversitesi Yayınları, Eskişehir, pp 694 s
14. Açikalin C (2011) "İçsel Ve Dışsal Doyumun Örgütsel Bağlılığa Etkisi: Eskişehir Üretim Sektörü Örneği", Dumlupınar Üniversitesi Sosyal Bilimler Enstitüsü Dergisi, Sayı: 31, Kütahya s. 237–254
15. Özen Kutanis, R. ve Muammer M (2010) Örgütsel Adaletin Çalışanların İş Tatmine Etkisi: Turizm Alanında Eğitim Veren Bir Yükseköğretim Kurumuna Yönelik Bir Örnek Olay Çalışması. Selçuk Üniversitesi İktisadi ve İdari Bilimler Fakültesi Sosyal ve Ekonomik Araştırmalar Dergisi, Cilt: 13, Sayı: 9, Konya, s. 527–552
16. Schappe P (1998) The influence of job satisfaction organizational commitment and fairness perceptions on organizational citizenship behavior
17. Colquitt JA (2001) On the dimensionality of organizational justice a construct validation of a measure. J Appl Psychol 86:386–400
18. Bakhshi A, Kumar K, Rani E (2009) Organizational justice perceptions as predictor of job satisfaction and organization commitment. Int J Bus Manag 4(9):145–154
19. Keskin H, Akgün AE, Günsel A (2008) Örgütsel Adaletsizlik Kavramı Boyutları ve Yönetimi. In: Özdevecioğlu M, Karadal H (eds) Örgütsel Davranışta Seçme Konular içinde. İlke Yayınevi, Ankara, pp 91–106
20. Hodgetts RM (1991) Organizational communication—a managerial perspective, 2nd edn. HarperCollins Publishers, New York
21. Bingöl D (1997) Personel Yönetimi, Beta Yayıncılık, 3.Baskı, İstanbul
22. Yaghoubi E, Mashinchi AS, Ahmad E, Hadi A, Hamid E (2012) An analysis of correlation between organizational justice and job satisfaction. African J Business Manage 6(3):995–1002, Iran
23. Dogan C (2009) Örgüt İkliminin İş Tatmini Üzerine Etkisi: Bir Uygulama, Marmara Üniversitesi Sosyal Bilimler Enstitüsü, Yayınlanmamış Yüksek Lisans Tezi, İstanbul-2009
24. Cropanzano R, Bowen DE, Gilliland SW (2007) The management of organazational justice. Acad Manage Perceptives 34–48
25. Judeh M (2011) Ethical decision making and its relationship with organizational justice. Int J Acad Res 3(4):215–220
26. Özgen H Ve, Yalçin A (2010) İnsan Kaynakları Yönetimi Stratejik Bir Yaklaşım. Nobel Kitapevi, Adana, p 452s
27. Hair J, Anderson R, Tatham R, Black W (1995) Multivariate data analysis. Prentice Hall, New Jersey
28. Papacharissi Z, Rubin AM (2000) Predictors of internet use. J Broadcast Electron Media 44(2):175–196
29. Herzberg F (2003) One more time: how do you motivate employees? Harvard Bus Rev 81(1):86
30. Barrett-Howard E, Tyler TR (1986) Procedural justice as a criterion in allocation decisions. J Personality Soc Psychol 50:296–304
31. Bektaş A (2012) "Yöneticilerin İş Tatmini Ve Motivasyon Düzeylerinin Yaşam Anlamı Düzeyi Üzerindeki Etkisi". (Yüksek Lisans Tezi) İstanbul
32. Dailey RC, Kirk DJ (1992) Distributive and procedural justice as antecedents of job dissatisfaction and intent to turnover
33. Folger R, ve Konovsky MA (1989) Effects of procedural and distributive justice on reactions. Acad Manage J 32(1):s.115–145
34. Ishfaq A, Usman A, Rana SL (2011) Jobs satisfaction mediates relationship between facets of job and citizenship behavior: a study of female employees of banking sector of Pakistan. Info Manage Business Rev (3)4:228–234
35. Leventhal GS (1980) What should be done with equity theory? new approaches to the study of fairness in social relationships In: Gergen KJ, Greenberg MS, Willis RH (ed) Social exchange: advances in theory and research, New York, Plenum, pp 27–55

36. Mcfarlin, ve Sweeney PD (1992) Distributive and procedural justice as predictors of satisfaction with personal and organizational outcomes. Acad Manage J 35:s626–637
37. Robinson KL (2004) The impact of individual differences on the relationship between employee perceptions of organizational justice and organizational outcome variables. California School of Organizational Studies Alliant International University, San Diego, Yayımlanmamış Doktora Tezi
38. Shaw JC, Wıld E, Colquitt JA (2003) To justify or excuse?: a meta-analytic review of the effects of explanations. J Appl Psychol 88(3):s.445
39. Yeşil S, Dereli S (2014) Örgütsel Adalet ve İş Tatmini Üzerine Bir Alan Çalışması. Kahramanmaraş Sütçü İmam Üniversitesi İktisadi Bilimler Fakültesi

Regenerative Supply Chain Through Digitalization in Diary

Peiman Alipour Sarvari, Sebastien-Augustin Martin, Gulcan Baskurt, Mohammad Nozari, and Djamel Khadraoui

Abstract Globally, enterprises are leveraging social media to promote their brands, monitor consumer trends, research new product ideas, drive business growth, and improve business processes. Integrating social media into existing supply chain networks is essential to provide instant access to real user data. This study defines tailored metrics by examining the current supply chain considering the data gathered from social media in order to have a re-designed supply chain based on the requirements defined by end-users alongside utilizing organizations' strategy, technology, process, and evaluation metrics. The target is to define a framework to take full advantage of intelligent automation in retail and consumer feedback for creating efficiency and creativity. As a case study, this study introduces a social data-driven causal analytics-based methodology that reflects Tweeter data for diagnosing supply chain management issues and determining its capabilities in a milk products company in Luxembourg.

Keywords Big data · Natural language processing · Support vector machines · Supply chain management · Sentiment analysis · Agro-food supply chain

---

P. Alipour Sarvari (✉) · S.-A. Martin
Trusted Security Services Unite, ITIS Department, Luxembourg Institute of Science and Technology, Belval, Luxembourg
e-mail: peiman.alipour@list.lu

S.-A. Martin
e-mail: sebastien.martin@list.lu

G. Baskurt
Mangement Department, Faculty of Management, Beykent University, Istanbul, Turkey

M. Nozari · D. Khadraoui
College of Art and Sciences, Drexel University, Philadelphia, USA
e-mail: mn468@drexel.edu

D. Khadraoui
e-mail: djamel.khadraoui@list.lu

F. Calisir (ed.), *Industrial Engineering in the Internet-of-Things World*,
Lecture Notes in Management and Industrial Engineering,
https://doi.org/10.1007/978-3-030-76724-2_28

Introduction

Success in the digital transformation process requires utilizing technologically improved processes to attain better products in the most accurate way [17]. Social Media (SM) refers to internet-based platforms, applications, and services. Social platforms are the most effective methods of bringing together customers with brands in the digital world. SM make it easier to manage all tiers of the internet, which was previously never possible. Customers can report positive or negative feedback, adverse feelings, ideas, and judgments about the product or service they receive to the brands through SM. Besides, the SM's supportive fabrication allows the interpretation to be broadcasted at a predictable speed. SM analytics implies accumulating data from various SM platforms and scrutinizing the data to help decision-makers deal with a specific issue. Nowadays, many companies use social networks in customer relationship management (CRM) and supply chain management (SCM) [2, 23].

Many companies want to enhance their supply chain (SC) to be promoted to a more consumer-oriented enterprise [14]. Thinking about SM that has about 2.5–3 billion users around the globe, reminds popular networking sites like Facebook, Instagram, Twitter, etc. SM has changed the approach businesses interact with their customers. This new approach not only zipped up the breach among customers but also businesses. As the transportation and logistics sector grew, SM craze began to move towards logistics and SCM. It is necessary to operate strategically to consolidate technology and communications to boost efficiency and link with core audiences and stakeholders. Data acquisition, mining, analysis, and sharing are just a few means through which SM can promote SC efficiency [28].

In the past, companies, businesses, and shareholders used to care about primary stakeholders' ideas, but the concerns of secondary stakeholders were mostly neglected. Primary stakeholders always had a dominant impact due to their power and legitimacy. Nevertheless, nowadays, thanks to the dynamism of SM, this is changing. Secondary stakeholders gather knowledge about the company and its practices, location and reveal ways to frame what they have experienced in the context of SM and convey their views to the authorities and peers. Using the SM revolution, secondary stakeholders can accumulate several years' worth of information from an organization. Facebook and Twitter's circle of friends makes it possible to gather more data and transfer them to blogs or websites. Another aspect of the revolution is fostering the process of framing and communication. Messages can reach impactors easier. The third influence of the social technology revolution is its capability to act as mobilizing structures for protest and action groups.

SM could assist SC and logistics management by first improving communication between suppliers and customers through receiving direct feedback leading to improved customer service. Next, SM provides better communication within the entire SC that contributes to the complete management system, increase productivity, and save time and money. Finally, SM networks are a great way to become more visible on digital media and promote business worldwide, reaching potential suppliers in the same or different markets. As an instance, in the logistics and shipping-forwarding industry, SM can add the following advantages:

1. Delivery schedules can be updated quicker; there is a rise in the tracking and visibility of the SC.
2. Freight transport roads covered by SM can receive critical information about accidents, to redirect deliveries.
3. Learning about weather conditions may affect commodity shipment or deliveries.
4. Being informed about the desired trends and other valuable information from customers and industry leaders.
5. Sharing or retrieving data for risk identification and identifying hazards in the SC.
6. Searching and collecting innovative schemes from different ideas to further improve the recruiting policies.
7. Reaching out to new employees, customers, partners, and opportunities.
8. Improving the networking capacity with the prospect of business partners and potential customers. SM provides feedback from your company's customers so you can improve your service.
9. The goal of each business is to make a profit. From the SC perspective, this goal can be achieved by maximizing increased productivity, better customer satisfaction, and lower operational costs.

Several studies have been conducted on SM in the domains and fields like disaster management [15], applications of SM by non-profit organizations [10], operation management [3], assimilating SM into the SCM [18], determining of the ranking of knowledge-creation [26], conventional knowledge management [12], lean thinking tools [27].

The following section introduces a methodology to combine an SM analytics phase with an SC regenerator tool. As a case study, we performed a social data-driven causal analytics-based approach that considered Tweeter data for the identification of SCM issues and capabilities in a milk products company in Luxembourg.

This paper looks at strategies and evaluation metrics, leading to more effective Supply Chain Management (SCM) activities thanks to the information available on Twitter, which reflects the customers' true opinions. The aim is to propose scaled and compliance metrics for establishing and monitoring the best-fitted transition strategies for enterprises and companies. The rest of this paper is organized as follows; Sect. 2 summarizes related works on applications of SM on SCM. Section 3 specifies details of the proposed methodology on spotting and defining the best-fitted evaluation metrics. Finally, Sect. 4 concludes this work and presents recommendations.

Sentiment Analysis on Social Media for Getting Supply Chain Insights

The development of the big data paradigm led to the further development of an existing trend with the availability of larger sets of structured data, for example through the uptakes of UAV and IoT sensors [11], also affecting the agro-food supply chain [8, 9]. Meanwhile, the extraction of meaningful insights from unstructured data, as well as their combination with data structured by design, also got a new wave of studies, linked to the increased availability of data and to the development of machine learning approaches. Unstructured data cover a large range of data, from video or pictures to text; this contribution is focusing on the latter.

Sentiment analysis (SA) is a task among the natural language processing approaches. From the linguistic perspective, it consists in identifying the polarity of a document, or a part of a document, which may be a message such as a Tweet. Most SAs implement a division based on positive, negative, or even neutral polarity, each relevant chain of characters being labeled following this model. Other researches intend to bring a finer-grained description of the sentiments, such as sarcasm or irony, but it is not explored in more detail in this work as it requires a large amount of labeled data to train the models. Affecting a label describing the polarity allows turning of this linguistic issue to a classification problem. In a recent literature review, [7] report that the most common approaches for sentiment analysis on social media are Machine Learning-Support Vector Machine (SVM), Naïve Bayesian, and ensemble methods. Sentiment analysis is covering a large range of use cases and is of paramount importance for research and solutions intending to capture customers' feedback, from restaurant reviews to social media content.

Reflecting this increasing interest, several literature reviews envisioning the contributions related to the analysis of unstructured textual data, including the social media, for the agro-food industry and its supply chain, were published (e.g., [5, 6, 13, 24]). Ref. [6] notes that sentiment analysis is an increasingly important approach in the domain of production and supply chain.

If sentiment analysis, and more generally natural language processing-based methods need to address challenges such as accuracy, granularity, or trust, literature is considering them as suitable to provide insights for various purposes, such as business intelligence or product development. Ref. [6] finds that it may lead to a more efficient and sustainable agro-food supply chain, as it is providing information on food consumption patterns, monitor food crises issues, and reduce food waste, which is an important concern as one-third of the food produces is discarded worldwide. A close related research of this work is the study on the capability of social media to be leveraged to reduce the wastes in the beef flesh supply chain. From a technical standpoint, [19] developed a hierarchical clustering of tweets through a multiscale bootstrap resampling approach.

In her literature review on the place of social media analysis in food innovation, [6] distinguishes between marketing and consumer research, supply chain management, sustainable food systems, food production, and food markets. Ref. [24] provide a similar and more developed framework, against which the contributions of this work are scored, including food safety and food fraud surveillance, consumer-opinion mining, dietary pattern characterization, new-product development, food knowledge discovery, food SCM, and online food services.

Methodology

The next-generation SC is more flexible and digital. To establish interactive trust and transparency, you need to optimize your retail SC and gain efficiency to handle changes and trends at any scale. In order to reach every touchpoint, you need to refine consumers' and customers' real-time insights with discovering and improving demand forecasting and order fulfillment with accurate insights. The following methodology is trying to reinvent consumer experience-based SC.

Figure 1 illustrates the proposed data-driven SC designer that is updating the traditional SC in three phases. The first phase, the information extraction, will turn data into knowledge. The second phase, the virtual experiment-based scenario generation tool, will provide all feasible supply scenarios and responses based on current and possible situations. Finally, the third phase is the tactical and operational support system tool for decision making.

Phase 1: Information extraction:

This engine searches various data sources, such as SM platforms, with different data structures and extracts the most accurate and reliable knowledge. Data mining and data analysis are supported by Artificial Intelligence (AI) and Machine learning (ML) algorithms. As a result of this phase, evaluation metrics are developed in shapes of numerical and categorical factors. This phase is composed of 4 stages:

The first stage recognizes the leading SM platforms such as Twitter. Next stage is data, which handles the ML-based text mining, image mining algorithms to realize the data creation. It prepares raw data by cleaning, categorizing, clustering, and visualizing. The third stage is listing the topic of interest by means of the list of appropriate keywords, including texts and hashtags.

This keywords list is used along with the Twitter streaming application program interfaces (APIs) to aggregate available datasets from the SM platform. These APIs empower data analysts to accumulate a certain percentage of the available SM datasets. The fourth stage of the knowledge extraction tool handles predictive analytics to generate inputs for scenario generation and optimization phases. This stage gets the comment, identifies the metrics, understands the demographics, measures the data legitimacy, identifies risks, diagnoses fault and infers weaknesses, and measures sentiments of people thanks to AI innovation.

Phase 2: The visual experiment tool:

This tool produces diverse types of situations and simulates them in the shapes of different combinations of soft windows constraints, hard windows constraints,

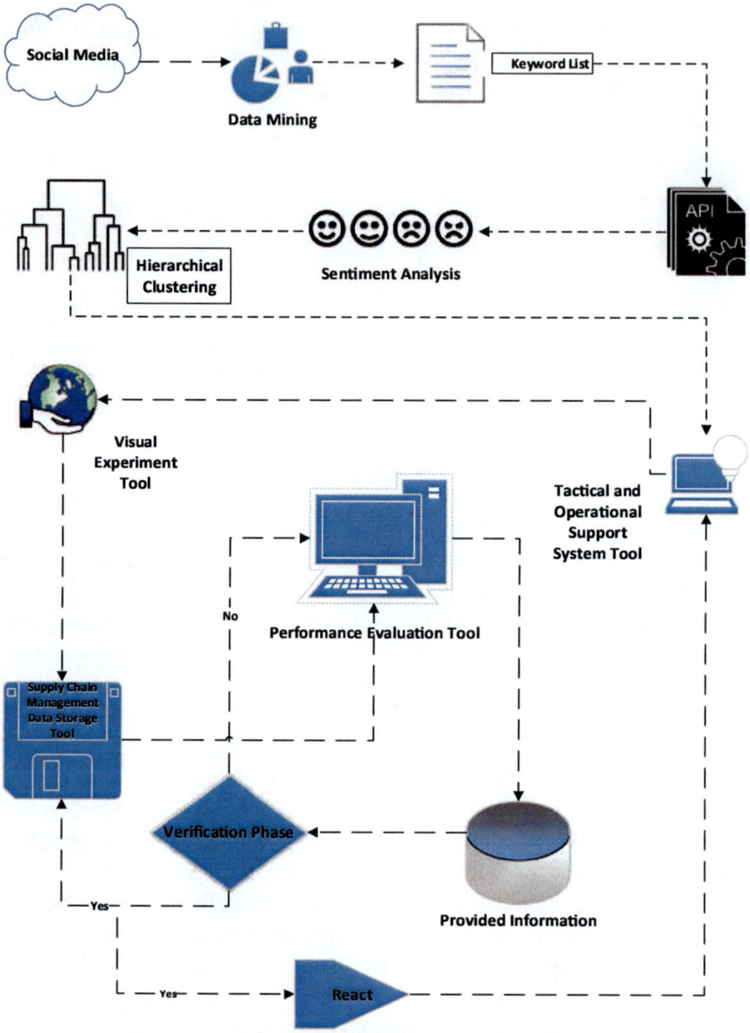

Fig. 1 The proposed methodology framework

and different product associations. The yields of this phase are stored in the data warehouses tool that utilizes file systems and programming models in association with optimization practices.

Phase 3: The tactical and operational support system tool:

This phase is comprised of three consequent and connected stages:

Table 1 Internal and external evaluation metrics of the regenerated SC using dynamic insights extracted from SM [25]

Assessed social performance and reports	Internal reconstruction/reengineering ability	Mutual trust in the SC networks
Conducted audits regarding social/operations/environment issues	Social network relationship ability	Modification flexibility in SC capability
Technical, managerial, and financial assistance to address social issues	Knowledge assessment/knowledge transfer to operational rate	Inbound/Outbound supplier flexibility
Identified possible social issues and prepare to response and response rate/period	SC partner development/collaborations	Information quality in building SC capability
Knowledge acquisition and absorptive capacity	SC re-conceptualization in sustainability	Internal manufacturing flexibility
Market-oriented perception/market growth rate ability	coevolving and Inter-firm information system capacity and usage	Inbound/Outbound logistics flexibility
Innovation ability and innovation	Inter-firm relational competency rate and perception	Mix flexibility in sustainable operations

Stage 1, a web-based optimization tool, is shaped by mathematical modeling-based solvers for SC cycle time, inventory turnover ratio, inventory velocity, and freight cost per unite problems.
Stage 2, the verification tool confirms or modifies the implemented SC model in tactical and strategical levels. This tool is an interface supervised by an expert to facilitate and support the accuracy of processes. Stage 3, the react tool that is feeding back the operation with factors to be used by the scenario creation tool. The output of this methodology from the SM-driven SC-point of view is summarized in Table 1.

Case Study

In order to implement the proposed SM data-based SC regenerator approach, we examine a milk products company's SC seated on consumer feedback. The SM analysis considers either positive or negative sentiments of milk customers and consumers tweeting in Luxembourg. The reasoning analysis will help to improve the company's SC and tackle problems.

The used data has been collected based on a 4-sided bounding approach for filtering tweets by location of Grand Duchy of Luxembourg, for the time range of 2019/02/15-2019/06/15. The collected data is handled and cleansed by CRISP data mining techniques. Python programming language is launched and used data scraping and deep learning tasks. In total, 1857 tweets have been analyzed by scikit-learn 0.23.1 package in three sets of training (70%), testing (20%), and validation (10%). For the performance evaluation of ML, the paper has used roc_auc_score, recall_score, and precision_score or sklearn.

The total number of tweets extracted for this research was 1857. They were captured from 21/05/2019 to 21/06/2019 using the main keywords of 'mil, 'lait,' 'mëllech,' and 'milch.' Table 2 represents some of the extracted keywords via data mining, which was used for sentiment analysis. The tweets were written in French, Luxembourgish, English, and German languages were considered, in the geographic limits of Luxembourg. Figure 2 illustrates the geolocation data on the Luxembourg map.

Table 2 A partial list of extracted keywords via data mining and used for sentiment analysis

#milk	#Esch-sur-Alzette	#Mersch	#Schengen	#Betzdorf	#Wormeldange
#lait	#Differdange	#Strassen	#Frisange	#Kopstal	#Redange
#mëllech	#Dudelange	#Kayl	#Mertert	#Contern	#Bettendorf
#milch	#Pétange	#Ettelbruck	#Habscht	#Rosport	#Beaufort
#price	#Sanem	#Walferdange	#Wincrange	#Remich	#Esch-sur-Sûre
#sour	#Hesperange	#Bertrange	#Rambrouch	#Sandweiler	#Vallée de l'Ernz
#lait	#Bettembourg	#Junglinster	#Dippach	#Parc Hosingen	#Koerich

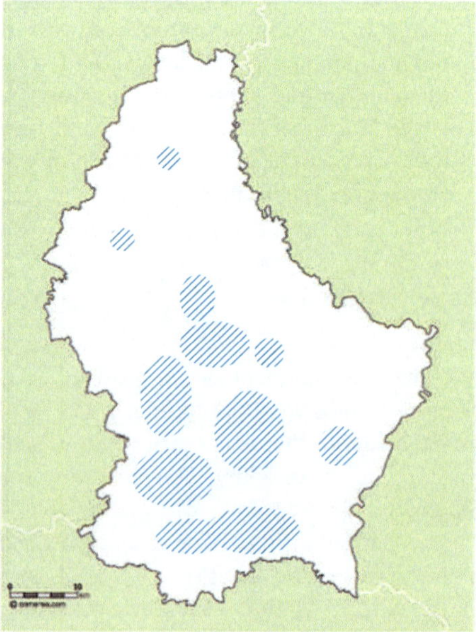

Fig. 2 Visualization of tweets with geolocation data (1857 tweets containing 'milk,' 'lait,' 'mëllech,' and 'milch.'

The set of training data for the Support Vector Machine (SVM) was executed based on sentiments. The training data were acquired by collecting 1374 (from all of the tweets with '#milk, #lait, #mëllech, and #milch') messages from the Twitter data. The self-propelled scoring process was terminated by generating 710 positive, 434 negative, and 70 discarded tweets.

The sentiment analysis based on the analyses on the selected feature sets with threefold cross-validation conducted on the Luxembourg-wise classification of tweets [16]. We used SVM and Naïve Bayes based classifier of scikit-learn v0.21.2. In order to identify meaningful issues and their content in the collected tweets, initially, we implemented sentiment analysis to understand the sentiments of each of the messages.

Understanding the Factors Affecting Milk Consumer Consent and Their Justification Through the SC

Throughout the study on consumer tweets, it is evident that there were diverse problems affecting consumer satisfaction, such as taste, packaging, less information on the packages, extra fat, low fat, color, and smell of milk. In tactical and operational support system tool, an Ishikawa diagram is evaluating the root cause of each issue, the output of this stage feeding the visual experiment tool and for visualizing the actual interrelated effects of SC components. Figure 3 elaborately illustrates the packaging error via an Ishikawa diagram in the tactical and operational support system tool.

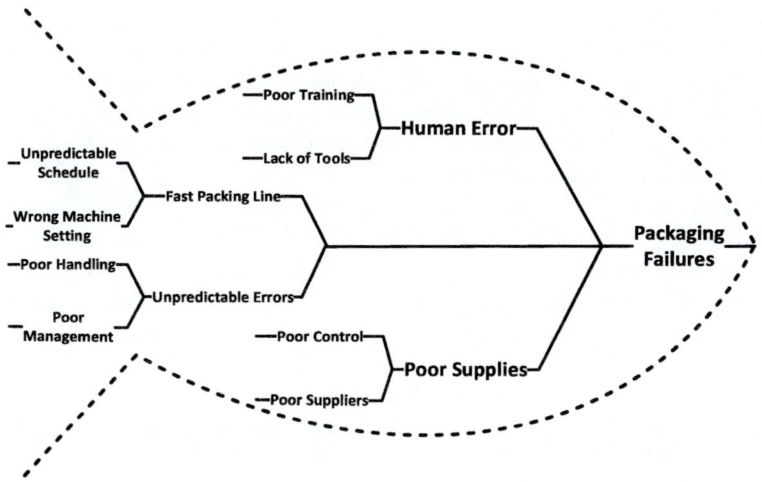

Fig. 3 Ishikawa diagram for packaging problem

Considering the packaging issue, the output of the visual experiment tool is summarizing six important possibilities of packaging failure like (1) Packaging design, (2) Contaminated packaging materials, (3) Sealing failures, (4) Physical defects during handling and shipping, (5) Label failures, and (6) Machine quality. After visualizing the possible reasons for packaging failure, the information is stored in data storage to feed the performance evaluation tool of different chains of the SC to spot the true reason for failure. The output of this stage is a virtually updated SC, which is tackling the problem that occurred during packaging. The provided new generation SC by performance evaluation tool is presented to the experts in the validation phase for the ultimate decision on performing related physical changes on the SC.

Tactical and Strategical Implications

This study contributes to various purposes. It confirms the results of [19] in another branch on the feasibility of applying natural language processing techniques to SC, especially with social media data, as well as their value in identifying products' issues or any customer concern at the downstream stages of the SC. From the sentiment analysis of tweets, [19] identified six root causes of customer dissatisfaction, namely bad flavor and unpleasant smell, traceability issues in beef products, extra fat, discoloration of beef products, hard texture, presence of foreign bodies. If some of these issues are very specific to the beef or meat sector, others are more generic to the agro-food supply chain and call for similar countermeasures, as are showing quite overlapping results. This study points out the specific aspect of packaging for milk-related products.

The relative abundance of geographic anchors should be confirmed on larger datasets. If so, it would be possible to combine different natural language processing tasks, such as named entity recognition [1], to identify the location at its finest degree of granularity, making the data more re-usable to address the potential issues. An important challenge to address is to get a finer description of the temporal information, which is essential for this kind of product.

It is showcasing the value of these data to address the issues arising at the upstream stages, with some limitations related to the inaccuracy of the references made by the customers. In particular, this shows the capability of SA on SM data to proceed descriptive and diagnostic analyses, while it demonstrates its interest to be combined with other structured data in the framework proposed in Fig. 1, for predictive and prescriptive approaches. This framework contributes to a better and seamless validation and integration of SM insights into the overall SC [4, 18].

Although this approach is strongly effective for the SC domain, yet it provides insights directly usable for other purposes mentioned in [24] because monitoring social networks may contribute to food safety monitoring relies directly on consumer-opinion mining and provides information for new product development.

The results of this study may also assist milk producing companies in promoting a customer-centric and dynamically regenerative SC. Through the analysis, we found that consumers were unsatisfied with the high price of milk products, the absence of synthetic milk for vegans, bad smell, lack of "enough" vitamins, and overall milk quality. The main part of milk waste was attributed to the close expiration date, lousy smell, and packaging issues. The milk packaging could be influenced by mishandling while the milk packs flow through the SC or by applying improperly designed packaging methods. Inadequate packaging affects color, taste, smell, and the quality of milk.

Recurrent diagnosis, maintenance, and monitoring of packaging robots and utilizing more sophisticated packaging methods, such as tweaked atmosphere packeting, might assist with targeting the above-mentioned problems. The milk production cost can be reduced by implementing a lean concept to improve the vertical and horizontal coordination within the milk SC. With increasing the coordination in the SC, the waste levels will decrease, which results in the low cost of milk products.

The problems linked to bad scent and taste can be imputed to bad pasteurization process and temperature violation of milk products. The effective cold chain control, all along with the SC, boosting awareness and decent coordination amongst different stakeholders, may resolve this issue. Studies on tweets coming from consumers disclosed that customers, especially the ones from Schuttrange, preferred local milk products better than the other brands. Another conclusion was related to the host farm of milk. Retailers can acquire consumer trust by incorporating the strictly interchained traceability platform (like blockchain technology) within the SC.

There are some shortcomings associated with the empirical analysis performed in this paper. First, larger data samples could be accumulated more extended time intervals to increase the representativeness of the gathered sample. What's more, annotation (using milk companies and retailers' names)-based approach brings in time and resources to manage a relevant examination of the use-case. Third, as most of the Luxembourg community are multilingual and communicate in different languages effectively, they may use various terms for the same problem, and inclusion of equivalent words could occur to high-grade visualization. It also represents a fruitful foundation for cross-lingual transfer learning approaches.

Results

SM is turning to a new platform for consumers to share their true feelings and thoughts openly and clearly. Using SM data, organizations may gain insights into their customers' concerns, needs, and preferences. As the focus of this work, we introduced a methodological approach to regenerate much more compliance SC based on the extracted insights and lessons learned from SM. Furthermore, we defined sophisticated and tailored metrics to examine the performance of a designed SC in strategical and tactical levels. The target was to define a framework to take full benefits of intelligent automation in production and retail for creating efficiency and creativity. As a case study, we performed a social data-driven causal analytics-based approach that considers Tweeter data for the identification of SCM issues and capabilities in a milk products company in Luxembourg. For future work, in order to have a fully automated management system, we will consider an artificial intelligence-based evaluation system for evaluating the regenerated SC components in the verification phase of our methodology.

References

1. Adams B, McKenzie G (2013) Inferring thematic places from spatially referenced natural language descriptions. In: Crowdsourcing geographic knowledge: Volunteered Geographic Information (VGI) in theory and practice, vol 9789400745. Springer, Netherlands, pp 201–221. https://doi.org/10.1007/978-94-007-4587-2_12
2. Alalwan AA, Rana NP, Dwivedi YK, Algharabat R (2017) SM in marketing: a review and analysis of the existing literature. Telematics Inform 34(7):1177–1190
3. Bhattacharjya S, Chandra R, Pareek N, Raverkar KP (2016) Biochar and crop residue application to soil: effect on soil biochemical properties, nutrient availability and yield of rice (Oryza sativa L.) and wheat (Triticum aestivum L.). Arch Agron Soil Sci 62(8):1095–1108
4. Chae B (2015) Insights from hashtag #supplychain and Twitter analytics: considering Twitter and Twitter data for supply chain practice and research. Int J Prod Econ 165:247–259. https://doi.org/10.1016/j.ijpe.2014.12.037
5. Drury B, Roche M (2019) A survey of the applications of text mining for agriculture. In: Computers and electronics in agriculture, vol 163. Elsevier B.V, p 104864. https://doi.org/10.1016/j.compag.2019.104864
6. Kosior K (2019) Social media analytics in food innovation and production: a review. Proc Food Syst Dyn 205–219. https://doi.org/10.18461/pfsd.2019.1921
7. Kumar A, Jaiswal A (2020) Systematic literature review of sentiment analysis on Twitter using soft computing techniques. Concurrency Comput Pract Experience 32(1). https://doi.org/10.1002/cpe.5107
8. Lezoche M, Panetto H, Kacprzyk J, Hernandez JE, Alemany Díaz MME (2020) Agri-food 4.0: a survey of the supply chains and technologies for the future agriculture. In: Computers in industry, vol 117, p 103187. Elsevier B.V. https://doi.org/10.1016/j.compind.2020.103187
9. Min W, Jiang S, Liu L, Rui Y, Jain R (2019) A survey on food computing. ACM Comput Surv 52(5). https://doi.org/10.1145/3329168
10. Muralidharan S, Dillistone K, Shin JH (2011) The Gulf Coast oil spill: extending the theory of image restoration discourse to the realm of SM and beyond petroleum. Public Relat Rev 37(3):226–232
11. Ni D, Xiao Z, Lim MK (2020) A systematic review of the research trends of machine learning in supply chain management. Int J Mach Learn Cybernet 11(7):1463–1482. https://doi.org/10.1007/s13042-019-01050-0
12. O'leary MB, Mortensen M, Woolley AW (2011) Multiple team membership: a theoretical model of its effects on productivity and learning for individuals and teams. Acad Manag Rev 36(3):461–478
13. Popovski G, Seljak BK, Eftimov T (2020) A survey of named-entity recognition methods for food information extraction. In: IEEE Access, vol 8, pp 31586–31594. Institute of Electrical and Electronics Engineers Inc. https://doi.org/10.1109/ACCESS.2020.2973502
14. Ross AR, Dumitrescu D (2019) 'Vox Twitterati': investigating the effects of SM exemplars in online news articles. New Media Soc 21(4):962–983
15. Sarvari PA, Nozari M, Khadraoui D (2019) The potential of data analytics in disaster management. In: Industrial engineering in the big data era, pp 335–348. Springer, Cham

16. Sarvari PA, Ustundag A, Takci H (2016) Performance evaluation of different customer segmentation approaches based on RFM and demographics analysis. Kybernetes
17. Sarvari PA, Ustundag A, Cevikcan E, Kaya I, Cebi S (2018) Technology roadmap for Industry 4.0. In: Industry 4.0: managing the digital transformation, pp 95–103. Springer, Cham
18. Sianipar CPM, Yudoko G, Adhiutama A, Dowaki K (2013) Community empowerment through appropriate technology: sustaining the sustainable development. Procedia Environ Sci 17:1007–1016
19. Singh A, Shukla N, Mishra N (2018) Social media data analytics to improve supply chain management in food industries. Transp Res Part E 114:398–415. https://doi.org/10.1016/j.tre.2017.05.008
20. Swani K, Brown BP, Milne GR (2014) Should tweets differ for B2B and B2C? An analysis of Fortune 500 companies' Twitter communications. Ind Mark Manage 43(5):873–881
21. Swani K, Brown BP, Mudambi SM (2019) The untapped potential of B2B advertising: a literature review and future agenda. Ind Mark Manag
22. Swani K, Milne GR, Brown BP, Assaf AG, Donthu N (2017) What messages to post? Evaluating the popularity of SM communications in business versus consumer markets. Ind Mark Manage 62:77–87
23. Taken Smith K, Smith LM (2019) SM usage by law firms: correlation to revenue, reputation, and practice areas. Serv Mark Quart 40(1):66–81
24. Tao D, Yang P, Feng H (2020) Utilization of text mining as a big data analysis tool for food science and nutrition. Compr Rev Food Sci Food Saf 19(2):875–894. https://doi.org/10.1111/1541-4337.12540
25. Tseng ML, Lim MK, Wu KJ, Peng WW (2019) Improving sustainable supply chain capabilities using SM in a decision-making model. J Clean Prod 227:700–711
26. Tyagi S (2016) An improved fuzzy-AHP (IFAHP) approach to compare SECI modes. Int J Prod Res 54(15):4520–4536
27. Tyagi SK (2015) Lean thinking for lead-time reduction and efficient knowledge creation in product development
28. Webster GB, Hume M (2018) Analysing the role of SM in dialogue marketing and management as a contemporary franchising local area marketing technique. In: SM marketing: breakthroughs in research and practice, pp 176–205. IGI Global

Healthcare Systems Engineering and Management

Metaheuristic Hybridization: A Case Study for Nurse Scheduling

Yakup Turgut and Zikriye Melisa Erdogan

Abstract This paper addresses a nurse scheduling problem frequently encountered in hospital management. To make nurses satisfied and use their best skills during the work process is a critical issue at the center of this problem. Besides, hospitals need to minimize personnel costs while keeping service quality at the highest level. We try to schedule nurses by considering their preferences and meet hospital management expectations at the same time. Our problem has hard and soft constraints that are faced in real-world case studies. Hard constraints are satisfied directly by applying the constraint programming method, and soft constraints are satisfied using a penalty cost applied in meta-heuristic algorithms. The initial model is structured using a Genetic algorithm (GA), then it is hybridized with the simulated annealing (SA) to obtain a nurse schedule. Results are compared with MIP solutions concerning the quality of solutions and the corresponding running time. Achievements are analyzed and discussed to make the proposed model applicable by hospital managers as well as researchers.

Keywords Nurse scheduling · Constraint programming · Genetic algorithm · Hybridization

Introduction

It is clear that the health sector has to give vital services, and it will continue to be one of the most demanding systems in the world. The main objectives of this industry are minimizing the costs and increasing the quality of work and productivity. To achieve these goals, it should be noted that all the available resources must be used in the right place at the right time [2]. Optimization applications in the field of healthcare

Y. Turgut (✉) · Z. Melisa Erdogan
Industrial Engineering Department, Management Faculty, Istanbul Technical University, Istanbul, Turkey
e-mail: turgut16@itu.edu.tr

Z. Melisa Erdogan
e-mail: erdoganzi16@itu.edu.tr

F. Calisir (ed.), *Industrial Engineering in the Internet-of-Things World*,
Lecture Notes in Management and Industrial Engineering,
https://doi.org/10.1007/978-3-030-76724-2_29

have attracted considerable attention in the past 30 years. Over time, these studies have expanded from high-level issues to operational issues, such as nurse scheduling, treatment planning, etc. [18].

Hospitals and other health care facilities strive to provide superior patient care at an affordable cost. It is already proven that efficient resource management is the key to the success of the goals. So, it is necessary to manage the workforce and other resources in those establishments efficiently. Moreover, hospitals have to make their patients satisfied with the health services, and this is only possible if the staff is fond of performing the job [14, 16]. Nurses are the fundamental workforce for health treatment organizations and have a direct impact on patients. This is why hospital administration must provide a reasonable timetable for nurses.

Creating a good timetable (or schedule) for nurses can be difficult and time consuming for organizations. Because scheduling is an NP-hard problem with a complex structure. A schedule is created to meet workforce requirements during a planning period [18]. Nurse scheduling is mainly the problem of assigning nurses with various skills and personal characteristics to various shifts under certain constraints [13]. Due to the constantly changing environmental conditions of hospitals, dynamic schedules or adjustments at certain intervals are requested. That is why the problem of nurse scheduling has been extensively studied in the literature by researchers [5]. The automatic and rapid creation of these schedules enables the hospitals to use their existing resources more effectively and helps to create balanced workloads among nurses [3].

After examining related researches in the field, we have seen that there are many techniques applied to Nurse Scheduling Problem (NSP). Yet, to our knowledge, there are few studies that compare the hybridized methods with the single meta-heuristic application. In this paper, we propose a hybrid meta-heuristic algorithm for the nurse scheduling problem. A sample application, including both hard and soft constraints, is performed. Initial solutions are created using constraint programming. Then, the Genetic Algorithm (GA) and the Hybridized Algorithm are used independently to improve the initial solutions, and the results are compared.

Problem Statement and Discrete Model in Scheduling

Our problem is nurse scheduling, and to address it, we propose a model that maximizes nurses' shift preferences while minimizing costs associated with understaffing and overstaffing. As is the case in real-world situations, the model is constructed under both hard and soft constraints. It is highly exceptional in practice to find a schedule that satisfies all the soft constraints. Therefore, the algorithm aims to minimize the real impact of constraint violations.

Assumptions:

- Consider a hospital that operates 24 h a day, 7 days a week.

- All instances start on Monday. Day index starts at zero so the first day in the planning period is day zero.
- All parameters are known. Time parameters such as the coverage requirements for each shift of each day, the minimum and the maximum values are given in minutes.
- The schedule is set for nurses who work in the same department of the hospital.
- The model does not see that all days are the same. There are weekdays and weekends.
- We have four shifts including a free shift. {Shift 1-8:00-16:00 (E), Shift 2-16:00-24:00 (D) Shift 3-24:00-8:00 (N), Shift 0: Day Off}.
- It is assumed that nurses comply with the assigned shift.

Model Notation for Sets and Parameters:

I: Set of employees.
h: Number of days in the planning horizon.
D: Set of days in the planning horizon = {1, 2...h}.
W: Set of weekends in the planning horizon = {1, 2...h/7}.
T: Set of shift types = {1, 2, ...t}.
l_t: The length of shift type t in minutes.
R_t: Set of shift types that cannot be assigned immediately after shift type t.
N_i: Set of days that employee i cannot be assigned a shift on.
R_{dt}: Required number of employees for shift type t on day d.
m_{it}^{max}: Maximum number of shifts of type t that can be assigned to employee i.
b_i^{max}: Maximum number of minutes that employee i can work.
b_i^{min}: Minimum number of minutes that employee i must work.
c_i^{max}: Maximum number of consecutive shifts that employee i can work.
c_i^{min}: Minimum number of consecutive shifts that employee i must work.
a_i^{max}: Maximum number of weekend days that employee i can work.
o_i: Minimum number of consecutive days offs that employee i can be assigned.
q_{idt}: Penalty if shift type t is not assigned to employee i on day d.
p_{idt}: Penalty if shift type t is assigned to employee i on day d.
w_{dt}^{min}: Weight if below the preferred cover for shift type t on day d.
w_{dt}^{max}: Weight if exceeding the preferred cover for shift type t on day d.
P_1: Penalty cost for under staffing.
P_2: Penalty cost for over staffing.

Decision Variables:

$$X_{idt} = \begin{cases} 1, \; if \; employee \; i \; is \; assigned \; to \; shift \; type \; t \; on \; day \; d \\ 0, \; otherwise \end{cases}$$

$$k_{iw} = \begin{cases} 1, \; if \; employee \; i \; works \; on \; weekend \; w \\ 0, \; otherwise \end{cases}$$

y_{dt} = Total number of nurses below the preferred coverage for shift type t on day d.
z_{dt} = Total number of nurse above the preferred coverage for shift type t on day d.

There are six hard constraints and three soft constraints. Hard and soft constraints that our model tries to satisfy are explained in the next section in detail.

Hard Constraints

Hard constraints are the constraints which cannot be violated in the model.

1. A nurse must be assigned to at most one shift on a single day.

$$\sum_{t \in T} X_{idt} \leq 1, \quad \forall i \in I, d \in D$$

2. In order to establish a minimum amount of idle time for each nurse (rest), certain shifts can not be assigned after the shift on the previous day. For example, an early shift cannot follow a late shift. This constraint can be called as shift rotation constraint. These are defined as follows: E|D, N|E, N|D.

$$X_{idt} + X_{i(d+1)u} \leq 1, \quad \forall i \in I, \quad d \in \{1, 2, \ldots, h-1\}, \quad t \in T, \quad u \in R_t$$

3. Each nurse has the maximum number of shifts of each type that can be assigned to them. For example, some nurses will say they can not work more than the maximum number of specific shift type.

$$\sum_{d \in D} X_{idt} \leq m_{it}^{\max}, \quad \forall i \in I, \quad t \in T$$

4. Maximum and minimum time work constraints determine the boundary of total time worked by each employee.

$$b_i^{\min} \leq \sum_{d \in D} \sum_{t \in T} l_t X_{idt} \leq b_i^{\max}, \quad \forall i \in I$$

5. A maximum consecutive shift constraint explains the maximum number of days a nurse can work without a day off. For example, part-time employees sometimes do not work as many consecutive days as full-time staff.

$$\sum_{j=d}^{d+c_i^{\max}} \sum_{t \in T} X_{ijt} \leq c_i^{\max}, \quad \forall i \in I, \quad d \in 1, 2 \ldots h - c_i^{\max}$$

6. The minimum consecutive shift constraint explains the minimum number of days a nurse can work without a day off.

$$\sum_{t\in T} X_{idt} + \left(s - \sum_{j=d+1}^{d+s} \sum_{t\in T} X_{ijt}\right) + \sum_{t\in T} X_{i(d+s+1)t} \geq 0,$$

$$\forall i \in I, \quad s \in 1 \ldots c_i^{\min} - 1, \quad d \in 1 \ldots h - s + 1$$

7. Minimum consecutive days off constraint explains the minimum number of consecutive days off that a nurse can be assigned.

$$\left(1 - \sum_{t\in T} X_{idt}\right) + \sum_{j=d+1}^{d+s} \sum_{t\in T} X_{ijt} + \left(1 - \sum_{t\in T} X_{i(d+s+1)t}\right) \geq 0,$$

$$\forall i \in I, \quad s \in 1 \ldots o_i - 1, \quad d \in 1 \ldots h - s + 1$$

8. The maximum number of weekend days constraint shows the maximum number of weekends which considered as being worked if the nurse has a shift on Saturday or Sunday.

$$k_{iw} \leq \sum_{t\in T} X_{i(7w-1)t} + \sum_{t\in T} X_{i(7w)t} \leq 2k_{iw}, \quad \forall i \in I, \quad w \in W$$

$$\sum_{w\in W} k_{iw} \leq a_i^{\max}, \quad \forall i \in I$$

9. Cover constraint ensures the relationship between X_{idt}, y_{dt}, z_{dt} decision variables, and the parameter, R_{dt}.

$$\sum_{i} X_{idt} + y_{dt} - z_{dt} = R_{dt}, \quad \forall d \in D, \quad t \in T,$$

10. To satisfy days off constraint, shifts must not be assigned to the specified nurse on the specified days. In other words, these are days that nurses can not work.

$$x_{idt} = 0, \quad \forall i \in I, \quad d \in N_i, \quad t \in T$$

Soft Constraints

Soft constraints are the ones that can be violated at the cost of incurring a penalty. The soft constraints we used in our problem are below:

1. If the required number of staff on the specified day for the specified shift is not assigned, then it is a soft constraint violation. If the assigned number (x) is less than or greater than the required number, a penalty cost is incurred as follows:

$$P_1 = (requirement - x) \times w_{dt}^{\min}$$
$$P_2 = (x - requirement) \times w_{dt}^{\max}$$

The parameters w_{dt}^{min} and w_{dt}^{max} are weights to represent the importance of minimizing under and over coverage.

2. Nurses can request certain shifts. If the schedule does not satisfy those requests, there will be a penalty cost, q_{idt}. If there is no request, then the parameter value is zero.
3. The nurses have some shift off requests. If the specified shift is assigned to the specified nurse on the specified day, then there is a penalty cost, p_{idt}.

Objective Function

The objective function of the model is to minimize the number of shifts that are incompatible with nurses' preferences while also minimizing the costs associated with under- and overstaffing. For example, an employee may request to work a certain shift-type on a particular day. The greater the weight assigned to a request, the more critical it is to the nurse.

$$Minimize \sum_{i \in I} \sum_{d \in D} \sum_{t \in T} q_{idt}(1 - x_{idt}) + \sum_{i \in I} \sum_{d \in D} \sum_{t \in T} p_{idt} x_{idt}$$
$$+ \sum_{d \in D} \sum_{t \in T} y_{dt} w_{dt}^{min} + \sum_{d \in D} \sum_{t \in T} z_{dt} w_{dt}^{max}$$

The variables y_{dt} and z_{dt} are the total numbers of staff below and above the preferred coverage level for each shift type t on each day d.

Methodology

Constraint satisfaction problem is a common problem in real life, for solving such problems as scheduling, vehicle routing, etc. Constraint satisfaction problem has three components: variables, domains, and constraints. The constraints restrict the value of each variable, which is determined by domain sets. The main goal of the problem is to find values for each decision variable without violating the constraints. Constraint programming is a reasonable way of solving constraint satisfaction problems using specific algorithms [4, 22].

GA is the most common evolutionary algorithm. GA is predicated on the idea that new individuals are created by applying a crossover operator to existing individuals and their diversity is increased through mutation. The GA's philosophy is based on the concept of survival of the fittest. Therefore, a probabilistic selection strategy is used to ensure the survival of the population's best individuals. The worst parents are replaced by better offsprings. In the crossover process, any point selected on the individual gene sequence is swapped with an individual to be crossed in a fixed or varying probability. On the other hand, in the mutation, the values of the genes that the individual possesses a flip with a fixed or random probability [20]. GA is commonly used in the literature to generate feasible nurse schedules that meet the constraints and specifications of the problem [1, 12, 15, 17, 21].

Simulated Annealing (SA) is a method for solving combinatorial optimization problems [6, 19]. SA operates in a manner analogous to how metal cools physically when it searches for a solution space. In traditional search algorithms, if the neighborhood solution is better than the current solution, the algorithm takes one step in that direction. Otherwise, the solution stays unchanged. This may cause being stuck in the local optima. On the other hand, SA's stochastic diversity enables it to avoid local optima. With some probability (acceptance probability), one can move towards the neighborhood solution, even if the neighborhood solution is worse than the current solution.

Case Dataset

Data is used by extraction from http://www.schedulingbenchmarks.org. The test instances range in size from eight to sixty staff. Table 1 gives the list of the instances and their sizes. Although larger instances are available on the website, we did not include them in the table because they are not used in our computational experiments. As an illustration of the proposed method's application, Instance 6 is used. According to Instance 6, we have 18 nurses to be scheduled over a 28-day period. We attempted to address the problem's hard constraints using a constraint programming approach. Following that, we used a genetic algorithm and then hybridized GA and SA to minimize the cost associated with the soft constraints.

Table 1 Test instances

Instances	Planning horizon in weeks	# of Staff	# of Shift Types
1	2	8	1
2	2	14	2
3	2	20	3
4	4	10	2
5	4	16	2
6	4	18	3
7	4	20	3
8	4	30	4
9	4	36	4
10	4	40	5
11	4	50	6
12	4	60	10

Table 2 Variable coding

S[i, j]	1	2	3	4
1	1	3	3	0
2	0	1	1	2
3	1	0	3	3
4	2	2	0	3

Representation of Decision Variables

Real-valued costing is used in constructing the GA model, as shown in Table 2. The data is coded by a data matrix S (I, J), where ith row, i∈ I = {1, 2, …, n}, represents the ith nurse, the jth column, j∈ J = {1, 2, …, m}, represents the jth day and the cell S (i, j) represents the shift types. For example, the S (1, 1) indicates that nurse 1 works at shift 1 on day 1. In the real world, shift types include Day, Evening, Night, and Free shift, which are referred to as phenotypes in genetic algorithms. They are represented with numbers; The day is denoted by 1, the evening is denoted by 2, the night is denoted by 3, and the free shift is denoted by 0. In a genetic algorithm, these values (0, 1, 2, 3) denote the genotype.

Constrained Programming

We define Variables, Domains, and Constraints of the problem as follows:

Variable: X_{ij}, the shift type for nurse i on day j.
Domain: {Values, X_{ij}} indicates the values that can be assigned to nurse i on day j. For example, {{0,1,2}, X_{11}} indicates that Nurse 1 on day 1 can have shift types {0,1,2}.
The Constraints: All hard constraints are satisfied. After assigning all possible domain values randomly to each variable, each constraint is checked. If any violation occurs, the value for that variable is changed depending on domain value. A backtracking algorithm is used for search. The filtering method is also used for narrowing the search space.

Genetic Algorithm

Given N nurses to allocate, the following steps are executed:

1. Initial Population Creation: The initial population is created by satisfying hard constraints using the constraint programming approach. The size of the population equals to N multiplied by 2.

2. Fitness Value Evaluation: Fitness value is calculated for each individual in the population.
3. Selection Operation: Using the roulette wheel selection strategy, the best individual in the population is selected.
4. Crossover Operation: A crossover point between 1 and N is selected. To create a child solution, the first x rows of parent solution P1 are extracted and concatenated with the remaining N-x rows of parent solution P2.
5. Mutation: To begin the mutation operation, two random numbers are generated. The first one represents the nurse id, which corresponds to a row number in the timetable, and the second one represents the day, which corresponds to a column number in the timetable. If the selected cell contains a value other than zero, it will be forced to change in accordance with the available nurses in the shift data. The impact of the mutation rate is analyzed in the next section named Parameter Impacts.
6. Objective function: It is calculated by minimizing the total penalty cost.
7. Stopping criteria: The number of iterations is used to maintain control. The impact of the number of iterations is reported in the Parameter Impacts Section.

Simulated Annealing and Genetic Algorithm Integrated

The initial population is generated randomly by satisfying hard constraints with constrained programming. Two individuals with the highest fitness values are chosen from the population, and offspring are generated through the crossover and mutation operations. By utilizing the acceptance probability concept of simulated annealing, the population's diversity is increased. All of these steps are repeated until the specified temperature is reached. The pseudo-code of the proposed hybrid algorithm is given in Fig. 1. The following section, titled Parameter Impacts, examines the effect of temperature.

Computational Experiments and Results

Parameter impacts

The performance of evolutionary algorithms is highly dependent on the parameters being set correctly. Certain parameters must be specified for both the genetic algorithm and the simulated annealing algorithm. Genetic algorithm parameters were classified into two categories: quantitative and qualitative by Eiben and Smith [9]. The qualitative category includes the parent selection strategy, the type of crossover, and the chromosome representation, while the quantitative category includes the crossover rate, mutation rate, and population size. Initial temperature ($T_{initial}$) and alpha are the parameters of simulated annealing that is used in the hybridized method. The response surface design method is used to determine the appropriate values for

```
Input  : Initial population,number of nurses, number of days,
         population size,T, Tmin,α
Output: Nurse Schedule Table and Fitness Value
while T>Tmin do
    Fitness value evaluation for each individual and select two of
     them with the minimum fitness value.
    Apply crossover operator to these two individual and obtain an
     offspring
    Apply mutation operator to the offspring
    /* comments on code,fv:fitness value.                        */
    if Offspring_fv < WorstInidividualofPopulation_fv then
        Replace the worst individual with offspring
    else
        Replace the worst individual when the acceptance probabiliy
         condition was satisfied in simulated annealing.
    end
    T← T*α
end
return best schedule and fitness value
```

Fig. 1 Hybrid algorithm

the parameters in Table 3. (see Kucukkoc et al. [11]) for the details of the method). Table 3 summarizes the level of each parameter used in the experiments. These values are derived from suggestions made in peer-reviewed articles by subject-matter experts.The best value of all parameters, which was obtained by applying the method, is given in the last column of Table 3.

Table 3 Level and values of GA and SA parameters

Parameter name	Level 1	Level 2	Level 3	Level 4	Level 5	Best value
Population size (P)	72	144	288	576	1152	288
Crossover rate (P_c)	0.1	0.3	0.5	0.7	0.9	0.70
Mutation rate (P_m)	0.05	0.1	0.15	0.20	0.25	0.10
$T_{initial}$	500	1000	10,000	50,000	100,000	10,000
Alpha	0.25	0.35	0.55	0.75	0.95	0.75

Computational Results

The instances in Table 1 were used to benchmark the performance of the hybridized method with other methods (genetic algorithm and MIP solution). The mathematical model in Problem Statement and Discrete Model in Scheduling Section has been coded in the GAMS software version 24 and solved using CPLEX solver for all instances. We utilized the Neos Servers, which is a free internet-based service for solving numerical optimization problems, to solve the model coded in GAMS [7, 8, 10].

The results of CPLEX solver for all instances are given in Table 4. Seven of twelve instances are solved by CPLEX solver optimally in a reasonable runtime. On the other hand, other instances require about one hour of computational time for a near-optimal solution. We implemented the genetic algorithm and hybrid algorithm in MATLAB R2015A with a computer that has the following CPU features: Intel® Core™ i5-2450 M CPU @ 2.5 GHz. The values in Table 4 represent the average of twenty iterations for the genetic algorithm and hybrid algorithm solutions.

Running time and the quality of solutions of the hybridized model, genetic algorithm, and MIP model were compared for the set of test instances given in Table 2.

The results show that the method is promising both in terms of solution quality and running times, especially for large instances (see Table 4). The results also indicate that the performance of the hybridized model is more desirable than a genetic algorithm in terms of solution quality.

We have applied two different methods (genetic algorithm and hybrid algorithm) for our nurse scheduling problem so far. Independent Samples t-test was used to compare the means of two solution sets. The p-value in the analysis is 0, which is

Table 4 The comparison of the proposed method with the MIP solution

Instances	MIP solution (objective value)	Time (s)	Genetic solution (fitness value)	Time (s)	Hybrid solution (fitness value)	Time(s)
1	607	0.59	912	3.856	654	5.698
2	828	0.97	1235	3.123	924	5.784
3	1001	10.08	1400	4.128	1126	6.324
4	1716	46.98	1963	5.132	1815	8.784
5	1143	101.91	1347	9.125	1236	12.326
6	1950	60.71	3218	8.367	2152	16.258
7	1056	1019.73	1321	24.147	1092	36.548
8	1320	3600.04	1658	75.236	1478	100.698
9	440	3600.06	652	72.458	478	95.478
10	4631	916.92	5300	50.266	5002	67.582
11	3443	83.96	3758	10.586	3655	22.547
12	4040	2831.41	4318	56.789	4156	77.896

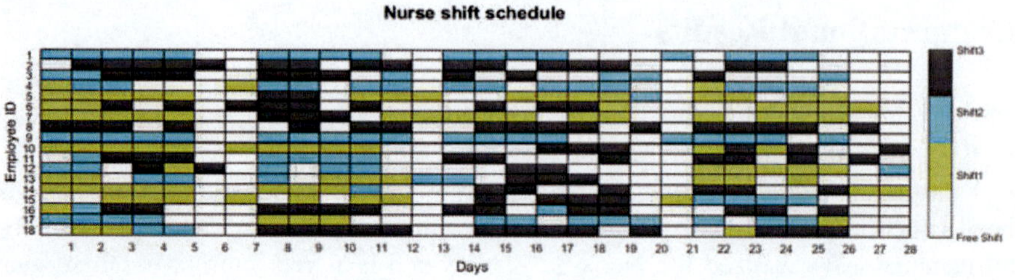

Fig. 2 Nurse schedule

less than 0.05, indicating that the null hypothesis being rejected. Consequently, the difference between the means of the two methods is significant, and we can say that they are statistically different from each other.

Figure 2 illustrates a nurse schedule for Instance 6. Additionally, Fig. 3 depicts the scheduled nurses and required nurses for each shift type. As illustrated in Fig. 3, the greatest penalty cost is incurred due to over and understaffing.

Conclusions and Further Research

Scheduling of nurses is generally a complex problem. A schedule that will satisfy all the parties (i.e., hospitals, nurses, and patients) is in request at reasonable costs. In this study, a hybridized meta-heuristic algorithm is proposed to find a less costly nurse schedule. We have implemented GA all alone and hybridization of GA and SA and then analyzed the results. The objective function values obtained by the hybridized method are found to be better. Moreover, we compared the results of the hybrid model with MIP model solutions. Although the fitness value of the hybrid model is about 10% worse than the MIP solution, it provides results in a substantially shorter time than the MIP solution, especially for larger instances.

The results contributed to the field in two-fold: First, we show that we can obtain a schedule which suits the nurse preferences while meeting the hospital constraints using hybridization; and second, we can enable the planners to make schedules in a reasonable time.

We recommend that future studies include additional constraints related to nurse preferences. Testing the proposed method with larger data sets will be our future study. Additionally, the proposed method will be compared to other hybridized methods, such as memetics.

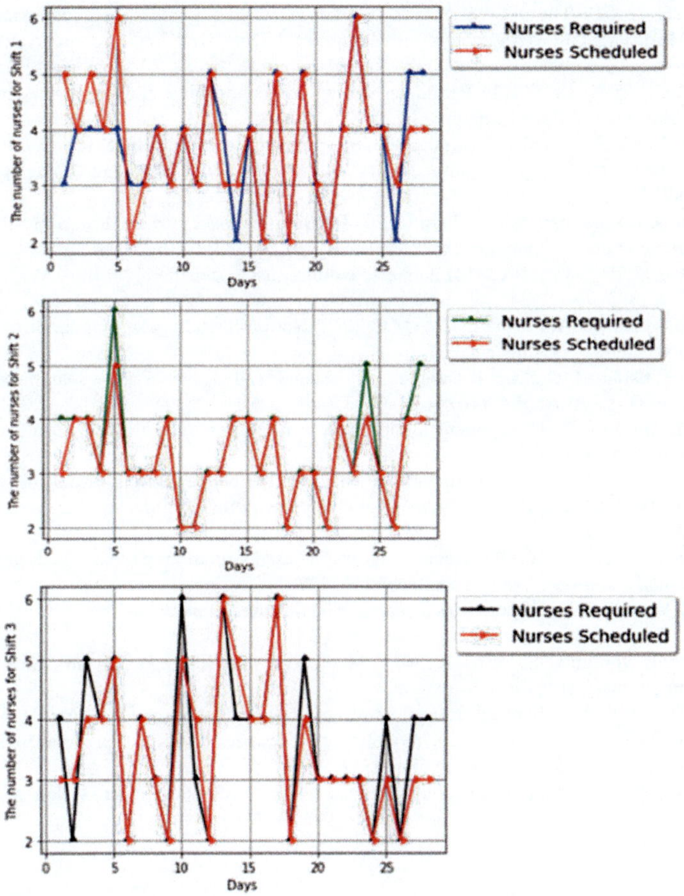

Fig. 3 The scheduled number of nurses versus the required number of nurses for each shift

References

1. Aickelin U, Dowsland KA (2000) Exploiting problem structure in a genetic algorithm approach to a nurse rostering problem
2. Atmaca E, Pehlivan C, Aydogdu CB, Yakıcı M (2012) Hemsire cizelgeleme problemi ve uygulaması. Erciyes Universitesi Fen Bilimleri Enstitusu Dergisi 28(4):351–358
3. Burke EK, De Causmaecker P, Berghe GV, Van Landeghem H (2004) The state of the art of nurse rostering. J Sched 7(6):441–499
4. Burke EK, Kendall G et al (2005) Search methodologies. Springer, Berlin
5. Cheang B, Li H, Lim A, Rodrigues B (2003) Nurse rostering problems-a bibliographic survey. Eur J Oper Res 151(3):447–460
6. Cheng M, Ozaku HI, Kuwahara N, Kogure K, Ota J (2008) Simulated annealing algorithm for scheduling problem in daily nursing cares. In: IEEE International conference on systems, man and cybernetics, 2008. SMC 2008. IEEE, pp 1681–1687
7. Czyzyk J, Mesnier MP, Moré JJ (1998) The NEOS Server. IEEE J Comput Sci Eng 5(3):68–75

8. Dolan E (2001) The NEOS Server 4.0 administrative guide. Technical Memorandum ANL/MCS-TM-250, Mathematics and Computer Science Division, Argonne National Laboratory
9. Eiben AE, Smit SK (2011) Parameter tuning for configuring and analyzing evolutionary algorithms. Swarm Evol Comput 1(1):19–31
10. Gropp W, Moré JJ (1997) Optimization environments and the NEOS server. In: Buhmann MD, Iserles A (eds) Approximation theory and optimization. Cambridge University Press, pp 167–182
11. Kucukkoc I, Karaoglan AD, Yaman R (2013) Using response surface design to determine the optimal parameters of genetic algorithm and a case study. Int J Prod Res 51(17):5039–5054
12. Leksakul K, Phetsawat S (2014) Nurse scheduling using genetic algorithm. Math Probl Eng 2014
13. Lim GJ, Mobasher A, Kardar L, Cote MJ, (2012) Nurse scheduling. In: Handbook of healthcare system scheduling. Springer, pp 31–64
14. Martin S, Ouelhadj D, Smet P, Berghe GV, ÖZcan E (2013) Cooperative search for fair nurse rosters. Expert Syst Appl 40(16):6674–6683
15. Moz M, Pato MV (2007) A genetic algorithm approach to a nurse rerostering problem. Comput Oper Res 34(3):667–691
16. van Oostveen CJ, Mathijssen E, Vermeulen H (2015) Nurse staffing issues are just the tip of the iceberg: a qualitative study about nurses perceptions of nurse staffing. Int J Nurs Stud 52(8):1300–1309
17. Podgorelec V, Kokol P (1997) Genetic algorithm based system for patient scheduling in highly constrained situations. J Med Syst 21(6):417–427
18. Rais A, Viana A (2011) Operations research in healthcare: a survey. Int Trans Oper Res 18(1):1–31
19. Rosocha L, Vernerova S, Verner R (2014) Medical staff scheduling using simulated annealing. Qual Innov Prosperity 19(1):1–8
20. Talbi EG (2009) Metaheuristics: from design to implementation, vol 74. Wiley
21. Tsai CC, Li SH (2009) A two-stage modeling with genetic algorithms for the nurse scheduling problem. Expert Syst Appl 36(5):9506–9512
22. Tsang E (2014) Foundations of constraint satisfaction: the classic text. BoD–Books on Demand

Home Health Care Services Management: Districting Problem Revisited

Onur Ozturk, Mehmet A. Begen, and Gregory S. Zaric

Abstract In this paper, we study optimal partitioning of home health care (HHC) units among different subregions of a city. Given each HHC crew is in charge of patients in a single subregion, the problem aims to group locations (districts) where patients live in larger subregions such that the total workloads of HHC crews are balanced. Moreover, due to geographical reasons, some locations cannot be assigned to the same subregion, which gives rise to feasible assignment problem of locations. First, we tackle the complexity of these problems and show that the feasible assignment problem is NP-complete. Then we propose a new objective function for the HHC workload balancing problem and develop a heuristic method to deal with the NP-hardness of the optimal partitioning problem. Compared to a mixed-integer linear model from the literature, the heuristic quickly finds an optimal solution in most of the problem instances and gives efficient and competitive results for instances which are not optimally solved.

Keywords Home health care services · Districting problem · Complexity analysis · Mathematical modeling · Heuristic

Introduction

Home health care (HHC) represents a wide range of health care services that can be given to a patient at home for illness or injury. Examples of services provided include wound care for pressure sores or a surgical wound, intravenous or nutrition therapy,

O. Ozturk
Telfer School of Management, University of Ottawa, Ottawa, ON, Canada
e-mail: ozturk@telfer.uottawa.ca

M. A. Begen (✉) · G. S. Zaric
Ivey Business School, University of Western Ontario, London, ON, Canada
e-mail: mbegen@ivey.uwo.ca

G. S. Zaric
e-mail: gzaric@ivey.uwo.ca

F. Calisir (ed.), *Industrial Engineering in the Internet-of-Things World*,
Lecture Notes in Management and Industrial Engineering,
https://doi.org/10.1007/978-3-030-76724-2_30

injections, monitoring serious illness, and unstable health status. HHC services may also be provided to assist with activities of daily living such as bathing, dressing, transferring, using the toilet, eating, and walking. HHC is an integral component of the post-hospitalization recovery, especially during the initial weeks after discharge, when patients still require some physical assistance [16].

HHC services are typically provided by a crew composed of nurses, caregivers, and, in some cases, physicians. The crew composition depends on patient needs. HHC services are believed to provide high-quality service for patients, and the demand for HHC services is expected to rise in the future [18]. For instance, in Canada, the government sees HHC as an opportunity to shift more health care into the home and out of expensive hospitals and long-term-care facilities [10].

The applications of operations research in the HHC literature has been flourishing in recent years such as resource dimensioning, i.e., finding the optimal composition of crews and medical equipment [7], districting, i.e., regrouping districts of a city into subregions [11], scheduling of human resource activities [3, 8], and finding optimal routing for the visits respecting treatment times [4, 6, 13, 17, 19]. In terms of methods, we see stochastic programming [5], multi-objective optimization [15], column generation [20], and also heuristic approaches [13, 14]. Recently, Benzarti et al. [2] studied a districting problem in HHC. There are locations within a city, and the aim is to group locations into larger subregions. Patients in each subregion are served by the crew(s) within that subregion. Some locations are incompatible with each other, i.e., they cannot be assigned to the same subregion due to geographical reasons. In Benzarti et al. [2], the authors proposed two mixed-integer linear programming (MILP) models: the first model minimizes the maximum workload difference among subregions, and the second model minimizes the maximum traveled distance of crews in each subregion. These MILP models can handle only moderate size instances. Furthermore, the incompatibility of locations gives rise to a feasible assignment problem, which is as important as the optimal assignment of locations.

In this paper, we determine the complexity of the feasible assignment problem, propose a new objective function for the workload balancing problem, and develop a competitive and efficient heuristic approach and demonstrate its performance.

Problem Description and Modeling

The districting problem aims to divide a given region into subregions so that patients living in different locations are regrouped in larger clusters. Patients of a subregion are only served by the crew(s) of that subregion. Crews are multi-skilled, i.e., any crew can give service to any patient. Patients have different profiles according to their medical needs. Each profile requires a different treatment or service, and thus service times vary depending on the patient profile. There are N locations called basic units (locations, districts) where patients live. The aim is to partition the N basic units into M subregions. It is known if two locations j and j' cannot be in the same subregion, and two basic units are said to be incompatible if they cannot be assigned to the same subregion. We next present the districting problem studied by Benzarti et al. [2], first the notation and then two models: The first model balances the workload among subregions, whereas the second model minimizes the maximum distance between districts within a subregion. We denote the number of basic units as N, the number of subregions as M, the number of patient profiles as H, index of basic units as i ($i = 1, \ldots, N$), index of subregions as j ($j = 1, \ldots, M$), index of patient profile as h ($h = 1, \ldots, H$), number of visits required by a patient having profile h as b_h, the average duration of a visit for patient profile h as T_h, number of patients living in location i and having profile h as P_{ih}, average workload per subregion as $\overline{wl} = \sum_{i=1}^{N}\sum_{h=1}^{H} P_{ih}b_hT_h/M$, the distance between basic units i and k ($i, k = 1, \ldots, N$) as d_{ik}, admissible percentage deviation of the workload as τ. The decision variables are $x_{ij} = 1$ if location i is assigned to subregion j, 0 otherwise; and $wl_j =$ workload of subregion j.

Model 1: Minimize the maximum workload gap.

$$\text{minimize } gap_{\max}$$

subject to

$$wl_j \geq \sum_{i=1}^{N}\sum_{h=1}^{H} P_{ih}b_hT_hx_{ij} \quad j = 1, \ldots, M \tag{1}$$

$$gap_{\max} \geq \left(wl_j - \overline{wl}\right)/\overline{wl} \quad j = 1, \ldots, M \tag{2}$$

$$gap_{\max} \geq \left(\overline{wl} - wl_j\right)/\overline{wl} \quad j = 1, \ldots, M \tag{3}$$

$$\sum_{j=1}^{M} x_{ij} = 1 \quad i = 1, \ldots, N \tag{4}$$

$$x_{ij} + x_{kj} \leq 1 \quad \forall i, k \text{ incomp. b. units}, \ j = 1, \ldots, M \tag{5}$$

The objective function together with (2) and (3) minimizes the maximum deviation of the workload among all subregions. Constraint set (1) defines the total workload of subregions, and (4) ensures the assignment of each district to a subregion. Constraint set (5) prevents the assignment of incompatible units.

Model 2: Minimize the total distance traveled

$$\text{minimize } distance_{\max}$$

subject to

$$distance_{\max} \geq d_{ik} * \left(x_{ij} + x_{kj} - 1\right) \quad \forall i, k, j \tag{6}$$

$$wl_j \geq (1 - \tau) * \overline{wl} \quad j = 1, \ldots, M \tag{7}$$

$$wl_j \leq (1 + \tau) * \overline{wl} \quad j = 1, \ldots, M \tag{8}$$

$$\text{constraint sets (1), (4) and (5) from Model 1} \tag{9}$$

Model 2 is an extension of model 1 with additional constraints. The objective function and the constraint set (6) guarantee the minimization of the maximum distance between districts within the same subregions. Constraint sets (7) and (8) allow feasible workload levels of subregions set by τ.

Problem Complexity

We formally determine the complexity of models 1 and 2. We show for the HHC problem that even determining the feasibility of a problem instance is NP-complete in the presence of incompatible basic units. Given M subregions and a set of basic units incompatible with each other, a feasible assignment may not be possible if M is small relative to the number of incompatibility relations among basic units. We can reduce the feasibility problem to a graph coloring problem such that incompatible basic units are represented as vertices connected with an edge. The connected vertices must be colored with different colors meaning that incompatible basic units must be assigned to different subregions. Let G be a graph where each node represents a basic unit that is incompatible with at least another basic unit (i.e., G does not contain any basic unit if it is compatible with all other basic units). We can then determine if a feasible assignment of incompatible basic units to M subregions is possible by answering the following question: Does G admit a proper vertex coloring with M colors? Since this problem is known to be NP-complete [9], deciding if N incompatible basic units can be feasibly assigned to M subregions is also NP-complete for $M > 2$. Last but not least, the optimal assignment problems solved by models 1 and 2 are NP-hard, since these problems can be reduced to assignment problems which are shown as NP-hard [1, 9].

Workload Balancing Problem Revisited

We revisit the HHC workload balancing problem (Model 1). We first propose a modification in the objective to guarantee the minimization of workload difference among subregions. Then we develop a heuristic approach and test its performance. Since there are incompatible basic units in the problem, the heuristic contains a pre-assignment procedure (PAP) that initiates the assignment of basic units without deteriorating the optimal value of the objective function. We show that the PAP can also be used to find a constraint set that increases the performance of MILP models.

A New Objective Function for Workload Balancing

Model 1 measures the maximum gap by comparing the workload of each subregion to the average workload. However, this type of measure does not always guarantee a balance between different subregions. Consider the following example, as shown in Fig. 1: three subregions indexed from 1 to 3 with the following initial workloads: 10 h, 40 h, and 10 h, respectively. Suppose that the second subregion is assigned two additional patients, each requiring a 5 h treatment (other assignments are not divisible except those two).

Model 1 would partition the two additional patients of subregion 2 among subregions 1, 2, and 3 to minimize the maximum workload difference compared to the average workload (the average workload is simply the total workload divided by the number of subregions, thus not a variable but a parameter). Thus, both solutions shown in Fig. 2 would be optimal for model 1 since they have the same value

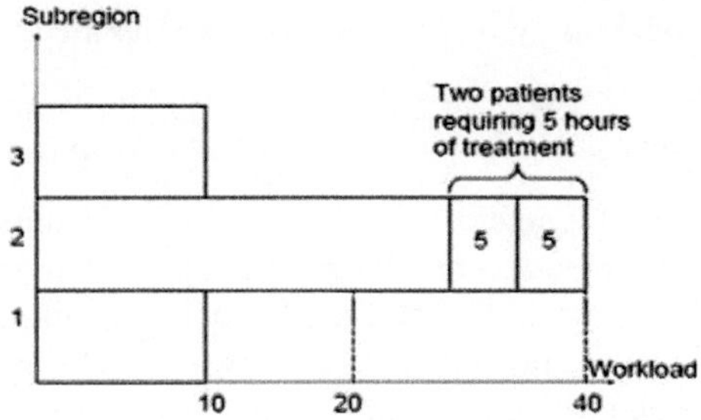

Fig. 1 Illustrative example

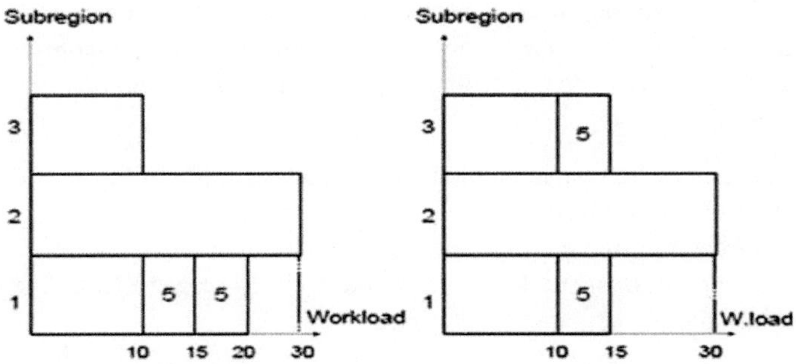

Fig. 2 Illustrative example (balanced version)

of $gap_{\max}$. However, the solution on the right-hand side depicts a more balanced configuration compared to the one on the right-hand side. The maximum difference among subregions is 15 h, and subregions 1 and 3 have the same amount of workload. Nevertheless, the objective function, $gap_{\max}$ is equal to 10 both solutions in Fig. 2. Thus, there is no guarantee that the right configuration will be provided by model 1.

To produce more balanced workloads, we define the maximum gap as the difference of workloads between subregions instead of using the average workload. We propose (10) to replace the constraint sets (2) and (3) in model 1. In the example above, $gap_{\max}$ for the solution on the left side of Fig. 2 is 20, and on the right side is 15. Thus, the one on the right would be preferred.

$$gap_{\max} \geq \left| wl_j - wl_k \right| \quad \forall j \neq k \tag{10}$$

The Heuristic Method for Workload Balancing

The heuristic method is composed of three phases. In the first phase, we make a pre-assignment of locations, which are the most incompatible ones with other locations. In the second phase, the remaining locations are partitioned among subregions to obtain an initial solution. Finally, an iterative process is applied to improve the initial solution in the third phase.

Phase 1: Pre-assignment procedure for incompatible locations

The re-assignment procedure determines a subset *S* of basic units with cardinality *M* (number of subregions) such that *S* has the greatest total incompatibility index. Here total incompatibility index means the total number of basic unit couples incompatible with each other. Afterward, among all elements of subset *S*, the basic unit, which is the most incompatible one with other elements of *S* is assigned to the first subregion. The other basic units in *S* are assigned to different subregions if they are incompatible with previous assignments. We give an example:

Instance: There are 3 subregions and 10 basic units. The first basic unit is incompatible with the second, third, and fourth ones. The second one is also incompatible with the third and fourth ones. The rest from the fifth to the tenth is not incompatible with any basic unit.

Generating set S: The first basic unit has the highest incompatibility index since it is incompatible with 3 other basic units. Thus, we put basic unit 1 to set *S*. The second basic unit also has the highest incompatibility index since it is incompatible with basic units 1, 3, and 4. Basic unit 2 is thus put to set *S*. Since set *S* has 3 elements (*M* = 3), either basic unit 3 or 4 can be put in set *S*.

Pre-assignment: We start from the first element of set *S* and assign basic unit one, w.l.o.g, to subregion 1. Then, checking that the basic unit 2 cannot be assigned to subregion 1, it is assigned to the following subregion, i.e., subregion 2. Reasoning the same way, basic unit 3 is tested on subregions 1 and 2. Because of the incompatibility of basic unit 3 with the previous assignments, it is assigned to a new subregion, i.e., subregion 3. After the pre-assignment procedure, phase 2 of the heuristic decides how unassigned basic units must be partitioned among subregions.

We present here an algorithm for the pre-assignment procedure. In the algorithm, E is the incompatibility matrix such that $E_{jj'} = 1$ if basic units j and j' are incompatible, 0 otherwise.

Preassignment assignment procedure: PAP

1. Find basic unit k in the incompatibility matrix E such that $\sum E_{k,l} > \sum E_{k',l} \forall k, k', l$ and $k \neq k'$.
2. Set $m' \leftarrow 1$. Assign basic unit k to subregion 1 (thus $x_{k1} = 1$).
3. Select row k in matrix E.
4. For all basic units k' incompatible with k.

    4.1. Set $loop \leftarrow false$.
    4.2. While $m' \leq M$,
        4.2.1 If basic unit k' is incompatible with the one assigned to m', set $m \leftarrow m' + 1$, $loop \leftarrow true$. Else, $loop \leftarrow false$ and $m' \leftarrow M + 1$ (i.e., exit while).
        4.2.2 $m' \leftarrow m' + 1$
    4.3. If $loop = true$, assign basic unit k' to subregion m (thus $x_{k'm} = 1$).
    4.4. Set $m' \leftarrow 1$.

5. For all unassigned basic units i and for all subrcgions m, set $x_{im} = 0$ if i is incompatible with any of the basic units assigned to subregion m.

Besides being the first step of the heuristic method, the PAP can also be used in the mathematical models to improve their solution times.

Phase 2: Workload balancing heuristic

This heuristic follows the principle of minimizing the workload difference between the least and the most loaded sub-regions. After initially running the PAP, the heuristic calculates the total workloads of every basic unit and then creates a list containing those basic units sorted in decreasing order of total workloads. Afterward, starting from the first element of the list, a basic unit is selected and is assigned to the least loaded subregion containing no other basic unit incompatible with the first element of the list. Then the same procedure is applied to the rest of the basic units list.

Workload balancing heuristic: WBH

1 Run PAP.

2 Calculate the total workload of every basic unit and create a list sorted in decreasing order of total workloads: L_{bu}
3 While $L_{bu} \neq \emptyset$,

3.1 Sort subregions in increasing of workloads $L_{subregion}$
3.2 Select the first element bu_{first} of L_{bu},
3.3 Starting from the first subregion in $L_{subregion}$, assign bu_{first} complying with the incompatibility constraint.
3.3.1 If bu_{first} cannot be assigned to any subregion, terminate the algorithm. Otherwise, $L_{bu} \leftarrow L_{bu} - bu_{first}$.

4 Run ImPr.

Phase 3: Solution Improvement Process

Once an initial solution is obtained with phase 2, an iterated improvement procedure is applied in phase 3. At each iteration, this procedure determines the least and the most loaded subregions. Then it selects a basic unit from the most loaded subregion such that the total load of the selected basic unit is closer to the difference between the workloads of the least and the most loaded subregions. Then respecting the incompatibility of basic units, it replaces the basic unit from the most loaded to the least loaded subregion. In case of incompatibility, the basic unit whose total workload is closer to the workload difference of the least and the next most loaded subregions is selected. The procedure is stopped if there is no improvement at the end of 1000 iterations or if all the basic units on the most loaded subregion are tried to be replaced.

Improvement process: ImPr

1 Set iteration number $iter \leftarrow 0$, maximum number of iterations $mni \leftarrow 1000$
2 While $iter < mni$,

2.1 Sort subregions in increasing order of total workloads
2.2 Set $index_{\min} \leftarrow 1$, $index_{\max} \leftarrow M$
2.3 Set $diff_{wl}$ the difference of workloads between subregions $index_{\min}$ and $index_{\max}$
2.4 Find the basic unit bu on subregiou $index_{\max}$ whose total load is closest to $diff_{wl}$
2.5 If bu is compatible with all other basic units of subregion $index_{\min}$, remove bu from $index_{\max}$ and assign to $index_{\min}$ Otherwise,
2.5.1 If $index_{\min} + 1 < index_{\max}$, set $index_{\min} \leftarrow index_{\min} + 1$, go to step 2.4. Otherwise, break.
2.6 If the objective function is improved, go to step 2.1. Otherwise set iteration number $iter \leftarrow iter + 1$

Numerical Experiments

Heuristic Performance

Test instances are inspired by Benzarti et al. [2]. The number of patient profiles H is equal to 2. The number of visits b_h and the average duration of the visits T_h with respect to the profile h are generated randomly from uniform distributions U[0, 2] and U[0, 5], respectively. Time is measured in hours. The number of patients P_{ih} having the profile h and living in the basic unit i is randomly generated from a discrete uniform distribution U[0, 20]. The number of subregions, M, varies from 2 to 6. The number of basic units, N, varies as 30, 60, 90, 120, 150. For each combination of M and N 10 test instances were generated. The uniform distribution U[0, N] is used to determine the incompatibility index. An additional 250 instances were also generated, considering there is no incompatibility between basic units (Table 2). CPLEX was used to implement the MILP model, and the heuristic coded in Java and solved with an Intel Core i3 CPU 2.27 GHz computer.

Table 1 presents differences between solutions, i.e., maximum workloads, given by the heuristic and model 1. We report the workload difference, which is calculated as $gap_{heuristic}$—gap_{model1} where gap_{model1} and $gap_{heuristic}$ are the maximum workload gap among subregions given by model 1 and heuristic, respectively. We measure the maximum, minimum, average, and the standard deviation between the workloads found with model 1 and the heuristic. Table 2 reports the same performance measures for instances containing only compatible basic units, i.e. when there is no incompatibility among basic units. We observe that the performance of the heuristic is satisfactory, especially with the increasing number of basic units. With the small number of basic units, there are relatively large gaps in workload per basic unit.

We also test how many times the heuristic finds the same solution as the MILP. Figure 3 shows that for almost 40% of compatible instances and 30% of all incompatible instances, the heuristic is able to find the optimal solution. Among all feasible instances (around 230), there are only fourteen that the heuristic cannot find a feasible assignment, although instances were feasible. The pre-assignment assignment procedure always provides a feasible initial partitioning into subregions in the presence of incompatible basic units. Thus, it could be used as an accelerating procedure for linear integer models.

While the heuristic performance is quite good both in terms of finding the optimality solution and optimality gap, it is a bit weak when it is about solving problem instances with incompatible basic units. The heuristic method does not always guarantee a solution even though a problem instance can be optimally solved if the incompatibility index is high. In Fig. 4, we report the percentage of infeasible solutions returned by the heuristic, although those instances admit feasible solutions that we can verify, thanks to the MILP model.

Table. 1 Difference between heuristic and optimal results

# of subregions	Performance measure	Incompatible					
		30	60	90	120	150	180
2	min.	0	0	2	2	0	0
	max.	46	14	18	14	15	1
	avg.	10.9	3.6	6.5	7.1	6.6	0.1
	std. dev.	13.3	4.5	4.6	4.2	5.1	0.7
3	min.	0	0	0	0	2	0
	max.	8	7	5	15	14	3
	avg.	1.2	2.4	2.1	3.6	5.6	1.5
	std. dev.	3.7	2.2	2.2	4.6	4.4	1.04
4	min.	0	0	0	0	0	0
	max.	6	5	4	4	7	4
	avg.	0.1	1.8	1.5	2.6	3.3	1.5
	std. dev.	2.2	1.4	1.6	1	2.5	1.19
5	min.	0	0	0	0	0	1
	max.	6	3	6	4	5	5
	avg.	0.6	1.8	1.7	1.7	2.1	2.5
	std. dev.	2.2	1.2	2.4	1.1	1.9	1.8
6	min.	0	0	0	0	0	1
	max.	2	4	3	4	6	6
	avg.	0.2	1.3	0.6	1.5	1.8	2.2
	std. dev.	1.4	1.1	1	1.4	1.6	1.8

Performance of the PAP

We now show how much the PAP can improve the solution times of mathematical models. Let S be the set of basic units initially assigned to a subregion with the PAP. Then, S can be interpreted as a constraint set since some decision variables x_{ij} are assigned a value with the procedure PAP. Recall that the pre-assignment procedure does not violate optimality since subregions are not specific to basic units.

We used model 2 to test problem instances since it was harder to solve than model 1 (model 1 instances are quickly solved with or without constraint set). The additional parameter we included for the generation of model 2 instances is the distance between two basic units i and k are generated as $d_{ik} \in U[0, 300]$.

The number of subregions, M, varies as 3 and 4 (instances with 2 subregions are quickly solved, and the impact of PAP is not significant when the number of subregions is greater than 4). Twenty problem instances are generated for each row in Table 3. For both cases of model 2 with and without constraint set S, Table 3 shows the number of instances solved in 300 s, number of times one case is faster than the other over 20 instances, and the average solution time of instances which are

Table. 2 Difference between heuristic and optimal results for only compatible units

# of subregions	Performance measure	Compatible					
		30	60	90	120	150	180
2	min.	0	0	0	0	0	0
	max.	12	4	2	4	4	2
	avg.	2.2	1	0.8	0.8	1	0.9
	std. dev.	3.8	1.6	1	1.6	1.4	0.5
3	min.	1	0	0	0	0	0
	max.	11	5	6	6	3	5
	avg.	5.2	1.8	1.2	2.4	1.5	1.6
	std. dev.	3.6	1.6	1.8	2	1.1	1.6
4	min.	0	0	0	0	0	0
	max.	7	11	10	5	10	2
	avg.	3.2	3.9	3	1.6	2.9	1.2
	std. dev.	2.5	3.2	2.9	1.7	2.9	0.6
5	min.	2	0	0	0	0	0
	max.	23	11	5	6	6	8
	avg.	9.5	4.6	1.7	2.3	1.9	2.5
	std. dev.	6.9	3.2	1.4	2	2	2.3
6	min.	2	0	0	0	0	1
	max.	22	16	9	3	3	5
	avg.	8.3	4.4	3.4	1.3	1.5	2.7
	std. dev.	5.8	5	2.5	0.9	1.1	1.6

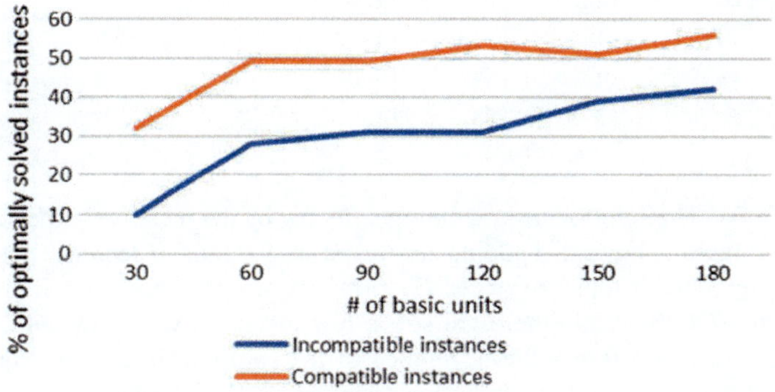

Fig. 3 Percentage of optimally solved instances with the heuristic

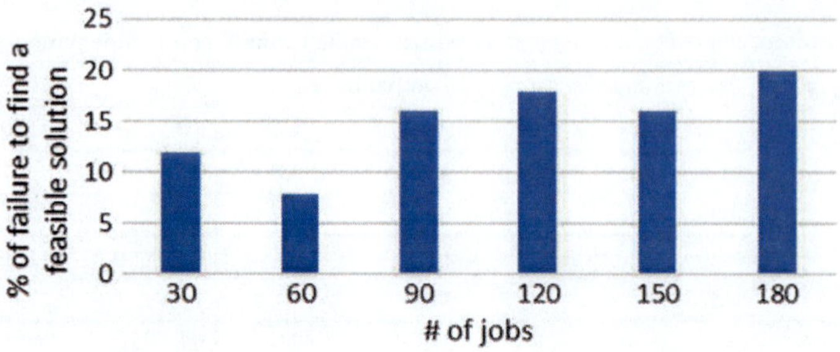

Fig. 4 Percentage of failure to find a feasible solution with the heuristic

Table. 3 Numerical results in the presence/absence of constraint set S

# of basic units	# solved in 300 s	# fastest	Average time (s)	# of basic units	# solved in 300 s	# fastest	Average time (s)
MILP with set S				MILP without set S			
$M = 3$				$M = 3$			
100	20	20	<1	100	20	0	8
150	20	20	3	150	20	0	24
200	20	16	25	200	20	4	87
250	17	15	13	250	15	2	175
300	6	6	17	300	0	0	> 300
$M = 4$	MILP with set S			$M = 4$	MILP without set S		
100	20	19	14	100	20	1	39
110	20	18	33	110	20	2	64
120	20	16	55	120	20	4	100
130	20	14	113	130	20	6	115
140	13	13	42	140	13	0	120
150	5	5	85	150	5	0	195

solved in 300 s. The solution time is mainly affected by the increase in the number of subregions. Table 3 shows that adding constraint set S is able to decrease the solution time considerably for most instances. The impact of constraint set S is bigger for cases where $M = 3$. While there are still a few instances where model 2 without set S finds a solution faster, the difference in the solution time is not significant. There are some instances for which constraint set S does not help to decrease the solution time. Those instances have a low incompatibility index, i.e., not so many basic units incompatible with each other. Otherwise, the results presented in this section show that the pre-assignment procedure we propose, improve the solution time by approximately more than 50% for almost all tested instances.

On the Use of PAP as an Acceleration Procedure

The decrease in the solution time, thanks to PAP, could be advantageous in many different solution methodologies. A first example would be the re-optimization models that have to relaunch the mathematical model(s) and/or algorithmic procedure(s) each time there is a parameter update. This is clearly a case we can count on in a districting problem. The partition of crews needs to be updated if the treatment of some patients takes longer than expected. Then, a new partition must be done to balance the workload of crews in different subregions. Another use of PAP would be in a nested solution procedure, as in [12]. The authors use a three-phase meta-heuristic to find a schedule for nurses. The first phase partitions nurses among patients. The second phase determines a schedule for each nurse. The final phase refines the solution value by resolving the second phase. Clearly, models 1 and 2 do not find a schedule nor a routing for HHC crews. Thus, the solution procedure can be extended by considering a second phase in which crews are assigned to different subregions (thus to patients in those subregions). After finding a schedule and/or a routing for crews, if the total workload of crews is different than one another, then some patients may be assigned to a crew from another subregion, and mathematical models 1 and 2 are resolved. Thus, a similar refining phase, as in [12], can be used by recalling those models, and even small decreases in solution time are important since the same models are solved many times.

Conclusion

We revisited the home health care (HHC) districting problem, which consists of partitioning locations where patients live into larger clusters (or subregions). The assignment of locations to subregions generates different workloads due to patient needs in each location. We first showed that the feasible assignment problem in the presence of incompatible locations is NP-complete. Afterward, for the HHC workload balancing problem, we proposed a slight modification in the objective function, aiming to provide a better balance between subregions. Then based on the idea of the new objective function, we provided a powerful heuristic method that finds solutions to either most the time-optimal or close to optimal. Besides providing good numerical results, another advantage of our heuristic method is its ease of implementation without requiring a commercial solver. Moreover, when instances get larger (e.g., more than 10 subregions and 500 locations), the solution time of the mathematical models increases considerably while the heuristic finds a solution quickly. The first step of the heuristic method can be used as an acceleration procedure for the mathematical model minimizing the maximum distance among locations assigned to the same subregion. Results showed that the acceleration procedure helps to decrease the solution time by more than 50% when the number of subregions is low.

Many extensions of the studied problem can be considered for future work. Considering some cancellations or last-minute visits that occur, the solution procedure can be integrated into a re-optimization model. Or instead of handling the problem with a deterministic point of view, some stochastic approaches can be applied. Another extension would be to consider the optimal assignment of locations to subregions as the first step of an HHC planning process. Then the scheduling of patient visits can be integrated as a second step in the planning process.

References

1. Aloise D, Deshpande A, Hansen P, Popat P (2009) NP-hardness of Euclidean sum-of-squares clustering. Mach Learn 75(2):245–248
2. Benzarti E, Sahin E, Dallery Y (2013) Operations management applied to home care services: analysis of the districting problem. Decis Support Syst 55(2):587–598
3. Bertels S, Fahle T (2006) A hybrid setup for a hybrid scenario: combining heuristics for the home health care problem. Comput Oper Res 33(10):2866–2890
4. Castillo-Salazar JA, Landa-Silva D, Qu R (2012) A survey on workforce scheduling and routing problems. In: Proceedings of the 9th international conference on the practice and theory of automated timetabling, pp 283–302
5. De Angelis V (1998) Planning home assistance for AIDS patients in the city of Rome, Italy. Interfaces 28(3):75–83
6. Duque PM, Castro M, Sörensen K, Goos P (2015) Home care service planning. The case of Landelijke Thuiszorg. Eur J Oper Res 243(1):292–301
7. Eaton SC (2000) Beyond 'unloving care': linking human resource management and patient care quality in nursing homes. Int J Hum Resour Manag 11(3):591–616
8. Eveborn P, Flisberg P, Rönnqvist M (2006) Laps care—an operational system for staff planning of home care. Eur J Oper Res 171(3):962–976
9. Garey MR, Johnson DS (1979) Computers and intractability: a guide to the theory of np-completeness. Freeman, San Francisco, LA
10. Grant K, Church E (2015) Ontario moves to fix plan to give personal support workers raises. URL http://www.theglobeandmail.com/news/national/ontario-tightens-wage-increase-rules-for-personal-support-workers/article24921222/. Retrieved 15 April 2015
11. Guti´errez EV, Guti´errez V, Vidal CJ (2013) Home health care logistics management: framework and research perspectives. Int J Ind Eng Manag 4(3):173–182
12. Issaoui B, Zidi I, Marcon E, Ghedira K (2015) New multi-objective approach for the home care service problem based on scheduling algorithms and variable neighborhood descent. Electron Notes Discrete Math 47:181–188
13. Liu R, Xie X, Augusto V, Rodriguez C (2013) Heuristic algorithms for a vehicle routing problem with simultaneous delivery and pickup and time windows in home health care. Eur J Oper Res 230(3):475–486
14. Mankowska DS, Meisel F, Bierwirth C (2014) The home health care routing and scheduling problem with interdependent services. Health Care Manag Sci 17(1):15–30
15. Milburn AB, Spicer J (2013) Multi–objective home health nurse routing with remote monitoring devices. Int J Plann Sched 1(4):242–263
16. Modin S, Furhoff AK (2002) Care by general practitioners and district nurses of patients receiving home nursing: a study from suburban Stockholm. Scand J Prim Health Care 20(4):208–212
17. Rasmussen MS, Justesen T, Dohn A, Larsen J (2012) The home care crew scheduling problem: preference-based visit clustering and temporal dependencies. Eur J Oper Res 219(3):598–610

18. Sieniuc K (2014) Canadian Hospitals see decline in overnight stays. http://www.theglobeandmail.com/life/health-and-fitness/health/canadian-hospitals-see-decline-in-overnight-stays/article18981290/. Retrieved 15 April 2015
19. Wirnitzer J, Heckmann I, Meyer A, Nickel S (2015) Patient-based nurse rostering in home care. Oper Res Health Care
20. Yuan B, Liu R, Jiang Z (2015) A branch-and-price algorithm for the home health care scheduling and routing problem with stochastic service times and skill requirements. Int J Prod Res 53(24):7450–7464

Meta-Analysis Study on the Effect of Managers' Leadership Behaviors on Work Performance of Employees

Baris Evcin and Cemil Ceylan

Abstract In today's world, where technological developments accelerate and great competition is experienced, managers play an important role in achieving the goals of the business. Managers must ensure the continuity of the enterprises. In order to ensure continuity, managers are trying to keep the work efficiency and quality of the work done as high as possible. In order to bring these factors to the level they want, managers knowingly or unknowingly exhibit various leadership behaviors. These leadership behaviors of managers are of great importance for employee performance. The culture and systems that leaders create within the organization have a significant impact on employee performance. For this reason, a meta-analysis was conducted to explain the relationship between leadership behaviors and employee performance and to determine the parameters that leaders indirectly affect employee performance. Within the scope of the meta-analysis study, the relationship between leadership behaviors and employee performance has been prioritized, and other parameters that have an indirect effect on the employee performance of the leaders have been examined. A synthesis by analyzing applications made in different regions and sectors in Turkey are formed. This research aims to explain the relationship between leadership behaviors and employee performance, to contribute to the researches about employee performance, and to serve as an example for those who will conduct a meta-analysis.

Keywords Leadership behaviors · Employee performance · Meta-analysis method

Introduction

Today, the number of scientific studies is increasing rapidly. In independent studies conducted on a specific subject, different results are often obtained. No matter how

B. Evcin · C. Ceylan (✉)
Industrial Engineering Department, Management Faculty, Istanbul Technical University, Istanbul, Turkey
e-mail: ceylan@itu.edu.tr

B. Evcin
e-mail: evcin17@itu.edu.tr

F. Calisir (ed.), *Industrial Engineering in the Internet-of-Things World*,
Lecture Notes in Management and Industrial Engineering,
https://doi.org/10.1007/978-3-030-76724-2_31

well designed, the results of a single study are limited to time, sample, and context characteristics [1]. In order to interpret this mass of information and lead to new studies, comprehensive and reliable studies are needed. A meta-analysis, which has frequently appeared in scientific journals in recent years, offers this opportunity.

Meta-analysis is a method of combining the findings of separate studies and reviewing criticism. Its specificity is that it uses quantitative methods rather than relying on judgment alone. This approach makes it different from other classic review methods.

Conducting a meta-analysis generally involves the following steps:

1. Defining the problem,
2. Identify criteria for including individual studies in the meta-analysis,
3. Obtaining individual research,
4. Coding and classification of each study according to the characteristics associated with meta-analysis,
5. Combining the findings of individual studies,
6. To relate the characteristics of the meta-analysis to the findings.
7. Report the findings of the meta-analysis.

Job Performance

Job performance is defined as an effort that employees should make for a salary [2]. Performance is defined as the quantitative and qualitative expression of an individual, a group, or an undertaking where he/she can reach for the intended purpose, in other words, what he/she can provide [3].

Evaluating and improving the performance of its employees is a very important task for all managers. Managers and team leaders, who value the job satisfaction expectations of employees, will see that they work much better. Managers and team leaders who do not pay enough attention to this need will only achieve short-term results and will have to deal with employees who are inadequate in terms of quality and quantity [4].

Perceived Organizational Support

Although support is a concept expressed as an informational, emotional, appreciation and material aid perceived by an individual [5] organizational support, which is a perception-based concept, means that an individual feels and thinks how much support is given to him by the organization [6]. Employees perceive the following attitudes in the organization as the support given to them:

- Consideration and implementation of ideas and suggestions,
- Tolerance of mistakes and feeling safe,

- Having a positive organizational climate in their relations,
- Believing that it is a fair administration and knowing that it will get what it deserves,
- Appreciated and honored by its organization.

Hypothesis 1: Perceived organizational support positively and significantly affects the work performance of employees.

Organizational Ethics

Organizational ethics is defined as the most valid perceptions of organizational procedures and activities based on correct and honest behavior in an organization. For example, when a decision affecting others is encountered, how does the organization determine the right alternative from the organization's point of view? The most important source of this information is the business climate that determines ethical behaviors in the business [7].

As a result, in today's economy where competition is inevitable and ruthless, on the one hand, and on the other hand, it is possible to make the right decisions and act according to objective measures in the ethical problems faced by the enterprises responsible for their environment.

Hypothesis 2: Ethical climate has a positive effect on individual performance.

Leadership Behaviors

Leader literally; is a creative person who guides, enlightens, shows forward, teaches, senses the needs and wishes of the people he works within time. A leader can also be defined as the person who is effective in determining or changing the aims and behaviors of the group he/she is a member of [8].

The main task of a manager is to use the human and material resources in the organization in the most efficient way to manage the organization by its objectives. Therefore, the manager must have some leadership qualities. The leadership qualities that a manager should have are as follows; it must have higher personality traits than its followers, interact with staff frequently enough, behave according to the conditions of the environment, and use its strength to motivate staff to work at all times and allow staff to participate in the decision. Expert writers of the subject also emphasized features that should not be present. These features are listed as follows; bias, lack of understanding, indecisiveness, being under the influence, fear, foresight.

The concept of leadership and management should not be confused. The leader does not need to take the biggest share in the decision process, and the fact that this share is low does not harm the leadership. Leaders are the creator and initiator of great plans. Executives enable these plans to be realized [9].

When the literature is examined, various concepts come to the forefront as leadership behaviors. These concepts are development culture, human-oriented leadership, task-oriented leadership, employee-oriented leadership, vision-oriented leadership, relationship-oriented leadership, risk-oriented leadership, control-oriented leadership, transformational leadership, interactional leadership, entrepreneurial leadership, and innovative leadership.

Hypothesis 3: Leadership behaviors positively and significantly affect the work performance of employees.

Organizational Motivation

The main purpose of motivation practices in organizations is to harmonize the objectives of the employees with the objectives of the organization and thus provide benefits both for themselves and for the organization while carrying out activities in line with the objectives of the organization. For this reason, various motivating factors are used in organizations, but although there is not always a motivating factor showing the same effect in each organization, an incentive tool for one employee may not have the same effect on the other employee. The impact of these encouraging factors depends on the individual's needs as well as the social level, education level, value judgments, and environmental factors.

Hypothesis 4: Organizational motivation positively and significantly affects the work performance of employees.

Rewarding

Colin Pitts describes the award as the gains that result from performing a task, providing a service, or performing a responsibility. According to Pitts, the main reward is the wage for employees who have been contracted to perform these tasks regularly. However, the prize is a package that includes wages and includes bonuses, pensions, health insurance, profit-sharing, and so on. According to Lundy and Cowling, the concept of rewards includes both direct wages and indirect employee gains. These are payments made by the employer to employees for their efforts and skills and are called external rewards. However, organizational rewards include internal rewards such as status, the definition of appreciation, company membership, security, career, development, sense of value given to the individual, and sense of accomplishment [10]. In the literature, the concept of reward is much higher than punishment. In the most common sense, "The reward is the boosters that enable the person to be encouraged to carry out the task assigned to him in a way that exceeds what is expected of him. "The award refers to tangible or non-tangible benefits that are provided on an uneven basis, in return for the employee's extraordinary achievements and contributions.

Hypothesis 5: Rewarding affects employees' job performance positively and significantly.

Organizational Democracy

Organizational democracy is the way that superiors and subordinates think together how to manage the organization, which methods will be used in this administration, how to achieve success [11].

Organizational democracy has some advantages as well as disadvantages. Advantages include: (i) organizational commitment may increase due to increased employee vocality, (ii) involvement in the decision-making process may facilitate the implementation of the decisions taken; (iii) democratic practices may pave the way for employees to feel and act more responsibly towards their organizations) adaptation to innovation and change can be saved together with the provision of participation climate; (v) joint decision-making and critical thinking can play an important role in enhancing the skills of employees; and (vi) democracy can be considered as a form of management in accordance with moral principles and rules. On the other hand, some disadvantages of democratic practices that give employees more power and authority are expressed as follows: (i) employees who do not have the ability to make decisions can sign decisions that are not suitable for the organization, (ii) democracy is a time-consuming set of practices and can reduce effectiveness, (iii) the application of the principles and rules of democracy may necessitate some changes, which may hinder organizational functioning, (iv) middle and upper-level managers may not want to share their power with lower-level employees and lower-level employees may oppose democratic practices due to their increased responsibilities, (v) democracy may not be a valid form of governance in all cases, for example, it may not meet expectations in a situation where rapid decisions are needed and, (vi) the implementation of democracy in the organization may not be morally correct, since if the performance of democracy decreases the stakeholders of the enterprise will be adversely affected [12].

Hypothesis 6: Organizational democracy positively and significantly affects the work performance of employees.

Workplace Roughness

Workplace rudeness is voluntary behavior that violates institutional norms, thus threatening the peace of the institution or employees [13].

Vulnerable behaviors exposed to working life cause individuals to lose confidence in their abilities and decrease their self-esteem and reputation [14]. As a result of the roughness of the workplace, which results in employees being withdrawn and drawn to their shells, they are deprived of the gains that can be obtained from the ideas and creativity of these individuals [15]. It can be said that these problems caused by roughness have the potential to decrease individual performance.

Hypothesis 7: Roughness in the workplace negatively affects the work performance of employees.

Leader-Member Interaction

The Leader-Member Interaction theory takes a different approach from other leadership theories by making it clear that the leaders do not interact with all group members within the working group in a similar leadership style and focusing on the mutual relations between the leader and members. The dual concept here refers to a leader and an audience, or a manager and a member/subordinate [16, 17].

Members who are in a high-quality relationship with the leader to whom they are affiliated tend to exert a high level of effort and personal commitment to their superiors. Thus, they contribute to improving the performance of the leader and the group. Leaders tend to respond to such subordinates by giving them more social support, offering organizational resources, and rewarding them [18].

The subordinate, who interacts at a high level with him, take on additional tasks and responsibilities in return for achieving higher status. The subordinate is asked to work more, to be attached to the leader, and to share some of the managerial work of the leader. Higher interaction develops slowly over time with mutual empowerment behaviors as the interaction cycle repeats. Unless this cycle is interrupted, the relationship develops to a point where there is a high degree of mutual trust, commitment, and support. The benefits of the leader from high-quality interaction are evident. For leaders who do not have enough time and energy, the help of dedicated subordinates is very important in the process of fulfilling the management responsibility. However, a high level of interaction also imposes certain obligations and constraints on the leader. In order to maintain these relationships, the leader must pay attention to his subordinates, respond to his needs and feelings, and devote more time to time-consuming methods of influence, such as persuading and consulting them.

Hypothesis 8: The level of Leader-Member interaction positively and significantly affects the work performance of employees.

Ethical Leadership

Key characteristics that characterize ethical leadership are character and accuracy, ethical awareness, community/individual focus, motivation, encouragement and empowerment, and ethical accountability management [19]. O'Connell and Bligh stated that the ethical leader has four main characteristics. Accordingly, the ethical leader first serves as a lens by reflecting on exactly what is happening. Secondly, the ethical leader creates clever awareness of his behavior and takes into account the effects of his behavior on his followers. Thirdly, the ethical leader exhibits a high level of virtue and acts within the framework of fair and correct behaviors in her professional and private life. Finally, the ethical leader is guiding his followers with value and justice because he believes it is the right thing to do [20].

Hypothesis 9: Ethical leadership affects employees' job performance positively and significantly.

Methodology

Study Design

In this study, the relationship between the concepts of leadership behaviors, organizational support, organizational ethics, organizational motivation, rewarding, organizational democracy, workplace roughness, leader-member interaction, and ethical leadership was tested using the meta-analysis method. Google Scholar was used for the literature review.

The criteria for this study are defined as follows:

- The surveys were conducted for 11 years between 2008 to 2019.
- Studies include correlation coefficient and sample size values for meta-analysis.
- Correlation value should be the correlation value of the concepts studied.
- Research should be carried out in Turkey.

Search Strategy

The research includes data on the work done together with examples from Turkey. Twenty of the 30 articles reviewed were evaluated for providing the necessary conditions (correlation coefficient, sample value, etc.) for the meta-analysis study. Ten articles are not included because they do not meet the requirements.

Coding Process

Coding is the process of obtaining all the necessary information from the selected studies. All the distinctive features of the studies are found in the coding form. The encoding form contains the following components:

- Name of Authors
- Year of publication
- Characteristics of the sample (sector, region)
- Correlation values
- Sample Dimensions.

Table 1 shows the characteristics of the collected studies.

Table 1 Summary of studies

	Researchers and publication year	Sample size	Province	Sector
1	Akkoç et al. 2012	348	Ankara	Service
2	Örücü and Kanbur 2008	270	Bursa	Industry
3	Altaş and Kuzu 2013	135	Sakarya	Service
4	Doğanay and Şen 2016	163	İstanbul	Service
5	Yılmaz and Karahan 2010	110	Uşak	Industry
6	Çınar and Yeşil 2016	306	Kahramanmaraş	Service
7	Altındağ and Akgün 2015	200	İstanbul	N/A
8	Kesen 2015	174	Aydın	N/A
9	Küçük and Çakıcı 2018	477	Mersin	N/A
10	Koçak and Özüdoğru 2012	340	Ankara	Service
11	Özutku et al. 2008	221	Afyonkarahisar	Industry
12	Akman et al. 2015	51	N/A	Industry
13	Bıyık et al. 2017	172	N/A	Industry
14	Alper and Keleş 2017	122	Sivas	Industry
15	Büte 2011	298	Gaziantep	N/A
16	Çalışkan and Akkoç 2012	543	Ankara	Service
17	Yılmaz and Tanrıverdi 2017	163	Antalya	Service
18	Demirel 2017	396	Antalya	Service
19	Aktuna and Kılıçlar 2019	397	Ankara	Service
20	Ayan 2015	112	Ankara	Service

Findings

Data, Heterogeneity and Publication Bias

Comprehensive Meta-Analysis Software (CMA) was used to analyze the data provided by the articles. The data sets are shown below:

1. data set: correlation coefficients between organizational support and employee performance
2. data set: correlation coefficients between organizational ethics and employee performance
3. data set: correlation coefficients regarding the relationship between leadership behaviors and employee performance (Table 2)

When the results are examined, it is seen that the 1st and 3rd data sets are heterogeneous at the 0.01 significance level, but the second data set is not heterogeneous at the same significance level. Compared to the variance of the sample, the I^2 statistic, which shows the percentage of variability due to differences between the studies, is above 97% in the 1st and 3rd data sets. This value ranges from 1 to 100%, and higher rates indicate higher heterogeneity. Due to the heterogeneity, a random-effects model will be used during the analyses. The I^2 statistic shows that there is no heterogeneity in the second data set. Furthermore, the significance level ($p = 0.836$) is not acceptable in heterogeneity analysis.

Publication bias can be defined as the probability that all research related to the subject of the missing data or meta-analysis study will not be published. It is one of the important limitations of meta-analysis studies. The funnel plot method is a method used to observe publication bias. The asymmetric distribution in the funnel plot is considered as a sign of publication bias, and publication bias was observed in the first dataset. The reasons for this bias may be reasons such as the fact that only published studies are taken into account, only Turkish publications are taken into account, bias due to citation, the low number of studies, and studies with negative findings are not taken into account.

There is no evidence of bias in the graph of the 3rd dataset. The fact that the number of studies is higher than the first data set may be a factor in this. At the same time, a more symmetrical graph can be obtained if more studies are included in the research. The reason for the lack of a funnel plot of the 2nd data set is that the number of studies is limited to 2. At least three studies are needed for the funnel plot.

Table 2 Heterogeneity analysis results

	Q	I^2
1st set	189.176	98.946*
2nd set	0.043	0.000
3rd set	285.389	97.197*

*Significant level at 0.01

Effect Size

The fixed-effect model and the random effect model are the two main models of a meta-analysis study. Using the random-effects model is one way to address heterogeneity. Heterogeneous studies are well suited for meta-analysis. The main assumption of the random effects model is that the actual impact may vary between studies. Therefore, the random-effects model was used.

The standard measure of meta-analysis is the impact dimension and is used to determine the strength and direction of the relationship being examined. Pearson correlation coefficients are used to calculate the effect size. In the hypothesis tests, the '*p*' value tends to shrink as the number of observations increases, whereas the effect size is not affected by the number of observations. The effect sizes of the data sets are shown in the following figures.

The effect size of the first data set is positive, and the H1 hypothesis is accepted. The impact of perceived organizational support on employee performance is 0.480 (r = 0.480) and reveals that the impact of perceived organizational support on employee performance is close to high [21].

The effect size of the second data set is positive, and the H2 hypothesis is accepted. The effect of organizational ethics on employee performance is 0.279 (r = 0.279), and the effect of organizational ethics on employee performance is moderate (see Cohen 1988).

The effect size of the third data set is positive, and the H3 hypothesis is accepted. The effect of leadership behaviors on employee performance is 0.617 (r = 0.617), and the effect of leadership behaviors on employee performance is high (see Cohen 1988).

Since the other studies subject to meta-analysis were not sufficient to form a cluster in the data set, the effect sizes were not calculated. The following table shows the values of these studies (Table 3).

When we look at other studies, it is seen that the relationship between organizational motivation and employee performance is high (r = 0.637), but it is not

Table 3 Other concepts affecting employee performance

Authors and publication year	The dependent variable	Sample size	Correlation coefficient
Örücü and Kanbur 2008	Organizational motivation	270	0.637
Altındağ and Akgün 2015	Rewarding	200	0.524*
Kesen 2015	Organizational democracy	174	0.104*
Küçük and Çakıcı 2018	Workplace roughness	477	0.247*
Özutku et al. 2008	Leader-member interaction	221	0.125*
Bıyık et al. 2017	Ethical leadership	172	0.478*

*Significant level is 0.05

significant at $p < 0.05$ significance level. In this case, the H4 hypothesis is rejected. There is no positive and significant relationship between organizational motivation and employee performance. On the other hand, when we look at other studies, we see rewarding 0.524, organizational democracy 0.104, workplace roughness 0.247, leader-member interaction 0.125, ethical leadership 0.478 correlation coefficients. Although there are strong, moderate, and weak relationships among them, these concepts are significant at $p < 0.05$ level. Hypotheses H5, H6, H7, H8, and H9 are accepted. There is a positive and significant relationship between employee performance and the concepts of reward, organizational democracy, workplace roughness, leader-member interaction, and ethical leadership.

Further studies are needed to evaluate the other concepts involved in the review in the meta-analysis format. When this happens, more realistic results and more consistent interpretations can be made.

Before starting the analysis of variables, the studies were grouped according to their geographical region and the business sector where the study was conducted. Studies are divided into Central Anatolia, Marmara, Mediterranean, and Aegean regions according to geographical regions. According to the business sector, the services and industry sectors were separated. In the following tables, the heterogeneity resulting from this grouping will be analyzed.

As shown in Table 4, high heterogeneity ($I^2 > 80\%$) was observed in all regions without distinguishing geographical regions. The reason for this heterogeneity is difficult to explain. However, reasons such as inter-study variability, quality of the study, bias from study selection, other bias effects, and the use of different statistical methods may be the cause of heterogeneity.

When the effect sizes are examined, high effect size values are observed with 0.639 and 0.549 in Central Anatolia and Marmara regions, respectively (see Cohen 1988). In the Mediterranean and Aegean regions, low impact magnitude values are observed with 0.275 and 0.359, respectively (see Cohen 1988). These results show that; In Central Anatolia and Marmara regions, 1 unit change in the dependent variable compared to Mediterranean and Aegean regions affects employee performance more. This change in the region may be due to the culture of the region, the business sector, the climate, or the educational level and marital status of the people.

As can be seen in Table 5, heterogeneity is high in both sectors without changing according to economic sectors ($I^2 > 80$). The reason for this heterogeneity is difficult to explain. However, reasons such as inter-study variability, quality of the study, bias

Table 4 Heterogeneity analysis by geographical regions

	Q	I^2	Point estimate
Central Anatolia	125.491	96.016	0.639*
Marmara	24.036	87.519	0.549*
Mediterranean	41.586	92.786	0.275*
Aegean	53.149	96.237	0.359*

*Significant level is 0.01

Table 5 Heterogeneity analysis by economic sectors

	Q	I^2	Point estimate
Service	372.613	97.316	0.531*
Industry	67.069	94.036	0.482*

*Significant level is 0.01

from study selection, other bias effects, and the use of different statistical methods may be the cause of heterogeneity.

When the effect sizes are examined, it is seen that there is a moderate relationship with 0.531 in the service sector and 0.482 in the industrial sector. The fact that the values are close to each other shows that the relationship between the dependent variable and employee performance does not change according to the sector.

Results

The results of the meta-analysis study showed that perceived organizational support and leadership behaviors have a significant effect on employee performance. In particular, the effect size of leadership behaviors has the highest impact value (ES = 0.617) of this study. However, organizational ethics does not affect employee performance. The impact value of organizational ethics (ES = 0.279) is very low.

The variable analysis showed that geographical regions might affect the relationship between dependent variables and independent variables. According to the heterogeneity analysis, according to geographical regions, although the effect size value was high in Central Anatolia and Marmara regions, the effect size value remained low in the Mediterranean and Aegean regions. This result is thought to be caused by various regional differences, especially cultural differences. When we analyzed heterogeneity according to economic sectors, no significant difference was found. There is no evidence that the relationship between dependent and independent variables is affected by sectoral differences.

In light of the findings of this study, H1, H2, H3, H5, H6, H7, H8, and H9 are supported, H4 is not supported.

Limitations and Future Research

The main purpose of this study is to examine the leadership behaviors and other components that are effective in the performance of the employees and to show how effective they are. The study contributes to the issues mentioned above. Therefore, it has some limitations. In this study, it was tried to reach enough studies with limited human resources in a limited time. In the future, some other authors may review local

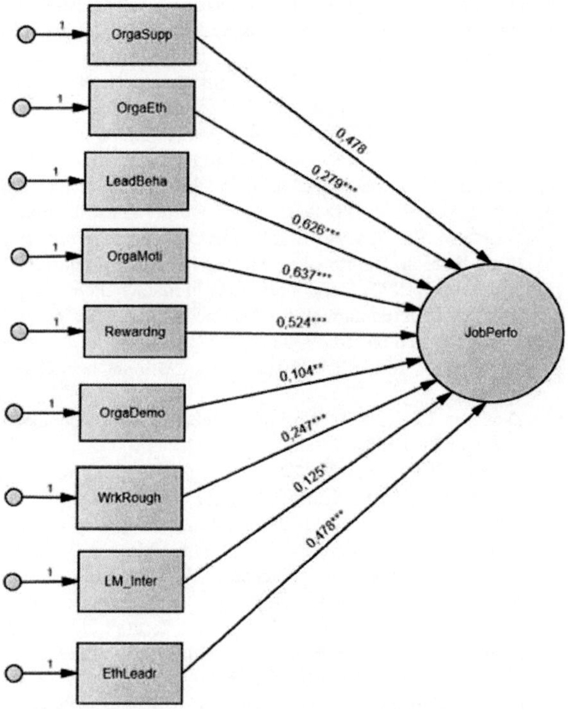

Fig. 1 Formal representation of the model with correlation coefficients. *Significant level is 0.05, **Significant level is 0.01, ***Significant level is 0.001

or national meta-analysis. Collaborations between authors can help authors conduct better research. This study may also serve as an example for managers and executive candidates.

Appendix

See Fig. 1.

References

1. Davies P (2000) The relevance of systematic reviews to educational policy and practice. Oxford Rev Edu 26(3–4):365–378
2. Rousseau DM, McLean PJ (1993) The contracts of individuals and organizations. In: Cummings LL, Staw BM (ed) Research in organizational behavior, vol 15. pp 1–43
3. Tınaz P (1999) Importance of performance evaluation system and a study of the application in Turkey. Journal of Celal Bayar University, Faculty of Economics and Administrative Sciences, vol 5. pp 389–496
4. Barutçugil İ (2004) Strategic human resources management. Kariyer Yayınları s, Istanbul 391

5. Parasuman S, Greenhaus JH, Granrose CS (1992) Role stressors, social support and wellbeing among two-career couples. J Organ Behav 13(4):339–356
6. Yoshimura KE (2003) Employee traits, perceived organizational support, supervisory communication, affective commitment, and intent to leave: group differences. Degree of Masters of Science, North Carolina State University, USA
7. Victor B, Cullen JB (1988) The organizational bases of ethical work climates. Administrative Science Quarterly (Psychological reports), pp 101–125
8. Saruhan SC, Yıldız ML (2009) Contemporary management science. Beta Basım A.Ş, İstanbul
9. Sünbül MA (2005) Teacher's world. Odun Pazarı Belediyesi Yayınları, Eğitim Dizisi, Eskişehir
10. Yürür Ş (1998) Analysis of relations between rewarding systems and organizational justice and an application. Unpublished Master's Thesis Cumhuriyet University, Sivas
11. Başaran İE (1992) Human relations in management. Kadıoğlu Matbaası, Ankara
12. Harrison JS, Freeman RE (2004) Special topic: democracy in and around organizations. Acad Manage Executive 18:49–53
13. Robbins SP, Judge TA (2012) Organizational behavior, (Çev. İ. Erdem). Nobel Yayınları, İstanbul
14. Zauderer DG (2002) Workplace incivility and the management of human capital. Public Manager 31(1):36–42
15. Pearson CM, Andersson LM, Wegner JW (2001) When workers flout convention: a study of workplace incivility. Hum Relat 54(11):1387–1419
16. Liden RC, Maslyn JM (1998) Multidimensionality of leader-member exchange: an empirical assessment through scale development. J Manage 24(1):43–72
17. Liden R, Graen G (1980) Generalizability of the vertical dyad linkage model of leadership. Acad Manage J 23(3):451–465
18. Schriesheim CA, Castro SL, Zhou XT, Yammarino FJ (2001) The folly of theorizing "A" but testing "B" a selective level-of-analysis review of field and a detailed leader member exchange illustration. Leadersh Q 12:515–551
19. Resick CJ, Hanges PJ, Dickson MW, Mitchelson JK (2006) A cross-cultural examination of the endorsement of ethical leadership. J Bus Ethics 63(4):345–359
20. O'Connell W, Bligh M (2009) Emerging from ethical scandal: can corruption really have a happy ending? Leadership 5(2):213–235
21. Cohen J (1988) Statistical power analysis for the behavioral sciences lawrence erlbaum associates, pp 77–83

Next-Day Operating Room Scheduling with Time-Dependent Stochastic Surgery Durations

Enis Kayış, Tuğçe Karataş, and Refik Güllü

Abstract Operating rooms (ORs) are the most costly part of hospitals, thus a priority for hospital administrations. In this paper, we consider the next-day OR scheduling problem for multiple operating rooms. We assume that surgeries have uncertain durations, and distributions of surgery durations are time-dependent. Our aim is to find the assignment of surgeries to the available ORs, the sequence, and the planned starting times of surgeries in order to minimize the weighted sum of expected waiting time of patients, idle time of ORs, and overtime of the hospital staff. In order to find solutions to the problem, we propose an L-Shaped method, customized to our problem formulation. We quantify the penalty of ignoring the time-dependency of surgery durations within a numerical study. We find that the penalty of ignoring the time-dependency increases with the overtime cost, average surgery durations, and decreases with surgery variability.

Keywords Operating rooms · OR scheduling · L-shaped method

Introduction

Healthcare expenditures are increasing with an ever-increasing rate [14]. According to [7], the proportion of healthcare expenditures in total GDP in the USA increased from 5.0% in 1960 to 17.8% in 2015. Hospital expenditures are responsible for a third of this spending. To reduce costs, many hospitals have been looking for ways to increase the efficient use of their resources. According to the Health Care Financial

E. Kayış (✉)
Industrial Engineering Department, Ozyegin University, Istanbul, Turkey
e-mail: enis.kayis@ozyegin.edu.tr

T. Karataş
Industrial Engineering and Operations Research Department, Columbia University, New York, NY, USA

R. Güllü
Industrial Engineering Department, Bogazici University, Istanbul, Turkey
e-mail: refik.gullu@boun.edu.tr

F. Calisir (ed.), *Industrial Engineering in the Internet-of-Things World*,
Lecture Notes in Management and Industrial Engineering,
https://doi.org/10.1007/978-3-030-76724-2_32

Management Association, operating rooms (ORs) account for more than 40% of a hospital's revenues and costs [18]. Therefore, it is crucial to increase the efficiency of ORs in order to better manage the increasing hospital costs.

OR planning and scheduling problems are especially challenging. First, many factors, including the ability of hospital staff, type of surgeries, and possible complications encountered during surgeries, result in the inherent uncertainty of surgery durations. Yet, the accuracy of surgery duration predictions is critical: If surgeries last shorter than predicted, ORs remain idle, and utilization of ORs decreases. If surgeries last longer than predicted, patients will wait for their surgeries, and this will result in dissatisfaction of patients. Moreover, the computational burden of the problem increases drastically as the number of ORs and thus the number of the surgeries increase, since there are too many possible assignments of surgeries to ORs and sequences of surgeries within ORs.

OR planning problems are divided into three categories in [1, 17]: strategic, tactical, and operational. Strategic level problems have long-term planning horizons. The long-term capacity of operating rooms and the allocation of operating rooms over different specialties are determined at the strategic level. The temporary capacity of operating rooms and the allocation of surgeries to different days in a week are determined at the tactical level. At the operational level, daily schedules of operating rooms are determined. Our research problem lies in this last category: What is the best next-day OR schedule for a list of surgeries given the time-dependent stochastic surgery durations?

In this chapter, we consider a next-day OR scheduling problem for multiple ORs. Emergent surgeries are ignored, and only elective surgeries are considered. It is assumed that surgeries have uncertain durations, and distributions of surgery durations are time-dependent. Furthermore, it is assumed that a surgery cannot start before its planned time, even if the previous surgery is completed earlier than its planned time. On the other hand, it may start later than the planned time if the previous surgery is completed after the planned time. The objective is to minimize the expected weighted sum of waiting time of patients, idle time of operating rooms, and overtime of the hospital staff. We provide a mathematical model formulation of the problem, and the L-Shaped method is used to solve the resulting formulation.

The vast literature on OR planning assumes that the duration of surgery does not depend on the scheduled time of the day. However, recent studies show that the performance of hospital staff (surgeons, nurses, anesthesiologists, technicians) could change according to the time of the day (e.g., [19, 19]). Most of the hospital staff are energetic in the morning, and this may result in a decrease in surgery durations of early surgeries. Moreover, hospital staff would not want to stay beyond planned working hours and be more effective in parallel processing and communication with other team members, which translates into reduced durations for late surgeries. We consider time-dependent surgery durations and quantify the effect of ignoring the time-dependency of stochastic surgery durations in OR planning.

A related avenue of research is scheduling papers that consider sequence-dependent setup times. In these papers, job duration is assumed to depend on the preceding job(s). In our study, we model time dependency such that the distribution of surgery duration depends on whether it is the first, middle, or last scheduled surgery. Apart from [35], which study the deterministic OR scheduling problem with sequence-dependent setup times, our paper is the first study that considers time-dependent stochastic surgery durations in next-day OR scheduling.

Literature Review

OR planning and scheduling problems have drawn great attention during the last 50 years. For a comprehensive review, we refer the reader to [6, 6]. In this section, we briefly discuss papers on stochastic OR planning grouped into two categories: papers that study only scheduling and those that consider both sequencing and scheduling of surgeries. Then, we highlight papers on machine scheduling with sequence-dependent setup times, which is a related stream for our study.

Papers in the first category assume that sequence of surgeries is known at the beginning and aims to find the optimally planned surgery durations. Reference [33] finds that the problem is of a "Newsvendor" type when there are two surgeries. If the number of surgeries is more, a myopic heuristic is proposed. Reference [31] aims to minimize the weighted customer delay time and system completion time. Under exponentially and independently distributed surgery durations, it proves the convexity of the objective function. It is argued that the optimal durations exhibit a dome-shaped pattern: the planned surgery durations first increase and then decrease as one moves toward the later surgeries in the sequence. Reference [8] formulate the problem as a two-stage stochastic linear program. It finds the lower and upper bounds of the objective function and then employs the L-Shaped method with sequential bounding to solve this problem. Reference [4] study the case where the surgery durations are distributed with a joint discrete probability function. It is shown that the optimal surgery durations can be calculated in polynomial time. Since the exact solution is hard to find, some papers focus on developing effective heuristics. Reference [29] propose a robust heuristic to minimize the weighted sum of the waiting time of the patients and the idle time of surgeons. Finally, [21] develop a method to find the exact solution to the problem and propose a hybrid heuristic motivated by real practices.

The problem of finding the optimal sequence as well as the planned starting times of surgeries within a single OR, is even more challenging. Reference [32] proposes a sequential two-phase method in which first the sequence, then the planned starting times are calculated. It finds that under exponentially distributed service times, sequencing jobs in order of decreasing means gives the optimal sequence. Reference [12] consider elective case scheduling to maximize the utilization of the operating rooms. Two heuristics for scheduling of surgeries are used: Earliest Start Time Heuristic and Latest Start Time Heuristic. The first heuristic is good at predicting start time, and the second heuristic can eliminate overtime. Reference [25] studies OR scheduling with the aims of maximizing the throughput and minimizing waiting time and overtime. It is argued that short procedures contain less variability than long procedures. Thus, scheduling short-duration procedures first maximize on-time performance. A two-stage stochastic programming model is presented in [10] to sequence surgeries and schedule surgery durations simultaneously. The paper employs the Sample Average Approximation (SAA) method to find solutions using four different sequencing heuristics. It is concluded that sequencing surgeries using Shortest-Variance-First (SVF) rule outperform other heuristics. Reference [28] prove that the finite scenario SAA method is NP-Complete. They develop a heuristic method based on Bender's decomposition, and the results are compared with the solutions found by the SAA method. It discusses that this heuristic has significantly better results than SVF heuristic when the unit costs are unequal. Reference [2] studies a single machine stochastic scheduling problem and finds that when the job processing times are distributed normally, and earliness and tardiness costs are the same for each job, the optimal sequence is SVF. Reference [15] extends this result and proves that the SVF rule is optimal under the assumption of dilation ordering of processing durations.

When there are multiple available ORs, another dimension of the assignment of surgeries to the ORs is introduced into an already challenging problem. Reference [9] develop a simulation model to handle uncertainties related to the intake process, surgical procedure, and recovery process. A simple simulated annealing method is used in order to improve the patient arrival schedule. Reference [22] develop a stochastic model for OR scheduling with elective and emergent surgeries under the assumption that surgery durations of elective surgeries are known and deterministic. A Monte Carlo optimization is proposed that combines Monte Carlo simulation and mixed-integer program. In this model, elective surgeries are assigned into different periods over the planning horizon in order to minimize the sum of elective patient-related costs and overtime cost of operating rooms. Reference [23] assume that operating rooms are identical, and the SAA method is used as a solution method. They find that the Monte Carlo optimization method converges exponentially to a real optimal solution. Reference [11] develop two methods to minimize the total sum of the fixed cost of opening operating rooms and the variable cost of overtime relative to a fixed length of a day. The first method is a two-stage stochastic linear program with binary decisions in the first stage, and simple recourse in the second stage. The binary decisions in the first stage are the number of operations to be opened, and the assignment of surgeries to operating rooms. The second method tries to minimize the maximum cost associated with the uncertainty of surgery durations. The paper finds that the second method is faster, and it benefits from limiting the worst-case outcome of the recourse problem. Reference [3] employ the L-Shaped method for multiple operating rooms and multiple surgeries problem. In the first stage, the number of operating rooms to be opened, the assignment of surgeries into operating rooms, and the sequence of the surgeries within ORs are determined. Therefore, this stage only includes binary variables. In the second stage, the weighted sum of the waiting time, idle time, and overtime are calculated by using the values of binary variables obtained in the first stage. Finally, [35] develop a method for solving a problem with sequence-dependent setup times of surgeries. The goal is to decide on the number of operating rooms to open, the assignment of surgeries to operating rooms and the sequence of surgeries within an operating room. It is assumed that surgeries belong to different types, each operating room allows only a set of surgery types and setup time, and surgery durations are deterministic. It is found that the Constraint Programming model gives more efficient results in terms of computation time and solution quality than the Mixed Integer Nonlinear Programming model.

There are several studies on machine scheduling with sequence-dependent setup times that are relevant to our problem. As in [35], the majority of studies on machine scheduling problems with sequence-dependent setup times focus on deterministic setup times due to the computational burden of stochastic problems. Reference [16] proposes an assignment algorithm, which is an extension of the Hungarian algorithm for scheduling jobs when there exist many identical parallel machines. Reference [26] propose a three-phase heuristic for scheduling of jobs when there exists only one machine. It is assumed that setup times are sequence-dependent, and all jobs are available for processing at time zero. In the first phase, several parameters that characterize the problem are calculated. By using these parameters, the schedule of jobs is found based on a priority rule in the second phase. In the last phase, the schedule obtained in the second phase is improved by using a local improvement procedure. Reference [27] extends the same problem to many parallel and identical machines. The same heuristic is used for the solution of the problem, and the simulated annealing method is applied in the third phase.

In conclusion, there are several studies on the scheduling and sequencing of the surgeries. It is nearly impossible to solve these problems exactly, and different heuristics are developed in order to solve these problems. However, there is no study on the scheduling and sequencing of ORs under time-dependent stochastic surgery durations. The problem of finding the optimal schedule under time-dependent setup times have been studied assuming deterministic setup and surgery durations. We consider stochastic surgery durations, and the computational complexity increases significantly as the number of surgeries and the number of ORs increases.

Problem Formulation

We study a stochastic next-day OR planning problem: Given the list of the surgeries to be performed the next day, one must assign each surgery in this list an OR, a sequence within this OR, and a planned starting time. Let $S = \{1, \ldots, N\}$ be the set of surgeries that need to be scheduled for the next day. There are m identical ORs, and each OR k is assigned with n_k surgeries for $k = \{1, \ldots, m\}$. Trivially we have $\sum_{k=1}^{m} n_k = N$. We assume that the number of surgeries assigned to each OR is determined in advance by hospital management.

All surgeries are assumed to be elective. Surgery durations are independently and non-identically distributed. We measure the surgery duration from wheels-in to wheels-out, i.e., surgery duration includes setup time before the surgery, actual surgery duration, and clean-up time after the surgery. $T_{i,j,k}$ is the random duration of surgery j, with a cumulative distribution function $F_{i,j}(x)$ and a probability density function $f_{i,j}(x)$, when it is assigned to i-th sequence in k-th OR. We assume that ORs are identical; hence surgery durations do not depend on the assigned OR. The first surgery of the OR starts without any waiting time, but surgeries later in the sequence should wait for the preceding surgery to be completed and the planned start time to arrive.

Our objective is to find a schedule that minimizes the expected weighted cost of waiting time of the patients, idle time of the operating rooms between two successive surgeries, and overtime of the ORs beyond planned opening times. We use weights for these cost components to reflect the fact that their priorities may not be equal. α_1 is defined as the unit cost of the idle time, α_2 is defined as the unit cost of the waiting time, and α_3 is defined as the unit cost of the overtime.

There are two types of decision variables in our model: continuous (planned surgery duration, waiting time, idle time and overtime) and binary (sequence and assignment of surgeries). $D_{i,k}$ is the assigned duration of i-th scheduled surgery in k-th OR. $w_{i,j,k}$ is the waiting time of the patient who is waiting for surgery j, which is assigned to i-th sequence in k-th OR. It takes positive value only if (i-1)-th scheduled surgery in k-th operating room lasts longer than its planned time. $s_{i,j,k}$ is the idle time of the OR right after i-th scheduled surgery in k-th OR. It is the time interval between the completion time of i-th scheduled surgery and the planned starting time of (i + 1)-th scheduled surgery in k-th OR. o_k is the excessive time over the total planned time of all surgeries assigned to k-th OR. The binary variable $x_{i,j,k}$ takes the value of 1 if surgery j is assigned to i-th sequence in k-th operating room.

Our problem could be formulated as a Stochastic Mixed-Integer Problem (SMIP). However, it is nearly impossible to find an exact solution. Instead, we will provide a formulation based on the SAA method, which is one of the most common approaches to solve SMIPs. SAA generates a number of scenarios (samples) for the random variables in the problem and estimates the objective function using the average weighted cost over these samples. In our context, a scenario is one particular sample path realization of all surgeries' random durations. Clearly, the approximation error reduces as the number of scenarios increases.

Some of the parameters and decision variables used in this method have an index p representing the scenario number. Let parameter $T^p_{i,j,k}$ be the random duration of surgery j in p-th scenario when it is assigned to i-th sequence in k-th OR. Since we assume that surgery durations are time-dependent, this parameter is sampled from the associated distribution with completing surgery j in the i-th sequence. $w^p_{i,j,k}$ is the waiting time of the patient who is waiting for surgery j, which is assigned to i-th sequence in k-th OR in p-th scenario. $s^p_{i,j,k}$ is the idle time of the operating room right after i-th scheduled surgery in k-th OR in p-th scenario. o^p_k is the excessive time over the total planned time of all surgeries assigned to k-th OR in p-th scenario.

One can formulate the problem to be solved with the SAA method as follows:

$$\text{Min } \frac{1}{P}\sum_{p=1}^{P}(\sum_{k=1}^{m}(\sum_{j=1}^{n}(\sum_{i=1}^{n_k}\alpha_1.s^p_{i,j,k}+\sum_{i=2}^{n_k}\alpha_2.w^p_{i,j,k})+\alpha_3.o^p_k)) \tag{1}$$

s.t.

$$-\sum_{j=1}^{n} w_{i,j,k}^{p} + \sum_{j=1}^{n} w_{i+1,j,k}^{p} - \sum_{j=1}^{n} s_{i,j,k}^{p} + D_{i,k} = \sum_{j=1}^{n} T_{i,j,k}^{p} \cdot x_{i,j,k} \tag{2}$$

i = 1,…,n_k - 1, k = 1,…,m and p = 1,…,P

$$-\sum_{j=1}^{n} w_{n_k,j,k}^{p} - \sum_{j=1}^{n} s_{n_k,j,k}^{p} + D_{n_k,k} + o_k^{p} = \sum_{j=1}^{n} T_{n_k,j,k}^{p} \cdot x_{n_k,j,k}. \tag{3}$$

k = 1,…,m and p = 1,…,P

$$\sum_{k=1}^{m} \sum_{i=1}^{n_k} x_{i,j,k} = 1 \quad j = 1,\ldots,N \tag{4}$$

$$\sum_{j=1}^{n} x_{i,j,k} = 1 \quad i = 1,\ldots,n_k \text{ and } k = 1,\ldots,m \tag{5}$$

$$s_{i,j,k}^{p} \leq M_1 \cdot x_{i,j,k} \quad i = 1,\ldots,n_k \text{ and } j = 1,\ldots,N \text{ and } k = 1,\ldots,m \text{ and } p = 1,\ldots,P \tag{6}$$

$$w_{i,j,k}^{p} \leq M_2.x_{i,j,k} \quad i = 2,\ldots,n_k \text{ and } j = 1,\ldots,n,\ k = 1,\ldots,m \text{ and } p = 1,\ldots,P \tag{7}$$

$$\sum_{k=1}^{j} \sum_{i=1}^{n_k} x_{i,j,k} = 1 \quad j = 1,\ldots,m \tag{8}$$

$$\sum_{k=r}^{\min\{j,m\}} \sum_{i=1}^{n_k} x_{i,j,k} - \sum_{a=r-1}^{j-1} \sum_{i=1}^{n_k} x_{i,a,r-1} \leq 0 \quad j = 2,\ldots,N \text{ and } r = 2,\ldots,\min(j,m) \tag{9}$$

$$s_{i,j,k}^{p}, w_{i,j,k}^{p} \geq 0 \quad i = 2,\ldots,n_k \text{ and } j = 1,\ldots,N \text{ and } k = 1,\ldots,m \text{ and } p = 1,\ldots,P \tag{10}$$

$$o_k^{p} \geq 0 \quad k = 1,\ldots,m \text{ and } p = 1,\ldots,P \tag{11}$$

$$D_{i,k} \geq 0 \quad i = 1,\ldots,n_k \text{ and } k = 1,\ldots,m \tag{12}$$

$$x_{i,j,k} \in \{0, 1\} \quad i = 1,\ldots,n_k \text{ and } j = 1,\ldots,N \text{ and } k = 1,\ldots,m \tag{13}$$

The objective function (1) is the weighted average cost of the waiting time of the patients, the idle time of the ORs, and the overtime of the surgeons. Constraints (2) and (3) provide the relationship between the waiting time of the patients, the idle time of the ORs, and the overtime of the surgeons. Constraints (4) and (5), respectively, ensure that surgery is assigned to exactly one sequence in exactly one OR, and each sequence in each OR is assigned to exactly one surgery. Constraints (6) and (7) guarantee that the waiting time and the idle time variables are equal to 0 unless j-th surgery is assigned to i-th sequence in k-th OR. In these constraints, M indicates a large number. Constraints (10), (11), and (12) define the non-negativity restrictions of the waiting time, the idle time, the overtime, and the assigned duration variables, respectively. Constraint (13) is the integrality constraint.

Constraints (8) and (9) are symmetry breaking constraints: In terms of sequencing of surgeries, there are N! possible allocation of surgeries. However, it is possible to decrease the number of allocations that should be considered, since we assume that ORs are identical. This constraint set prevents obtaining the same solution again. Constraint (8) helps assigning the surgeries to operating rooms in the lexicographical order, since assigning surgeries lexicographically will give a feasible solution to the problem (see, for example, [30]). According to this constraint, surgery 1 should be assigned to OR 1. Surgery 2 can be assigned to the OR 1 or OR 2. By following this pattern, surgery (m − 1) can be assigned to the first (m − 1) ORs, and surgery m can be assigned to any OR. Constraint (9) prevents assigning surgeries to ORs that have a larger OR number index than the smallest OR number index of empty ORs (see, for example, [11]). For example, surgery 4 cannot be assigned to the third OR if there exists no surgery in the second OR.

Solution Methodology

As the number of operating rooms and the number of surgeries increase, the problem size increases drastically. For these cases, the SAA method cannot provide a solution within a reasonable computation time. As an alternative, we propose to solve our problem using the L-Shaped method (see, for example, [3], and [5, 24, 34]).

The L-shaped method computes a solution in two stages, defined as a master problem and a subproblem. In the master problem, the assignment of surgeries to ORs and their sequences within the OR are determined by solving a mixed-integer problem. The objective of this problem is to minimize the value of a variable, denoted by Θ, which is the expected recourse function. Θ is the expected value of the objective function of the subproblem given the sequences of surgeries in their assigned ORs. In the second stage, the total weighted sum of the waiting time of the patients for the surgeries, the idle time of the operating rooms, and the overtime of the ORs is minimized. A slightly modified version of the SAA model defined in the previous section is used to solve the subproblem. Since the values of the binary variables are assigned in the first stage, the second stage of the L-Shaped method is solving a simple linear programming problem. Figure 1 presents the overall algorithm of the L-Shaped method.

As an input, the algorithm requires $\alpha_1, \alpha_2, \alpha_3$ weights, the number of operating rooms (m), and the total number of surgeries (N). It is assumed that each operating room will have an equal number of surgeries. Furthermore, parameters of time-dependent surgery duration distributions for each surgery should be provided.

Require $\alpha_1, \alpha_2, \alpha_3$, m and N;
Require Parameters of distributions of surgery durations for i=1,2,...,N/m and j=1,2,...,N;
$w \Leftarrow 0$;
$\Theta \Leftarrow -\infty$;
while $w > \Theta$ **do**
    solve the master problem to obtain the values of x decision variables and the corresponding objective value Θ;
    substitute x into the subproblem;
    generate scenarios and obtain T matrix;
    solve the subproblem;
    $w \Leftarrow \pi^T * (h - T * x)$;
    add $\Theta \geq \pi^T * (h - T * x)$ to the master problem;
end while
w is the optimal value of the problem.

Fig. 1 L-Shaped method algorithm

The algorithm starts by initializing w and Θ variables: w is set to 0, and Θ is set to $-\infty$ in this step. w is the objective value of the subproblem, and Θ is the objective value of the master problem. At each step of this algorithm, Θ and w values are obtained, respectively. The algorithm continues to iterate as long as the value of w is larger than the value of Θ. At each iteration, an optimality cut is added to the master problem, and these cuts help Θ converge to the optimal value of the problem.

In the algorithm, the value of w is calculated as $\pi^T \times (h - T \times x)$, π is the matrix of optimal simplex multipliers, and h is the vector of constants of each constraint. In our problem, we take $h = 0$ since there exist no constants in all constraints. T is the coefficient matrix of the values of the $x_{i,j,k}$ binary variables. This matrix consists of the surgery duration values generated in different scenarios. Once the subproblem is solved, if the value of w is larger than Θ, an optimality cut, $\Theta \geq \pi^T \times (h - T \times x)$, is generated and added to the master problem.

The master problem formulation starts with the objective function, which is simply the minimization of the variable Θ. The constraints in the first iteration are constraints (4), (5), (8), (9), and (13) of the SAA problem formulation, as defined in the previous section. In the subsequent iterations, additional optimality cuts are added to the master problem, as explained before.

Once the assignment of surgeries to ORs and their sequences within the ORs are determined in the master problem, the subproblem aims to find the planned starting times of the surgeries. The objective function is exactly that of the SAA problem formulation as defined in the previous section, which is the average total weighted cost of idle time, waiting time, and overtime across the scenarios. The constraints of the subproblem are Eqs. (2–3), (6–7), and (10–12) from the SAA problem formulation.

Numerical Results

To understand the computational performance of the L-shaped method and quantify the value of considering time-dependent surgery durations, we conduct a computational study. We investigate the relationship between different parameter settings and the penalty of ignoring time-dependent surgery durations, which is calculated as the per-centage difference between the optimum objective value of the problem with and without the assumption of time-dependent surgery durations, as shown below:

$$PI = \frac{Obj^i - Obj}{Obj} \times 100\%$$

In this equation, Obj^i is the objective value obtained when time-dependent surgery durations are ignored, and Obj is the optimum objective value obtained when time-dependent surgery durations are incorporated.

In order to decrease the computation time of the L-shaped method further, the SVF rule is used to sequence the surgeries. Particularly, we include the following constraint to the master problem in order to sequence the surgeries according to the SVF rule:

$$\sum_{j=1}^{N} x_{i-1,j,k} \times V_{i-1,j} - \sum_{j=1}^{N} x_{i,j,k} \times V_{i,j} \leq 0 \quad i = 2, \ldots, n_k \text{ and } k = 1, \ldots, m$$

In this constraint $V_{i,j}$ is the variance of the distribution of the surgery duration of surgery j when it is assigned to i-th sequence. Finally, we use mean surgery durations as the planned surgery durations as a heuristic in the initial iteration of the L-shaped method to give an initial lower bound to the objective function of the master problem. The weighted cost obtained from this heuristic is used as a lower bound for the objective function of the master problem.

Experimental Setting

In our computational study, we vary a few parameters to study the extent of PI and the relationship between PI and our problem parameters. The number of operating rooms is chosen from the set $\{2, 3, 4\}$, and the number of surgeries for each operating room is taken as 3. Following the literature on surgery duration estimation, we use the log-normal distribution family to model surgery duration distributions. We take the time unit as an hour. We consider two different modes for the surgery duration distributions: short duration and long duration. For each mode, two different values for mean and coefficient of variation (CV) of surgeries durations are generated. The mean of surgery durations is generated randomly from U[0.5, 1] to model short duration surgeries and from U[1, 4] to model long duration surgeries. CV of surgery durations is generated randomly from U[0.4, 0.8] for short duration surgeries, and from U[0.8, 1.2] for long-duration surgeries.

We consider the time-dependency of surgery durations in our computational study as follows. The mean and standard deviation of the surgery durations of surgeries assigned to the first sequence in each operating room and surgeries assigned to the last sequence in each operating room is adjusted by multipliers. The value of the coefficient of variation will remain the same since both mean and standard deviation will be multiplied by the same value. The multiplier of surgeries assigned to the first sequence in each operating room is chosen from the set $a_{\text{first}} = \{0.9, 0.95, 0.983\}$, and it is called the first-surgery multiplier. The multiplier of surgeries assigned to the last sequence in each operating room is chosen from the set $a_{\text{last}} = \{0.8, 0.9\}$, and it is called the last-surgery multiplier. These parameter values are chosen following the empirical study of [19].

Furthermore, there will be four different combinations of $\{\alpha_1, \alpha_2, \alpha_3\}$: {0.2, 0.2, 0.6} is used to observe the effect of high overtime weight, {0.2, 0.6, 0.2} is used to observe the effect of high waiting time weight, {0.6, 0.2, 0.2} is used to observe the effect of high idle time weight, and {0.33, 0.33, 0.33} is used to model equal weights settings.

There are three options for the number of ORs, three options for the first-surgery multiplier, three options for the last-surgery multiplier, four different combinations for $\{\alpha_1, \alpha_2, \alpha_3\}$, two options for the mean of surgery durations, and two options for coefficient of variation of surgery durations. In total we consider $3\times3\times2\times4\times2\times2 =$ 288 different parameter combinations.

Analysis of the Results

We summarize the minimum, average, and maximum percentage penalty of ignoring time dependency, PI, as we change problem parameters one at a time in Table 1.

The average penalty decreases as the first-surgery multiplier increases, as expected. The slope of the decrease is smaller as one gets closer to 1. Similarly, the average penalty decreases as the last-surgery multiplier increases from 0.8 to 0.9. Remember that when the last surgery duration is overestimated, only the idle time cost due to the last surgery being completed before planned time increases. None of the preceding surgeries are affected. Hence the penalty could be limited. However, it is possible to modify OR assignment and sequence in order to take advantage of time-dependency. These results support our insights.

Next, we investigate the effect of having long surgeries or surgeries with higher duration variability on the penalty of ignorance. When the list includes relatively long surgeries, the average penalty is higher. On the other hand, when the list includes surgeries with relatively more variable durations, the average penalty decreases. We believe that when there is sufficient variability in the surgery durations, the advantage of considering time-dependency disappears.

Table 1 Statistics of PI as problem parameters vary

Parameter	Value	Min (%)	Avg (%)	Max (%)
a_{first}	0.9	0.06	1.96	6.52
	0.95	0.04	1.56	6.78
	0.983	0	1.59	4.88
a_{last}	0.8	0.06	2.04	6.52
	0.9	0	1.36	6.78
Mean mode	Small	0	1.36	5.70
	Large	0.05	1.99	6.78
CV mode	Small	0.09	2.15	6.78
	Large	0	1.12	3.27
Unit cost comb	{0.2, 0.2, 0.6}	0.12	2.01	6.52
	{0.2, 0.6, 0.2}	0.09	1.52	6.78
	{0.33, 0.33, 0.33}	0.06	1.7	5.44
	{0.6, 0.2, 0.2}	0	1.54	4.74

Keeping the unit cost of idle time as 0.2, we observe that the average penalty of ignoring time-dependent surgery durations decreases when the unit cost of waiting time increases, and the unit cost of overtime decreases. Moreover, when the unit cost of waiting time is kept as 0.2, we find that the average penalty of ignoring time-dependent surgery durations decreases as the unit cost of idle time increases, and the unit cost of overtime decreases. Finally, we observe that the average penalty is high when the overtime cost has significant weight or all unit costs are equal. The former observation is due to the effect of the last-surgery multiplier.

Computational performance of our L-shaped method deteriorates as the number of ORs increases due to the increasing number of binary variables: 36 binary variables are used when there are 2 ORs. This number is 81 with 3 ORs and 144 with 4 ORs. Table 2 shows the number of iterations in the L-shaped algorithm as well as the total solution time in seconds. There is a superlinear increase in the average number of iterations as the number of ORs increases. However, the increase in the average solution time is more superlinear. This observation highlights the fact that the master problem requires significantly more time as the number of ORs increases. To check the effect of using SVF heuristic in the master problem, we computed the solution

Table 2 Solution time and the number of iterations as the number of ORs varies

	# of iterations			Time (seconds)		
# of ORs	Min	Avg	Max	Min	Avg	Max
2	2	2.61	14	12.27	25.16	99.52
3	2	4.90	104	16.63	174.70	1471.56
4	2	8.77	315	24.52	1315.37	8853.20

with 2 ORs without this heuristic. We find that the optimum values are very close, yet the average (maximum) solution time increases from 25.16 (99.52) seconds to 167.57 (2650.69) seconds. Therefore, we conclude that using SVF heuristic in the master problem is quite effective.

Conclusions

ORs are responsible for the majority of a hospital's revenues, and also the costliest part of healthcare services. In this paper, we consider the next-day OR scheduling problem with multiple ORs. We study this problem under time-dependent and stochastic surgery durations. We formulate the problem as a Stochastic Mixed Integer Program and propose an L-shaped algorithm to solve this problem within a reasonable time limit. Through a computational study, we quantify the penalty of ignoring time-dependency in creating OR schedules and investigate the effects of problem parameters on this penalty. Results show that penalty of ignoring time-dependent surgery durations increases as the unit cost of idle time or the unit cost of waiting time decreases. Furthermore, the penalty of ignoring time-dependent surgery durations is directly proportional to surgery durations, whereas inversely proportional to surgery duration variability.

Further research is needed to develop heuristics to solve problems with larger numbers of surgeries and ORs within reasonable computational times. Another avenue for research is to incorporate other factors that effect surgery duration, such as surgery team and other resources. Finally, integration of pre and post-operative processes and their resource usage could create additional insights for the OR managers.

Acknowledgements This work was supported by the Scientific and Technological Research Council of Turkey (TÜBİTAK), Grant No: 114M388.

References

1. Abdelrasol ZY, Harraz N, Eltawil A (2013) A proposed solution framework for the operating room scheduling problems. Proc World Congr Eng Computs Sci 2:23–25
2. Baker KR (2014) Minimizing earliness and tardiness costs in stochastic scheduling. Eur J Oper Res 236(2):445–452
3. Batun S, Denton BT, Huschka TR, Schaefer AJ (2011) Operating room pooling and parallel surgery processing under uncertainty. INFORMS J Comput 23(2):220–237
4. Begen MA, Queyranne M (2011) Appointment scheduling with discrete random durations. Math Oper Res 36(2):240–257
5. Birge JR, Louveaux F (2011) Introduction to stochastic programming. Springer Science & Business Media
6. Cardoen B, Demeulemeester E, Beliën J (2010) Operating room planning and scheduling: a literature review. Eur J Oper Res 201(3):921–932

7. Centers for Medicare and Medicaid Services (2015) NHE summary including share of GDP, CY 1960–2015. https://www.cms.gov/Research-Statistics-Data-and-Systems/Statistics-Trends-and-Reports/NationalHealthExpendData/Downloads/NHEGDP15.zip. Accessed at March 2016
8. Denton B, Gupta D (2003) A sequential bounding approach for optimal appointment scheduling. IIE Trans 35(11):1003–1016
9. Denton BT, Rahman AS, Nelson H, Bailey AC (2006) Simulation of a multiple operating room surgical suite. In: Simulation conference, 2006. WSC 06. Proceedings of the winter, IEEE, pp 414–424
10. Denton B, Viapiano J, Vogl A (2007) Optimization of surgery sequencing and scheduling decisions under uncertainty. Health Care Manag Sci 10(1):13–24
11. Denton BT, Miller AJ, Balasubramanian HJ, Huschka TR (2010) Optimal allocation of surgery blocks to operating rooms under uncertainty. Oper Res 58(4):802–816
12. Dexter F, Traub RD (2002) How to schedule elective surgical cases into specific operating rooms to maximize the efficiency of use of operating room time. Anesth Analg 94(4):933–942
13. Erdogan SA, Denton BT (2011) Surgery planning and scheduling. Wiley encyclopedia of operations research and management science
14. Erixon F, Van der Marel E (2011) What is driving the rise in health care expenditures?: an inquiry into the nature and causes of the cost disease. ECIPE working papers 5
15. Guda H, Dawande M, Janakiraman G, Jung KS (2016) Optimal policy for a stochastic scheduling problem with applications to surgical scheduling. Prod Oper Manag 25(7):1194–1202
16. Guinet A (1993) Scheduling sequence-dependent jobs on identical parallel machines to minimize completion time criteria. Int J Prod Res 31(7):1579–1594
17. Hans EW, Van Houdenhoven M, Hulshof PJ (2012) A framework for healthcare planning and control. Handbook of healthcare system scheduling. Springer, Berlin pp 303–320
18. Healthcare Financial Management Association (2003) Achieving operating room efficiency through process integration. Healthc Financial Manage J Healthc Financial Manage Assoc 57(3):1–8
19. Kayış E, Khaniyev TT, Suermondt J, Sylvester K (2015) A robust estimation model for surgery durations with temporal, operational, and surgery team effects. Health Care Manag Sci 18(3):222–233
20. Kc DS (2013) Does multitasking improve performance? Evidence from the emergency department. Manuf Serv Oper Manag 16(2):168–183
21. Khaniyev T, Kayış E, Güllü R (2020) Next-day operating room scheduling with uncertain surgery durations: exact analysis and heuristics. Eur J Oper Res, forthcoming
22. Lamiri M, Xie X, Dolgui A, Grimaud F (2008) A stochastic model for operating room planning with elective and emergency demand for surgery. Eur J Oper Res 185(3):1026–1037
23. Lamiri M, Grimaud F, Xie X (2009) Optimization methods for a stochastic surgery planning problem. Int J Prod Econ 120(2):400–410
24. Laporte G, Louveaux FV (1993) The integer L-shaped method for stochastic integer programs with complete recourse. Oper Res Lett 13(3):133–142
25. Lebowitz P (2003) Schedule the short procedure first to improve OR efficiency. AORN J 78(4):651–659
26. Lee YH, Bhaskaran K, Pinedo M (1997) A heuristic to minimize the total weighted tardiness with sequence-dependent setups. IIE Trans 29(1):45–52
27. Lee YH, Pinedo M (1997) Scheduling jobs on parallel machines with sequence-dependent setup times. Eur J Oper Res 100(3):464–474
28. Mancilla C, Storer R (2012) A sample average approximation approach to stochastic appointment sequencing and scheduling. IIE Trans 44(8):655–670
29. Robinson LW, Chen RR (2003) Scheduling doctors' appointments: optimal and empirically-based heuristic policies. IIE Trans 35(3):295–307
30. Sherali HD, Smith JC (2001) Improving discrete model representations via symmetry considerations. Manage Sci 47(10):1396–1407

31. Wang PP (1993) Static and dynamic scheduling of customer arrivals to a single-server system. Nav Res Logist 40(3):345–360
32. Wang PP (1999) Sequencing and scheduling N customers for a stochastic server. Eur J Oper Res 119(3):729–738
33. Weiss EN (1990) Models for determining estimated start times and case orderings in hospital operating rooms. IIE Trans 22(2):143–150
34. Van Slyke RM, Wets R (1969) L-shaped linear programs with applications to optimal control and stochastic programming. SIAM J Appl Math 17(4):638–663
35. Zhao Z, Li X (2014) Scheduling elective surgeries with sequence-dependent setup times to multiple operating rooms using constraint programming. Oper Res for Health Care 3(3):160–167

Value of MRI and Ultrasound Screening for Breast Cancer in Non-High-Risk Populations

Burhaneddin Sandikçi and Mücahit Çevik

Abstract Mammography is the gold standard for breast cancer screening in the vast majority of the world, but it is known to be less accurate for women with dense breasts. To improve cancer detection accuracy, supplemental ultrasound and Magnetic Resonance Imaging (MRI) screening have been recently introduced and are actively recommended for high-risk populations by many agencies. This chapter studies the value of supplemental tests in non-high-risk populations using a partially observable Markov decision process model alongside a simulation model. A numerical study using these models driven by clinical data reveals that supplemental tests may not cause any meaningful improvement in the quality-adjusted life expectancy for non-high-risk women, and they may indeed be harmful if used routinely after biennial or annual mammographies for non-high-risk women. However, they are associated with significant improvements in overall cancer detection rate, the time to detect cancer, fraction of in situ cases that deteriorate to the invasive stage as well as the fraction of women who die with undiagnosed cancer. While MRI is generally more effective than ultrasound on several performance metrics, it also suffers from significantly increased false positives, hindering its viability for this population.

Keywords Medical decision making · Breast cancer screening · Mammography · Magnetic resonance imaging · Ultrasound · Markov decision process · Simulation

Introduction

Breast cancer is a disease characterized by the uncontrolled growth of cells in the breast tissue. These abnormal cells can spread over time to other organs in the body—a phenomenon known as metastasis—and cause death. Breast cancer is the most

---

B. Sandikçi (✉)
Booth School of Business, University of Chicago, Chicago, IL, USA
e-mail: burhan@chicagobooth.edu

M. Çevik
Department of Mechanical and Industrial Engineering, Ryerson University, Toronto, ON, Canada
e-mail: mcevik@ryerson.ca

F. Calisir (ed.), *Industrial Engineering in the Internet-of-Things World*,
Lecture Notes in Management and Industrial Engineering,
https://doi.org/10.1007/978-3-030-76724-2_33

frequently diagnosed cancer and the leading cause of cancer death in women worldwide: an estimated 2.09 million new cases were diagnosed (accounting for more than 25% of all new female cancers), and 626,679 women died due to breast cancer (accounting for 15% of cancer-related female deaths) in 2018 [11].

Global Cancer Observatory (http://gco.iarc.fr) data indicate that incidence rate (i.e., new cases per 100,000 population) is 46.3 worldwide, but it varies significantly across the globe: it is highest in continental Oceania (Australia and New Zealand), Europe (with Belgium having the highest global rate of 113.2) and North America. Risk factors associated with developing breast cancer include personal and family history of breast or ovarian cancer, inherited gene mutations such as BRCA1/2 genes, breast density, early age of menarche, late age of menopause, postponement of childbearing and having fewer children, use of oral contraceptives and hormone replacement therapy, unhealthy diet and obesity, tobacco and alcohol use, decreased levels of physical activity and breastfeeding [3, 12, 17, 33].

Mortality rates, however, display much less variability: worldwide average in 2018 is estimated at 13.0 per 100,000 population, while it is lowest in Asia (11.3) and highest in Africa (17.2). Survival rates improve over time across the globe. Early detection is critical to improving survival since stage at diagnosis is one of the most important factors affecting prognosis [3]. Mammography screening is identified as the most cost-effective method for the early detection of breast cancer [15]. Screening refers to the examination of asymptomatic individuals to detect a disease before clinical symptoms are revealed, hence prevent the spread of the disease from helping maximize the likelihood of treatment success. Simply stated, mammography screening involves taking an X-ray image of the breast and the interpretation of this image by an experienced radiologist. While most of the developed world (including the United States and Western European countries) use mammography as the gold standard for breast cancer screening, there are disagreements about at what age to start/stop screening and the frequency of screening. The imperfect nature of mammography (i.e., false positives indicating false alarms in healthy women and false negatives indicating missed cancer cases) contribute to such disagreements. The degree of its accuracy decreases even further for women with dense breasts [20].

The need to improve the detection accuracy of screening has led to supplementing mammography, particularly for women who are at high risk of breast cancer, with additional screening modalities. Among the available modalities for evaluation of the breast, ultrasound and MRI have been found to have the highest incremental effectiveness for the detection of breast cancer [2, 10, 20]. American Cancer Society, for example, recommends breast MRI as a supplement to mammography for patients at high risk defined as those with a lifetime risk of at least 20%. The main rationale for limiting MRI screening to those at greatest risk is in part due to its relatively high cost [8, 25].

In this study, we focus on quantifying the value of supplemental MRI and ultrasound screening for breast cancer in non-high-risk women. For this purpose, we ignore financial costs and focus purely on the incremental life benefits of supplemental screening. To optimize the use of available screening modalities (i.e., mammography, supplemental ultrasound, and supplemental MRI) for a woman, we use a partially observed Markov decision process (POMDP) model that maximizes an individual's quality-adjusted life expectancy (QALE).

POMDPs provide a natural framework to optimize stochastic and dynamic systems, where the system state evolves stochastically over time, but the decision-maker cannot directly observe its evolution. Lacking direct access to the actual system state (also known as the core state), the decision-maker receives signals that (partially) informs about the actual state. For instance, in a manufacturing system, a machine may be going through various deterioration states. Opening up the machine to directly observe the machine state may be too costly, but the system controller may infer the state of the machine by assessing the quality of the products obtained from the machine. We refer the interested reader to [13, 17, 20, 29] for detailed reviews of the POMDP methodology, solution algorithms, and applications.

Several papers have used partially observed Markov process models to study screening optimization in breast cancer detection. Maillart et al. [23] are among the first to formulate a partially observed Markov chain model to evaluate a given set of mammography screening policies through sample-path enumeration of the underlying Markov chain. Madadi et al. [22] extend the work of Maillart et al. [23] to also incorporate patient adherence. Ayer et al. [4] formulate and solve a POMDP model to optimize mammography screening decisions based on an individual's risk of breast cancer, which is later extended by Ayer et al. [5] to also incorporate patient adherence. Cevik et al. [14] propose a constrained POMDP model to optimize the mammography screening schedule when patients are allowed a limited number of mammographies due to resource constraints. All of the cited literature, however, consider mammography as the only screening modality. The only exception, to the best of our knowledge, is the work of Sandıkçı et al. [28], which offers a POMDP model to optimize mammography plus supplemental ultrasound and MRI screening for high-risk women. We utilize the model proposed in Sandıkçı et al. [28] to study the value of supplemental ultrasound and MRI screening in non-high-risk populations.

Methodology

The POMDP Model

We start with simplifying assumptions to capture the complex breast cancer screening decision process in a mathematically tractable way. Accordingly, we assume that the screening decisions for a woman start when she reaches a certain age (e.g., 40 years) and continues at regular intervals (e.g., annually) until she reaches a certain age (e.g., 80 years). At each decision interval, for the woman under consideration, the physician assesses her risk of breast cancer, recommends whether or not mammography screening should take place and, depending on the result of the mammography, recommends whether or not supplemental ultrasound or MRI screening should take place. Since missing a cancer case (i.e., a false negative conclusion) is arguably the most important concern, we consider supplemental tests only after a negative mammography test. We also assume that the patient fully adheres to the recommendations of her physician. In addition to imperfect screening tests (namely, mammography, ultrasound, and MRI), the physician has access to a perfect diagnostic test (e.g., biopsy) that can reveal the actual cancer state of the woman. The biopsy option, however, is costly/painful/stressful, and, therefore, it is performed only if the woman tests positive from any of the screening tests. If the biopsy confirms the positive screening test, then the screening decision process concludes, and the patient immediately starts cancer treatment. If the biopsy finding reveals that the screening outcome is a false positive, then the physician revises the cancer risk assessment for the patient, and the screening process continues with this updated risk assessment in the next decision interval.

We use a discrete-time finite-horizon nonstationary POMDP model, first proposed in Sandıkçı et al. [28], to optimize breast cancer screening recommendations of a physician. For completeness, we recall the technical details of this model. The planning horizon is denoted by $\tau < \infty$ and decision points are indexed by $t = 1, 2, \ldots, \tau$. Let $i \in S$ denote the actual state of the patient. The state-space S includes an absorbing state for 'patient death' as well as many 'surviving' states, which can track all relevant information for making the screening decisions. A subset of the surviving states (e.g., whether or not the patient has cancer and the stage of cancer) may not be directly observable to the decision-maker. This issue is circumvented by constructing a belief vector π, whose ith component $\pi(i)$ denotes the probability that the patient's actual state is $i \in S$. This construct is a widely accepted modeling approach for partially observable systems as it is known to be a sufficient statistic for the entire history of the process [24], and therefore regains the Markovian property of the decision process. The set of all possible actions available to the physician is denoted $A = \{W$ (watchful waiting), M (mammography), U (ultrasound), R (MRI), N (no followup)$\}$.

Disease onset and progression is modeled through state transition probabilities $p_t^a(j|i)$, which denotes the probability that the actual state of the patient is $j \in S$ at time $t + 1$ given that her actual state is $i \in S$ and she takes action $a \in A$ at time t. This construct captures the stochastic evolution of the patient's actual state over time and emphasizes that the state evolution is modeled in a Markovian fashion. The actual state of the process is not observable to the decision-maker, but taking action $a \neq N$ yields a signal $o \in O$ with probability $q_t^a(o|i)$ given the actual underlying state is $i \in S$ at time t. The set O of all observations can be broad and include all relevant observations that the physician can make. For simplicity, we restrict it to the screening outcomes; that is, $O = \{$'$-$', '+'$\}$, with '+' ('$-$') indicating abnormalities were (were not, respectively) observed in the test result. The parameter $q_t^a(o|i)$ is referred to as observation probability, which reflects the accuracy of the screening action a. For example, when the mammography (M) action is taken, q_t^M('+'| *patient has cancer*) would be the sensitivity of mammography, while q_t^M('+'| *patient has cancer*) would be the specificity of mammography.

The set of all belief states is the probability simplex $\Pi(S) = \left\{ \boldsymbol{\pi} \in \Re^{|S|} : \sum_{i \in S} \pi(i) = 1, \pi(i) \geq 0, \forall i \in S \right\}$. The decision-maker updates a belief vector $\boldsymbol{\pi}_t \in \Pi(S)$ at time t to $\boldsymbol{\pi}_{t+1}$ at time $t+1$ in a Bayesian manner. In particular, upon observing signal $o \in O$ and taking action $a \in A$, the jth component of the updated belief vector $\boldsymbol{\pi}_{t+1}$ can be found by

$$\pi_{t+1}(j) = \frac{\sum_{i \neq death} \pi_t(i) \cdot p_t^a(j|i) \cdot q_t^a(o|i)}{\sum_{j' \neq death} \sum_{i \neq death} \pi_t(i) \cdot p_t^a(j'|i) \cdot q_t^a(o|i)}, \quad \forall j \in S \backslash \{death\}.$$

The 'death' state is excluded in this updating formula since 'death' is an observable state. The numerator of this updating equation denotes the probability of observing signal o and the actual system moves to state j conditional on the current belief $\boldsymbol{\pi}_t$, while the denominator is a normalization constant to ensure $\sum_{j \in S} \pi_{t+1}(j) = 1$ and can be interpreted as the probability of observing signal o conditional on the current belief $\boldsymbol{\pi}_t$. We have chosen this concise presentation due to space limitations; more explicit expressions for belief updating are available in Sandıkçı et al. [28].

The physician's goal is to maximize the QALE for the patient. Let $\lambda_t(i)$ denote the quality-adjustment factor and $\omega_t^a(i)$ denote the quality-adjusted intermediate reward for occupying state $i \in S$ during decision epoch t and taking action $a \in A$. Using the commonly used half-cycle correction method in medical decision-making literature to account for a potential death during the decision epoch, we can write $\omega_t^a(i) = \lambda_t(i) \cdot \left[0.5 p_t^a(death|i) + \left[1 - p_t^a(death|i)\right]\right]$. Furthermore, let $R_\tau(i)$ denote the expected total quality-adjusted remaining life when patient occupies state $i \in S$ at time τ and $\psi_t(i)$ denote the lump-sum QALE from initiating cancer treatment in state $i \in S$ at time t upon diagnosing the patient with cancer—note that the screening decision process concludes when cancer is diagnosed. Screening tests and positive test outcomes can cause considerable stress and anxiety for the patient. To account for these effects, let δ^a, δ^{fp}, and δ^{tp} denote the disutility associated with action $a \in A$, having a false positive result, and having a true positive result, respectively.

Let $V_t^*(\boldsymbol{\pi}_t)$ be the maximum QALE for belief state $\boldsymbol{\pi}_t \in \Pi(S)$ at time $t \leq \tau$. For convenience, define $q_t^a(o|i) = 1$ for $o = \emptyset$ and all t, i, a, and for $a \in A, o \in O \cup \emptyset$, and $\boldsymbol{\pi}_t \in \Pi(S), t < \tau$, let

$$f(a, o) = \sum_{i \in S} \pi_t(i) \cdot \omega_t^a(i) + \sum_{i,j \in S} \sum_{o \in O} \pi_t(i) \cdot q_t^a(o|i) \cdot p_t^a(j|i) \cdot V_{t+1}^*(\pi_{t+1}),$$

$$g(a, v(a)) = -\delta^a + \sum_{i \neq cancer\ free} \pi_t(i) \sum_{o \ni \prime + \prime} q_t^a(o|i)\left[-\delta^{tp} + \psi_t(i)\right]$$

$$+ \sum_{i \ni cancer\ free} \pi_t(i) \sum_{o \ni / +'} q_t^a(o|i) \left[-\delta^{fp} + \omega_t^a(i) + \sum_{i \in S} p_t^a(j|i) \cdot V_{t+1}^*(\pi_{t+1}) \right]$$

$$+ \sum_{i \in S} \pi_t(i) \sum_{o \ni / -/} q_t^a(o|i) \cdot v(a) \text{ , and}$$

$$v(a) = \begin{cases} V_{t^+}^*(\pi_{t^+}) & if\ a = M, \\ \omega_t^a(i) + \sum_{i \in S} p_t^a(j|i) \cdot V_{t+1}^*(\pi_{t+1}) & if\ a \in \{U, R\}. \end{cases}$$

(The notation t^+ indicates the time point t, where a negative mammography outcome is observed, and the supplemental screening decision is to be made following this negative mammography outcome). Then the optimal value function $V_t^*(\boldsymbol{\pi}_t)$ can be found by recursively solving the following Bellman optimality equations:

$$\begin{array}{ll} V_t^*(\boldsymbol{\pi}_t) = \max\{f(W, \cdot), g(M, v(M))\} & \forall \boldsymbol{\pi}_t \in \Pi(S), t < \tau \\ V_{t^+}^*(\boldsymbol{\pi}_{t^+}) = \max\{f(N, \emptyset), g(U, v(U)), g(R, v(R))\} & \forall \boldsymbol{\pi}_t \in \Pi(S), t < \tau \\ V_\tau^*(\boldsymbol{\pi}_\tau) = \sum_{i \in S} \pi_\tau(i) \cdot R_\tau(i) & \forall \boldsymbol{\pi}_\tau \in \Pi(S). \end{array}$$

The first set of equations chooses between watchful waiting (W) and mammography screening (M). Given a negative outcome from mammography, the second set of equations optimizes the supplemental screening decisions: no followup (N), ultrasound (U), and MRI (R). The last set of equations calculates the expected terminal reward since no decision is made in the final decision epoch of the process.

POMDPs are well-known for being computational challenging since the belief simplex $\Pi(S)$ is a continuum. Increasing the dimension of the belief simplex (i.e.,$|S| - 1$) contributes to the intractability and, therefore, POMDPs are rarely solved to optimality except when $|S|$ is small (typically, less than 5). In our numerical study, we solve the optimality equations approximately using a grid-based approximation, as described in Sandıkçı et al. [28]. Simply stated, this method samples a finite number of belief points from $\Pi(S)$, reducing the infinitely many equations down to a finite number, and solves the resulting finite set of equations.

The Simulation Model

While the screening policy recommended by the POMDP model described in the previous section maximizes QALE, it is not guaranteed to be optimal on other important outcomes such as false positive rate, the fraction of cancer patients who die before detecting cancer, mean sojourn time (i.e., time from cancer onset until diagnosis or death, whichever occurs first). Therefore, in addition to optimizing the screening policy using a POMDP model, we also use a simulation model to assess the resulting policy with respect to outcomes other than QALE. We also utilize this simulation model to compare the QALE-maximizing POMDP policy with commonly encountered population-based screening guidelines and report their comparative performances in the next section.

In this model, we simulate the health evolution of a given patient using the underlying transition probabilities. The patient's demographic, as well as clinical attributes, are used in simulating her health evolution. Each patient enters the simulation at age 40 and leaves at age 80 unless she dies or is diagnosed with cancer sooner. To assess the performance of any given screening policy σ, we superimpose the screening prescriptions given by σ on the simulated life of the patient. We make 1000 independent replications of the simulation for each simulated patient and use the common random numbers [19] to facilitate comparisons across the different screening policies.

Cohort Selection

National Cancer Institute [26] offers an online breast cancer risk assessment tool to estimate a woman's risk (i.e., probability) of developing invasive breast cancer over the next five years. The tool has been validated for white, black/African American, Hispanic, and for Asian and Pacific Islander women in the United States. It uses the following risk factors in its estimation for a given woman: her age, her race/ethnicity, whether or not she ever had a breast biopsy, if yes, how many breast biopsies and whether or not she ever had a biopsy with atypical hyperplasia, her age at the time of her first menstrual period, her age when she gave birth to her first child, and the number of her first-degree relatives with a history of breast cancer. Relying on the features utilized by this tool, we define our study population as those woman without any family history or previous biopsy, since these two factors are associated with significant increases in breast cancer risk assessment [3]. NCI's estimated risk of breast cancer for our resulting cohort averages 0.23% and ranges between 0.11 to 0.54%.

An important consideration in breast cancer screening is the breast density of the woman, which is not only associated with screening accuracy, but also it is an important risk factor contributing to breast cancer. Breast Imaging Reporting and Data System (BI-RADS) classifies breast densities into four groups [29]: fatty, scattered fibrodengular, heterogeneously dense, and extremely dense breast. Since supplemental tests are generally argued for women with dense breasts, we assume our cohort is composed of women with extremely dense breast density and adjust screening accuracies and cancer development probabilities to reflect extremely dense breast density.

Table 1 Model parameters and data sources

Parameter	Explanation	Value	Source
$p_t^a(\cdot)$	Core state transition probabilities	*	Alagoz et al. [1]
$\psi_t(\cdot)$	QALE post-cancer diagnosis	*	SEER data
$R_\tau(\cdot)$	Expected remaining life at age 80	*	SEER data
δ^W	Disutility of watchful waiting	0 days	Assumption
δ^M, δ^U	Disutility of mammography, ultrasound	0.5 days	Ayer et al. [4], Berg et al. [9]
δ^R	Disutility of MRI	2 days	Ayer et al. [4], Berg et al. [9]
δ^{fp}, δ^{tp}	Disutility of biopsy	2 weeks	Ayvaci et al. [6]
Sensitivity and specificity data used in estimating observation probabilities $q_t^a(\cdot)$			
	(Sensitivity, specificity) of W	(44%, 92)	Barton et al. [7]
	(Sensitivity, specificity) of M	*	Stout et al. [32], Alagoz et al. [1]
	(Sensitivity, specificity) of U action	(55%, 94)	Sprague et al. [31]
	(Sensitivity, specificity) of R action	(90%, 72)	Peters et al. [16], Knuttel et al. [27]

* Changes by age and/or health state of the woman

Parameter Estimation

We set the length of each decision epoch to one year, and so, allow only for annual screening decisions for patients starting at age 40 until age 80 (i.e., $\tau = 40$). All model parameters and their estimates, along with data sources used are summarized in Table 1. Note that some parameter values change by patient's age and/or health state, which is cumbersome to present in a table; therefore, we indicate such parameter values in Table 1 using the '*' symbol.

Results

Optimal Screening Policies

QALE-maximizing screening policies obtained from solving the POMDP model for a randomly selected patient from our cohort are visualized in Fig. 1. Each triangle in

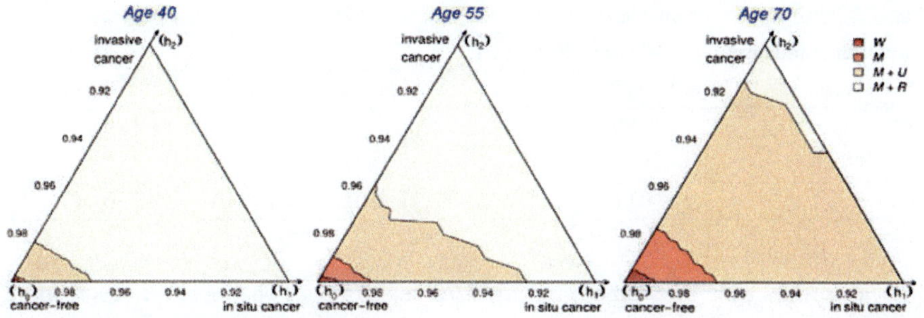

Fig. 1 QALE-maximizing screening policy

this figure denotes the belief simplex corresponding to three health states (namely, the patient is cancer-free, the patient has in situ cancer, and the patient has invasive cancer). Different grades of shading inside the triangles visualize the optimal actions in different zones of the belief simplex.

There is a similar pattern in each one of these screening recommendations: W (watchful waiting) is optimal when physician's assessment of patient's health is highly likely to be cancer-free (darkest shading on the lower-left corner); as the assessment leans towards having some form of cancer, M (screening with mammography only), followed by $M+U$ (mammography supplemented with ultrasound), followed by $M+R$ (mammography supplemented with MRI, lightest shading) becomes optimal. This pattern implies that supplemental tests should only be considered when the patient's estimated breast cancer risk is sufficiently high.

Figure 1 also displays that the switching points from W to M to $M+U$ to $M+R$ depend on the age of the woman undergoing screening. The switching points increase as the patient gets older, which implies less aggressive screening for older, but otherwise identical, women. This interesting finding may be explained by the fact that there is a greater upside potential of detecting cancer at younger ages.

Simulation Results

Using the simulation model, we first evaluate the utilization of available screening modalities under the screening policies prescribed by the POMDP model. Table 2 summarizes these results, which are averaged over 100 patients sampled at random. These results suggest that increases in woman's age, on average, cause more frequent use of screening tests, although it is not necessarily in the form of supplemental tests. In particular, mammography utilization, which is defined as the fraction of periods the simulated woman takes the M action, is 72.67% when they start their simulated lives at 40 years of age. A practical interpretation of this number would be to employ roughly 3 mammography tests per woman every 4 years. This fraction jumps to 86.04% when women start their simulated lives at 50 years of age and progressively

Table 2 Utilization of available screening modalities under QALE-maximizing policies

Screening modality	Woman's age (years)			
	40 (%)	50 (%)	60 (%)	70 (%)
M	72.67	86.04	88.44	89.83
$M+U$	5.99	3.49	4.06	7.27
$M+R$	0.00	0.01	0.01	0.04

increases to 89.84% for women starting at 70 years of age. The increase in screening intensity with the age of the woman is related to an increased risk of developing breast cancer as a woman ages [3]. In contrast, supplemental ultrasound screening is rarely used (although at an increased rate as woman ages), while supplemental MRI screening is almost never utilized (Table 2).

We next simulate several commonly encountered screening recommendations and compare their outcome performance to policies obtained by the POMDP model. The screening policies included in our comparisons are biennial and annual mammography-only, biennial and annual mammography supplemented with ultrasound, and biennial and annual mammography supplemented with MRI. Furthermore, we re-solve the POMDP model for each patient in our cohort after eliminating the ultrasound and MRI options to obtain QALE-maximizing mammography-only screening strategies. When interpreting our results, we choose the 'no screening' strategy as a common baseline and interpret the performance of each screening policy relative to this baseline.

To compare the performance of these screening policies, we re-simulate the lives of 100 patients under each policy and summarize results, averaged over these 100 patients, in Table 3. These results suggest that screening improves the average QALE of the cohort compared to no screening. QALE improvement associated with mammography-only screening policies averages to 58.9, 60.2, and 60.7 weeks for 'biennial,' 'annual,' and 'POMDP without Supplemental tests' policies, respectively. The numerical (but statistically insignificant) superiority of the POMDP-based policies can be attributed to the POMDP model's ability to provide dynamic, personalized screening recommendations that diminish the disutilities associated with unnecessary screenings.

On the other hand, introducing ultrasound or MRI screening to supplement mammography does not necessarily further improve QALE. On the contrary, supplementing biennial and annual mammographies with an ultrasound cause a reduction of 0.4 weeks and 5.2 weeks, respectively, in average QALE. Using MRI to supplemental biennial and annual mammographies causes an even larger loss in average QALE: 7.8 weeks and 22.2 weeks, respectively. These losses are attributed to accumulated disutilities due to aggressive screening.

While supplemental tests may cause reductions in QALE, they help improve several other performance metrics:

(i) Not only cancer detection rate, but also the stage at which cancer is detected improves significantly with supplemental test: for example, supplementing

Table 3 Simulation results comparing various screening policies

	Noscreening	Biennial	Biennial + Ultrasound	Biennial + MRI	Annual	Annual + Ultrasound	Annual + MRI	POMDP without suppl.	POMDP
QALE (years)	38.960	40.092	40.084	39.941	40.117	40.016	39.690	40.126	40.129
Cancer cases diagnosed (%)	—	90.73	93.16	94.46	97.80	99.10	99.90	96.21	96.21
at in situ stage	—	56.55	59.50	60.99	63.61	65.33	66.34	61.82	61.80
at invasive stage	—	34.18	33.66	33.47	34.18	33.77	33.56	34.39	34.41
by mammography	—	90.73	79.12	73.19	97.80	83.39	76.73	96.21	94.33
by ultrasound	—	—	14.05	—	—	15.71	—	—	1.88
by MRI	—	—	—	21.27	—	—	23.17	—	—
Duration with undiagnosed cancer (years)	9.45	1.83	1.57	1.43	1.29	1.11	1.03	1.50	1.50
in situ	4.84	1.22	1.04	0.95	0.87	0.75	0.69	1.01	1.01
invasive	4.61	0.61	0.52	0.48	0.42	0.37	0.34	0.49	0.49
In situ cases deteriorated to invasive before diagnosis (%)	55.40	9.02	5.88	4.36	3.16	1.25	0.28	5.03	5.04
Cases died with undiagnosed cancer (%)	72.93	6.24	4.34	3.25	1.73	0.69	0.08	3.26	3.29
False positive outcomes (%)	—	10.81	9.68	18.84	10.89	9.72	18.91	10.91	10.76
by mammography	—	10.81	5.75	5.75	10.89	5.77	5.78	10.91	10.34
by ultrasound	—	—	3.93	—	—	3.95	—	—	0.42
by MRI	—	—	—	13.09	—	—	13.13	—	—

biennial mammographies with ultrasound improves detection rate from 90.73 to 93.16% and detection rate at the in situ stage from 56.55 to 59.50%; using MRI instead of ultrasound further improves these rates to 94.46 and 60.99%, respectively.

(ii) Supplementing mammographies by ultrasound (MRI) shortens the time to detect cancer by about 14% (21%, respectively).

(iii) The fraction of in situ cases that deteriorate to the invasive stage reduces by 35% and 60% when ultrasound supplements biennial and annual mammographies, respectively, and by 52% and 91% when MRI supplements biennial and annual mammographies, respectively.

(iv) The fraction of cancer patients who die before any diagnosis reduces by 31% and 60% when ultrasound supplements biennial and annual mammographies, respectively, and by 48 and 95% when MRI supplements biennial and annual mammographies, respectively.

It should, however, be emphasized that while supplementing mammographies with ultrasound reduces the false positive rate by about 10%, using MRI instead of ultrasound causes the false positive rate to increase drastically by about 74%, which reveals an important concern regarding using MRI as a supplemental screening test.

Considering the POMDP-based screening policies, results in the last two columns of Table 3 indicate that cautiously utilizing supplemental tests, as suggested by the screening recommendations of the 'POMDP' policy, can eliminate the potential QALE losses associated with supplementing routine (biennial or annual) mammography screening with ultrasound or MRI. Nevertheless, since the POMDP policy rarely utilizes supplemental tests (as noted above in our discussion surrounding Table 2), the incremental benefit of using supplemental tests as suggested by the POMDP policies over the mammography-only POMDP policies is negligibly small (0.1 weeks). In fact, differences between the values in the last two columns of Table 3 are all negligibly small, indicating that QALE-maximization objective results in such a conservative utilization of supplemental tests that using them do not translate to any meaningful improvement in any of the performance metrics measured.

Concluding Remarks

Mammography screening is the gold standard for breast cancer screening in the vast majority of the world, but it is known to be less accurate for women with dense breasts. To improve cancer detection accuracy, supplemental ultrasound and MRI screening have been recently introduced and are actively recommended by many agencies (e.g., American Cancer Society, American College of Radiology) for high-risk populations such as those women that have a lifetime breast cancer risk of at least 20%, those with a family history, or those with inherited gene mutations of BRCA1/BRCA2. Numerous studies have demonstrated that breast MRI has an increased cancer detection rate when compared to ultrasound and other available imaging modalities. Nevertheless, widespread utilization of breast MRI in clinical practice is hindered due to its relatively higher cost.

In this study, we have focused on quantifying the value of supplemental MRI and ultrasound screening for breast cancer in non-high-risk women, ignoring the financial costs to exclusively focus on incremental life benefits of supplemental screening. We have used a POMDP model to optimize the use of available screening modalities (i.e., mammography, mammography supplemented with ultrasound and mammography supplemented with MRI) for a woman with the objective of maximizing her QALE. We have further scrutinized the screening recommendations of this POMDP model and compared its performance to population-based routine screening recommendations using a simulation model.

Detailed numerical study driven by clinical data reveals that QALE-maximization objective results in very conservative use of supplemental tests, and almost exclusively prefer ultrasound instead of MRI. On the other hand, using supplemental tests as part of routine (biennial or annual) mammographies causes significant reductions in QALE since accumulated disutility from frequent supplemental tests outweigh their benefits for non-high-risk women. However, they are associated with significant improvements in overall cancer detection rate, the time to detect cancer, fraction of in situ cases that deteriorate to the invasive stage as well as the fraction of cases who die with undiagnosed cancer. While MRI is generally more effective than ultrasound on several performance metrics, it also suffers from significantly increased false positives, hindering its viability for this population.

Routine screening for breast cancer is not without controversy. When to start and stop screening, as well as the frequency of screening, have been intensely debated. Recently added to these debates is whether or not, and in what frequency, supplemental tests should be used. Decades-long controversies are largely rooted in the assumption that there exists a general population, in which everyone has the same risks. However, it is known that breast cancer is a heterogeneous disease, and there is a significant variation in individuals' risk of developing breast cancer. While the focus of this chapter was on quantifying the value of supplemental screening tests, the POMDP model used in this paper can also be utilized to offer personalized screening recommendations, adjusting for risk factors for the woman under consideration, and making it feasible to relax the one-size-fits-all assumption.

Future research can explore several directions. While our model assumed the decision-makers are risk-neutral, which may be justifiable when searching for screening guidelines for a homogeneous population, it is true that such homogeneous populations rarely exist and, therefore, risk attitudes of individual groups should be considered. It is generally agreed that most patients take a risk-averse attitude when faced with life-and-death related decisions. Furthermore, while we suppressed financial aspects of screening in our study to focus on the value –as measured by QALE– of supplemental screening, the insights gained from this study should be tested under financial constraints. Another interesting direction of research would be to incorporate parameter uncertainty in the models to examine the robustness of current findings and identify robust solutions. Finally, this study and studies of this kind should ultimately be complemented with pilot studies that test the insights from these models in small populations to pave the road for larger implementations.

References

1. Alagoz O, Ergun MA, Cevik M, Sprague BL, Fryback DG, Gangnon RE, Hampton JM, Stout NK, Trentham-Dietz A (2018) The University of Wisconsin breast cancer epidemiology simulation model: an update. Med Decis Making 38(1_suppl):99S–111S
2. American Cancer Society (2016) Breast cancer early detection and diagnosis. www.cancer.org/content/dam/CRC/PDF/Public/8579.00.pdf
3. American Cancer Society (2019) Breast cancer facts and figures 2019–2020. Atlanta, GA
4. Ayer T, Alagoz O, Stout NK (2012) A POMDP approach to personalize mammography screening decisions. Oper Res 60(5):1019–1034
5. Ayer T, Alagoz O, Stout NK, Burnside ES (2016) Heterogeneity in women's adherence and its role on optimal breast cancer screening policies. Manage Sci 62(5):1339–1362
6. Ayvaci MUS, Alagoz O, Burnside ES (2012) The effect of budgetary restrictions on breast cancer diagnostic decisions. Manuf Serv Oper Manag 14(4):600–617
7. Barton MB, Harris R, Fletcher SW (1999) Does this patient have breast cancer? The screening clinical breast examination: should it be done? how? J Am Med Assoc 282(13):1270–1280
8. Berg WA (2014) How well does supplemental screening magnetic resonance imaging work in high-risk women? J Clin Oncol 32(21):2193–2196
9. Berg WA, Blume JD, Adams AM, Jong RA, Barr RG, Lehrer DE, Pisano ED, Evans WP 3rd, Mahoney MC, Hovanessian Larsen L, Gabrielli GJ, Mendelson EB (2010) Reasons women at elevated risk of breast cancer refuse breast MR Imaging screening. Radiology 254(1):79–87
10. Berg WA, Zhang Z, Lehrer D, Jong RA, Pisano ED et al (2012) Detection of breast cancer with addition of annual screening ultrasound or a single screening MRI to mammography in women with elevated breast cancer risk. J Am Med Assoc 307(13):1394–1404
11. Bray F, Ferlay J, Soerjomataram I, Siegel RL, Torre LA, Jemal A (2018) Global cancer statistics 2018: GLOBOCAN estimates of incidence and mortality worldwide for 36 cancers in 185 countries. CA Cancer J Clin 68(6):394–424
12. Brinton LA, Gaudet MM, Gierach GL (2018) Breast cancer. In: Thun MJ, Linet MS, Cerhan JR, Haiman CA, Schottenfeld D (eds) Cancer epidemiology and prevention, 4th edn. Oxford University Press, New York, pp 861–888
13. Cassandra AR (1998) Exact and approximate algorithms for partially observable Markov decision processes. Ph.D. thesis, Brown University, Providence, RI
14. Cevik M, Ayer T, Alagoz O, Sprague BL (2018) Analysis of mammography screening policies under resource constraints. Production Oper Manag 27(5):949–972
15. Feig S (2010) Cost-effectiveness of mammography MRI and ultrasonography for breast cancer screening. Radiol Clin North Am 48(5):879–891
16. Knuttel FM, Menezes GL, van den Bosch MA, Gilhuijs KG, Peters NH (2014) Current clinical indications for magnetic resonance imaging of the breast. J Surg Oncol 110(1):26–31
17. Komen Foundation (2020). https://ww5.komen.org/BreastCancer/LowerYourRisk.html
18. Krishnamurthy V (2016) Partially observed Markov decision processes: from filtering to controlled sensing. Cambridge University Press, Cambridge, United Kingdom
19. Law AM (2015) Simulation modeling and analysis, 5th edn. McGraw-Hill, New York, NY
20. Lee CH, Dershaw DD, Kopans D, Evans P, Monsees B, Monticciolo D, Brenner RJ, Bassett L, Berg W, Feig S, Hendrick E, Mendelson E, D'Orsi C, Sickles E, Burhenne LW (2010) Breast cancer screening with imaging: recommendations from the society of breast imaging and the ACR on the use of mammography, breast MRI, breast ultrasound, and other technologies for the detection of clinically occult breast cancer. J Am Coll Radiol 7(1):18–27

21. Lovejoy WS (1991) A survey of algorithmic methods for partially observed Markov decision processes. Ann Oper Res 28(1):47–66
22. Madadi M, Zhang S, Henderson LM (2015) Evaluation of breast cancer mammography screening policies considering adherence behavior. Eur J Oper Res 247(2):630–640
23. Maillart LM, Ivy JS, Ransom S, Diehl K (2008) Assessing dynamic breast cancer screening policies. Oper Res 56(6):1411–1427
24. Monahan GE (1982) A survey of partially observable Markov decision processes: theory, models, and algorithms. Manage Sci 28(1):1–16
25. Moore SG, Shenoy PJ, Fanucchi L, Tumeh JW, Flowers CR (2009) Cost-effectiveness of MRI compared to mammography for breast cancer screening in a high risk population. BMC Health Serv Res 9:9
26. National Cancer Institute (2020) Breast cancer risk assessment tool. https://bcrisktool.cancer.gov/. Accessed on 10 Jan 2020
27. Peters NH, Borel Rinkes IH, Zuithoff NP, Mali WP, Moons KG, Peeters PH (2008) Meta-analysis of MR imaging in the diagnosis of breast lesions. Radiology 246(1):116–124
28. Sandıkçı B, Cevik M, Schacht DV (2019) Screening for breast cancer in high-risk women: the role of supplemental tests and breast density information. Working paper, University of Chicago Booth School of Business
29. Sickles EA, D'Orsi CJ, Bassett LW, et al. (2013) American College of Radiology BI-RADS Atlas. Breast Imaging Reporting and Data System. 5th edn. Reston, VA
30. Spaan MTJ (2012) Partially observable Markov decision processes. In: Wiering M, van Otterlo M (eds) Reinforcement learning. Adaptation learning and optimization, vol 12. Springer, Berlin, Heidelberg, pp 387–414
31. Sprague BL, Stout NK, Schechter C, van Ravesteyn NT, Cevik M, Alagoz O, Lee CI, van den Broek JJ, Miglioretti DL, Mandelblatt JS, de Koning HJ, Kerlikowske K, Lehman CD, Tosteson ANA (2015) Benefits, harms, and cost-effectiveness of supplemental ultrasonography screening for women with dense breasts. Ann Intern Med 162(3):157–166
32. Stout NK, Lee SJ, Schechter CB, Kerlikowske K, Alagoz O, Berry D, Buist DSM, Cevik M, Chisholm G, de Koning HJ, Huang H, Hubbard RA, Miglioretti DL, Munsell MF, Trentham-Dietz A, van Ravesteyn NT, Tosteson ANA, Mandelblatt JS (2014) Benefits harms, and costs for breast cancer screening after US implementation of digital mammography. J National Cancer Inst 106(6):dju092
33. World Health Organization (2020). https://www.who.int/news-room/fact-sheets/detail/cancer

What Drives the Turkish Government's Response to COVID-19 Pandemic—Daily Cases or Daily Deaths

Basak Cetinguc and Fethi Calisir

Abstract COVID-19 has been changed the way of each others' lifestyles. After the outbreak of this virus all over the world, governments have taken precautions to prevent the spread of it. In some countries, schools are closed, working conditions are changed, international flights are canceled, lockdowns are announced, public gatherings are banned, and so on. Basically, our daily routines and lives have altered into a new pattern. Knowing how contiguous COVID-19, governments had to take serious precautions and develop policies. Government Response Index has been introduced by Oxford University, evaluating the stringencies of 165 countries' government response actions on COVID-19 [8]. The main aim of this study is to figure out while governments are developing these policies, whether daily cases and/or daily deaths of COVID-19 drive the Turkish Governmnent's Response. Regression analysis is conducted via Jamovi by gathering data from Oxford's Government Response Index. Findings showed that the Turkish government's responses to COVID 19 are affected by the daily deaths, not daily cases.

Keywords COVID-19 · Government response · Daily COVID-19 cases · Daily COVID-19 deaths · Developing policies

Introduction

While waiting for technological developments to change our worlds, a virus called COVID-19 has shifted the paradigms of our daily lives. The cases have started by reported atypical pneumonia illnesses in Wuhan, China. By January 9, 2020, the World Health Organization (WHO) announced that Chinese authorities reported the

B. Cetinguc (✉)
Industrial Engineering Department, Engineering Faculty, Yalova University, Yalova, Turkey
e-mail: basak.cetinguc@yalova.edu.tr

F. Calisir
Industrial Engineering Department, Management Faculty, Istanbul Technical University, Istanbul, Turkey
e-mail: calisirfet@itu.edu.tr

F. Calisir (ed.), *Industrial Engineering in the Internet-of-Things World*,
Lecture Notes in Management and Industrial Engineering,
https://doi.org/10.1007/978-3-030-76724-2_34

illnesses are caused by a novel coronavirus [13]. Furthermore, close neighbors of China respectively announced the first cases in their countries, beginning with Thailand (January 13), Japan (January 15), and South Korea (January 20). Unfortunately, the epidemic without retarding has continued to spread all over the world in a short time. Since the daily average number of commercial flights in the whole world was above 100.000, the spread of the virus was inevitable. The first case outside Asia was reported in North America on January 21 in the United States, followed by Oceanica on January 25 in Australia, on January 28 in Germany in Europe, on February 15 in Egypt in Africa, and finally on February 26 in Brazil in South America. This virus was not the first epidemic, and recently the world has dealt with Severe Acute Respiratory Syndrome (SARS) and Middle East Respiratory Syndrome (MERS). However, COVID-19 infect many more people than SARS and MERS. The number of cases diagnosed by SARS has been 8096 since the first occurrence in 2002 (World Health Organization) [14] and by MERS has been 2494 since 2012, the first appearance in 2012 [15]. Furthermore, WHO announced COVID-19 as a pandemic on March 11, 2020, when the cumulative number of COVID-19 cases reached 126,547 [11]. While understanding the pandemic, which has been a new term for many people, has not been easy, humans have to need to fight with it. One of the first research to understand COVID-19 symptoms and risk groups has done by Zhou and his colleagues in China. Their sample consists of 191 patients diagnosed with COVID-19 (age between 18 and 87 years) [17]. Moreover, they found out that elder age groups are under risk more than young generation and people with comorbidity (e.g. hypertension, diabetes, coronary heart diseases, chronic lung diseases etc.) are affected more by COVID-19. Additionally, the symptoms are associated with high fever, cough, fatigue, myalgia, increased sputum production, etc. in admission [6, 16]. Fu and his colleagues conducted a systematic literature review regarding the characteristics of COVID-19. They figured out that 25.6% of patients are severe cases, and 3.6% of cases, unfortunately, end up with fatality [5].

As of July 18, there have been almost 14 million cases and 600 thousand deaths in all continents except Antarctica so far. Although the virus transmission is very rapid and its consequences are known to be fatal, no definitive evidence for the curative treatment has been obtained. There has been plenty of research funded by governments and firms to discover a vaccine for COVID-19. But at the end of the day, COVID-19 is still an enigma for human beings. Knowing how contagious this virus is, the essential thing to do is propounding preventive actions as a first step. As precautions, governments from all over the world develop policies to defeat the spread of pandemic with non-pharmaceutical interventions such as lockdowns, travel restrictions, closing schools, working home-office, banning public gatherings, etc. [9]. The main purpose of these actions is to keep social distance enough among people to prevent spreading the COVID-19 [1]. Yet these restrictions come with the price of economic problems [4]. In response, governments announced economic support packages for their citizens. Hale and his colleagues proposed a government response index based on their research on 165 countries related to governments' responses to COVID-19 [8]. Government actions differ based on their perceptions of the seriousness of the pandemic and their capabilities. The government response index is comprised of 13 different indicators as a school closing, workplace closing, public events cancellations, restrictions on crowded populations, public transport restrictions, stay at home requirements, internal movement limitations, international travel restrictions, income support, debt/contact relief of households, starting a public information campaign, testing policy, and contact tracing. Under these indicators, data from 165 governments were collected, and an index was developed. Each government has a score between 0 and 100 for each day.

The Turkish government is one of the governments included in the Government Response Index. The first case was declared by the Minister of Health at the late hours of March 11 in Turkey. Since the first day, there have been known 218,717 COVID-19 cases and 5475 death. Turkey has a 2.5% fatality rate in registered cases. Before the first case occurred in Turkey, international flights have already been limited, and public information campaigns regarding the severity of COVID-19 have been started [2]. After the first case announcement, schools were closed on March 16 and home office working conditions adapted to daily lives. Moreover, partial lockdowns have been declared at the weekends and official holidays by the Ministry of Interior. While taking these precautions, the research question is what the reference point of the decision is. In this regard, the main purpose of this study is whether the Turkish government policy developing process against COVID-19 is affected by the daily COVID-19 cases or deaths. In the following section, data and methodology are introduced. The results are given in the third section. Respectively, discussion and further research and limitations are mentioned in the following parts.

Data and Methodology

The data were derived from the Oxford Coronovirus Government Response Tracker (Blavatnik School of Government and University 2020) between March 12 and July 3, a total of 115 observations [3]. Variables are daily cases, daily deaths, and government response scores of Turkey. As seen in Fig. 1, daily cases and daily deaths have reached their peak points in April. The highest daily case occurred on April 12, with 5138 cases, when the highest death number was on April 20 with 127.

Turkey's score in government response index varies between 25 and 76, 28 out of 100 points [2].

Firstly, data were examined to check if there are wrong, biased, or missing values. Yet, this data set is derived from OxGRT that is an open and reliable source. Each observation is precious for giving an idea for COVID-19. In our data set, daily cases above 4285 are determined as outliers, but we did not eliminate these cases from our data sets since the infectious disease may show erratic patterns [12]. Especially in peak times of daily cases and daily deaths, government responses are important to explore.

To answer the research question, two hypotheses were developed, and causal relationships are tested among variables. These hypotheses are supported by the idea

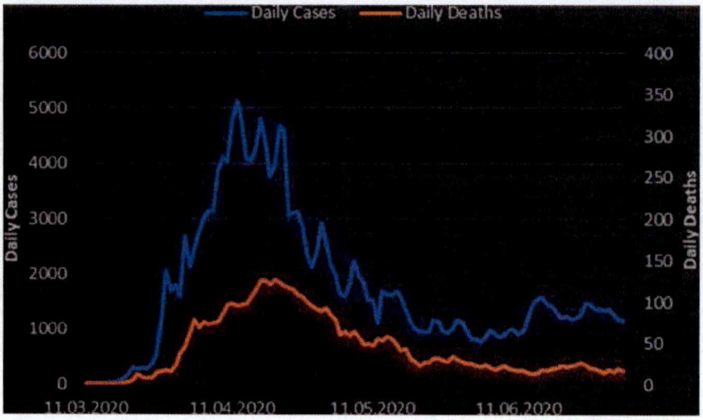

Fig. 1 Number of daily cases and daily deaths

of making policies by governments to prevent or reduce the spread and damage of COVID-19 [1, 9].

H1: Turkish government policy development process is positively affected by the number of daily cases.

H2: Turkish government policy development process is positively affected by the number of daily deaths.

The research model is comprised of two independent variables, which are daily cases and daily deaths in Turkey, and one dependent variable that is government response score of Turkish government as given in Fig. 2.

Linear regression is employed to test the causal relationships between these variables via Jamovi, which is an open and free statistical software made by and for the scientific community [10]. As shown in Eq. 1, a prediction model is proposed (Y: Government response scores predictions, X1: Daily Cases, X2: Daily Deaths). Equation 3 shows the error terms after subtracting predicted values from real values, which is given in Eq. 2.

$$\widehat{Y_i} = b_0 + b_1 * X_1 + b_2 * X_2 \tag{1}$$

$$Y_i = b_0 + b_1 * X_{1(i-1)} + b_2 * X_{2(i-1)} + \varepsilon_i \tag{2}$$

Fig. 2 The model of the study

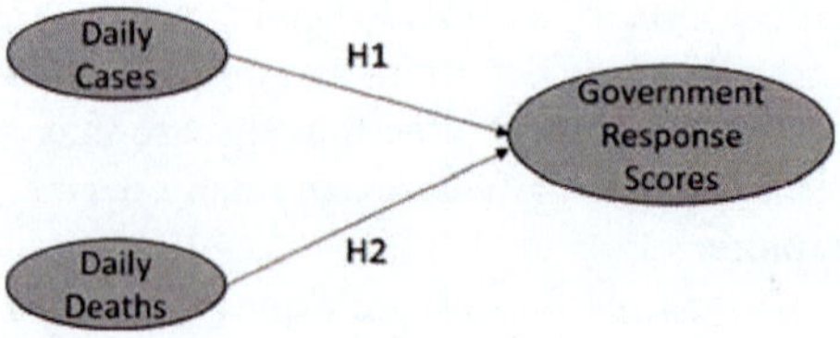

$$\varepsilon_i = Y_i - \widehat{Y_i} \quad (3)$$

Results

Before beginning the regression analysis, assumptions of the test are checked. First, having ratio scale variables in this study satisfies the initial condition of regression analysis that is employing only continuous variables. Then, the residuals expect to distribute normally [7]. To check the normality of the residuals, Q-Q plots, as seen in Fig. 3, are examined. Moreover, the constant variance of error terms is another assumption.

Moreover, assessing multicollinearity is one of the other issues in regression analysis. It is expected to have a higher correlation between dependent variables and independent variables but has lower correlations between independent variables. Hair expressed the cut off value for variance inflation factor (VIF) as 10, and as shown in Table 1, our findings are below this cut-off value [7].

After satisfying assumptions, firstly, it is important to check if our prediction model is significant or not. Based on the overall model test results given in Table 2, our prediction model is significant, with a p-value below 0.01.

Fig. 3 Q-Q plot

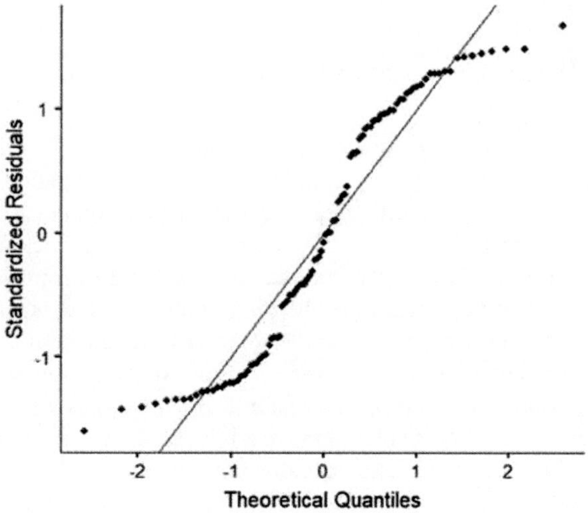

Table 1 Collinearity statistics

	VIF	Tolerance
Daily deaths	5.53	0.181
Daily cases	5.53	0.181

Table 2 Model fit measures

			Overall model test			
Model	R	R^2	F	df1	df2	p
1	0.470	0.221	15.8	2	112	<0 .001

Table 3 Model coefficients—government response scores

Predictor	Estimate	SE	t	p
Intercept	61.329	1.55753	39.376	<0 .001
Daily deaths	0.116	0.05791	1.999	0.048
Daily cases	7.24e−4	0.00169	0.429	0.669

After this step, whether coefficients are meaningful or not is checked. Table 3 shows that intercept and daily deaths are meaningful, while daily cases are not. In other words, the government response index is affected by daily deaths, not by daily cases. Therefore, H2 is supported while H1 is rejected based on these results. Additionally, interpretations of coefficients show that ten death of COVID-19 induces governments to develop more stringent policies and raise their government responses 1 point more. Furthermore, government response scores are estimated based on government policy regarding COVID-19, and daily deaths explain 22% of government response scores, as seen in Table 2.

Discussion and Conclusion

This study was conducted by the objective of determining whether government non-pharmaceutical policy decision to prevent the spread of COVID-19 is made based on daily deaths or daily cases. The Turkish government constituted a science committee consisting of physicians from various branches weeks before the first case occurred in Turkey. This committee has been working on informing the public about what COVID-19 is, what are the symptoms are, and what to do if symptoms appear. Moreover, for preventing the spread of this pandemic, the scientific committee has given advice to the government about limiting international air traffic, lockdowns, school closures, how to transfer Turkish citizens who live abroad to Turkey, and so on. Their duty is solely to give advice to the government and inform the public, and policies are made and developed by the Turkish Government. Our findings clarified that daily deaths are a concern while making and developing COVID-19 policies and are not affected by daily cases. The Turkish government has been confident with the Turkish health system and the capability of Turkish physicians since the beginning of the pandemic. Additionally, the Turkish government has started a project called "City Hospitals" before the pandemic, and they accelerated the construction of these hospital buildings to have extra space for COVID-19 patients. Therefore, there has not been any over-load crisis or equipment shortages in Turkey so far during the pandemic. Moreover, the young population percentage is relatively higher than the older ones' in Turkey. Since the older age group is affected more severely than younger age groups by COVID-19, the numbers have never been as high as the numbers observed in other countries such as Italy and Spain, which have higher older population percentages than Turkey. However, there is still not any proven medicine or vaccine for this pandemic in the whole world. Serious vaccination studies have been ongoing in very respected institutes and organizations.

On the other hand, COVID-19 deaths are the cruelest and the most irreparable side of this pandemic. Even funerals have not been proceeded properly due to infectious issues. The life of loved ones matters much more than restrictions. That is why the Turkish government develops policies based on COVID-19 deaths. Especially, the obligations of lockdowns during national holidays and the weekends were very helpful to reduce the COVID-19 deaths.

For further research, the other countries might be included in the analysis and examined if the COVID-19 deaths and/or cases affect the stringency of governments' actions. Moreover, panel-data analysis can be employed by including other countries. Additionally, government response actions might be divided into subsections based on their purpose and analysis regarding their aims. The progress of this pandemic still unclear, thus our predictions are based on what we have been through so far. Hence, limited information is the biggest limitation of this study.

References

1. Ashraf BN (2020) Economic impact of government interventions during the COVID-19 pandemic: international evidence from financial markets. J Behav Exp Financ Article in: 100371. https://doi.org/10.1016/j.econlet.2020.109231
2. Blavatnik School of Government, University of Oxford (2020a) Coronavirus Government Response Tracker. https://www.bsg.ox.ac.uk/research/research-projects/coronavirus-government-response-tracker
3. Blavatnik School of Government, University of Oxford (2020b) Oxford COVID-19 Government Response Tracker. https://covidtracker.bsg.ox.ac.uk/
4. Eichenbaum M, Rebelo S, Trabandt M (2020) The macroeconomics of epidemics. Natl Bur Econ Res. https://doi.org/10.3386/w26882
5. Fu L, Wang B, Yuan T et al (2020) Clinical characteristics of coronavirus disease 2019 (COVID-19) in China: a systematic review and meta-analysis. J Infect 80:656–665. https://doi.org/10.1016/j.jinf.2020.03.041
6. Guan W, Ni Z, Hu Y et al (2020) Clinical Characteristics of coronavirus disease 2019 in China. N Engl J Med. https://doi.org/10.1056/NEJMoa2002032
7. Hair JF, Black WC, Babin BJ, Anderson RE (2010) Multivariate data analysis, 7th edn. Pearson Prentice Hall, New Jersey
8. Hale T, Angrist N, Kira B et al (2020a) Variation in government responses to COVID-19
9. Hale T, Hale AJ, Kira B, et al (2020b) Global assessment of the relationship between government response measures and COVID-19 Deaths. medRxiv 2020.07.04.20145334. https://doi.org/10.1101/2020.07.04.20145334
10. Jamovi (2020) The Jamovi Project

11. Johns Hopkins University (JHU) (2020) COVID-19 dashboard by the center for systems science and engineering (CSSE) at Johns Hopkins University (JHU). https://gisanddata.maps.arcgis.com/apps/opsdashboard/index.html#/bda7594740fd40299423467b48e9ecf6
12. Weisent J, Seaver W, Odoi A, Rohrbach B (2014) The importance of climatic factors and outliers in predicting regional monthly campylobacteriosis risk in Georgia, USA. Int J Biometeorol 58:1865–1878. https://doi.org/10.1007/s00484-014-0788-6
13. World Health Organization (2020) Timeline of WHO's response to COVID-19. https://www.who.int/news-room/detail/29-06-2020-covidtimeline
14. World Health Organization summary of probable SARS cases with onset of illness from 1 November 2002 to 31 July 2003. https://www.who.int/csr/sars/country/table2004_04_21/en/
15. World Health Organization (2019) Middle East respiratory syndrome coronavirus (MERS-CoV). https://www.who.int/emergencies/mers-cov/en/
16. Yang W, Cao Q, Qin L, et al (2020) Clinical characteristics and imaging manifestations of the 2019 novel coronavirus disease (COVID-19): a multi-center study in Wenzhou city, Zhejiang, China. J Infect. https://doi.org/10.1016/j.jinf.2020.02.016
17. Zhou F, Yu T, Du R, et al (2020) Clinical course and risk factors for mortality of adult inpatients with COVID-19 in Wuhan, China: a retrospective cohort study. Lancet 395

Taxonomy of Mathematical Modeling Studies for Hepatitis C Among Injection Drug Users

Emine Yaylali and Sahincan Ucler

Abstract Hepatitis C (HCV) is one of the significant public health problems, and around 71 million people have chronic Hepatitis C infections globally. Among infected, people who inject drugs (PWID) have an increased risk of transmission, and it is a key risk population. In the last decade, the introduction of direct-acting antivirals (DAA) has started a new era for HCV treatment, and elimination of HCV globally by 2030 became a World Health Organization (WHO) target. Besides clinical and medical studies on HCV infection, modeling and economic studies have been conducted to analyze the spread of this disease, to prevent and eliminate HCV, to determine interventions and programs that are effective and cost-effective, and to optimally allocate resources among high-risk and general populations. In this paper, we survey modeling and cost-effectiveness studies of HCV among injection drug users, and we provide a taxonomy of 73 studies published from 2000 to 2020 that represent 15 countries. We categorized studies based on modeling type, whether it contained cost-effectiveness analysis, the focus of study, and treatment type included. Our taxonomy could be used by modelers, public health policy-makers, and researchers to review past HCV models and to develop new models and cost-effectiveness studies.

Keywords Mathematical modeling · Taxonomy · Hepatitis C · Injection drug users · Cost-effectiveness analysis

Introduction

Hepatitis C (HCV) is a significant liver disease infection that is caused by the Hepatitis C virus. Hepatitis C infection could lead to cirrhosis and hepatocellular carcinoma. Hepatitis C infection has two phases acute and chronic hepatitis. According to the World Health Organization (WHO)'s estimation, there were 399,000 people died

E. Yaylali (✉) · S. Ucler
Industrial Engineering Department, Management Faculty, Istanbul Technical University, Istanbul, Turkey
e-mail: emineyaylali@itu.edu.tr

F. Calisir (ed.), *Industrial Engineering in the Internet-of-Things World*, Lecture Notes in Management and Industrial Engineering,
https://doi.org/10.1007/978-3-030-76724-2_35

from Hepatitis C in 2016, a major part of deaths caused by cirrhosis and hepatocellular carcinoma. Hepatitis C is an important public health problem because a person who is infected may not have any symptoms or mild symptoms during initial infection. For that reason, Hepatitis C is known as a silent epidemic. There is no vaccine for Hepatitis C infection. It is an illness that affects not only people who infected it also has an effect on society.

Literature Review

In our study, we conducted a literature review on PubMed using the following Medical Subject Headings (MeSH) search terms: (Hepatitis C OR HCV) AND transmission AND (model OR modeling OR modeling) AND (inject OR drug) during the period 2000–2019 [54]. We expanded our search with keywords for 2020. In this paper, literature about HCV transmission, elimination, and prevention modeling among persons who inject drugs (PWID) are analyzed. A taxonomy study has been conducted and categorized with five main groups, which are study type, model type, the focus of study, related treatment, and cost-effectiveness analysis. Articles' classification is represented in Table 5.

Study Type

The study type section contains two subgroups, which are theoretical and practical (Table 1). The theoretical group included studies that did not use real-world data; that apply model or analysis on hypothetical examples, or that did not use any case. There is only one article on theoretical research among 73 articles. Due to the lack of existing work on the theoretical part, we did not further divide into subgroups. The research in the practical group split into two subgroups, which are empirical data and empirical and simulation. The empirical data section contains Hepatitis C patients' surveillance data from the real world. Researchers used data for parameterization and calibration of the parameters. There are fifty-five articles that utilized empirical data for their work. Another subgroup is empirical data and simulation that applied real-world data in their research and also developed a simulation model for estimation and scenarios such as vaccine impact, prevention studies, and cost-effectiveness analysis,

Table 1 Study type classification table

1	Study type
1.1	Theoretical
1.2	Practical
1.2.1	Empirical data
1.2.2	Empirical data and simulation

Fig. 1 Practical group percentages in sub-groups

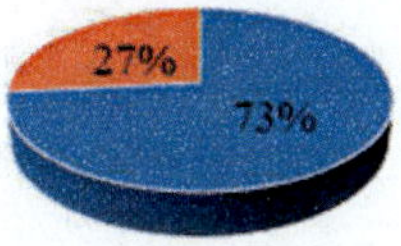

etc. In our taxonomy work, 19 of the article is categorized as the empirical data and simulation subgroup. While 27% of articles utilized empirical data and simulation for their research, 73% used empirical data (Fig. 1).

Model Type

The main idea of modeling is to help people understand the system, to analyze systems, or to determine solutions to existing problems in systems. Therefore, models are tools that could enable to reduce the complexity of systems. In health care modeling applications, models are useful for decision-makers to analyze the consequences of health care decisions (Caro, Briggs, et al. 2012). In the context of infectious disease modeling and epidemiology, Keeling et al. argue, *"In epidemiology, models allow us to translate between behavior at various scales, or extrapolate from a known set of conditions to another. As such, models allow us to predict the population-level epidemic dynamics from an individual-level knowledge of epidemiological factors, the long-term behavior from the early invasion dynamics, or the impact of vaccination on the spread of infection"* [38]. Hence, researchers investigate and develop models to understand the spread of infectious diseases such as measles, malaria, COVID-19, HIV, and TB, to prevent the public from these infectious diseases and to determine the best and most cost-effective methods for eliminating these diseases. Similarly, there are many studies for exploring the dynamics of HCV infection.

In the literature, there exist different modeling methods. Our taxonomy suggested that the most common modeling methodologies for Hepatitis C are dynamical modeling and statistical modeling. We also included a separate section on literature review studies on HCV modeling and cost-effectiveness analysis. Thus, we divided the model type category into three main groups (Table 2).

Dynamic modeling is the most common modeling type that has been used by 96% of the articles surveyed. After dynamical modeling, the literature review group follows with 3%, and statistical modeling is 1% (Fig. 2).

We further classified modeling methodologies into two subgroups, which are stochastic and deterministic modeling. Deterministic modeling studies include 75% of the total of 73 articles. The rest of the articles can be classified as stochastic modeling for Hepatitis C (Fig. 3). The deterministic model is defined as models in which the values for their dependent variables are determined by the model's

Table 2 Model type classification table

2	Model type
2.1	Dynamic
2.1.1	Deterministic
2.1.1.1	Compartmental
2.1.1.2	Individual based
2.1.2	Stochastic
2.1.2.1	Compartmental
2.1.2.2	Agent-based
2.1.2.3	Network-based
2.1.2.4	Markov chain
2.2	Statistical
2.3	Literature review

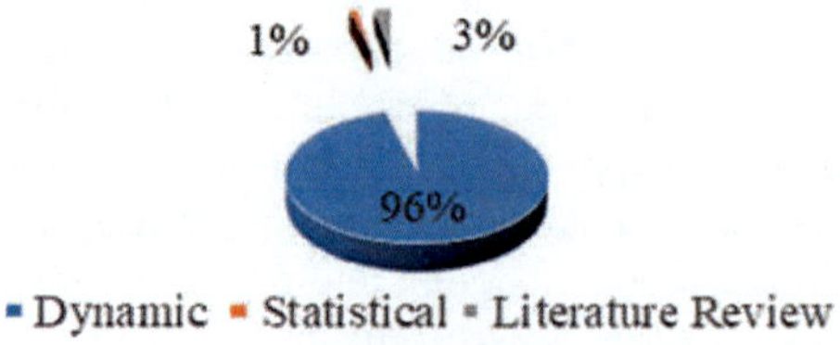

Fig. 2 Model type

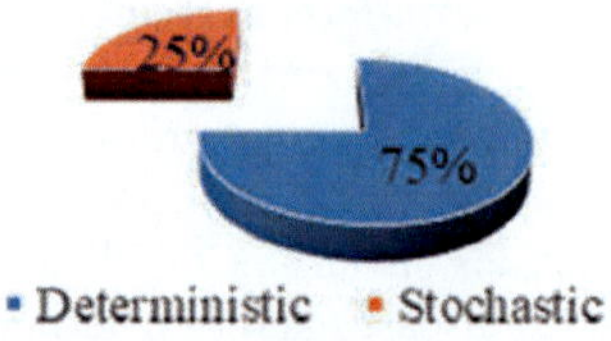

Fig. 3 Dynamic model type

parameters, and parameters do not include randomness or chance [56]. On the other hand, stochastic models consider input parameters that involve uncertainty.

According to our taxonomy, among deterministic models, the most common method in the literature is compartmental models, which are also known as SIR models. In the most basic SIR model, the population is represented by three compartments in which S represents the Susceptible population, I represents the Infected population, and R represents the Recovered population (or removed). In this type of model, populations divided into homogenous compartments that imitate the transmission of an infectious disease from an infected person to a susceptible individual by contact, and the model can determine transmission risk and spread of the disease over a period of time. Compartmental models could be tailored based on characteristics of infectious diseases, and they could be named as SI, SIR, SIRS, SEIR, MSEIR, etc. depending on the type of compartments included. For instance, if a person could be immune to infectious disease due to maternally inquired antibodies such as measles,

Fig. 4 Deterministic model type

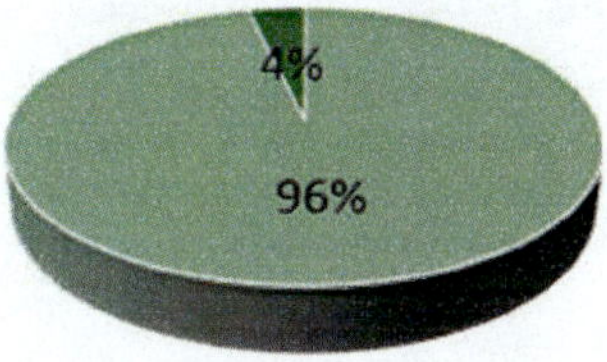

M (maternally derived immunity) compartment is added to the model. Similarly, if there is no recovery from this infectious disease such as HIV, the Recovered compartment could be eliminated from the model. In terms of modeling HCV, the SIR model is often preferred. Among all studies included in this paper, the deterministic compartmental model was used by 96% under the dynamical modeling group. On the other hand, individual-based models, also called agent-based models, could include heterogeneity of the population and individual characteristics compared to population-based compartmental models. As a result, they often enable the modelers to understand the high-level complexity of interactions among individuals. However, they require more parameters and data as well as computational resources compared to the compartmental models [16]. Due to their complexity, this approach is not preferred in most of the studies included in this paper, and only four percent of these deterministic models include an agent-based or individual model (Fig. 4).

In contrast to deterministic models, stochastic models consider randomness in the system rather than certainty. Stochastic modeling could be more realistic than deterministic models since real-life systems usually have uncertain elements. However, it is not easy to construct and implement stochastic models, the complexity level is higher, and the workload and parameter requirement is more than the deterministic approach.

In our taxonomy, we separated the stochastic modeling section into five subgroups, which are compartmental models, agent-based models, network-based models, Bernoulli models, and Markov models. Compartmental models and agent-based models could be deterministic or stochastic. Hence we included this categorization in both subgroups. If a model of these types included probabilistic nature or parameters, they are analyzed under the stochastic model group. Network-based models in epidemiology consider individuals as a node that can be reached from any other by network links, which are contacts that enables disease transmission among individuals. With the help of such a network, who infects whom could be demonstrated [37]. Due to the complexity of network models, sophisticated interventions, and different strategies could be tested in an artificial environment. Only one study of the 19 studies in the stochastic modeling category was a network model. In stochastic modeling, one of the most well-known methodologies is Markov models. In Markov models, the natural progression of the disease is divided into states, and probabilistic transitions between disease acquisition and progression are provided. These models should also satisfy the Markovian property [42]. Markovian property is based on the conditional distribution of any future state depends only on the present. Six percent of 19 articles in this category included Markov models for the transmission of the

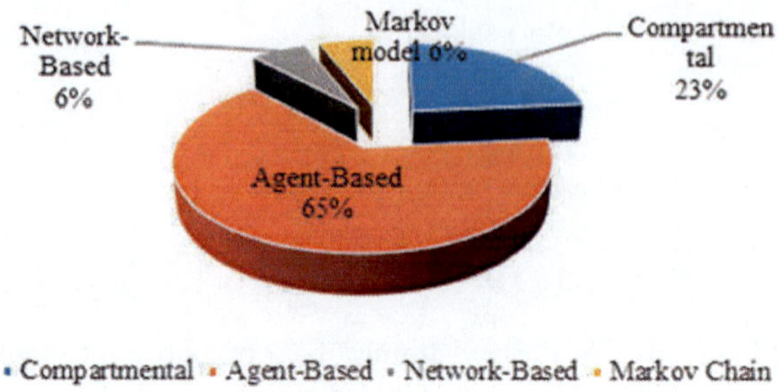

Fig. 5 Deterministic model type

Hepatitis C. Agent-based model, also known as individual-based models, was the most common method in the stochastic modeling group, which contributed 65% of the articles. Stochastic compartmental models were the second most common method for stochastic modeling and included 23% of 19 articles (Fig. 5).

Focus of Study

Modeling studies could focus on several important aspects of Hepatitis C, such as the spread of the disease, prevention methods, and ultimately elimination of HCV. We classified the focus of studies included in our work into three subgroups, which are Hepatitis C elimination, Hepatitis C prevention, and Hepatitis C transmission (Table 3). The papers included in the Hepatitis C elimination category are defined as studies that evaluated elimination strategies for Hepatitis C according to WHO HCV elimination goals. WHO elimination targets are specified as a 65% reduction in mortality and a 90% reduction in new infection by 2030 and 90% coverage for treatment with direct-acting antivirals for those who have Hepatitis C [31]. Fourteen percent of the articles contributed to the literature about the elimination strategies of HCV. Hepatitis C prevention subgroup contained articles on prevention strategies such as needle exchange programs, hand hygiene, training of health care workers, improving diagnosis, scaling up treatment, access to treatment, and blood safety programs. The ratio of this subgroup was 41%. All modeling studies include HCV transmission before analyzing prevention and elimination strategies because models developed in these studies initially forecast the future of the HCV epidemic in the status quo and then, they include further analysis of prevention and elimination strategies to determine the number of infections prevented as well as quality-adjusted

Table 3 Classification focus of study

3	Focus of study
3.1	HCV elimination
3.2	HCV prevention
3.3	HCV transmission

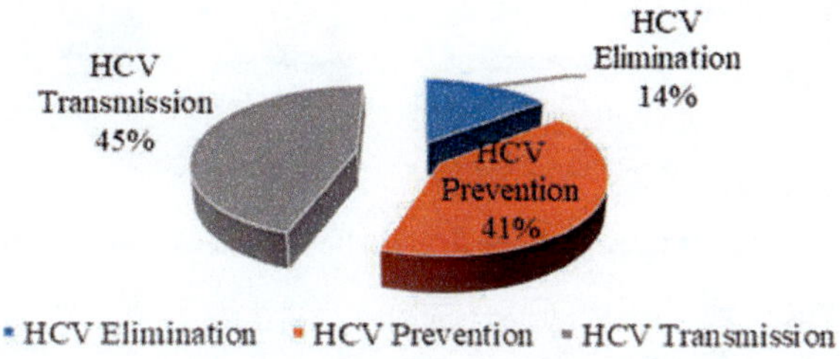

Fig. 6 Pie chart of focus of study

life-years saved. Based on our analysis, this subgroup had the highest number of articles, with 45% among 73 articles (Fig. 6).

Treatment Strategies

Recently, the treatment of Hepatitis C infection has drastically changed due to the development of direct-acting antiviral agents (DAA). New medications, compared to the old standard of care treatment with antivirals, which include PEGylated interferon and ribavirin, have a higher success rate, lower side effects, and shorter treatment duration. Unfortunately, they also have a quite higher cost compared to interferon-based treatment. As a result, many studies analyzed the effect and cost-effectiveness of DAAs and often compared them with older treatment methods in the last decade. DAAs and interferon-based therapy have different durations of sustained virologic response. *"Virologic response means that the Hepatitis C virus is not detected in the blood during treatment. When the virus continues to be undetectable 12 weeks or more after completing treatment, a "sustained" virologic response (SVR) has been achieved"* [21]. Based on the major changes in the treatment guidelines that moved towards DAAs from antivirals, we categorized treatment strategies included in the articles. We presented four subgroups related to treatment strategies, which are DAAs, antivirals, not defined, and comparison studies (Table 4). Comparison studies included both DAAs and antivirals and conducted a comparison study in terms of old and new treatment strategies. Our results suggested that 34% of 73 articles compared treatments, 36% did not define treatment included in the study, 27% focused on direct antiviral agents (DAA) as a treatment strategy and finally, 3% included interferon-based therapies (Fig. 7).

Table 4 Classification of treatment strategies

4	Treatment strategies
4.1	DAA
4.2	Antivirals (PEG/RBV interferon based)
4.3	Not defined
4.4	Comparison studies

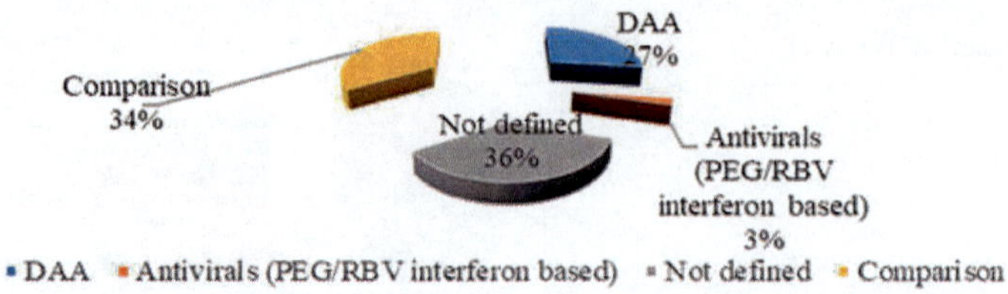

Fig. 7 Pie chart of related treatment

Cost-Effectiveness Analysis

An important dimension of infectious disease modeling is the inclusion of economic analysis. With the help of economic analysis techniques such as cost-benefit analysis, cost-utility analysis, and cost-effectiveness analysis, modelers can answer the questions such as which interventions are cost-effective, which programs should be implemented before others and what is the best allocation of health-care resources. Among economic analysis methods, cost-effectiveness analysis (CEA) is the most common and widely used technique. CEA compares prevention interventions and treatment methods to determine if an intervention or treatment is cost-effective compared to other interventions or treatments. As a result, policymakers could choose the best interventions or treatments that would give the best health outcomes with the least cost. Since many modeling studies include CEA, we ask the following question for our taxonomy work "Is there any cost-effectiveness analysis?" Thirty-seven percent of 73 articles contained CEA. The remaining 67% did not contain CEA in their study (Fig. 8).

Conclusion

Mathematical modeling has been widely used in the prediction of infectious disease outbreaks, and modeling methodologies such as compartmental models and agent-based simulation have been applied in many infectious diseases including malaria, measles, HIV, gonorrhea, whooping cough, SARS, MERS, pandemic influenza and Ebola. Hepatitis C is one of the infectious diseases that is an important public health problem, and it is a leading cause of death worldwide. There are several modeling studies that focus on the transmission, prevention, and elimination of HCV in the literature. In this study, we examined 73 papers about Hepatitis C modeling published in the last 20 years, 2000–2020. There is a significant growth in the number of studies in the last decade due to the introduction of DAAs for the treatment of HCV. Based on our taxonomy, we presented that modeling techniques that are mostly used were compartmental modeling and individual-based modeling. The majority of the interest of articles is in economic analysis, particularly cost-effectiveness analysis and in understanding the best prevention methods and determining the impact of

Table 5 Classification of articles

Article	1	1.1	1.2	1.2.1	1.2.2	2	2.1	2.1.1	2.1.1.1	2.1.1.2	2.1.2	2.1.2.1	2.1.2.2	2.1.2.3	2.1.2.4	2.2	2.3	3	3.1	3.2	3.3	4	4.1	4.2	4.3	4.4	5	5.1	5.2
[63]					×				×										×				×						×
[47]					×				×											×						×			×
[13]					×							×								×					×			×	
[57]					×								×								×				×				×
[15]					×							×									×				×				×
[59]					×				×										×				×						×
[20]				×					×											×					×				×
[25]				×						×										×					×			×	
[80]				×									×							×						×		×	
[50]				×					×											×				×					×
[48]				×					×										×				×					×	
[22]				×					×												×					×			×
[72]				×					×												×				×				×
[7]				×					×												×				×				×
[35]				×									×								×				×				×
[33]				×									×							×			×						×
[2]				×					×												×					×		×	
[8]				×					×												×				×				×
[46]				×					×												×					×		×	
[12]				×									×								×					×			×
[10]				×									×						×				×					×	

(continued)

Table 5 (continued)

Article	1	1.1	1.2	1.2.1	1.2.2	2	2.1	2.1.1	2.1.1.1	2.1.1.2	2.1.2	2.1.2.1	2.1.2.2	2.1.2.3	2.1.2.4	2.2	2.3	3	3.1	3.2	3.3	4	4.1	4.2	4.3	4.4	5	5.1	5.2
(P [70]				×									×							×					×				×
[41]				×					×											×			×						×
[39]				×					×												×				×			×	
[79]				×					×											×					×				×
[23]				×					×											×									×
[55]				×					×												×					×		×	
[5]				×													×			×			×					×	
[45]				×					×												×					×		×	
[9]				×													×				×				×				×
[24]				×					×												×					×			×
[27]				×								×							×				×						×
[32]				×					×											×						×			×
[52]				×					×											×				×				×	
[17]				×					×											×					×				×
[58]				×									×												×				×
[3]				×					×												×					×		×	
[19]				×					×											×			×					×	
[11]				×					×											×						×			×
[65]				×					×											×			×						×
[69]				×						×										×						×		×	
[36]		×							×												×				×				×

(continued)

Table 5 (continued)

Article	1	1.1	1.2	1.2.1	1.2.2	2	2.1	2.1.1	2.1.1.1	2.1.1.2	2.1.2	2.1.2.1	2.1.2.2	2.1.2.3	2.1.2.4	2.2	2.3	3	3.1	3.2	3.3	4	4.1	4.2	4.3	4.4	5	5.1	5.2
[29]				×												×				×					×				×
[51]				×					×											×						×		×	
[44]				×					×											×						×		×	
[49]					×				×											×						×		×	
[11]				×									×							×			×						×
[61]					×				×											×			×						×
[74]					×				×										×				×						×
[66]					×				×											×						×		×	
[40]				×																	×				×				×
[71]					×				×												×				×				×
[14]					×							×								×						×		×	
[77]					×				×												×				×				×
[34]				×										×						×			×						×
[73]				×					×												×				×				×
[60]					×				×										×				×					×	
[75]					×				×										×							×			×
[62]					×				×											×						×		×	
[64]				×					×										×				×						×
[76]				×					×												×					×		×	
[43]				×					×											×					×				×
[30]				×					×												×				×				×

(continued)

Table 5 (continued)

Article	1	1.1	1.2	1.2.1	1.2.2	2	2.1	2.1.1	2.1.1.1	2.1.1.2	2.1.2	2.1.2.1	2.1.2.2	2.1.2.3	2.1.2.4	2.2	2.3	3	3.1	3.2	3.3	4	4.1	4.2	4.3	4.4	5	5.1	5.2
[53]				×					×												×				×				×
[28]				×									×								×				×				×
[18]				×					×										×				×					×	
[31]					×				×										×							×			×
[4]					×								×								×		×						×
[26]					×				×												×		×						×
[1]				×					×												×					×		×	
[67]					×				×												×					×			×
[68]				×					×											×			×					×	
[78]				×					×												×				×				×
[6]				×													×				×					×		×	

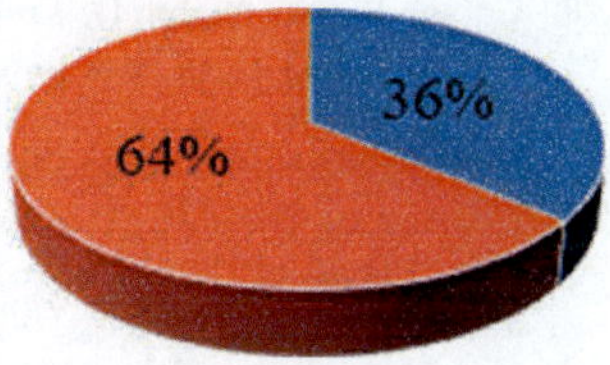

Fig. 8 Pie chart of cost-effectiveness analysis

these methods on the spread and future of HCV infections. All modeling techniques mentioned above have advantages and disadvantages in terms of data and computational requirements, simplicity and implementation, however often their main objective is to understand the dynamics of this disease, to predict the future transmission of the disease better, to find the most cost-effective intervention for prevention of HCV and to find efficient public health strategies for the ultimate elimination of HCV. The taxonomy proposed here could potentially be a useful tool for public health policymakers in developing new HCV models via understanding existing modeling studies, providing the most common modeling techniques, and assessing treatment strategies included.

References

1. Barbosa C, Fraser H, Hoerger TJ, Leib A, Havens JR, Young A, Kral A, Page K, Evans J, Zibbell J, Hariri S, Vellozzi C, Nerlander L, Ward JW, Vickerman P (2019) Cost-effectiveness of scaling-up HCV prevention and treatment in the United States for people who inject drugs. Addiction 114(12):2267–2278. https://doi.org/10.1111/add.14731
2. Bennett H, Gordon J, Jones B, Ward T, Webster S, Kalsekar A, Yuan Y, Brenner M, McEwan P (2017) Hepatitis C disease transmission and treatment uptake: impact on the cost-effectiveness of new direct-acting antiviral therapies. Eur J Health Econ 18(8):1001–1011. https://doi.org/10.1007/s10198-016-0844-8
3. Bennett H, McEwan P, Sugrue D, Kalsekar A, Yuan Y (2015) Assessing the long-term impact of treating Hepatitis C Virus (HCV)-infected people who inject drugs in the UK and the relationship between treatment uptake and efficacy on future infections. PLoS ONE 10(5). https://doi.org/10.1371/journal.pone.0125846
4. Buchanan R, Meskarian R, van der Heijden P, Grellier L, Parkes J, Khakoo SI (2020) Prioritising Hepatitis C treatment in people with multiple injecting partners maximises prevention: a real-world network study. J Infect 80(2):225–231. https://doi.org/10.1016/j.jinf.2019.12.010
5. Chhatwal J, He T, Lopez-Olivo MA (2016). Systematic review of modelling approaches for the cost effectiveness of Hepatitis C treatment with direct-acting antivirals. In: Pharmaco economics, vol 34(6). Springer International Publishing, pp 551–567. https://doi.org/10.1007/s40273-015-0373-9
6. Chhatwal J, Kanwal F, Roberts MS, Dunn MA (2015) Cost-effectiveness and budget impact of Hepatitis C virus treatment with sofosbuvir and ledipasvir in the United States. Ann Intern Med 162(6):397–406. https://doi.org/10.7326/M14-1336
7. Corson S, Greenhalgh D, Hutchinson SJ (2013) A time since onset of injection model for Hepatitis C spread amongst injecting drug users. J Math Biol 66(4–5):935–978. https://doi.org/10.1007/s00285-012-0577-y

8. Corson S, Greenhalgh D, Taylor A, Palmateer N, Goldberg D, Hutchinson S (2013) Modelling the prevalence of HCV amongst people who inject drugs: an investigation into the risks associated with injecting paraphernalia sharing. Drug Alcohol Depend 133(1):172–179. https://doi.org/10.1016/j.drugalcdep.2013.05.014
9. Cousien A, Tran VC, Deuffic-Burban S, Jauffret-Roustide M, Dhersin J-S, Yazdanpanah Y (2015) Dynamic modelling of Hepatitis C virus transmission among people who inject drugs: a methodological review. J Viral Hepatitis 22(3):213–229. https://doi.org/10.1111/jvh.12337
10. Cousien A, Tran VC, Deuffic-Burban S, Jauffret-Roustide M, Mabileau G, Dhersin J-S, Yazdanpanah Y (2018) Effectiveness and cost-effectiveness of interventions targeting harm reduction and chronic Hepatitis C cascade of care in people who inject drugs: the case of France. J Viral Hepatitis 25(10):1197–1207. https://doi.org/10.1111/jvh.12919
11. Cousien A, Leclerc P, Morissette C, Bruneau J, Roy É, Tran VC, Yazdanpanah Y, Cox J (2017) The need for treatment scale-up to impact HCV transmission in people who inject drugs in Montréal, Canada: a modelling study. BMC Infect Dis 17(1). https://doi.org/10.1186/s12879-017-2256-5
12. Cousien A, Tran VC, Deuffic-Burban S, Jauffret-Roustide M, Dhersin JS, Yazdanpanah Y (2016) Hepatitis C treatment as prevention of viral transmission and liver-related morbidity in persons who inject drugs. Hepatology 63(4):1090–1101. https://doi.org/10.1002/hep.28227
13. de Vos AS, Kretzschmar MEE (2014) Benefits of Hepatitis C virus treatment: a balance of preventing onward transmission and re-infection. Math Biosci 258:11–18. https://doi.org/10.1016/j.mbs.2014.09.006
14. de Vos AS, Prins M, Kretzschmar MEE (2015) Hepatitis C virus treatment as prevention among injecting drug users: who should we cure first? Addiction 110(6):975–983. https://doi.org/10.1111/add.12842
15. De Vos AS, Kretzschmar MEE (2013). The efficiency of targeted intervention in limiting the spread of HIV and Hepatitis C virus among injecting drug users. J Theor Biol 333:126–134. https://doi.org/10.1016/j.jtbi.2013.05.017
16. DeAngelis DL, Grimm V (2014) Individual-based models in ecology after four decades. F1000Prime Rep 6. https://doi.org/10.12703/P6-39
17. Durier N, Nguyen C, White LJ (2012) Treatment of Hepatitis C as prevention: a modeling case study in Vietnam. PloS One 7(4). https://doi.org/10.1371/journal.pone.0034548
18. Echevarria D, Gutfraind A, Boodram B, Layden J, Ozik J, Page K, Cotler SJ, Major M, Dahari H (2019) Modeling indicates efficient vaccine-based interventions for the elimination of Hepatitis C virus among persons who inject drugs in metropolitan Chicago. Vaccine 37(19):2608–2616. https://doi.org/10.1016/j.vaccine.2019.02.081
19. Echevarria D, Gutfraind A, Boodram B, Major M, Del Valle S, Cotler SJ, Dahari H (2015) Mathematical modeling of Hepatitis C prevalence reduction with antiviral treatment scale-up in persons who inject drugs in metropolitan Chicago. PLoS One 10(8). https://doi.org/10.1371/journal.pone.0135901
20. Esposito N, Rossi C (2004) A nested-epidemic model for the spread of Hepatitis C among injecting drug users. Math Biosci 188(1):29–45. https://doi.org/10.1016/j.mbs.2003.11.001
21. FAQs—Viral Hepatitis and Liver Disease. (n.d.). Retrieved 30 July 2020 from https://www.hepatitis.va.gov/hcv/patient/faqs/
22. Fraser H, Martin NK, Brummer-Korvenkontio H, Carrieri P, Dalgard O, Dillon J, Goldberg D, Hutchinson S, Jauffret-Roustide M, Kåberg M, Matser AA, Matičič M, Midgard H, Mravcik V, Øvrehus A, Prins M, Reimer J, Robaeys G, Schulte B, Hickman M et al (2018). Model projections on the impact of HCV treatment in the prevention of HCV transmission among people who inject drugs in Europe. J Hepatol 68(3):402–411. https://doi.org/10.1016/j.jhep.2017.10.010
23. Fraser H, Mukandavire C, Martin NK, Goldberg D, Palmateer N, Munro A, Taylor A, Hickman M, Hutchinson S, Vickerman P (2018) Modelling the impact of a national scale-up of interventions on Hepatitis C virus transmission among people who inject drugs in Scotland. Addiction 113(11):2118–2131. https://doi.org/10.1111/add.14267

24. Fraser H, Zibbell J, Hoerger T, Hariri S, Vellozzi C, Martin NK, Kral AH, Hickman M, Ward JW, Vickerman P (2018) Scaling-up HCV prevention and treatment interventions in rural United States-model projections for tackling an increasing epidemic. Addiction 113(1):173–182. https://doi.org/10.1111/add.13948
25. Fu R, Gutfraind A, Brandeau ML (2016) Modeling a dynamic bi-layer contact network of injection drug users and the spread of blood-borne infections. Math Biosci 273:102–113. https://doi.org/10.1016/j.mbs.2016.01.003
26. Gicquelais RE, Foxman B, Coyle J, Eisenberg MC (2019) Hepatitis C transmission in young people who inject drugs: insights using a dynamic model informed by state public health surveillance. Epidemics 27:86–95. https://doi.org/10.1016/j.epidem.2019.02.003
27. Gountas I, Sypsa V, Anagnostou O, Martin N, Vickerman P, Kafetzopoulos E, Hatzakis A (2017) Treatment and primary prevention in people who inject drugs for chronic Hepatitis C infection: is elimination possible in a high-prevalence setting? Addiction 112(7):1290–1299. https://doi.org/10.1111/add.13764
28. Gountas I, Sypsa V, Papatheodoridis G, Paraskevis D, Kalamitsis G, Anagnostou O, Antaraki A, Fotiou A, Hatzakis A (2019) A Hepatitis C outbreak preceded the HIV outbreak among persons who inject drugs in Athens, Greece: insights from a mathematical modelling study. J Viral Hepatitis 26(11):1311–1317. https://doi.org/10.1111/jvh.13178
29. Hagan H, Pouget ER, Des Jarlais DC, Lelutiu-Weinberger C (2008) Meta-regression of Hepatitis C virus infection in relation to time since onset of illicit drug injection: the influence of time and place. Am J Epidemiol 168(10):1099–1109. https://doi.org/10.1093/aje/kwn237
30. Hancock E, Ward Z, Ayres R, Neale J, Hussey D, Kesten JM, Hickman M, Vickerman P (2020) Detachable low dead space syringes for the prevention of Hepatitis C among people who inject drugs in Bristol, UK: an economic evaluation. Addiction 115(4):702–713. https://doi.org/10.1111/add.14849
31. Heffernan A, Cooke GS, Nayagam S, Thursz M, Hallett TB (2019) Scaling up prevention and treatment towards the elimination of Hepatitis C: a global mathematical model. Lancet 393(10178):1319–1329. https://doi.org/10.1016/S0140-6736(18)32277-3
32. Hellard ME, Jenkinson R, Higgs P, Stoové MA, Sacks-Davis R, Gold J, Hickman M, Vickerman P, Martin NK (2012) Modelling antiviral treatment to prevent Hepatitis C infection among people who inject drugs in Victoria, Australia. Med J Aust 196(10):638–641. https://doi.org/10.5694/mja11.10981
33. Hellard M, McBryde E, Sacks Davis R, Rolls DA, Higgs P, Aitken C, Thompson A, Doyle J, Pattison P, Robins G (2015) Hepatitis C transmission and treatment as prevention—the role of the injecting network. Int J Drug Policy 26(10):958–962. https://doi.org/10.1016/j.drugpo.2015.05.006
34. Hellard M, Rolls DA, Sacks-Davis R, Robins G, Pattison P, Higgs P, Aitken C, Mcbryde E (2014) The impact of injecting networks on Hepatitis C transmission and treatment in people who inject drugs. Hepatology 60(6):1861–1870. https://doi.org/10.1002/hep.27403
35. Hutchinson SJ, Bird SM, Taylor A, Goldberg DJ (2006) Modelling the spread of Hepatitis C virus infection among injecting drug users in Glasgow: Implications for prevention. Int J Drug Policy 17(3):211–221. https://doi.org/10.1016/j.drugpo.2006.02.008
36. Imran M, Rafique H, Khan A, Malik T (2014) A model of bi-mode transmission dynamics of Hepatitis C with optimal control. Theor Biosci 133(2):91–109. https://doi.org/10.1007/s12064-013-0197-0
37. Keeling MJ, Eames KTD (2005) Networks and epidemic models. J Royal Soc Interface 2(4):295–307. https://doi.org/10.1098/rsif.2005.0051
38. Keeling MJ, Rohani P, Pourbohloul B (2008) Modeling infectious diseases in humans and animals: modeling infectious diseases in humans and animals. Clin Infect Dis 47(6):864–865. https://doi.org/10.1086/591197
39. Kwon JA, Anderson J, Kerr CC, Thein H-H, Zhang L, Iversen J, Dore GJ, Kaldor JM, Law MG, Maher L, Wilson DP (2012) Estimating the cost-effectiveness of needle-syringe programs in Australia. AIDS 26(17):2201–2210. https://doi.org/10.1097/QAD.0b013e3283578b5d

40. Kwon JA, Iversen J, Maher L, Law MG, Wilson DP (2009) The impact of needle and syringe programs on HIV and HCV transmissions in injecting drug users in Australia: a model-based analysis. J Acquir Immune Defic Syndr 51(4):462–469. https://doi.org/10.1097/QAI.0b013e3181a2539a
41. Lim AG, Qureshi H, Mahmood H, Hamid S, Davies CF, Trickey A, Glass N, Saeed Q, Fraser H, Walker JG, Mukandavire C, Hickman M, Martin NK, May MT, Averhoff F, Vickerman P (2018) Curbing the Hepatitis C virus epidemic in Pakistan: the impact of scaling up treatment and prevention for achieving elimination. Int J Epidemiol 47(2):550–560. https://doi.org/10.1093/ije/dyx270
42. Markov Model (2013) Encyclopedia of biophysics. https://doi.org/10.1007/978-3-642-16712-6_100568
43. Marks C, Borquez A, Jain S, Sun X, Strathdee SA, Garfein RS, Milloy MJ, DeBeck K, Cepeda JA, Werb D, Martin NK (2019) Opioid agonist treatment scale-up and the initiation of injection drug use: a dynamic modeling analysis. PLoS Med 16(11). https://doi.org/10.1371/journal.pmed.1002973
44. Martin NK, Foster GR, Vilar J, Ryder S, Cramp ME, Gordon F, Dillon JF, Craine N, Busse H, Clements A, Hutchinson SJ, Ustianowski A, Ramsay M, Goldberg DJ, Irving W, Hope V, De Angelis D, Lyons M, Vickerman P, Hickman M (2015) HCV treatment rates and sustained viral response among people who inject drugs in seven UK sites: real world results and modelling of treatment impact. J Viral Hepatitis 22(4):399–408. https://doi.org/10.1111/jvh.12338
45. Martin NK, Hickman M, Hutchinson SJ, Goldberg DJ, Vickerman P (2013a) Combination interventions to prevent HCV transmission among people who inject drugs: modeling the impact of antiviral treatment, needle and syringe programs, and opiate substitution therapy. Clin Infect Dis 57(Suppl.2). https://doi.org/10.1093/cid/cit296
46. Martin NK, Hickman M, Hutchinson SJ, Goldberg DJ, Vickerman P (2013b) Combination interventions to prevent HCV transmission among people who inject drugs: modeling the impact of antiviral treatment, needle and syringe programs, and opiate substitution therapy. Clin Infect Dis 57(suppl_2):S39–S45. https://doi.org/10.1093/cid/cit296
47. Martin NK, Pitcher AB, Vickerman P, Vassall A, Hickman M (2011) Optimal control of Hepatitis C antiviral treatment programme delivery for prevention amongst a population of injecting drug users. PLoS One 6(8). https://doi.org/10.1371/journal.pone.0022309
48. Martin NK, Vickerman P, Brew IF, Williamson J, Miners A, Irving WL, Saksena S, Hutchinson SJ, Mandal S, O'Moore E, Hickman M (2016) Is increased Hepatitis C virus case-finding combined with current or 8-week to 12-week direct-acting antiviral therapy cost-effective in UK prisons? A prevention benefit analysis. Hepatology 63(6):1796–1808. https://doi.org/10.1002/hep.28497
49. Martin NK, Vickerman P, Dore GJ, Grebely J, Miners A, Cairns J, Foster GR, Hutchinson SJ, Goldberg DJ, Martin TCS, Ramsay M, Hickman M (2016) Prioritization of HCV treatment in the direct-acting antiviral era: an economic evaluation. J Hepatol 65(1):17–25. https://doi.org/10.1016/j.jhep.2016.02.007
50. Martin NK, Vickerman P, Foster GR, Hutchinson SJ, Goldberg DJ, Hickman M (2011) Can antiviral therapy for Hepatitis C reduce the prevalence of HCV among injecting drug user populations? A modeling analysis of its prevention utility. J Hepatol 54(6):1137–1144. https://doi.org/10.1016/j.jhep.2010.08.029
51. Martin NK, Vickerman P, Grebely J, Hellard M, Hutchinson SJ, Lima VD, Foster GR, Dillon JF, Goldberg DJ, Dore GJ, Hickman M (2013) Hepatitis C virus treatment for prevention among people who inject drugs: Modeling treatment scale-up in the age of direct-acting antivirals. Hepatology 58(5):1598–1609. https://doi.org/10.1002/hep.26431
52. Martin NK, Vickerman P, Hickman M (2011) Mathematical modelling of Hepatitis C treatment for injecting drug users. J Theor Biol 274(1):58–66. https://doi.org/10.1016/j.jtbi.2010.12.041
53. Miller-Dickson MD, Meszaros VA, Almagro-Moreno S, Brandon Ogbunugafor C (2019) Hepatitis C virus modelled as an indirectly transmitted infection highlights the centrality of injection drug equipment in disease dynamics. J Royal Soc Interface 16(158). https://doi.org/10.1098/rsif.2019.0334

54. Pitcher AB, Borquez A, Skaathun B, Martin NK (2019) Mathematical modeling of Hepatitis C virus (HCV) prevention among people who inject drugs: a review of the literature and insights for elimination strategies. J Theor Biol 481:194–201. https://doi.org/10.1016/j.jtbi.2018.11.013
55. Platt L, Sweeney S, Ward Z, Guinness L, Hickman M, Hope V, Hutchinson S, Maher L, Iversen J, Craine N, Taylor A, Munro A, Parry J, Smith J, Vickerman P (2017) Assessing the impact and cost-effectiveness of needle and syringe provision and opioid substitution therapy on Hepatitis C transmission among people who inject drugs in the UK: an analysis of pooled data sets and economic modelling. Public Health Res 5(5):1–118. https://doi.org/10.3310/phr05050
56. Rey SJ (2015) Mathematical models in geography. In: International encyclopedia of the social and behavioral sciences, 2nd ed. Elsevier Inc, pp 785–790. https://doi.org/10.1016/B978-0-08-097086-8.72033-2
57. Rolls DA, Sacks-Davis R, Jenkinson R, McBryde E, Pattison P, Robins G, Hellard M (2013) Hepatitis C transmission and treatment in contact networks of people who inject drugs. PLoS One 8(11): https://doi.org/10.1371/JOURNAL.PONE.0078286
58. Rolls DA, Sacks-Davis R, Jenkinson R, McBryde E, Pattison P, Robins G, Hellard M (2013b) Hepatitis C transmission and treatment in contact networks of people who inject drugs. PLoS One 8(11). https://doi.org/10.1371/journal.pone.0078286
59. Rozada I, Coombs D, Lima VD (2016) Conditions for eradicating Hepatitis C in people who inject drugs: a fibrosis aware model of Hepatitis C virus transmission. J Theor Biol 395:31–39. https://doi.org/10.1016/j.jtbi.2016.01.030
60. Scott N, Doyle JS, Wilson DP, Wade A, Howell J, Pedrana A, Thompson A, Hellard ME (2017) Reaching Hepatitis C virus elimination targets requires health system interventions to enhance the care cascade. Int J Drug Policy 47:107–116. https://doi.org/10.1016/j.drugpo.2017.07.006
61. Scott N, McBryde E, Vickerman P, Martin NK, Stone J, Drummer H, Hellard M (2015) The role of a Hepatitis C virus vaccine: modelling the benefits alongside direct-acting antiviral treatments. BMC Med 13(1). https://doi.org/10.1186/s12916-015-0440-2
62. Scott N, Mohamed Z, Rwegasha J, Mbwambo J, Lemoine M, Hellard M (2019) Upscaling prevention, testing and treatment to control Hepatitis C as a public health threat in Dar es Salaam, Tanzania: a cost-effectiveness model. Int J Drug Policy. https://doi.org/10.1016/j.drugpo.2019.102634
63. Scott N, Ólafsson S, Gottfreðsson M, Tyrfingsson T, Rúnarsdóttir V, Hansdottir I, Hernandez UB, Sigmundsdóttir G, Hellard M (2018) Modelling the elimination of Hepatitis C as a public health threat in Iceland: a goal attainable by 2020. J Hepatol 68(5):932–939. https://doi.org/10.1016/j.jhep.2017.12.013
64. Stocks T, Martin LJ, Kühlmann-Berenzon S, Britton T (2020) Dynamic modeling of Hepatitis C transmission among people who inject drugs. Epidemics 30. https://doi.org/10.1016/j.epidem.2019.100378
65. Stone J, Martin NK, Hickman M, Hellard M, Scott N, McBryde E, Drummer H, Vickerman P (2016) The potential impact of a Hepatitis C vaccine for people who inject drugs: is a vaccine needed in the age of direct-acting antivirals? PLoS One 11(5). https://doi.org/10.1371/journal.pone.0156213
66. Stone J, Martin NK, Hickman M, Hutchinson SJ, Aspinall E, Taylor A, Munro A, Dunleavy K, Peters E, Bramley P, Hayes PC, Goldberg DJ, Vickerman P (2017) Modelling the impact of incarceration and prison-based Hepatitis C virus (HCV) treatment on HCV transmission among people who inject drugs in Scotland. Addiction 112(7):1302–1314. https://doi.org/10.1111/add.13783
67. Sweeney S, Ward Z, Platt L, Guinness L, Hickman M, Hope V, Maher L, Iversen J, Hutchinson SJ, Smith J, Ayres R, Hainey I, Vickerman P (2019) Evaluating the cost-effectiveness of existing needle and syringe programmes in preventing Hepatitis C transmission in people who inject drugs. Addiction 114(3):560–570. https://doi.org/10.1111/add.14519
68. Trickey A, Fraser H, Lim AG, Walker JG, Peacock A, Colledge S, Leung J, Grebely J, Larney S, Martin NK, Degenhardt L, Hickman M, May MT, Vickerman P (2019) Modelling the potential prevention benefits of a treat-all Hepatitis C treatment strategy at global, regional and country levels: a modelling study. J Viral Hepatitis 26(12):1388–1403. https://doi.org/10.1111/jvh.13187

69. Van Santen DK, De Vos AS, Matser A, Willemse SB, Lindenburg K, Kretzschmar MEE, Prins M, De Wit GA (2016) Cost-effectiveness of Hepatitis C treatment for people who inject drugs and the impact of the type of epidemic; extrapolating from Amsterdam, the Netherlands. PLoS One 11(10). https://doi.org/10.1371/journal.pone.0163488
70. Vickerman P, Hickman M, Judd A (2007) Modelling the impact on Hepatitis C transmission of reducing syringe sharing: London case study. Int J Epidemiol 36(2):396–405. https://doi.org/10.1093/ije/dyl276
71. Vickerman P, Martin NK, Hickman M (2012) Understanding the trends in HIV and Hepatitis C prevalence amongst injecting drug users in different settings—implications for intervention impact. Drug Alcohol Depend 123(1–3):122–131. https://doi.org/10.1016/j.drugalcdep.2011.10.032
72. Vickerman P, Martin NK, Roy A, Beattie T, Jarlais DD, Strathdee S, Wiessing L, Hickman M (2013) Is the HCV–HIV co-infection prevalence amongst injecting drug users a marker for the level of sexual and injection related HIV transmission? Drug Alcohol Depend 132(1):172–181. https://doi.org/10.1016/j.drugalcdep.2013.01.020
73. Vickerman P, Martin N, Turner K, Hickman M (2012) Can needle and syringe programmes and opiate substitution therapy achieve substantial reductions in Hepatitis C virus prevalence? Model projections for different epidemic settings. Addiction 107(11):1984–1995. https://doi.org/10.1111/j.1360-0443.2012.03932.x
74. Virlogeux V, Zoulim F, Pugliese P, Poizot-Martin I, Valantin MA, Cuzin L, Reynes J, Billaud E, Huleux T, Bani-Sadr F, Rey D, Frésard A, Jacomet C, Duvivier C, Cheret A, Hustache-Mathieu L, Hoen B, Cabié A, Cotte L, Lamaury I et al (2017) Modeling HIV-HCV coinfection epidemiology in the direct-acting antiviral era: the road to elimination. BMC Med 15(1). https://doi.org/10.1186/s12916-017-0979-1
75. Walker JG, Kuchuloria T, Sergeenko D, Fraser H, Lim AG, Shadaker S, Hagan L, Gamkrelidze A, Kvaratskhelia V, Gvinjilia L, Aladashvili M, Asatiani A, Baliashvili D, Butsashvili M, Chikovani I, Khonelidze I, Kirtadze I, Kuniholm MH, Otiashvili D, Vickerman P et al (2020) Interim effect evaluation of the Hepatitis C elimination programme in Georgia: a modelling study. Lancet Global Health 8(2):e244–e253. https://doi.org/10.1016/S2214-109X(19)30483-8
76. Ward Z, Campbell L, Surey J, Platts S, Glass R, Hickman M, Story A, Vickerman P (2019) The cost-effectiveness of an HCV outreach intervention for at-risk populations in London, UK. J Antimicrob Chemother 74(5):v5–v16. https://doi.org/10.1093/jac/dkz451
77. Ward Z, Platt L, Sweeney S, Hope VD, Maher L, Hutchinson S, Palmateer N, Smith J, Craine N, Taylor A, Martin N, Ayres R, Dillon J, Hickman M, Vickerman P (2018) Impact of current and scaled-up levels of Hepatitis C prevention and treatment interventions for people who inject drugs in three UK settings—what is required to achieve the WHO's HCV elimination targets? Addiction 113(9):1727–1738. https://doi.org/10.1111/add.14217
78. Ward Z, Reynolds R, Campbell L, Martin NK, Harrison G, Irving W, Hickman M, Vickerman P (2020) Cost-effectiveness of the HepCATT intervention in specialist drug clinics to improve case-finding and engagement with HCV treatment for people who inject drugs in England. Addiction. https://doi.org/10.1111/add.14978
79. Zeiler I, Langlands T, Murray JM, Ritter A (2010). Optimal targeting of Hepatitis C virus treatment among injecting drug users to those not enrolled in methadone maintenance programs. Drug Alcohol Depend 110(3):228–233. https://doi.org/10.1016/j.drugalcdep.2010.03.006
80. Zelenev A, Li J, Mazhnaya A, Basu S, Altice FL (2018) Hepatitis C virus treatment as prevention in an extended network of people who inject drugs in the USA: a modelling study. Lancet Infect Dis 18(2):215–224. https://doi.org/10.1016/S1473-3099(17)30676-X

Author Index

F. Calisir (ed.), *Industrial Engineering in the Internet-of-Things World*, Lecture Notes in Management and Industrial Engineering,
https://doi.org/10.1007/978-3-030-76724-2